Dispersal Ecology

Symposia of The British Ecological Society

41 Ecology: Achievement and Challenge
Press, Huntly & Levin, 2001

40 The Ecological Consequences of Environmental Heterogeneity
Hutchings, John & Stewart, 2000

39 Physiological Plant Ecology
Press, Scholes & Barker, 2000

37 Dynamics of Tropical Communities
Newbery, Prins & Brown, 1997

36 Multitrophic Interactions in Terrestrial Systems
Gange & Brown, 1997

33 Genes in Ecology
Berry, Crawford & Hewitt, 1992

32 The Ecology of Temperate Cereal Fields
Firbank, Carter, Darbyshire & Potts, 1991

31 The Scientific Management of Temperate Communities for Conservation
Spellerberg, Goldsmith & Morris, 1991

30 Toward a More Exact Ecology
Grubb & Whittaker, 1989

Dispersal Ecology

The 42nd Symposium of the British Ecological Society held at the University of Reading 2–5 April 2001

EDITED BY

JAMES M. BULLOCK
NERC Centre for Ecology and Hydrology
CEH Dorset, UK

ROBERT E. KENWARD
NERC Centre for Ecology and Hydrology
CEH Dorset, UK

AND

ROSIE S. HAILS
NERC Centre for Ecology and Hydrology
CEH Oxford, UK

CAMBRIDGE
UNIVERSITY PRESS

CAMBRIDGE UNIVERSITY PRESS
Cambridge, New York, Melbourne, Madrid, Cape Town, Singapore, São Paulo, Delhi

Cambridge University Press
The Edinburgh Building, Cambridge CB2 8RU, UK

Published in the United States of America by Cambridge University Press, New York

www.cambridge.org
Information on this title: www.cambridge.org/9780521839945

First published on behalf of the British Ecological Society by Blackwell Science Ltd 2002
First published on behalf of the British Ecological Society by Cambridge University Press 2008
Re-issued in this digitally printed version 2008

A catalogue record for this publication is available from the British Library

ISBN 978-0-521-83994-5 hardback
ISBN 978-0-521-54931-8 paperback

Contents

List of Contributors

H. P. Andreassen
Division of Zoology, Department of Biology, University of Oslo, PO Box 1050 Blindern, N-0316 Oslo, Norway

M. J. Bailey
Molecular Microbial Ecology, Institute of Virology and Environmental Microbiology, Centre for Ecology and Hydrology, Mansfield Road, Oxford OX1 3SR, UK

J. M. Bond
NERC Centre for Ecology and Hydrology, CEH Dorset, Winfrith Technology Centre, Dorchester, Dorset DT2 8ZD, UK

I. L. Boyd
NERC Sea Mammal Research Unit, Gatty Marine Laboratory, University of St Andrews, St Andrews, Fife KY16 8LB, UK

J. K. M. Brown
Department of Disease and Stress Biology, John Innes Centre, Colney Lane, Norwich NR4 7UH, UK

J. M. Bullock
NERC Centre for Ecology and Hydrology, CEH Dorset, Winfrith Technology Centre, Dorchester, Dorset DT2 8ZD, UK

C. Calogeropoulos
Department of Biology, Concordia University, Montreal, Quebec, Canada H3G IM8

H. Caswell
Biology Department MS-34, Woods Hole Oceanographic Institution, Woods Hole, MA 02543, USA

R. T. Clarke
NERC Centre for Ecology and Hydrology, CEH Dorset, Winfrith Technology Centre, Dorchester, Dorset DT2 8ZD, UK

A. N. Cohen
San Francisco Estuary Institute, 7770 Pardee Lane, Oakland, CA 94621-1424, USA

S. G. Compton
Centre for Biodiversity and Conservation, School of Biology, University of Leeds, Leeds LS2 9JT, UK

S. J. Coulson
The Northmoor Trust, Little Wittenham, Abingdon, Oxon OX14 4RA, UK

G. Dwyer
Department of Ecology and Evolution, University of Chicago, Chicago, IL 60637, USA

J. R. Freeland
Department of Biological Sciences, Open University, Milton Keynes MK7 6AA, UK

J. A. Gill
Tyndall Centre for Climate Change Research and School of Biological Sciences, University of East Anglia, Norwich NR4 7TJ, UK

C. J. Gliddon
School of Biological Sciences, Deiniol Road, University of Wales—Bangor, Bangor, Gwynedd LL57 2UW, UK

D. F. Greene
Department of Geography, Concordia University, Montreal, Quebec, Canada H3G IM8

R. S. Hails
NERC Centre for Ecology and Hydrology, CEH Oxford, Mansfield Road, Oxford OX1 3SR, UK

L. Hemerik
Subdepartment of Mathematics, Wageningen University, Dreijenlaan 4, 6703 AH Wageningen, the Netherlands

R. Hengeveld
ALTERRA, Department of Ecology and Environment, Droevendaalsesteeg 3, PO Box 47, 6700 AA Wageningen, the Netherlands

D. J. Hodkinson
Department of Animal and Plant Sciences, University of Sheffield, Sheffield S10 2TN, UK

M. S. Hovmøller
Department of Crop Protection, Danish Institute of Agricultural Sciences, Flakkebjerg, 4200 Slagelse, Denmark

R. A. Ims
Division of Ecological Zoology, Department of Biology, University of Tromsø, N-9037 Tromsø, Norway

K. Kawasaki
Department of Knowledge Engineering and Computer Sciences, Doshisha University, Kyo-Tanabe 610-0321, Japan

R. E. Kenward
NERC Centre for Ecology and Hydrology, CEH Dorset, Winfrith Technology Centre, Dorchester, Dorset DT2 8ZD, UK

A. K. Lilley
Molecular Microbial Ecology, Institute of Virology and Environmental Microbiology, Centre for Ecology and Hydrology, Mansfield Road, Oxford OX1 3SR, UK

H. D. Loxdale
Plant and Invertebrate Ecology Division, IACR-Rothamsted, Harpenden, Herts AL5 2JQ, UK

D. W. Macdonald
Wildlife Conservation Research Unit, Department of Zoology, South Parks Road, University of Oxford, Oxford OX1 3PS, UK

I. L. Moy
NERC Centre for Ecology and Hydrology, CEH Dorset, Winfrith Technology Centre, Dorchester, Dorset DT2 8ZD, UK

A. M. Nolan
Department of Geography, University of Southampton, Highfield, Southampton SO17 1BJ, UK

K. Norris
School of Animal and Microbial Sciences, University of Reading, Whiteknights, PO Box 228, Reading RG6 6AJ, UK

B. Okamura
School of Animal and Microbial Sciences, University of Reading, Whiteknights, PO Box 228, Reading RG6 6AJ, UK

J. L. Osborne
Plant and Invertebrate Ecology Division, IACR-Rothamsted, Harpenden, Herts AL5 2JQ, UK

C. M. Perrins
Edward Grey Institute, Department of Zoology, South Parks Road, University of Oxford, Oxford OX1 3PS, UK

R. F. Pywell
NERC Centre for Ecology and Hydrology, Monks Wood, Abbots Ripton, Huntingdon, Cambs PE28 2LS, UK

A. F. Raybould
Syngenta, Jealott's Hill International Research Centre, Bracknell, Berks RG42 6EY, UK

M. Rees
Department of Biology, Imperial College at Silwood Park, Ascot, Berks SL5 7PY, UK

L. C. Rickard
Department of Biology, Imperial College at Silwood Park, Ascot, Berks SL5 7PY, UK

S. P. Rushton
Centre for Life Sciences Modelling, University of Newcastle upon Tyne, Newcastle NE1 7RU, UK

N. Shigesada
Department of Information and Computer Sciences, Nara Women's University, Kita-Uoya Nishimachi, Nara 630-8506, Japan

A. B. South
Anatrack Ltd, Edward Grey Institute, Department of Zoology, South Parks Road, University of Oxford, Oxford OX1 3PS, UK

N. C. Stenseth
Division of Zoology, Department of Biology, University of Oslo, PO Box 1050, Blindern, N-0316 Oslo, Norway

W. J. Sutherland
School of Biological Sciences, University of East Anglia, Norwich NR4 7TJ, UK

C. D. Thomas
Centre for Biodiversity and Conservation, School of Biology, University of Leeds, Leeds LS2 9JT, UK

K. Thompson
Department of Animal and Plant Sciences, University of Sheffield, Sheffield S10 2TN, UK

A. R. Watkinson
Tyndall Centre for Climate Change Research and Schools of Biological and Environmental Sciences, University of East Anglia, Norwich NR4 7TJ, UK

R. E. Welters
NERC Centre for Ecology and Hydrology, CEH Dorset, Winfrith Technology Centre, Dorchester, Dorset DT2 8ZD, UK

M. Williamson
Department of Biology, University of York, York YO10 5DD, UK

R. J. Wilson
Centre for Biodiversity and Conservation, School of Biology, University of Leeds, Leeds LS2 9JT, UK

I. P. Woiwod
Plant and Invertebrate Ecology Division, IACR-Rothamsted, Harpenden, Herts AL5 2JQ, UK

R. A. Wyand
Department of Disease and Stress Biology, John Innes Centre, Colney Lane, Norwich NR4 7UH, UK

D. Yu
Plant Protection Institute, Hubei Academy of Agricultural Sciences, 430064 Wuchang, Wuhan, China

History of the British Ecological Society

The British Ecological Society is a learned society, a registered charity and a company limited by guarantee. Established in 1913 by academics to promote and foster the study of ecology in its widest sense, the Society currently has around 5000 members spread around the world. Members include research scientists, environmental consultants, teachers, local authority ecologists, conservationists and many others with an active interest in natural history and the environment. The core activities are the publication of the results of research in ecology, the development of scientific meetings and the promotion of ecological awareness through education. The Society's mission is:

To advance and support the science of ecology and publicize the outcome of research, in order to advance knowledge, education and its application.

The Society publishes four internationally renowned journals and organizes at least two major conferences each year plus a large number of smaller meetings. It also initiates a diverse range of activities to promote awareness of ecology at the public and policy maker level in addition to developing ecology in the education system, and it provides financial support for approved ecological projects. The Society is an independent organization that receives little outside funding.

British Ecological Society
26 Blades Court
Deodar Road, Putney
London SW15 2NU
United Kingdom
Tel.: +44 (0)20 8871 9797
Fax: +44 (0)20 8871 9779
E-mail: info@BritishEcologicalSociety.org
ULR: http://www.britishecologicalsociety.org

The British Ecological Society is a limited company, registered in England No. 15228997 and a Registered Charity No. 281213.

Preface

Dispersal has been recognized as an important subject throughout the relatively short history of ecology as a science. However, development of refined techniques for measuring dispersal, extension of ecological theory and attempts to solve current applied problems have all led to dispersal assuming a central role in modern ecology.

Population ecologists have long been able to obtain accurate measurements of survival and fecundity in plant and animal populations, and there is a good understanding of these processes even in microbial population. However, dispersal, the remaining element of many life cycles, has traditionally been measured only rarely. Theoretical advances in several areas of ecology have highlighted this omission, and have created a need for empirical analyses of dispersal. Dynamics of metapopulations centre on dispersal between habitat patches. Invasion ecology, island biogeography and colonization studies require data on how individuals reach virgin communities. Landscape ecology needs to incorporate dispersal to understand the effect of landscape change. The spatial modelling of metapopulations, invasions and even the persistence of individual populations requires a representation of the amount and form of dispersal between patches. Dispersal underpins ecological genetics, and is vital for understanding gene flow between populations and species. Life-history evolution and theories of the development of life-history strategies need to take more account of dispersal abilities and the costs and benefits of particular dispersal strategies. Biogeographical interest in the limits of species distributions and the formation of regional biota also needs information on dispersal. Studies of dispersal are fundamental to many problems in applied ecology: the spread of alien species, responses to habitat loss and fragmentation, re-establishment in species recovery programmes, the spread of diseases, species' ability to track climate change, and the escape of genetically modified organisms.

These considerations have caused an upsurge in interest in the study of dispersal, especially over the last 2–3 years. The 2001 Ecological Society of America meeting in Madison, Wisconsin had a special symposium on 'Long distance dispersal' (to be published in *Ecology*), and there have been two recent symposia with associated books on particular aspects of dispersal; the genetics and evolution of dispersal, especially in animals (Clobert *et al.* 2001) and dispersal in insects (Woiwod *et al.* 2001). What is missing, however, is a broad overview of dispersal in terms of its role in all areas of ecology and covering the range of taxa from microbes to vertebrates. This book aims to achieve this and is based on the BES Symposium on Dispersal which took place in April 2001 at Reading University, UK. As well as a review of the status of knowledge about dispersal, it provides suggestions as to where dispersal research is going and what gaps in knowledge need to be plugged.

To achieve this we commissioned 21 chapters by international experts in particular fields. Within the size constraints of this book, these represent as wide-ranging an overview of dispersal as is available. The chapters cover most taxa—baculoviruses, bacteria, fungi, plants, invertebrates and vertebrates—in order to consider similarities and differences in the problems and questions being tackled in traditionally separate disciplines. Despite the great differences in approach, the outcome is an emergence of many basic similarities (see Nathan 2001 for an insightful review of this symposium). These include the fundamental importance of dispersal for all taxa, the role of dispersal in linking ecological processes across different scales, the problems of accurately quantifying dispersal, and the importance yet difficulty in measuring rare long-distance dispersal events.

Many debates in ecology arise through semantic differences. To achieve clarity, we therefore asked authors to use a common definition of dispersal, as 'intergenerational movement'. So, for example, we exclude so-called 'dispersal in time' (e.g. seed banks) and foraging movement of animals. The authors have mostly kept to this definition, and any deviations are explained. Another term used often is 'migration'. The *Oxford English Dictionary* defines this both as 'The action . . . of moving from one place to the other' and as 'Change in . . . the distribution of a [species]'. We have not imposed a definition, so where 'migration' is used in passing, the context must be considered. However, where authors have used this word extensively they have provided a definition. The periodic migrations by many birds and other animals, e.g. from summer to winter grounds, are termed 'annual migrations' in this book.

The structure of the book is to consider the four fundamental questions in dispersal research: (i) *what techniques are available to measure dispersal*; (ii) *what is the evolutionary and behavioural basis of dispersal*; (iii) *what is the role of dispersal in spatial processes at all scales*; and (iv) *what do we need to know about dispersal to solve applied ecological problems*? Because of the biological differences among different taxa, the chapters on techniques are classified taxonomically, although molecular methods merit a separate overarching chapter. Otherwise, the chapters are designed to cover a full range of issues within each question, either using a specific taxon or a range of taxa to illustrate each issue. Generality is achieved by a consideration of links among chapters, but also by the literature reviews presented in each chapter and the modelling approaches which are at the core of many. Thus the second question is addressed by considering dispersal cues, dispersal behaviour, genetic structure at different scales, evolutionary trade-offs, and the effects of dispersal trait changes on fitness, in a range of taxa that includes baculoviruses, birds, plants, bryozonans and insects.

The scales of spatial processes considered in different chapters are within populations, metapopulations, landscapes and biogeographical regions. The explicit consideration of dispersal in each chapter means that none stick to one particular (anthropomorphically defined) scale, but rather the role of dispersal by individual bacteria, plants, insects or vertebrates in linking scales is a common theme. Applied problems are many, and the chapters in the fourth section show how dispersal is at the core of approaches to solving these. Thus, the chapters consider: the spread of non-native organisms, of pests and diseases and invasions in general; conservation re-introductions and responses to habitat fragmentation; and whether species will be able to track climate change. It is indicative of the increasing involvement of ecologists in environmental concerns that many of the other chapters in this book also consider applied questions. Finally, Mark Williamson provides a personal overview of the book in terms of the objectives we set ourselves.

We would like to thank the BES for funding both the symposium and the production of this book, particularly Hazel Norman whose organizational skills made the symposium such a success. Ian Keary and Paul Hatcher also helped enormously with the running of the symposium. The authors all worked very hard to produce the top-quality reviews for which they were asked, and we are grateful to the many ecologists who helped enhance the quality by critically reviewing the chapters.

It has been extremely exciting to be involved in production of a book in such a fast-developing and important field. This book describes the state of the art. It will be very interesting to come back in 10 years' time to review the leaps and bounds forward this subject will have taken—partly, we hope, as a result of this book.

James Bullock
Robert Kenward
Rosie Hails

References

Clobert, J., Danchin, E., Dhondt, A.A. & Nichols, J.D. (eds) (2001) *Dispersal.* Oxford University Press, Oxford.

Nathan, R. (2001) The challenges of studying dispersal. *TREE* **16**, 481–483.

Woiwod, I.P., Reynolds, D.R. & Thomas, C.D. (eds) (2001) *Insect Movement: Mechanisms and Consequences.* CAB International, Wallingford.

Part 1
Techniques for dispersal studies

Chapter 1

Measuring and modelling seed dispersal of terrestrial plants

David F. Greene and Catherine Calogeropoulos

Introduction

In this review we will focus on seed dispersal as an empirically verified function that could be coupled with other life history arguments to simulate spatially explicit population dynamics at any scale of interest. Our interest, therefore, is on the problems of observing and then expressing the dispersal curve. A review of techniques in the study of seed dispersal and recruitment is timely because it has become clear that there are two outstanding problems, and both are related to present limitations in field methods: (i) the shape of the dispersal curve near the maternal source as a function of source geometry and dispersal capacity, and (ii) the magnitude of seed deposition as a function of source strength (seeds per metre squared within the source) at very large distances.

We begin by defining these two scales of interest more concretely, and then offer a formal definition of a dispersal curve. Next, we summarize what is known about seed dispersal curves in relation to dispersal vectors for ideal situations where the seed or recruit can be attributed to a source without ambiguity. We then deal with models (primarily empirical; mechanistic models are dealt with cursorily), not merely because of their utility in spatially explicit simulators, but because of their recent importance as a tool for 'disentangling' overlapping dispersal curves. Finally, we will examine vector-specific methods for determining dispersal curves, analyse their merits and drawbacks, and, for the more tractable problems, make some recommendations for improvements.

Scales of interest

At one extreme, we know from the palynological literature that most trees, shrubs and herbs have (in the higher latitudes at least) migrated at a velocity (V) on the order of 200 m year^{-1} (MacDonald 1993). It is likely that the potential velocity is greater than this but migration rates were constrained by the rate at which the climate regime (affecting reproduction and growth rates) itself was moving latitudinally (e.g. MacDonald *et al.* 1993). The scalar for this minimal migrational velocity can be written as Vt_g where t_g is the characteristic generation time (in years). For annuals ($t_g = 1$ if they cannot produce more than one generation in a growing season), Vt_g is about 200 m. For trees, the age at which they can produce a large (we leave this undefined) crop is a function of the species-specific growth rate (not age). Unconstrained by any serious data set, let us assume this generation time is on the order of 20 years for trees, and therefore the scaled distance is $Vt_g = 4$ km. This great difference in the value of the scalar for herbs versus trees has obvious implications for field logistics. We note in passing that the scalar for migrational distance is also of interest in metapopulation studies given landscape fragmentation (Hanski 1999; Cain *et al.* 2000).

The shortest distance of interest is more difficult to define. A population that can seize vacant space from dying competitors or conspecifics will tend to persist across a few disturbance cycles, thus easing the required immigration within a metapopulation (Bolker & Pacala 1999). Let us arbitrarily assert that population persistence requires that some fraction of the crop achieves a dispersal distance (x) greater than $0.5H$ (where H is the plant height). Thus, we argue that population persistence requires that vectors disperse a large fraction (again, undefined) of the crop to a distance greater than about 0.1 m for a small herb such as *Taraxacum officinale* or 12 m for a tree.

In summary, short-term and long-term population persistence requires that the species available for study today have simultaneously satisfied the requirement that dispersal curves span about three orders of magnitude. Below, we examine how individual species have achieved this range using one vector (rarely) or more typically a large number of vectors.

Defining the dispersal curve

A point source is an individual plant where the lateral spread among the preabscission seeds is small compared to the median distance travelled. Obviously, widely spread clonal plants are not point sources. We define the dispersal curve as the number of seeds deposited at distance x from a point source bearing Q seeds:

$$Q_x = Q f(x) \tag{1.1}$$

Here $f(x)$ is the dispersal term. Alternatively, one might be more interested in the density of seeds at x (Q_{Dx}), and this density curve is merely

$$Q_{Dx} = Q_x / 2\pi x \tag{1.2}$$

For an area source (a very large array of point sources) the curve is given by

$$Q_{Dx} = (QN_D/2) f(x, y) \tag{1.3}$$

where Q is now the averaged seed production of a conspecific point source, N_D is the density of these point sources, $x = 0$ is defined as the edge of the area source, and $f(x,y)$ is an empirical function that accounts for the position of the point source in Cartesian space.

Any of these three equations can be changed into a recruitment curve by multiplying Q by the mean survivorship averaged across the area of interest if this survivorship is expected to be roughly independent of distance. In what follows, we will casually let Q_x and Q_{Dx} refer to recruitment as well as seed curves.

With lesser or greater difficulty, these equations can be changed to express density dependence or azimuthal bias in dispersal. Empirical demonstrations of azimuthal bias can be found in Harris (1969) (bias induced by the abscission response to the relative humidity of air masses), Yumoto (1999) (birds preferentially defecating into gaps) and McDonnel and Stiles (1983) (birds' defecation biased by structural elements of the vegetation).

Empirical dispersal curves (Q_x)

In this section we divide seed dispersal vectors into those that can merely satisfy the requirement for local dispersal and those that can also satisfy the migrational requirement. We further dichotomize vectors as primary (the initial postabscission event) or secondary (i.e. involving subsequent re-entrainment). These dispersal agents are listed in Table 1.1.

Local distance vectors

These include ants (secondary), small flightless mammals (primary or secondary), wind (primary but for seeds with very large terminal velocities) and ballistic (primary). These generally produce right-skewed Q_x curves (we know of two exceptions for ballistic), and most place the modal density of the Q_{Dx} curve at the point source (Figs 1.1 and 1.2). A large fraction of the crop is transported well beyond 50% of a plant height from the source. While distances for herbs (Fig. 1.1) are much less than for trees (Fig. 1.2), note that an ant dispersing seeds 1 m from a 0.3 m tall herb is, using the scalar x/H, equivalent to a rodent transporting tree ($H = 25$ m) seeds about 75 m. Maximum reported distances are several plant heights for ants, caching rodents, ballistic and species with large terminal velocities (Table 1.1; maxima for herbs are also tabulated in Willson 1993). Even for herbs with short generation times, however, these distances are too small to satisfy the migrational scale (Vt_g).

Long-distance vectors

These include wind as both a primary (plumes, wings or seeds less than about 0.02 mg) and secondary (on low-friction surfaces such as snow, ice or sand) vector, defecating large terrestrial or arboreal mammals (primary or secondary), defecating birds (primary), caching birds (primary or secondary), and epizoochory by both primary (the seed is attached to the animal via hooks or an adhesive substance) and secondary (a very small seed, say <10 mg, is attached to an animal in a fleck of mud) means.

At the migrational scale our assertion of the adequacy of these vectors is based completely on anecdotes or models. There are anecdotal reports for distances of many kilometres achieved by *Nucifraga* (Vander Wall & Balda 1977) and other caching birds, but of course we have no estimate of the proportion of the crop in-

Table 1.1 Seed dispersal vectors by mode: primary (p) or secondary (s). Our categorization is based on similarities in vector-specific field methods *and* in similarities in mechanistic modelling. Vectors are typified by logarithmic seed size (mg) classes (<0.05 is very small; 0.05–5 is small; 5–50 is large; >500 is very large) where rare exceptions are ignored. Maximum distances (Max. *x*) observed are only for full dispersal curves (from a source of any shape or size), not anecdotal reports. The empirical argument that a seed size <0.05 mg will have terminal velocities as low as many plumed or winged seeds is based on Cremer (1977), Woodall (1982) and Jongejans and Schippers (1999). (Based on Cremer 1966; Howe & Smallwood 1982; Willson 1993; Greene & Johnson 1995; Yumoto 1999.)

Vector and mode	Common vegetation and strata	Seed size range (mg)	Max. *x* (m)
Ants (s)	Short plants; sclerophyllous vegetation and forest understories	0.05–5	8
Small terrestrial vertebrates (p or s)	No limits on height; no typical vegetation	>0.05	60
Ballistic (p)	Mainly shorter plants; no typical vegetation	0.05–50	Herbs: 12 Trees: 30
Wind (large terminal velocity) (p)	Wide range of plant sizes; no typical vegetation	>0.05	Herbs: 5 Trees: 120
Wind (plumes or wings or very small size) (p)	All plant heights; mainly canopy-stratum plants	<50	Herbs: 150 Trees: 1600
Wind (s)	Low-friction substrates: sand, snow, ice; high-latitude vegetation and deserts	<50	100 m
Caching birds (p or s)	Mainly mid-latitude trees	>50	–
Defecating large animals (flying, arboreal or terrestrial) (p or s)	No limit to plant size or vegetation type	>0.05	700
Epizoochory (hooks, barbs, adhesive material) (p)	Small plants; grasslands and disturbed sites	0.05–5	Herbs: 300 Trees: –
Epizoochory (seeds in mud) (s)	Small plants: grasslands and hydric sites	<5	–

volved in this far tail. For epizoochory, essentially nothing is known at any scale with the lone exception of Yumoto (1999; but the sample size is quite small). For wind (secondary) entraining seeds on snow or sand there is no observed dispersal curve extending beyond 100 m (Greene & Johnson 1997; Vander Wall & Joyner 1998).

Given the empirical void at great distances, it is not surprising that there is no agreement on whether some vectors are more effective at achieving such distances than are others. Willson (1993) argued that wind (primary) is more effective at modest distances than animals (all vectors grouped), whereas Hewitt (1999) made

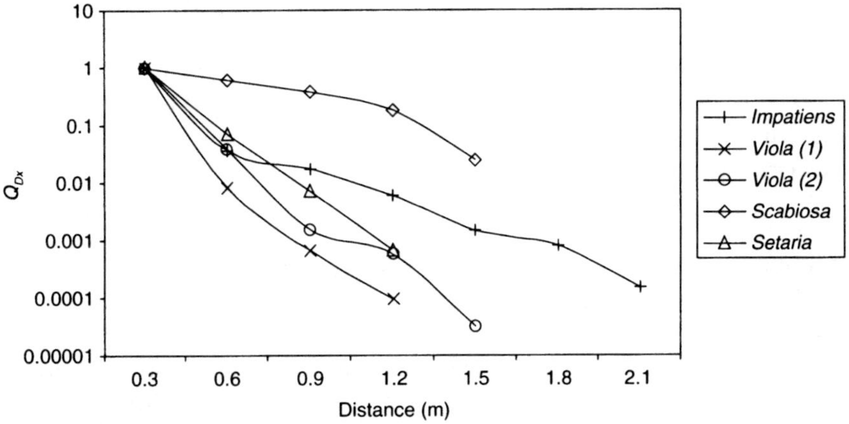

Figure 1.1 Local dispersal for herbaceous point sources dispersing ballistically (*Impatiens*, Schmitt *et al.* 1985), by ants (*Viola* 2, Ohkawara & Higashi 1994; *Viola* 1, Anderson 1988) and by wind (seeds with a high terminal velocity) (*Scabiosa*, Verkaar *et al.* 1983; *Setaria*, Ernst *et al.* 1992).

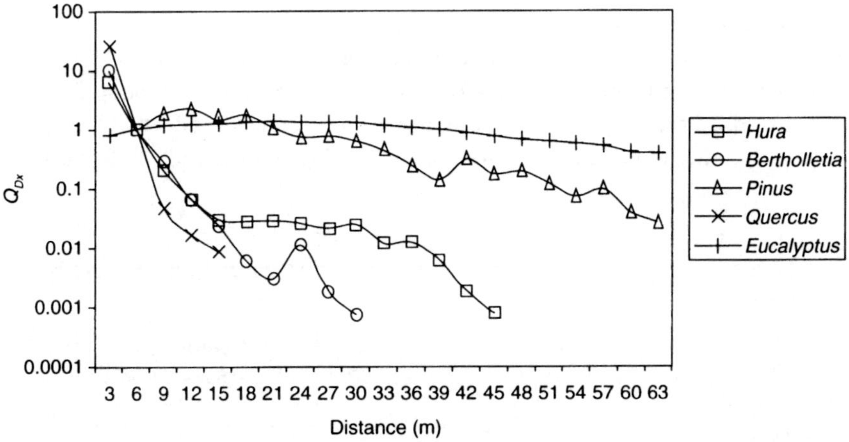

Figure 1.2 Local dispersal for tree point sources dispersing ballistically (*Hura*, Swaine & Beers 1977), by rodents (*Bertholletia*, Peres & Baider 1997; *Pinus*, Vander Wall 1993; *Quercus*, Sork 1984) and by wind (seeds with a high terminal velocity) (*Eucalyptus*, Cremer 1966). Note that the *Eucalyptus* is an 80 m tall tree.

the opposite argument for colonization of *Pinus* plantations by hardwood species. Wilkinson (1997) asserted that birds must be the unacknowledged long-distance vectors for even winged or plumed seeds because it was (somehow) clear that the wind was incapable of moving seeds many kilometres. Given that our best data sets at the scale of 1 km are at present provided by wind-dispersed species, this speculation seems odd.

In Fig. 1.3 we show the only long-distance point source dispersal curves for animal vectors (two defecating terrestrial bird species and a large arboreal mammal). As before, the Q_x curves are right skewed. We are not, of course, able to say if there is a characteristic Q_{Dx} curve shape.

Models, source geometry and the empirical curves of wind-dispersed trees

The contagious distribution of conspecific sources means that, almost invariably, dispersal curves overlap. There are two classes of models: (i) simple dispersal terms with one or more parameters that are fitted to a data set, or (ii) mechanistic terms whose parameters can be estimated independently of an empirical dispersal curve. We begin with the former class, showing their performance at various scales, and then we use one of them (the 2Dt) to demonstrate the effect of source geometry on the shape of the dispersal curve. Mechanistic models are treated here only briefly as an alternative although one might presume they will eventually be of more interest as this discipline tires of curve-fitting and begins to seek generality.

Empirically fitted models

Figure 1.4a and Table 1.2 show some commonly used functions in their Q_{Dx} forms with the median distance (x_m) set at 40 m and Q (seed crop size) at 1000 seeds. We have also added the simplest intuitive argument: a rectilinear model that has seed density (Q_{Dx}) declining linearly with distance. The log-normal and 2Dt forms are

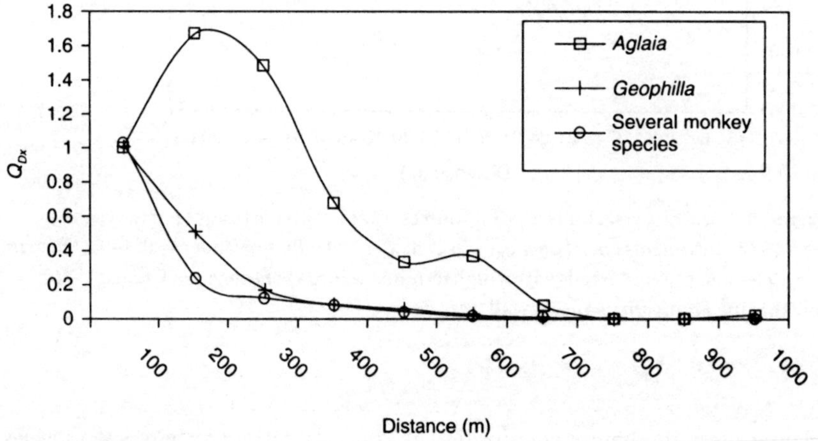

Figure 1.3 Long-distance seed dispersal of trees via defecation by a large terrestrial bird (*Aglaia*, Mack 1995), a flightless bird (*Geophilla*, Yumoto 1999) and monkeys (Stevenson 2000).

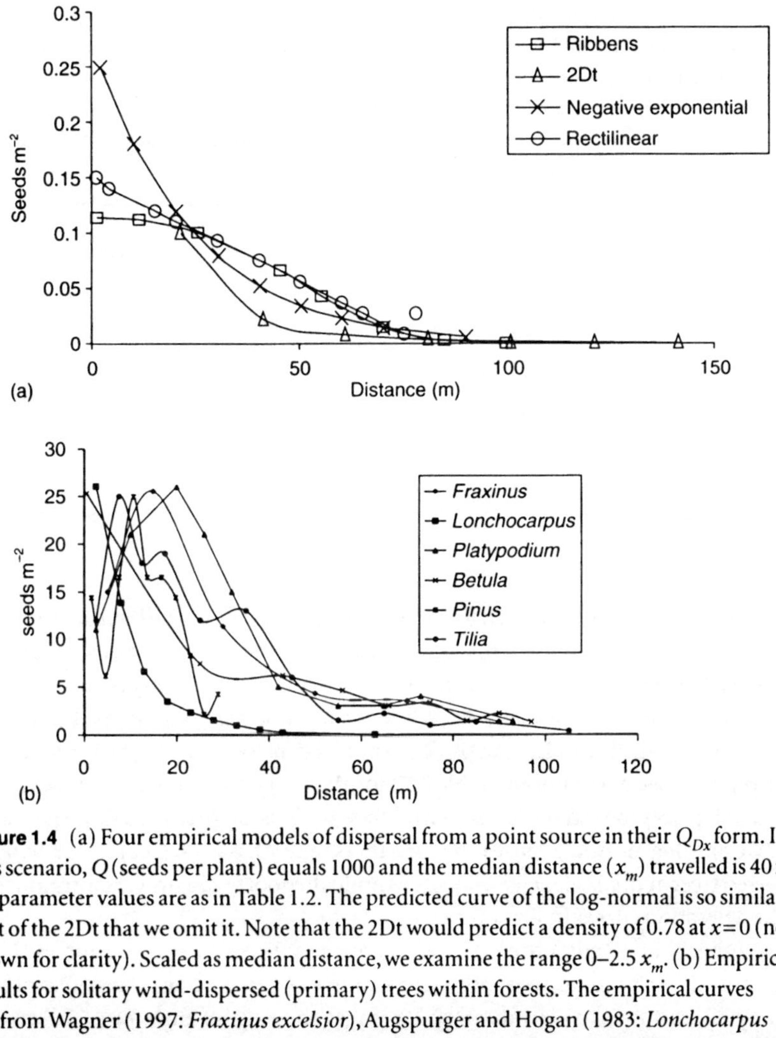

Figure 1.4 (a) Four empirical models of dispersal from a point source in their Q_{Dx} form. In this scenario, Q (seeds per plant) equals 1000 and the median distance (x_m) travelled is 40 m. All parameter values are as in Table 1.2. The predicted curve of the log-normal is so similar to that of the 2Dt that we omit it. Note that the 2Dt would predict a density of 0.78 at $x=0$ (not shown for clarity). Scaled as median distance, we examine the range 0–2.5 x_m. (b) Empirical results for solitary wind-dispersed (primary) trees within forests. The empirical curves are from Wagner (1997: *Fraxinus excelsior*), Augspurger and Hogan (1983: *Lonchocarpus pentaphyllus*), Augspurger (1983b: *Platypodium elegans*; deposition reduced by 100), Rudis *et al.* (1978: *Pinus strobus*) and *Tilia americana* (D.F. Greene, unpubl. data).

additionally constrained (based on the empirical evidence below) to have a deposited density at 1 km that is 2% of the density well inside an area source.

Because of the crucial role of recruitment in postharvest stocking success in forestry, there is, compared to herbs and shrubs, a wealth of studies on the dispersion of seeds or germinants of trees within forests and adjacent clearings. Figure 1.4b depicts observed curves of mid-latitude and tropical wind-dispersed tree species where: (i) we have a reasonable density of sampling points, (ii) the authors explicitly state that there are no other nearby conspecifics, and (iii) the tree is not in a clearing or at a forest edge. (Our 40 m median in Fig. 1.4a is based loosely on Fig. 1.4b.) In four of five cases, the mode of the Q_{Dx} curve is clearly displaced well away from the base of

Table 1.2 Empirical models and parameters. x_m (median distance travelled) is set at 40 m and Q (seed production) at 1000 seeds. The log-normal and 2Dt are given the additional constraint that the density of deposition at 1 km must be 2% of the density well within an area source. The sources for the functions include Willson (1993), Ribbens *et al.* (1994), Clark *et al.* (1999) and Greene and Johnson (2000). For the rectilinear model (introduced here) the single parameter is the maximum distance travelled (x_{max}).

Model	Dispersal term for seeds m^{-2} at distance x	Parameter values
Gaussian	$a\exp(-ax^2)/\pi$	$a = 0.000425$
Ribbens	$1.1d^{0.67}\exp(-dx^3)/\pi$	$d = 0.0000057$
Rectilinear	$3(x_{max}-x)/\pi x^3_{max}$	$x_{max} = 80$
Negative exponential	$g^2\exp(-gx)/2\pi$	$g = 0.0415$
2Dt	$p/[\pi u(1+(x^2/u)^{p+1})]$	$p = 0.25; u = 110$
Log-normal	$[(2\pi)^{1.5}\sigma_{ln}x^2]^{-1}\exp[-(0.5/\sigma_{ln}^2)(\ln(x/x_m))^2]$	$\sigma_{ln} = 2.4; x_m = 40$

the tree. Clark *et al.* (1999) argued that the modal density should be beneath the tree because the tree is not truly a point source (i.e. there is variation in release height as well as in lateral position within the crown) but we consider that this justification for locally convex models is not correct. While the log-normal (as an empirical argument or a mechanistic model) can place the modal density at a distance greater than 0, the σ_{ln} value we have chosen (Table 1.2) for the proper fit at great distances is so large that it places the modal deposit virtually at 0. Thus all these models (especially the 2Dt and log-normal) fail to capture the shape of the Q_{Dx} curve for wind-dispersed trees at this scale. Likewise, they would do poorly as a universal argument for the defecation curves in Fig. 1.3.

Two other source geometries of interest are line sources and patch sources (Fig. 1.5). Using the 2Dt model, we express a line source as a single linear array one plant (occupying a 10 × 10 m cell) deep, and $x = 0$ at the edge of the source. Likewise we show in Fig. 1.4 what we will term a *patch source* which is defined, necessarily vaguely, as a collection of conspecifics much smaller than an area source. Here we depict the 2Dt model for a patch source 80 × 40 m (32 individuals). For the line source the edge deposit is around 25% of the product QN_D (where N_D is the point source density as plants per metre squared), and for the patch source it is around 35%. Note that both source shapes produce a flatter curve than would a point source. (This effect would be even more dramatic if our co-ordinant system placed $x = 0$ in the middle of the line or patch source). Intrepretation of dispersal curves must take account of source geometry. For example, Johnson (1988) argued that *Fraxinus americana* was a better disperser than *Acer saccharum* because the empirical Q_{Dx} curve for the former (a line source along a fence row) was flatter than that of the latter (a single tree source) when plotted on a semilog graph. His many biomechanical speculations

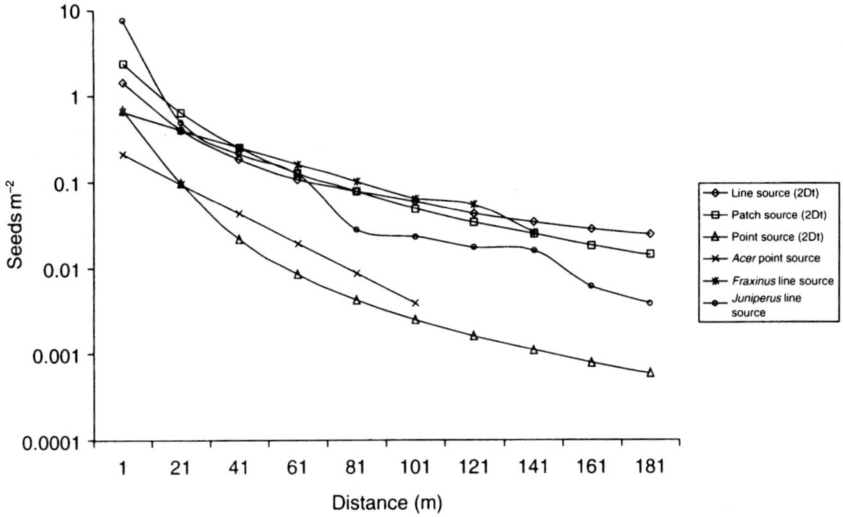

Figure 1.5 The 2Dt model as a line source or small patch source with the same point source parameter values as in Table 1.2. The line source is 4 km long, and has a width of one point source per 10 × 10 m, and this can be scaled as 0.25 × 100 x_m. The patch source is 80 × 40 m (scaled 2 × 1 x_m). We also show two empirical curves for *Fraxinus americana* (line source) and *Acer saccharum* (point source) from Johnson (1988) that are made identical to the simulated density at x= 20 m. For comparison, a line source of *Juniperus virginiana*, a bird-defecated species, is also shown.

notwithstanding, it is clear from Fig. 1.5 that no such intuitive interpretation was warranted.

Where the empirical models differ most is at larger spatial scales. In Fig. 1.6 we show the expected curves (again, for each point source, Q = 1000 seeds; x_m = 40 m) for an area source (an array of individual conspecific sources of extent ≫ x_m^2). In this case, the plants were uniformly distributed across a rectangle 4 × 2 km abutting an area with no conspecific sources; x = 0 marks the edge of the area source with increasing x as we sample further from the source. With a spacing (S) of 1 plant per 10 × 10 m contiguous cell (and the seeds located at a single point in the middle of that cell), we have a deposited seed density of 10 seeds m^{-2} (10 = QN_D where N_D, as in equation (1.3), is the individual source density ($1/S^2$)) well within the area source. This density declines slowly to 5 m^{-2}, i.e. half of QN_D, at x = 0. It subsequently declines more rapidly from x = 0 to x = 200. Only the log-normal and 2Dt permit appreciable dispersal at this scale.

Empirical studies of greater distances than 200 m for wind-dispersed trees abutting clearings are rare; the few available examples are shown in Fig. 1.7. The flattening hinted at in Fig. 1.6 is confirmed; at larger distances there is a remarkable reduction in the rate of decline of seed deposition. Of the models presented in Table

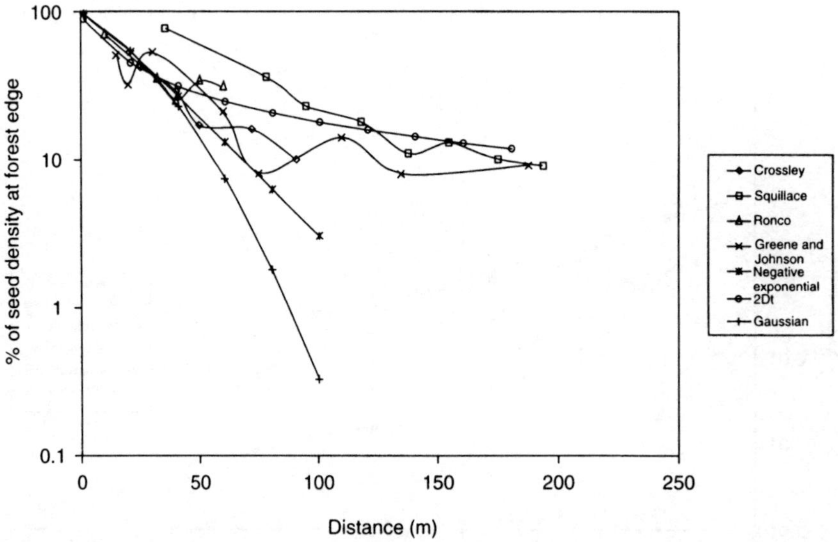

Figure 1.6 Area source curves (Q_{Dx}) for three of the models in Fig. 1.4b (Q and x_m remain the same). $x=0$ indicates the edge of the area source. Scaled as median distance, the range in distances is 0–5 x_m. We have presented the deposition as a percentage of the deposit at $x=0$. Also shown are four empirical curves for wind-dispersed (primary) area sources abutting clearings. The species are *Picea engelmannii* and the closely related *P. glauca*. (References for these studies as well as numerous other examples from the forestry literature are given in Greene and Johnson (1996).)

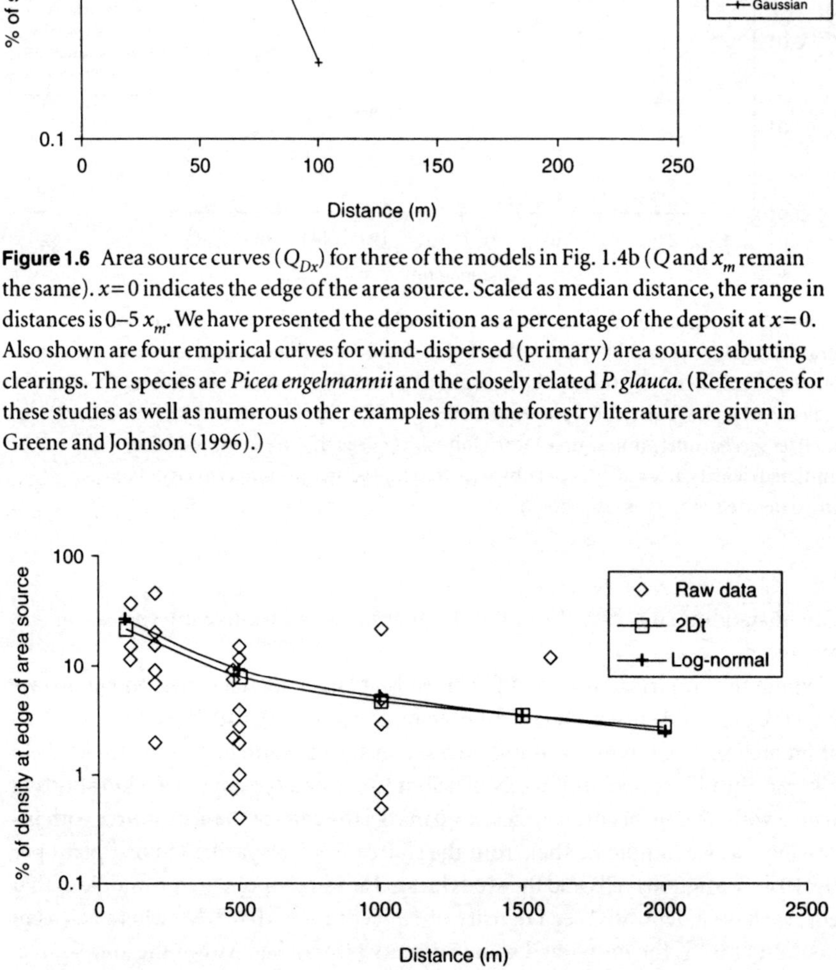

Figure 1.7 Observed long-distance dispersal up to 1.6 km from a forested area source for wind-dispersed (primary) trees (scaled 0–40 x_m). Original data sets can be found in Greene and Johnson (1995, 2000). Also shown are model curves using the parameter values in Table 1.2 for the 2Dt and the log-normal. The log-normal was not shown in Figs 1.4–1.6 because the curve was essentially identical to the 2Dt for these parameter values. Both models were forced to have a median (x_m) at 40 m and to have a deposit at 1 km that is 4% (mean of the empirical data) of the deposit at the area source edge.

1.2, only the 2Dt and log-normal can be sufficiently flat at these great distances. (Note: Higgins and Richardson (1999) ungenerously refer to the log-normal as a 'thin-tailed' function; this is an error.)

Generalizations regarding the effect of source geometry on dispersal curve shape

This exercise (Figs 1.4–1.7) allows us to make certain generalizations. First, as the size of the conspecific source aggregation becomes larger, the Q_{Dx} curve flattens strongly. (The bird-defecated line source curve for *Juniperus virginiana*, median distance unknown, was added to Fig. 1.5 to underscore the role of source geometry.) The shape of the curve is a function of source geometry. As we pointed out, considerations of source geometry are crucial when one takes empirical data, estimates some measure of central tendency (e.g. the mode or median in a Q_D curve) or rate of decline as a summarizing parameter for dispersal capacity, and then compares different species. This is especially problematic for herbs (examples include Okubo & Levin 1989; Willson 1993); a small patch or clone (say, 1 m^2 with a height of 0.2 m) of herbaceous conspecifics would, all else equal, appear to have a greater mean dispersal distance than an isolated conspecific point source.

Figures 1.3–1.7 make one wonder how we arrived at the commonplace assertion that 'most seeds travel short distances' (e.g. Cain *et al.* 2000). Certainly, this is wrong for the long-distance vectors, and is analogous to declaring that keys or contact lenses are typically lost near streetlights. Q is seldom estimated, and so our evaluation of the median distance for a Q_x curve is based on the 50th percentile of what we sampled; and, invariably, we sample primarily near the plant.

A final lesson from this exploration of empirically fitted models is that all of them will perform equally well at the scale of one or two medians away from the source. Whether one uses chi-square (e.g. Stoyan & Wagner 2001) or a regression of predicted versus observed (e.g. Ribbens *et al.* 1994) for a Q_{Dx} curve to test for significance, all of them will perform about equally well at short distances. Even the rather silly rectilinear model will achieve a correlation as good as the others in Fig. 1.4. With correlation, for example, a model might underpredict very badly at, say, 80 m but both the observed and predicted values will be near 0 at such a large distance and have little effect on the correlation or the regression coefficients. Taking the logs of the values is hampered by the fact that the invariably insufficient trap area examined leads to an abundance of 0 values at large distances. Likewise, with chi-square, the ubiquity of 0 values for the observed deposition at large distances will force one to lump cells, and thus the significance of the test will depend almost completely on the fit in the distance interval within about one median away from the source. The problem is exacerbated when one uses the inverse approach developed by Ribbens *et al.* (1994) and refined by Clark *et al.* (1999) for estimating the point source parameters of a model given empirical data from within an area source. Again, any of the models in Fig. 1.4a could be substituted for the one used by either of these groups of authors and would achieve a similar level of significance. Clearly, it is only at much larger distances (Figs 1.6 and 1.7) that we detect serious differences among them (e.g. Bullock & Clarke 2000).

We recommend that in modelling exercises (e.g. stand dynamics or metapopulation simulators) only the 2Dt and log-normal be used as they are the only ones that can capture the shape of the density curve across a large range of distances. However, both models will greatly overpredict within about 0.1 x_m if they are to predict well at other distances, and thus it would appear it is not possible to simultaneously perform well at all scales of interest if the dispersal term has no more than two parameters. As discussed by Higgins and Richardson (1999), a mix of two models or a three-parameter dispersal term may be the most sensible approach for dispersal subroutines within simulators.

Mechanistic models

Mechanistic models for wind (primary) and animal (primary or secondary) dispersal are reviewed by Turchin (1998), and treated here only briefly. All the mechanistic arguments we are aware of necessarily lead to the right-skewed Q_x curves so typical of the empirical literature. For example, Greene *et al.* (1999) argued that horizontal wind speeds are log-normally distributed, and that the skew increases with the magnitude of the vertical turbulence (which is the primary control on $\sigma_{\ln}$ (Table 1.2)). As we have seen, with a large standard deviation, observed wind (primary) curves can be reasonably expressed at all relevant scales (except within about one-tenth of a plant height of the source). The outstanding problem is understanding the interaction between vertical turbulence and the abscission bias (Greene and Johnson (2000) present a first attempt at this coupling) so that a $\sigma_{\ln}$ as large as in Table 1.2 can be justified mechanistically and can reduce the overestimation near the point source. One biomechanically realistic possibility is that vertical winds are not normally distributed during abscission; i.e. the seed is more likely to be removed from the ovulate cone or preduncle by an updraft than by a downdraft (Greene & Johnson 1992). Implementation of this suggestion would require a random walk model (e.g. Jongejans & Schippers 1999) as a platform.

A model of secondary entrainment and dispersal by wind on idealized surfaces (snow, ice, sandy desert substrates) with a uniform distribution of roughness elements has been developed by Greene and Johnson (1997). There is no data set available at present that would allow the model to be properly tested at even modest distances.

There are a number of mechanistic models available for animal dispersal of seeds (either primary or secondary) and they tend to have a minimum of four parameters: the mean and variance of the seed transit time on or within the animal, and the mean and variance of the net velocity of the animal away from the source (reviewed in Turchin 1998). For a single seed, the distance travelled is simply the product of net velocity (a function of territory size, foraging behaviour, etc.) and transit time on or within the animal (primarily a function of seed size and animal size). Given the multiplicative process, then, even if the distributions for both transit time and velocity are normal, then the Q_x curve will be right skewed. Thus, very simple models involving only these minimal parameters will produce right-skewed curves (e.g. Murray 1988) and should approach a log-normal form. Indeed, the shape (not necessarily the scale) of the curves generated by Murray (1988) for bird dispersal and Hickey *et al.* (1999) for dispersal by the marten are similar to the empirical wind-dispersal curves in Fig. 1.4b.

Field methods for observing dispersal curves

We begin by reviewing the more common field methods for developing empirical dispersal curves. The first two methods (traps; examination of seeds, germinants or seedlings *in situ*) share a major problem: the source cannot be known with certainty except in ideal situations (e.g. a single plant well-isolated from other conspecifics; a single area source or patch source). Another problem is that one cannot be sure which vector was responsible for the dispersal. The next two methods (following individual seeds; mark-recapture of labelled seeds) permit identification even when one has overlapping dispersal curves.

Seed traps

For wind or defecating flying vectors, seed traps are typically employed. There are four possibilities for trap designs. The most common approach is a box on the ground (Pickford 1929) or a suspended bag (Hughes *et al.* 1987). They may be screened on top to prevent granivory (with the added benefit that the reduction in litter input reduces the time examining the trap contents), and with screened holes in the bottom to permit water drainage. Other kinds of traps include a flat surface covered with a viscous substance (e.g. 'Tanglefoot': Werner 1975), a container filled with water (Walker *et al.* 1986) or a soil-filled tray which can later be brought into the lab or greenhouse for germination trials (Leadum *et al.* 1997).

One major problem with traps is that the investigator must decide in advance, well before the abscission season, how much trap area will be employed and how it will be arranged within or near the source. Obviously, one wants to avoid zero values for seed deposition but it is easy to guess incorrectly the crop size (seeds m^{-2}) within the source and the likely diminution in density with distance. Indeed, one of the most common features of empirical dispersal studies is the proliferation of zero values that begins, often abruptly, at one or two plant heights from the source.

Trapless methods: examining seeds, germinants and seedlings

If one can sample the ground freely for seeds then the logistical limitations of traps can be avoided. Granivores, of course, are not excluded. This approach can be useful for large seeds (e.g. Augspurger & Hogan 1983). However, for small seeds this becomes a very time-consuming method (see the quantification below) as one must sift through the leaf litter; worse, very small seeds (say 1 mg or less) often adhere to drying leaves and thus each leaf must be examined carefully.

Letting the granivores take their share, examination of germinants and seedlings has been widely used (e.g. MacArthur 1964; Ribbens *et al.* 1994). A major problem is that the investigator must be sure that the mix of seedbeds is relatively independent of distance from the source because, especially for small-seeded species, there can be more than an order of magnitude difference in seedbed-related juvenile survivorship (Greene & Johnson 1998). Taking a worst case example, with fires juvenile survivorship is as much as 50 times higher in a burnt area than in the adjacent intact vegetation. Further, the investigator must have reason to think that density-dependent or distance-dependent losses are unimportant. One way around this problem is to sow seeds at various sites to permit site-specific mortality estimates, but this potential correction procedure is seldom done. For seedlings, these potential seedbed-related problems are compounded by the subsequent light-mediated survivorship (e.g. Augspurger 1983). Now, the investigator must argue that canopy gaps are independent of distance to the source.

Determining the source

Where source curves overlap, there are two methods for determining the maternal source. One is strictly experimental: secondary dispersal can be studied by labelling the seeds. Small seeds can be radiolabelled, dusted with pigments or dyed (e.g. Radvanyi 1966; Vander Wall 1993). Very large seeds (e.g. *Quercus*) can have small bits of metal inserted in them and then subsequently relocated with a metal detector (Sork 1984; Mack 1995). A major problem is that, typically, a large fraction of the tagged seeds are not recovered during the detection procedure and one does not know if these represent insufficient sampling or a fraction that was carried beyond the maximum sampling distance. (A power-law regression of percentage recovered on maximum distance observed yielded a significant (negative) result ($r^2 = 0.58$) for 12 literature studies.)

The ideal method is one that determines both how far a seed travelled and how it arrived there. The Lagrangian method (i.e. following individual seeds during transit) has been tried with wind (primary) (e.g. Zasada & Lovig 1983; Greene & Johnson 1995) as well as with defecating (e.g. Yumoto 1999) and caching birds (e.g. Vander Wall & Balda 1977), but invariably the approach only works for the fraction of seeds travelling short distances. At longer distances the seed or the vector are soon out of sight. The Lagrangian approach is, however, well suited for small caching animals such as ants or, when coupled with labelling, large defecating mammals, flightless birds, and ballistic seeds (e.g. Anderson 1988; Passos & Ferreira 1996). The animal can be followed, and its caches or dung piles examined.

Statistical and genetic methods for overlapping curves

Statistical methods

An alternative approach to disentangling overlapping curves was pioneered by Ribbens *et al.* (1994). With this statistical method, one uses equation (1.1) in combination with the Cartesian co-ordinates and sizes of source plants and the co-ordinates of progeny to estimate the parameters within the dispersal term. Seed production is replaced by a measure of plant size (e.g. basal area for trees; or one might use some surrogate of plant mass for herbs: Shipley & Dion 1992). LePage *et al.* (2000) took the method further by including seedbed-specific mortality in the Cartesian space.

There are a number of problems with this approach. As pointed out earlier, for short distances almost any dispersal term will perform about equally well as long as it places the mode of the Q_{Dx} curve at or near the source; a significant correlation is not an occasion for opening an expensive bottle of Merlot. The models differ greatly at a distance of a few plant heights, however, and thus model choice will certainly affect the outcome of simulations of stand dynamics. For example, the model outcome of Pacala *et al.* (1996) led to some species forming virtually monospecific patches that tended to resist invasion by competitors. But this result may simply be an artefact of their strongly curtailed dispersal term (where distance has the exponent 3 within an exponential term: Table 1.2).

While some of the models in Table 1.2 may be useful for some short-distance vectors, none is useful for all vectors at all spatial scales. This is a major problem in, for example, the recruitment subroutine of Ribbens *et al.* (1994) where the same function is used for the dispersal of winged *Acer, Betula* entrainment on snow, birds defecating *Prunus,* rodents caching *Quercus,* and dispersion of the asexual root suckers of *Fagus.* Figures 1.1–1.4 argue that the world does not have this pleasing uniformity.

Genetic methods

Molecular markers represent a solution to the problem of deconstructing overlapping curves. Sequencing and analytical methods are reviewed in Cain *et al.* (2000) and Ouborg *et al.* (1999). While the approach holds great promise, it is not a panacea as logistical constraints will restrict its use to certain carefully chosen source configurations. Sampling within an area source, the task of assignment of dispersed seeds or recruits to just a few point sources becomes logistically formidable as we must sample thousands of progeny to find the small fraction that belong to the maternal parents of interest. This technique, then, is best applied to a situation where we have a few small patch sources with overlapping curves, and thus we can sample intensively at the larger distances knowing that these specimens cannot represent short-distance events by other, nearer sources.

Methods and outstanding problems in local dispersal

There are a number of reasons why it is of some interest to understand exactly what the dispersal curve looks like near the source. Initial models of selection for dispersal capacity (Hamilton & May 1977) or persistence of populations (Shmida & Ellner 1984) were simple algebraic arguments that divided a crop into fractions (e.g. seeds under the plant versus those not under; seeds within the patch versus those dispersing outside the patch). Such arguments would become more useful if they relied on a full dispersal curve rather than dichotomies. Second, a comparison of the efficaciousness of asexual lateral expansion and short-distance seed dispersal requires that the latter be adequately characterized.

Precisely what the Q_{Dx} curve looks like near a point source is an issue that has been needlessly muddied. The first source of confusion is mixing up point sources and patch sources. A patch source will certainly have the highest density within the source or, if $x = 0$ defines the patch edge, then at 0. A second source of confusion is that inverse methods coupled with some very unlikely curve shapes will give satisfactory significance. A third problem is that strictly empirical studies often sample very lightly near the source for the intuitive reason that they want to allocate the majority of their finite trap area to greater distances. One would choose an area where the species of interest is relatively rare.

The method of choice here is molecular markers, but only for the special source configuration of a few adjacent patch sources. Unlike inverse methods, it allows us to determine parentage directly. Further, it can show us, unambiguously, the curve form.

Methods and outstanding problems in long-distance dispersal

For both metapopulation and migrational studies, it is crucial that we test models with empirical data collected from the far tail; i.e. where densities will be quite low. This has never been done. Instead we have only inferences from the palynological record and anecdotes to tell us that this tail does indeed exist (Higgins & Richardson 1999; Cain *et al.* 2000; Nathan & Muller-Landau 2000). Below, we offer tentative solutions for enumerating long-distance dispersal by selected vectors.

Large terrestrial mammals (defecation or epizoochory)

The best method here is the one that is already being employed: labelled seeds of a locally rare species are fed to, or attached to, animals who are then followed. One major problem is sample size: Yumoto (1999), for example, obtained only 26 faecal pellets (two plant species) in 223 h of observation time. What is needed now is a larger collection of animals that are followed for larger distances than anyone has done so far. Sampling of the faeces or periodic resampling of attached seeds could permit the characterization of the dispersal curve to long distances. The right-censoring of the data (mentioned above) can be reduced by employing a large number of field assistants, but it can never be eliminated. The ideal is that the animals have already been habituated to the presence of the investigators (e.g. Yumoto 1999; but see Turchin (1998) for a discussion of the potential problems with habituation).

Secondary dispersal by wind on snow

One possibility for determining long-distance dispersal would be staggered lines of well-supported, fine mesh fences with wide, deep sills at the base. Snow and seeds will build up on the windward side of the fence. Following snow-melt, the seeds would end up in the wooden sills. The results for seeds per metre squared would, of course, represent cumulative deposition at x.

Epizoochory and caching by flying animals

Direct observation can be done but the result is a very small sample size and small range of observed distances. Further, where great distances are involved, it is not clear that any combination of labelling, radiocollaring and genetic analysis is logistically feasible. The only possibility where we might measure the dispersal curve would be via molecular markers in situations where there are limited numbers of source populations that need to be examined for parental tissue (for example, corvids dispersing *Pinus* in arid fault-block terrain where the conifer populations are limited to isolated ridges). This would not be cheap.

Primary dispersal by wind and by defecating flying or arboreal animals

Heroically following individual seeds or animal vectors invariably results in small sample sizes and an inability to observe the distributional tail. Some initial considerations allow us to see how the problem can best be solved. First, to maximize the observed density at great distances, we should be working with area sources rather than some smaller source configuration; use of an individual plant is out of the question. Second, while it makes little difference if we work with herbs or trees (because fields and forests supply about equal deposition densities (seeds m^{-2}), the scalar Vt_g (above) reminds us that long-distance dispersal will be much easier to study with herbs; that is, the absolute distances involved are much shorter. Third, it will be easier to work in the higher latitudes where the lower diversity ensures that, on average, the source strength of a species of interest is greater.

A costing example: trees dispersed by wind (primary)

In what follows we will make some rough calculations to see which of three field methods (screened traps; enumerating seeds in the litter; enumerating germinants) would be the cheapest way to collect data for trees. (This section can be regarded as the budget justification of a grant proposal that will probably be rejected as too expensive.) We will idealize a forested landscape as consisting of a very large area source, tens of kilometres in extent, that abuts another area void of conspecifics for a great distance. We will imagine sampling within the source and at 1, 2, 3 and 4 km away. The source species of interest has a 2 mg seed and would, as a monoculture enjoying a mast year, produce about 100 seeds m^{-2} (calculated from Greene & Johnson 1994). Let us assume that 40% of the areal coverage of this nominal source consists of river, lakes, very young stands, roads, etc. Additionally only one-third of the total basal area consists of the species of interest. Thus, at the landscape scale this source provides us with 18 seeds m^{-2} within the nominal source. We will counterpose this to a larger-seeded species (50 mg) that would, by the same reasoning as above, have a realized source strength of 2.5 seeds m^{-2}. Using the same parameter values for the dispersal term of the 2Dt model as in Table 1.2, we estimate that the density of seeds at 4 km will be only 0.066 seeds m^{-2} for the small-seeded species. To obtain at least 100 seeds (in case our extrapolation beyond the empirical limit in Fig. 1.7 is markedly overoptimistic) at each distance, we would need 0.08 ha of trap area at the farthest distance (4 km) alone.

Table 1.3 Field time required for a two-person team with three different field procedures for documenting the long-distance dispersal curve of wind (primary) or defecating flying vectors. Expected seeds per metre squared are based on the 2Dt model using the parameter values in Table 1.2. The small-seeded (2 mg) species has a realized source strength of 18 seeds m^{-2} well within the area source while the larger-seeded (50 mg) species has a source strength of 2.5 seeds m^{-2}. (Both arguments are based on Greene and Johnson (2000) for a mast year; compare with Shipley and Dion (1992) for herbs.) We seek to obtain 100 seeds or germinants per distance and we are sampling at 1, 2, 3 and 4 km from the area source. Deep within the area source we will try to obtain 400 seeds or germinants at widely scattered locations. For a screened trap we assumed a 0.28 m^2 bag appended from poles with wide mesh screening to exclude much of the litter. For seeds sampled in litter it is assumed that granivores have removed 50% of the deposited seeds regardless of seed size, while for germinants it is expected that granivores and seedbed-mediated mortality will reduce the deposited seed density by 95% for small seeds and 75% for large seeds (based on Greene & Johnson 1998). Note that transportation time (twice as great for traps as for the other two methods; i.e. two separate trips are involved) and material costs are ignored.

	Screened traps		Seeds in litter		Germinants	
Time	2 mg	50 mg	2 mg	50 mg	2 mg	50 mg
Installation (h m^{-2})	1.77	1.77	–	–	–	–
Examination (h m^{-2})	0.33	0.08	0.83	0.21	0.04	0.04
Total area examined (ha)	0.24	1.72	0.48	3.44	4.8	6.88
Weeks per team (40 h $week^{-1}$)	126	797	100	179	50	72

Table 1.3 shows the expected time allocations for the three field methods based on our own experience in recently burned sites and intact forests. This exercise makes a number of things clearer. First, there is no single method that is dramatically preferable to all others. The cheapest method is an examination of germinants. The worst method (especially since Table 1.3 does not include the cost of materials) is seed traps. Second, while larger-seeded species necessarily have reduced source strength, this is, except for traps, roughly cancelled by the ease with which these larger seeds can be sampled in the field and by their higher juvenile survivorship. Third, if we want to limit this exercise to say a month of field work then the total number of field personnel must be much larger than the two-person team envisioned in Table 1.3: in the best example (germinants of small-seeded species) we would require about 25 people for a month of field work. This would represent a major departure for this field-orientated discipline where we normally hire a small number of people for an entire field season. Fourth, we have imagined a species that comprised 30% of the individuals within the source area. As the species of interest becomes less common, or as we choose to examine more than just these five distances, the total time (Table 1.3) will increase proportionally.

Our idealized landscape meant that we could determine the source of the seeds or germinants. A more realistic, but still carefully selected, situation would have at least a few patch sources within 8 km of the single area source. In this scenario, the best method would be, as stressed by Cain *et al.* (2000), a study of germinants followed by a genetic analysis. However, now in addition to the time involved in Table 1.3, we must add the time for a complete sampling of the tissue of the adults in the patch sources so that we can declare with confidence that a germinant had not arrived from one of the patches (and therefore, by default, had arrived from the area source). The recruitment curves from the patch sources themselves are, of course, also of interest in understanding metapopulation dynamics and migrational velocity.

In summary, there is no royal road to the empirical characterization of the migrational scale.

References

Anderson, A.N. (1988) Dispersal distance as a benefit of myrmecochory. *Oecologia* **75**, 507–511.

Augspurger, C.K. (1983) Offspring recruitment around tropical trees: changes in cohort distance with time. *Oikos* **20**, 189–196.

Augspurger, C.K. & Hogan, K.P. (1983) Wind dispersal of fruits with variable seed number in a tropical tree (*Lonchocarpus pentaphyllus*, Leguminosae). *American Journal of Botany* **70**, 1031–1037.

Bolker, B.M. & Pacala, S.W. (1999) Spatial moment equations for plant competition: understanding spatial strategies and the advantages of short dispersal. *American Naturalist* **153**, 575–602.

Bullock, J.M. & Clarke, R.T. (2000) Long distance seed dispersal by wind: measuring and modelling the tail of the curve. *Oecologia* **124**, 506–521.

Cain, M.L., Milligan, B.G. & Strand, A.E. (2000) Long-distance seed dispersal in plant populations. *American Journal of Botany* **87**, 1217–1227.

Clark, J.S., Silma, M., Kern, R., Macklin, E. & RisLambers, J.H. (1999) Seed dispersal near and far: patterns across temperate and tropical forests. *Ecology* **80**, 1475–1494.

Cremer, K.W. (1966) Dissemination of seed from *Eucalyptus regnans*. *Australian Forester* **30**, 33–37.

Cremer, K.W. (1977) Distance of seed dispersal in eucalypts estimated from seed weights. *Australian Forest Research* **7**, 225–228.

Crossley, D.I. (1995) *The Production and Dispersal of Lodgepole Pine Seed*. Technical Note No. 25. Forest Research Division, Department of Northern Affairs and Nature Resources, Forestry Branch, Ottawa.

Ernst, W.H.O., Veenendaal, E.E. & Kebakile, M.M. (1992) Possibilities for dispersal in annual and perennial grasses in a savanna in Botswana. 1992. *Vegetatio* **102**, 1–11.

Greene, D.F. & Johnson, E.A. (1992) Fruit abscission in *Acer saccharinum* with reference to seed dispersal. *Canadian Journal of Botany* **70**, 2277–2283.

Greene, D.F. & Johnson, E.A. (1994) Estimating the mean annual seed production of trees. *Ecology* **75**, 339–347.

Greene, D.F. & Johnson, E.A. (1995) Long-distance wind dispersal of tree seeds. *Canadian Journal of Botany* **73**, 1036–1045.

Greene, D.F. & Johnson, E.A. (1996) Wind dispersal of seeds from a forest into a clearing. *Ecology* **77**, 595–609.

Greene, D.F. & Johnson, E.A. (1997) Secondary dispersal of tree seeds on snow. *Journal of Ecology* **85**, 329–340.

Greene, D.F. & Johnson, E.A. (1998) Seed mass and early survivorship of tree species in upland clearings and shelterwoods. *Canadian Journal of Forest Research* **28**, 1307–1316.

Greene, D.F. & Johnson, E.A. (2000) Tree recruitment from burn edges. *Canadian Journal of Forest Research* **30**, 1264–1274.

Greene, D.F., Zasada, J.C., Sirois, L. *et al.* (1999) A review of the regeneration dynamics of boreal forest tree species. *Canadian Journal of Forest Research* **29**, 824–839.

Hamilton, W.D. & May, R.M. (1977) Dispersal in stable habitats. *Nature* **269**, 578–581.

Hanski, I. (1999) *Metapopulation Ecology*. Oxford University Press, Oxford.

Harris, A.S. (1969) *Ripening and dispersal of a bumper western hemlock–sitka spruce seed crop in southeast Alaska.* United States Forest Service Note PNW-105. US Department of Agriculture, Forest Service, Portland, OR.

Hewitt, N. (1999) *Tree dispersal and colonization in fragmented forests.* PhD dissertation, Department of Geography, York University, Toronto.

Hickey, J.R., Flynn, R.W., Buskirk, S.W., Gerow, K.G. & Willson, M.F. (1999) An evaluation of a mammalian predator, *Martes americana*, as a disperser of seeds. *Oikos* **87**, 499–508.

Higgins, S.I. & Richardson, D.M. (1999) Predicting plant migration rates in a changing world: the role of long-distance dispersal. *American Naturalist* **153**, 464–475.

Howe, H.F. & Smallwood, J. (1982) Ecology of seed dispersal. *Annual Review of Ecology and Systematics* **13**, 201–228.

Hughes, J.W., Fahey, T.J. & Brown, B. (1987) A better seed and litter trap. *Canadian Journal of Forest Research* **17**, 1623–1624.

Johnson, W.C. (1988) Estimating dispersability of *Acer*, *Fraxinus*, and *Tilia* in fragmented landscapes from patterns of seedling establishment. *Landscape Ecology* **1**, 175–187.

Jongejans, E. & Schippers, P. (1999) Modeling seed dispersal by wind in herbaceous species. *Oikos* **87**, 362–372.

Leadum, C.L., Gillies, S.L., Yearsley, H.K., Sit, V., Spittlehouse, D.L. & Burton, P.J. (1997) *Field Studies of Seed Biology.* Research Branch, British Columbia Ministry of Natural Resources, Victoria.

LePage, P.T., Canham, C.D., Coates, K.D. & Bartemucci, P. (2000) Seed abundance versus substrate limitation of seedling recruitment in northern temperate forests of British Columbia. *Canadian Journal of Forest Research* **30**, 415–427.

MacArther, J.D. (1964) *A Study of Regeneration after Fire in the Gaspe Region.* Department of Forests Publication No. 1074. Department of Forests, Ottawa.

MacDonald, G.M. (1993) Reconstructing plant invasions using fossil pollen analysis. *Advances in Ecological Research* **24**, 67–110.

MacDonald, G.M., Edwards, T.W.D., Moser, K.A., Pienitz, R. and Smol, J.P. (1993) Rapid response of treeline vegetation and lakes to past climate warming. *Nature* **361**, 243–246.

Mack, A.L. (1995) Distance and non-randomness of seed dispersal by the dwarf cassowary *Casuarius bennetti. Ecography* **18**, 286–295.

McDonnell, M.J. & Stiles, E.W. (1983) The structural complexity of old field vegetation and the recruitment of bird-dispersed plant species. *Oecologia (Berlin)* **56**, 109–116.

Murray, K.G. (1988) Avian seed dispersal of three neotropical gap-dependent plants. *Ecological Monographs* **58**, 271–298.

Nathan, R. & Muller-Landau, H.C. (2000) Spatial patterns of seed dispersal, their determinants and consequences for recruitment. *Trends in Ecology and Evolution* **15**, 278–285.

Ohkawara, K. & Higashi, S. (1994) Relative importance of ballistic and ant dispersal in two diplochorous *Viola* species (Violaceae). *Oecologia* **100**, 135–140.

Okubo, A. & Levin, S.A. (1989) A theoretical framework for data analysis of wind dispersal of seeds and pollen. *Ecology* **70**, 329–338.

Ouborg, N.J., Piquot, Y. & Groenendael, J.M. (1999) Population genetics, molecular markers and the study of dispersal in plants. *Journal of Ecology* **87**, 551–568.

Pacala, S.W., Canham, C.D., Saponara, J., Silander, J.A., Kobe, R.K. & Ribbens, E. (1996) Forest models defined by field measurements: estimation, error analysis and dynamics. *Ecological Monographs* **66**, 1–43.

Passos, L. & Ferreira, S.O. (1996) Ant dispersal of *Croton priscus* (Euphorbiaceae) seeds in a tropical semideciduous forest in southeastern Brazil. *Biotropica* **28**, 697–700.

Peres, C.A. & Baider, C. (1997) Seed dispersal, spatial distribution and population structure of Brazilnut trees (*Bertholletia excelsa*) in southeastern Amazonia. *Journal of Tropical Ecology* **13**, 595–616.

Pickford, A.R. (1929) Studies in seed dissemination in British Columbia. *Forest Chronicle* **5**, 8–16.

Radvanyi, A. (1966) Destruction of radio-tagged seeds of white spruce by small mammals during summer months. *Forest Science* **12**, 1307–1315.

Ribbens, E.J., Silander, J.A. & Pacala, S.W. (1994) Seedling recruitment in forests: calibrating models to predict patterns of tree seedling dispersion. *Ecology* **75**, 1794–1806.

Ronco, F. (1970) *Engelmann Spruce Seed Dispersal and Seedling Establishment in Clearcut Forest*

Openings in Colorado—a Progress Report. Research Note No. RM-168. Rocky Mountain Forest and Range Experiment Station, Fort Collins, CO.

Rudis, V.A., Ek, A.R. & Balsiger, J.W. (1978) Within-stand seedling dispersal for isolated *Pinus strobus* within hardwood stands. *Canadian Journal of Forest Research* **8**, 10–13.

Schmitt, J., Ehrhardt, D. & Swartz, D. (1985) Differential dispersal of self-fertilized and outcrossed progeny in jewelweed (*Impatiens capensis*). *American Naturalist* **126**, 570–575.

Shipley, B. & Dion, J. (1992) The allometry of seed production in herbaceous angiosperms. *American Naturalist* **139**, 467–483.

Shmida, A. & Ellner, S. (1984) Coexistence of plant species with similar niches. *Vegetatio* **58**, 29–55.

Sork, V.L. (1984) Examination of seed dispersal and survival in red oak, *Quercus rubra* (Fagaceae), using metal-tagged acorns. *Ecology* **65**, 1020–1022.

Squillace, A.E. (1954) *Engelmann Spruce Seed Dispersal into a Clear-cut Area.* Research Note No. INT-II. Intermountain Forest and Range Experimental Station, Ogden, UT.

Stevenson, P.R. (2000) Seed dispersal by woolly monkeys (*Lagothrix iagothricha*) at Tinigua National Park, Columbia: dispersal distance, germination rates and dispersal quantity. *American Journal of Primatology* **50**, 275–289.

Stoyan, D. & Wagner, S. (2001) Estimating the fruit dispersion of anemochorous forest trees. *Ecological Modelling* **145**, 35–47.

Swaine, M.D. & Beers, T. (1977) Explosive seed dispersal in *Hura crepitans* L. (Euphorbiaceae). *New Phytologist* **78**, 695–708.

Turchin, P. (1998) *Quantitative Analysis of Movement.* Sinauer, Sunderland, MA.

Vander Wall, S.B. (1993) Cache site selection by chipmunks (*Tamias* spp.) and its influence on the effectiveness of seed dispersal in Jeffrey pine (*Pinus jeffreyi*). *Oecologia* **96**, 246–252.

Vander Wall, S.B. & Balda, R.P. (1977) Coadaptations of Clark's nutcracker and the pinon pine for efficient seed harvest and dispersal. *Ecological Monographs* **47**, 89–111.

Vander Wall, S.B. & Joyner, J.W. (1998) Secondary dispersal by the wind of winged pine seeds across the ground surface. *American Midland Naturalist* **139**, 365–373.

Verkaar, J., Schenkeveld, A.J. & van de Klashorst, M.P. (1983) The ecology of short-lived forbs in chalk grasslands: dispersal of seeds. *New Phytologist* **95**, 335–344.

Wagner, S. (1997) A model describing the fruit dispersal of ash (*Fraxinus excelsior*) taking into account directionality. *Allgemeine Forst und Jagdzeitung* **168**, 149–155.

Walker, L.R., Zasada, J.C. & Chapin, F.S. (1986) The role of life history processes in primary succession on an Alaskan floodplain. *Ecology* **67**, 1243–1253.

Werner, P.A. (1975) A seed trap for determining patterns of seed deposition in terrestrial plants. *Canadian Journal of Botany* **54**, 1189–1197.

Wilkinson, D.M. (1997) Plant colonization: are wind dispersed seeds really dispersed by birds at larger spatial and temporal scales? *Journal of Biogeography* **24**, 61–65.

Willson, M.F. (1993) Dispersal mode, seed shadows, and colonization patterns. *Vegetatio* **107/108**, 261–280.

Woodall, S.L. (1982) Seed dispersal in *Melaleuca quinquenervia. Florida Scientist* **45**, 81–93.

Yumoto, T. (1999) Seed dispersal by Salvin's curassow, *Mitu salvini* (Cracidae), in a tropical forest of Columbia: direct measurements of dispersal distance. *Biotropica* **31**, 654–660.

Zasada, J.C. & Lovig, D. (1983) Observations on primary dispersal of white spruce, *Picea glauca*, seed. *Canadian Field Naturalist* **97**, 104–106.

Chapter 2

Monitoring insect dispersal: methods and approaches

Juliet L. Osborne, Hugh D. Loxdale and Ian P. Woiwod

Introduction

Insects are generally small, often very numerous, frequently winged and can travel extremely long distances, either actively or passively in the airstream. Consequently, tracking their movements over space and time presents a great challenge. In this chapter we discuss why and how insects disperse, focusing on the peculiarities of studying insect movement as opposed to that of other organisms, and we review the variety of techniques used to quantify insect dispersal. Reynolds *et al.* (1997) and Southwood and Henderson (2000) provide comprehensive reviews of techniques used to study insect movement, and Turchin (1998) describes measurement techniques and associated analysis. To avoid repetition of these works, we concentrate on recently developed techniques.

In this volume, dispersal has been defined as 'intergenerational (spatial) movement', but since long-range insect movement is often termed 'migration', the difference between migration and dispersal should be noted (Woiwod *et al.* 2001a, 2001b). The most commonly used entomological definition of migration is that of Kennedy (1985), who concluded that 'Migratory behaviour is persistent and straightened-out movement effected by the animal's own locomotory exertions or by its active embarkation on a vehicle. It depends on some temporary inhibition of station-keeping responses, but promotes their eventual disinhibition and recurrence'. Migration, in insects, can thus be considered as a behaviour which usually results in dispersal, but not all dispersal events are migratory (Woiwod *et al.* 2001b). A plethora of literature tackles the phenomenon of insect migration and techniques associated with its study (Danthanarayana 1986; Dingle 1989, 1996; Gatehouse 1989; Drake & Gatehouse 1995; Reynolds *et al.* 1997; Woiwod *et al.* 2001a). Migration may be obligatory (e.g. in the gregarious form of the African armyworm moth), but is often facultative, and its incidence is highest in species whose habitats are spatially and/or temporally patchy (Drake & Gatehouse 1995). It has been studied most intensely in the acricoid Orthoptera, Hemiptera and Lepidoptera, for pest forecasting information (Pedgley 1981; Tatchell 1991; Day & Knight 1995; Krall *et al.* 1997; Tucker & Holt 1999; Rose *et al.* 2000).

Why do insects disperse?

Habitats and resources are ephemeral in time and space. Dispersal provides a mechanism for tracking changes in environmental conditions, whether it be over a few centimetres or over hundreds of kilometres (Roderick & Caldwell 1992). Dispersal may result from deterioration of the habitat, for example through physical conditions, resource depletion, competition for mates, or a combination of these factors (MacDonald & Smith 1990). By dispersing, an animal may increase its fitness if it arrives in a new habitat with an improved physical, competitive or genetic milieu (MacDonald & Smith 1990). Dispersal is often triggered by factors such as food availability or population density (Danthanarayana 1986), or other environmental factors which may act as 'surrogates' for habitat change (e.g. temperature, humidity and rain, photoperiod, sun's position, air pressure, chemical factors or wind) (Dingle 2001).

Why study insect dispersal?

Quantifying insect dispersal is key to understanding insect population dynamics, and also to answering questions related to insects' behavioural traits, and physiological or genetic constraints. But the purpose of studying dispersal is not only to increase our fundamental knowledge of insect populations *per se*, but also to develop forecasting systems to alert farmers and foresters to pest invasions. Large-scale insect movement causes considerable economic problems in crop production and forest systems. Crops provide a temporal and spatial concentration of resources which can attract or arrest large numbers of insects from long distances.

Features of insect dispersal

The flight capabilities and small size of insects enable them to disperse both actively and passively (i.e. being transported by an external 'vehicle' such as the airstream or water), in contrast to most vertebrates (generally active) and plants (mainly passive). (It should be noted that, although we use the term 'passive' for dispersal of wind-borne insects, these insects have suites of behavioural adaptations to enable them to enter and stay in high-altitude airstreams outside their flight boundary layer (Taylor 1974), so their movements are under some degree of control.) This combination of passive and active dispersal allows insects to travel over huge distances relative to their size, and to accurately target habitats specific to their requirements (Compton, this volume). Some of the characteristics of insects that are advantageous to their dispersal can be disadvantageous to the researcher attempting to study the phenomenon. For example, large populations of tiny winged individuals are carried upward by convection currents and conveyed by high winds, including jet streams, above their flight speed, over large geographical distances, making tracking very difficult. Remote sensing, netting or trapping are then necessary. Even if they are travelling at low altitude, their size and the speed at which they fly make it difficult to follow individuals.

Most insects have short life cycles (often with one or more generations per year), giving them the opportunity to complete a whole generation within a short-lived patch of habitat or resource. As their resources are often ephemeral, so are the insect populations: new populations, or subpopulations, can establish and go extinct over short time periods, making their metapopulation dynamics very fluid (see Wilson & Thomas, this volume). This provides a challenge for monitoring such population displacements and, as individuals do not live very long, then artificial markers can only be used over short periods (compared to vertebrate species).

Dispersal is usually by adult insects, although other life stages, notably larval, may have a dispersive phase, for example several forest Lepidoptera (McManus 1988). Many insects are crepuscular or nocturnal, making visual observation very difficult without disturbing their behaviour with the use of lamps.

Insects show physical and behavioural adaptations to dispersal, for example polymorphism and polyphenism (Dingle 1989). Polymorphism is the existence of different genotypes within a species. Polyphenism is a facultative phenotypic response to environmental conditions. Many orders of insects have flight polymorphisms, alary or polyphenisms. For example, aphid asexual lineages produce winged aphids in response to environmental cues, e.g. host plant quality, crowding, temperature or photoperiod (Dingle 1989; Loxdale & Lushai 1999). Winged and wingless morphs occur notably in the Orthoptera, Homoptera, Hemiptera and Coleoptera. *Spodoptera exempta* (Walker) moth larvae show density-dependent flight polyphenism, depending on whether larvae emerge alone or at high density (Gatehouse 1989).

Insects disperse either individually or in swarms, for example Orthoptera (Cooter 1989) and Hymenoptera. Some swarms are non-cohesive, resulting from numerous winged individuals taking off at once, triggered by an environmental cue such as food or weather, e.g. ants or butterflies. Others are actively cohesive, orientated towards an external feature or goal as the result of complex behavioural interactions, e.g. honey bees or locusts. Honey bees take this to the extreme with nest scouts from the swarm finding potential nest sites and communicating the location to the swarm (Winston 1987).

Techniques

Insect movement can be measured directly or indirectly. Direct methods involve tracking the movement of an individual, whilst indirect methods are often focused on monitoring displacement of populations. Reynolds *et al.* (1997) review techniques concentrating on long-distance, air-borne insects, primarily at high altitude; whilst Reynolds and Riley (2002) review recent remote-sensing and computer-based technologies. Jones *et al.* (1999) review methods for studying dipteran dispersal.

When choosing a technique to monitor insect movement, it should be borne in mind that observing behaviour (by marking or following an insect, for example) could in itself alter that behaviour. It is essential that the method of analysis is decided before the data are collected so that an experiment at a suitable spatial and temporal resolution is performed to ensure measurements are appropriate to the aims of the study.

Direct methods of studying dispersal involve either characterizing the actual path of the insect's movement, or individual mark–recapture experiments. Indirect methods include mass mark–recapture experiments, genetic markers, trapping, behavioural observations, scanning radar, vertical looking radar (VLR) and modelling (Turchin 1998; Turchin & Omland 1999). These methods are briefly summarized below. Three relatively novel techniques (harmonic radar, VLR and genetic markers) are then discussed in more detail. A literature search (ISI, Web of Science, Science Citation Index) for 1999–2001, specifying 'insects', 'dispersal' and/or 'migration' as keywords, resulted in a collection of 47 papers. Most researchers used mark–recapture, trapping or genetic markers to estimate insect movement. Lepidoptera, Homoptera and Heteroptera were, by far, the most commonly studied orders.

Tracking paths

Movement paths are usually characterized either by eye or by video (Wratten 1994). The insect's position at different intervals can be recorded digitally, on audio tape or with numbered flag markers. Although this is the most accurate and precise way of recording insect movement, it is time consuming and can only be done on a relatively small spatial scale. Recently developed harmonic radar technology has enabled the tracking of individual flying insects over longer distances (see below and Riley *et al.* 1996). In addition, although it has previously not been possible to track flying insects with radio telemetry due to the requirement for batteries attached to the insect, success has recently been achieved with large Scarabaeid beetles, *Osmoderma eremita* (Scopoli) (Hedin & Ranius 2002).

Mark–recapture

Mark–recapture is probably the most widely used technique for tracking insect movement (see Wilson & Thomas, this volume). Application of this technique and analysis of associated data are discussed in detail by Southwood and Henderson (2000). If the marker allows identification of individuals, then movement records are direct. If insects are mass marked, the results are an indirect measure of movement. A variety of elegant methods has been used to mark insects, depending on their size and mobility (Table 2.1). Artificial markers can be applied to the insect, or use can be made of natural markers such as phenotype, genotype and forensic evidence from pollen or gut contents (Reynolds *et al.* 1997). Southwood and Henderson (2000) list criteria essential for mark–recapture studies: the markers must not impose a disadvantage on an individual or population (e.g. by altering mating or predation chances, or ability to move!); markers must be recognizable for the duration of the experiment; and they must be easy to apply, with minimal interference. The least disturbing techniques involve marking the insects without capture, for example using treated bait.

The major disadvantage of marking techniques is the difficulty of relocating the marked individuals. Insects can travel over vast distances and the large dilution effect means the chances of finding many marked insects again are low (Urquhart

Table 2.1 Variety of markers used in mark–release–recapture experiments on insects.

	Insect examples	Reference
Artificial markers		
Paint dot or number*	Numerous	See Reynolds *et al.* 1997; Southwood & Henderson 2000
Numbered tags*	Bumble bees	Osborne & Williams 2001
Coloured tags*	*Danaus plexippus* L. (monarch butterfly)	Urquhart & Urquhart 1979
Wing notching, elytra punching or branding*	Orthoptera Coleoptera	Gangwere *et al.* 1964 Southwood & Henderson 2000
Powder dye or dust (or spray)	*Musca domestica* L. (house flies) *Bemisia tabaci* (whitefly) Coleoptera, Apoidea	Denholm *et al.* 1985 Byrne 1999 Southwood & Henderson 2000
Stain or dye	*Musca domestica* L. (house flies) *Heliothis virescens* (F.) (tobacco budworm)	See Jones *et al.* 1999 Schneider 1999
Ferrous discs and magnets	*Apis mellifera* L. (honey bees)	Gary *et al.* 1977
Dye in larval diet (adults laid coloured eggs)	*Pieris rapae* L. (cabbage white butterfly)	Jones *et al.* 1980
Glitter in caddis cases	*Gumaga nigricula* larvae (caddis fly)	Jackson *et al.* 1999
Stable isotopes, e.g. rubidium, caesium	*Heliothis zea* (Boddie) (corn earworm) *Spodoptera frugiperda* (J. E. Smith) (fall armyworm)	Graham *et al.* 1978; see also Reynolds *et al.* 1997
Natural markers		
Phenotype, e.g. colour, size	*Pectinophora gossypiella* (Saunders) (pink bollworm)	Bartlett & Lingren 1984
Genotype*	Many	See molecular markers section
Pollen on insects' bodies	Noctuid moths	Gregg 1993; Lingren *et al.* 1993
Gut contents	*Meligethes aeneus* (pollen beetle)	Free & Williams 1978
Isotope signature (δD and δC) or elemental composition	*Danaus plexippus* (monarch butterfly) *Entomoscelis americana* Brown (red turnip beetle)	Hobson *et al.* 1999 Turnock *et al.* 1980

*Suitable for marking individuals rather than mass marking.

1987; Showers 1997). Bias is introduced because only a certain area can be surveyed for marked insects, and this limits the number seen at long distances from the source (Wilson & Thomas, this volume). However, mark–recapture may be the only way to directly demonstrate that very long-distance movements occur in certain insect species, and landscape-scale studies have been attempted with some success. Schneider (1999) marked 7 million *Heliothis virescens* (F.) moths and recaptured some of them, in traps, over an area of 2000 km^2.

Natural markers provide a means by which the whole population can be identified, though the resolution depends on the marker. Genetic markers (phenotypic and genotypic) are an example (see below). Other natural markers involve taking 'forensic evidence' from the insect. This means examining the bodies of the insects for clues as to their previous whereabouts. For example, Gregg (1993) and Gregg *et al.* (1995) identified the pollen on *Helicoverpa* moths on the Australian coast. Some of the pollen came from plants that only grew inland, indicating that the moths had immigrated over thousands of kilometres. Lingren *et al.* (1993) also gave evidence of North American noctuid migration using pollen grains. Other examples of forensic evidence include examination of plant material ingested by insects and examination of isotope signatures (Hobson *et al.* 1999).

Behavioural observations

Reynolds *et al.* (1997) reviewed the variety of laboratory experiments performed to examine behavioural cues for migration, e.g. what stimulates flight and landing, and the determination of insects' physiological capabilities. Experiments are often conducted in wind tunnels or on flight mills (Dingle 1989), or with computerized video rigs (e.g. 'virtual reality' flight simulators; Reynolds & Riley 2002). In the field, behavioural observations are used to estimate indirectly where an insect, or swarm, may be heading. Gibo (1986) and Schmidt-Koenig (1993) used vanishing bearings to calculate the direction of monarch butterfly, *Danaus plexippus* L. migrations. Srygley *et al.* (1996) followed butterflies and moths flying over water, in relation to solar and celestial cues, to investigate whether the insects were compensating for wind drift, and consequently whether they were goal-orientated (also Srygley & Oliveira 2001). Seeley and Morse (1978) read the dances of honey bee nest scouts on a swarm to predict where those swarms would decide to nest.

Traps

A great variety of traps has been utilized at a range of spatial scales to examine changes in insect population densities, and these may then be interpreted in terms of movement patterns, when some information is available on local population dynamics. Southwood and Henderson (2000) comprehensively review trap types and other methods for scoring insect population densities. Reynolds *et al.* (1997) focus on aerial sampling for migration studies, often involving the use of nets attached to balloons or aircraft. Traps are often used in combination with other techniques, such as meteorology and mark–recapture, to determine dispersal patterns in more detail.

Networks of traps, run over long periods of time, at fixed positions, can provide useful information on insect dispersal. Such networks are often set up to study migrant species of economic importance, for example the pheromone/light trap network operating in eastern Africa to monitor the movements of the African armyworm, *Spodoptera exempta* (Rose *et al.* 2000). The Rothamsted Insect Survey illustrates another approach, where long-term fixed-position networks of light and suction traps, monitoring moths and aphids, respectively, provide daily information on a wide range of species throughout the UK. These data have been used for a variety of purposes including the study of population dynamics and insect movement (Woiwod & Harrington 1994). The Survey's aphid monitoring network uses standard suction trap sampling at a height of 12.2 m, designed for the efficient sampling of small mobile wind-borne insects such as aphids (Macaulay *et al.* 1988). Because of the mixing of the insects by the movement of the air, samples are representative of the aerial insect fauna over a large source area.

The first '12-metre' high suction trap began operating at Rothamsted in 1965; there are currently 15 traps in the UK, and altogether 73 such traps operating in 19 countries throughout Europe. An EU-funded database is now being constructed (under the acronym EXAMINE: see http://www.iacr.bbsrc.ac.uk/examine), to co-ordinate the information from all these sites across Europe. This will provide the most comprehensive standardized set of data for any mobile terrestrial invertebrate group anywhere in the world and should greatly increase our understanding of the dispersal and the impacts of global change on this important group of insects.

Mathematical approaches

Turchin (1998) and Turchin and Omland (1999) describe and review methods for analysing and interpreting insect dispersal data, so we do not attempt to do so here. Models and analytical approaches can be separated into those centred on a moving individual (Langrangian approach) and those centred on a point in space (Eulerian approach), measuring population densities and fluxes. Both have advantages and disadvantages, which depend on the insect being studied and the spatial and temporal scale. Individual approaches lead to a more mechanistic understanding of dispersal, whilst those centred on a point in space lead to further understanding of the population dynamics of the species and the resultant patterns of distribution.

Forecasting

The forecasting of pest invasions (Day & Knight 1995) usually depends on data from a combination of sources, for example large-scale trap surveys (like the Rothamsted Insect Survey), behavioural observations, insect density counts, meteorological data (Wellington *et al.* 1999), geographical information systems (GIS) (Liebhold *et al.* 1993; Robinson 1995) and modelling (Stinner *et al.* 1986). Raulston *et al.* (1986) used a combination of pheromone traps, behavioural laboratory experiments, weather forecasting and developmental observations to study *Heliothis* spp. and *Spodoptera* spp. migrations in the USA. Gregg *et al.* (1995) and Rochester *et al.*(1996) have created mathematical models to forecast *Helicoverpa* spp. migrations in Australia, using data from pheromone and light traps, sweep-netting larvae, pollen analysis, rain and wind records (and meteorological backtracking) and vegetation indices from satellite data.

Novel techniques

We will describe three of the more recently developed techniques for measuring insect movement. Firstly, the harmonic radar provides a direct method for tracking individual insects flying at low altitude. Secondly, the vertical looking radar samples the air column for insects flying at a variety of altitudes and separates them according to flight speed, direction, body mass and shape. It focuses on a point in space rather than an individual. Thirdly, molecular genetic markers provide an indirect, highly versatile method for monitoring movement at a variety of spatial and temporal scales, with different degrees of resolution.

Harmonic radar

Mascanzoni and Wallin (1986) were the first to use a detection system, based on the harmonic radar principle (originally developed for finding avalanche victims), to locate ground-dwelling beetles. It has subsequently been used on other carabids (Wallin & Ekbom 1988, 1994; Kennedy & Young 1993; Kennedy 1994), butterflies (Roland *et al.* 1996) and caddis flies (R. Briers *et al.*, unpubl. data). The insect is tagged with a diode and aerial which, when illuminated with microwaves, re-radiates the signal at a harmonic frequency of the transmitted one. The aerial trails behind the insect and its length affects the detection range. Mascanzoni and Wallin (1986) and Wallin and Ekbom (1988) used 3–5 cm aerials on ground-dwelling carabids. The microwaves are emitted by portable detection equipment, used to locate the tagged insect, and a marker is placed in the ground where the insect is detected. After a time gap (15 min in Wallin and Ekbom (1988)), another search is made and another positional fix is taken. The carabids could be detected to a depth of 20–30 cm below ground and over a range of 10 m in vegetation.

Although described as a 'radar', this system gives no range information and is, rather, a portable direction-finding device: alone it cannot produce geometrically accurate maps. It is useful for slow-moving or stationary objects, but is not suited to tracking 'real-time' trajectories or use with fast-moving insects. The scanning harmonic radar system, developed by Professor Joe Riley and colleagues at the Natural Resources Institute (NRI) Radar Unit, is currently the only technique which enables accurate tracking of the complete flight paths of individual, large, low-flying insects over hundreds of metres (Riley *et al.* 1996). It has yet to be used on insects engaged in dispersal, but holds great potential for future studies of this behaviour. Details of experiments performed so far using this technique are described in Riley *et al.* (1998, 1999), Osborne *et al.* (1999), Capaldi *et al.* (2000) and Riley and Osborne (2001).

The radar apparatus (3.2 cm wavelength, 25 kW peak power) is described in Riley *et al.* (1996). The tracked insect carries a small transponder consisting of a vertical dipole aerial, 16 mm long, with a diode at the centre (Fig. 2.1). Preliminary investigations need to be carried out to ensure that attachment of the transponder does not affect the insect's flight behaviour. For example, bumble bees were found to collect similar quantities of nectar and pollen with and without the transponder in place (Osborne *et al.* 1999), and calculations indicated that the increased drag imposed by the transponder was negligible compared to the drag from the bee's own body (Riley *et al.* 1996). The transponder captures some of the energy in the radar transmissions and re-radiates part of it at double the transmitted frequency. This returned signal is easily distinguished from even strong echoes reflected by ground features. Since the illuminating radar delivers the energy to operate the transponder, no 'on-board' battery is required and extreme miniaturization is therefore possible.

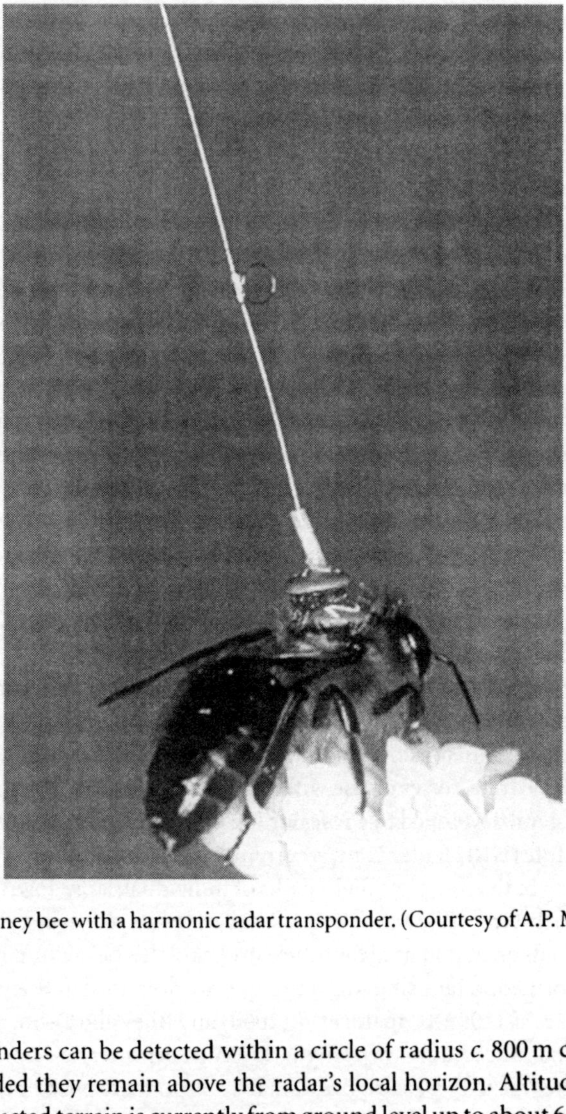

Figure 2.1 Honey bee with a harmonic radar transponder. (Courtesy of A.P. Martin.)

The transponders can be detected within a circle of radius *c.* 800 m centred on the radar, provided they remain above the radar's local horizon. Altitude coverage in flat, unobstructed terrain is currently from ground level up to about 6–7 m. The two radar antennae rotate in azimuth at 20 revolutions per minute, so position fixes (precision ± 3 m) from the transponder are received approximately every 3 s. These are digitized to provide a temporally and spatially explicit, or geometrically accurate, track of the insect's flight path.

For individual insects, the direction, distance and straightness of flight can be characterized, and even destination if it is within range. The insect's course control and navigational performance can be investigated by examining flight speed and direction, and the relationship with wind speed and direction.

The constraints of the system include the size of the transponder. The original versions weighed *c.* 12 mg (Osborne *et al.* 1999), but lighter versions (*c.* 3 mg) have been used on moths (Riley *et al.* 1998) and honey bees (J.R. Riley, unpubl. data), and even lighter ones (*c.* 1 mg) have been designed for use on tsetse flies (J.R. Riley, pers. comm.). Individual insects with transponders cannot be distinguished from each other, so only one or two can be tracked at a time. A relatively flat landscape, devoid of tall vegetation, is required to utilize the maximum spatial range of the radar. The system can not be used in forests, tree-filled landscapes or hilly areas, and altitudinal range of the present system limits the variety of insects that can be studied. The radar equipment is large and, so far, has not been mobile during studies, although that would increase its horizontal range. Riley's team is working to reduce the weight of the transponders, and to improve the precision of the position fixes. They have also developed a 3 cm wavelength, handheld direction finder, which can be used to find the tagged insect at close range, and can therefore pinpoint more precisely the destination of the tracked insect (for example a bee which has disappeared into a patch of forage).

The technique has been used to investigate flight behaviour of bumble bees (Osborne *et al.* 1999; Riley *et al.* 1999), honey bees (Carreck *et al.* 1999; Capaldi *et al.* 2000) and moths (Riley *et al.* 1998), and a pilot study was performed on tsetse flies (J.R. Riley, unpubl. data). In the first study of bumble bee foraging flight, 65 bees were tracked flying to and from their colony over a 700 m range (Fig. 2.2) (Osborne *et al.* 1999). The results showed that bumble bees do not necessarily forage close to their nest. They fly along fast, straight paths and show route constancy between trips. They also actively compensate for wind speed and direction (Fig. 2.3) (Riley *et al.* 1999). In the first honey bee study using this technique, naive bees were tracked when they first left the nest. They were observed to make characteristic orientation flights that were very different from those of experienced foragers (Capaldi *et al.* 2000).

This technique would be most suitable for studying the dispersal of large insects, flying at low altitudes over a range of hundreds of metres. Butterflies or moths, particularly those grassland species thought to be localized in their movement, would be an ideal choice. Hymenopterans, such as large solitary bees or wasps, would also be suitable. It would also be interesting to try the technique on Coleoptera, Orthoptera or Odonata: orders containing large, sturdy insects capable of carrying the transponder.

Vertical looking radar (VLR)

Scanning radars have been used to study the migratory behaviour of insects for over 25 years (Schaefer 1976; Drake 1993; Reynolds & Riley 1997). Indeed, much of our current knowledge about frequency, altitude and orientation behaviour associated with high-altitude movement of migrant pests has been derived from studies using such radars (Reynolds *et al.* 1997). The main limitations of the technique are the difficulty of target identification and the labour-intensive extraction of data. For these two reasons, scanning radars have tended to be used in relatively short intensive studies where the species under investigation have been numerically dominant (e.g. Drake 1993; Reynolds & Riley 1997).

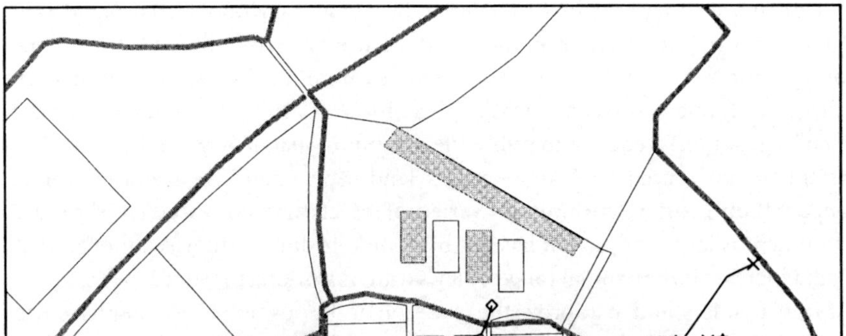

Figure 2.2 Outward tracks ($n = 35$) of nine bumble bees (indicated by different end symbols to the track) flying away from their colony (black diamond) to forage over the experimental arena. Range rings are at 200 and 400 m from the radar (open circle). Shading is as follows: field boundaries (thin lines), flowering crops (▨), hedges (thick lines) and wooded areas (▦). Radar visibility was reduced by buildings (■) and hedges. Flowering crops included oilseed rape, spring field beans, winter field beans and lupins. Gardens over the northeast hedges contained unknown forage. Tracks have been joined where missing points occurred for less than 60 s. When one bee stopped for 5 min 54 s and then continued, the continuation is plotted as a dashed line. (From Osborne *et al.* 1999, with permission.)

Further progress in the automation of 'radar entomology' has come from the development of inexpensive radars that have fixed, upward-pointing beams and rely on the insect's movement to come into the beam, rather than utilizing a moving beam scanning an area of sky at regular intervals. Using these new types of radar, it has been possible to control their operation from a personal computer and, perhaps more importantly, to fully automate signal analysis and data capture. Hence, for the first time, it is possible to contemplate the use of insect radars for long-term studies

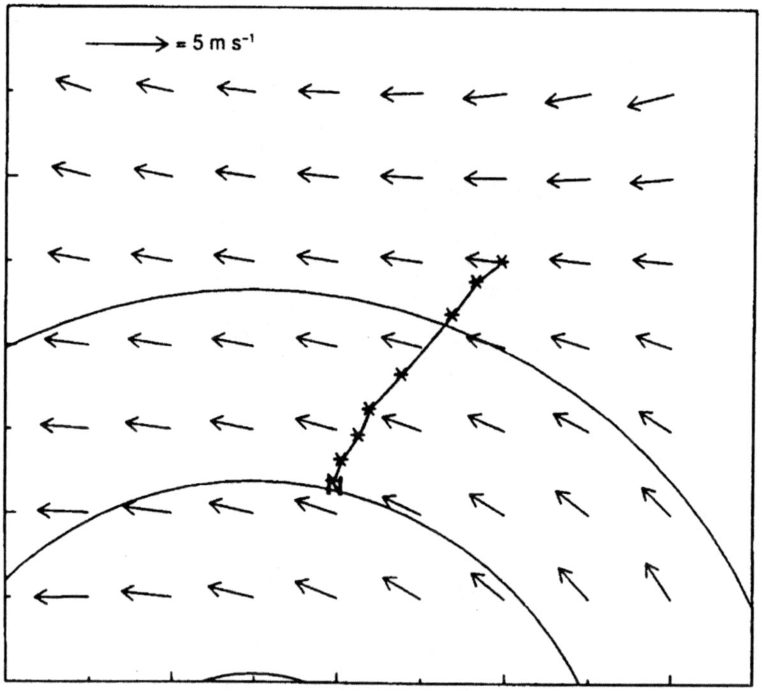

Figure 2.3 Track of a bumble bee flying back to the colony (N) in a strong crosswind. The area represented is 450 m (E to W) by 400 m (N to S). Radar rings are at 115 m intervals. The grid of arrows indicates wind speed and direction (ESE at *c.* 2.4 m s^{-1}) at 2.7 m above ground level. The bee is travelling from the NE at 6.9 m s^{-1}. (Adapted from Riley & Osborne 2001, with permission.)

and to develop their use for routine monitoring of migrant pest species. The VLR, developed by the NRI Radar Unit, is of this type (Smith *et al.* 1993, 2000; Chapman *et al.* 2002). Others using the same principles, but developed independently, are now in operation in Australia for monitoring the dispersal of Australian plague locusts, *Chortoicetes terminifera* (Walker), and native budworm moths, *Helicoverpa punctigera* (Wallengren) (Drake *et al.* 2001).

There are currently two VLRs in operation in the UK that are being used to develop the radar's potential for studying fundamental aspects of insect dispersal behaviour, and to assess the equipment's value for long-term monitoring and forecasting (Chapman *et al.* 2002). They are set up to record insects passing through the beam at altitudes from 150 to 1166 m above the radar in 15 bands (or 'range gates'). Each band is 45 m deep and there is a 26 m gap between adjacent bands (Fig. 2.4). The radar is automatically switched on once every 15 min throughout the day and night, and collects data for 5 min. The remaining 10 min are available for signal processing. A variety of parameters is automatically extracted from each individual target passing through the radar beam. These are height, displacement speed and direction,

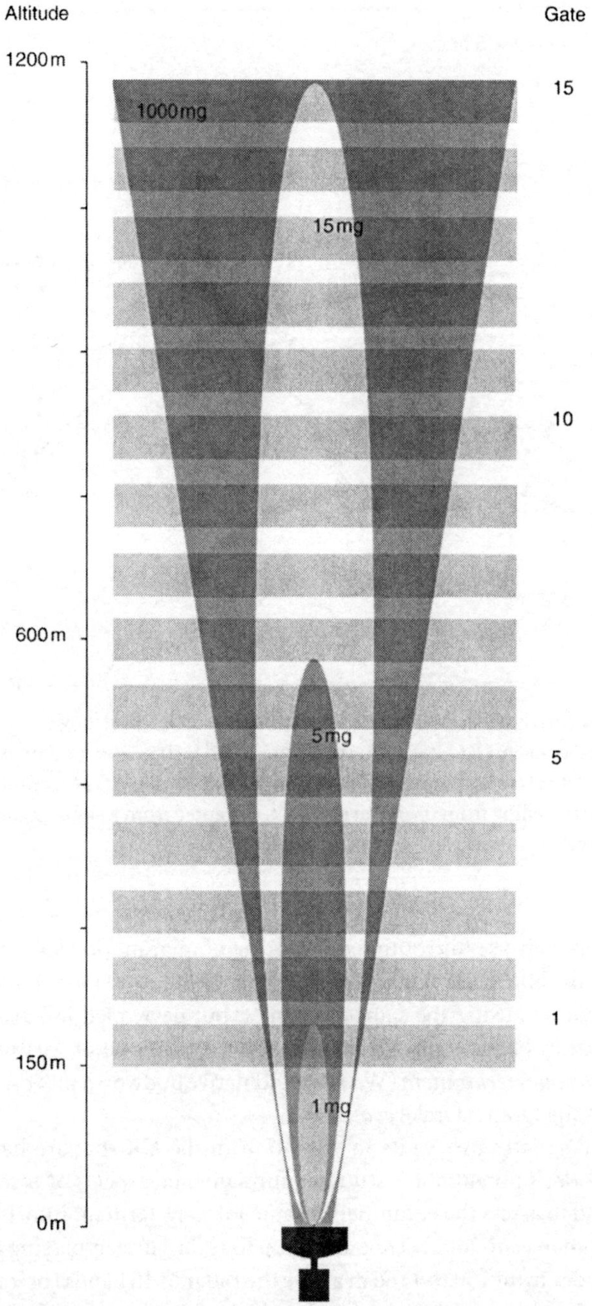

Figure 2.4 Diagrammatic representation of a VLR sampling envelope (not to scale), for insects in four size categories (1, 5, 15 and 1000 mg).

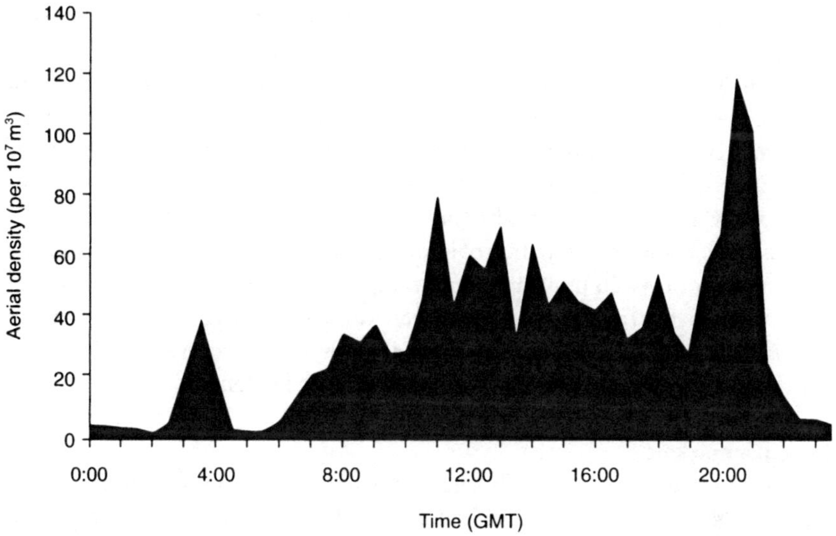

Figure 2.5 Diurnal periodicity of insect aerial density between 150 and 195 m altitude, obtained from the Rothamsted VLR in June 1999.

orientation and three scattering terms related to the insect's mass and shape. The height at which insects can be detected is dependent on their radar cross-section, which is strongly dependent on their mass. With the current equipment, all insects 15 mg or heavier can be detected at all heights. Insects of 1 mg can only be detected at the lowest level (Fig. 2.4).

Most insects that reach the heights monitored by the VLR are likely to be windborne individuals involved in some form of migration or dispersal behaviour. The radars currently in operation are 150 km apart (at Rothamsted and Malvern, UK) and are already providing interesting results with such insects. For example, plots of diurnal periodicity, summed across whole months, show clear dawn and dusk peaks of movement through the lowest recording height bands, as insects take off and descend during the periods of transition between diurnal and nocturnal species (Fig. 2.5).

With the VLR, density–height profiles can be produced to indicate the peak densities of aerial insects at different times throughout the day and night. Stratification of nocturnal insects, with well-defined dense accumulations at particular heights, is well known from radar studies in other regions of the world and was found in some preliminary VLR data from Malvern (Smith *et al.* 2000). Rather more unexpected has been the discovery that a similar phenomenon also occurs in day-flying insects in the UK. When summed over a month, there is very little fall off in aerial density with altitude until about 800 m, after which there is a rapid decline. The reason for this decline after 800 m is probably temperature related and is currently under investigation (Chapman *et al.* 2002).

As with scanning radar, the VLR has the problem, common to many remote-sensing devices, of target identification. Some form of direct sampling is very important in this respect, and the current VLR studies incorporate sampling at 200 m with an aerial net suspended from a balloon, and also make use of samples from the 12.2 m high suction traps and standard light traps of the Rothamsted Insect Survey (Chapman *et al.* 2002). This strategy can be quite successful, as shown in a recent migration study of the diamond-backed moth, *Plutella xylostella* (L.) (J.W. Chapman, unpubl. data). Using a combination of information, such as time of flight, insect size and occurrence of insects in all three direct sampling methods, it has been possible to demonstrate that particular radar peaks were caused by a large-scale migration of this relatively small (*c.* 1 mg) species. Further work will be required in this critical area of target identification for the VLR to deliver its full potential.

Molecular markers

Molecular markers may be broadly categorized as non-recombinant or mitochondrial (mtDNA) markers or nuclear genomic (nDNA) markers, including protein markers (Avise 1994; Hoy 1994; Loxdale 1994; Loxdale *et al.* 1996; Loxdale & Lushai 1998, 1999, 2001). From the mid-1960s to the present day, protein markers, more especially allozymes (Richardson *et al.* 1986), have been used extensively to study population structure and dynamics of animal and plant populations, including insects (Loxdale 1994). They provide good markers in species that display a reasonable amount of polymorphism, although they rarely have more than 10 alleles per locus, and, in some instances, they are under direct selection (e.g. insecticide-resistance alleles).

DNA markers are now generally preferred and many are considered to be predominantly selectively neutral, particularly hypervariable sequences from non-coding regions, such as microsatellites. Unlike protein markers, they involve more complex isolation and purification procedures. However, they usually have the advantage of providing more variation, thereby increasing the power to distinguish individuals, populations and levels of divergence up to species level, especially for insects with a dearth of allozyme polymorphism, such as aphids.

The range of DNA markers now available for population studies include RFLPs (restriction fragment length polymorphism, both nuclear and mitochondrial), RAPDs (random amplified polymorphic DNA), AFLPs (amplified fragment length polymorphism), SSCPs (single-stranded conformation polymorphism), introns, single and multilocus probes (including for rDNA), mini- and microsatellites and, potentially, even transposable elements (Tu 2001). They have been reviewed extensively elsewhere in relation to entomological studies (Hoy 1994; Roderick 1996; Loxdale & Lushai 1998). Vos *et al.* (1995) discuss AFLP markers specifically (see also Reineke *et al.* 1999), Sunnucks *et al.* (2000) explain SSCPs (see also Vaughn & Antolin 1998) and Goldstein and Schlötterer (1999) give a detailed account of microsatellite markers. Each marker has advantages or disadvantages due to the extent of its variation and dominance. For example, RAPDs provide a lot of bands at different loci so are potentially very variable. Since the markers are dominant it is difficult to ascribe bands to particular loci without formal crossing experiments. Even so, they have proved very informative for examining intraclonal (Lushai *et al.* 1997, 1998) as well as interclonal variation (De Barro *et al.* 1995).

Microsatellites are especially useful because they are often found to be highly polymorphic at a range of loci and produce single locus Mendelian banding patterns (like allozymes, but unlike the majority of other DNA markers). However, because most designed primers tend to be species specific, they may not be very useful for cross taxa studies (although see Goldstein & Schlötterer 1999).

Flying insects are especially rich in mitochondrial DNA due to the large number of mitochondria in the very energetically demanding flight muscles (Cullen 1974). Mitochondrial DNA (essentially maternally inherited and non-recombinant, but see Wallis 1999) may be used to give RFLP profiles or directly sequenced (Roehrdanz & Degrugillier 1998). Both nuclear and mitochondrial markers provide valuable genetic data useful in population genetic and taxonomic studies (Roehrdanz & Degrugillier 1998, and see below). However, for the various reasons outlined, use of molecular markers to study movement is rather a matter of 'horses for courses' in terms of which marker is chosen and why.

The use of such markers can first and foremost help in the determination of species integrity, i.e. uniformity in terms of species identity (Hawksworth 1994; Claridge *et al.* 1997). They can also provide the means of discerning the level of gene flow and, by inference, movement between subpopulations of differing genetic composition (demes) (Slatkin 1985, 1987). Usually the latter is achieved by comparing population allele and genotype frequencies (including heterozygote frequencies) within and among populations and thereby deriving statistical estimates of gene flow using conventional measures such as *F*-statistics, and rare allele and other derived parameters; and when population divergence is numerically large enough, estimates of genetic similarity or distance (Slatkin 1985, 1987; Daly 1989; Avise 1994; Roderick 1996; Mallet 2001; Rousset 2001; also see Raybould *et al.*, this volume). *F*-statistics may be described in a hierarchical fashion in terms of gene flow/dispersal ability (Daly 1989), whilst similarity/distance matrices are often presented as dendrograms of relatedness. If levels of differentiation are very small, such measures may not be particularly meaningful for within-species population studies, unless temporal data are available (Loxdale 1994). More meaningful in such circumstances are correlations of the data in relation to geographical parameters: for example, multivariate presentations (Napolitano & Descimon 1994), spatial autocorrelation (Stone & Sunnucks 1993) and UPGMA (unweighted pair group method with arithmetic averages) cluster analysis (Peterson 1996). In the case of mitochondrial genes, interpopulation nucleotide divergence data and maximum parsimony methods can be exploited to represent population identity relative to geographical location, i.e. demographic assessments (Chapco *et al.* 1994). Some studies have plotted *F*-statistical parameters (F_{ST}) and genetic distance against geographical distance (Loxdale *et al.* 1985; Rousset 2001; Raybould *et al.*, this volume). When *Nm* (estimate of gene flow derived from *F*-statistics) was plotted against geographical distance for a range of winged phytophagous species of varying mobility,

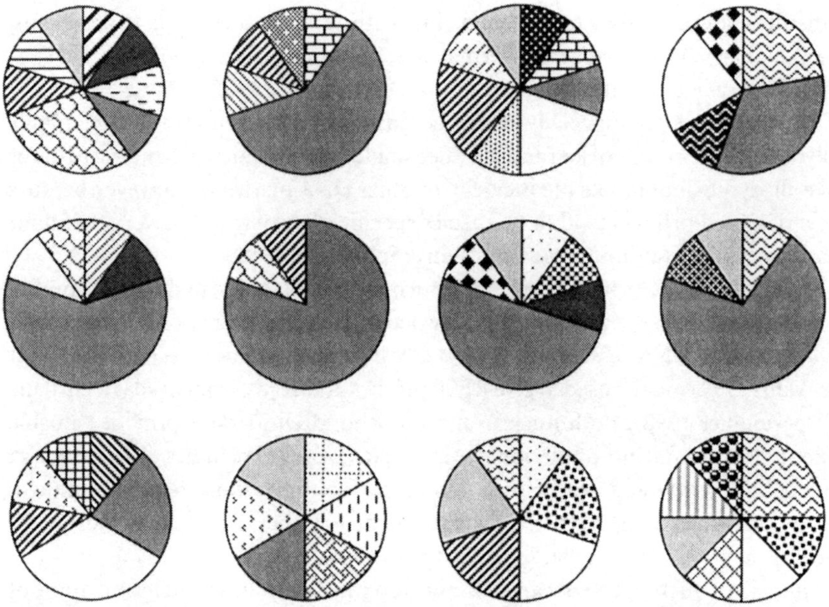

Figure 2.6 Pie diagrams showing genotypic frequency distributions at the dr283 microsatellite locus in field-collected, 'mummy'-reared females of the primary aphid parasitoid, *Diaeretiella rapae* M'Intosh (Hymenoptera: Braconidae) attacking the cabbage aphid, *Brevicoryne brassicae* L. (Hemiptera: Aphididae). Mummies from 12 racemes of a single oilseed rape plant, *Brassica napus* L, were sampled. Each pie diagram is a sample of *c.* 10 insects from one raceme (H.D. Loxdale *et al.*, unpubl. data). At this very small spatial scale, population genetic structure is highly heterogeneous, reflecting the searching behaviour of original individual 'progenitor' female parasitoids (each of which is diploid and maximally produces two genotypes following a single mating with a haploid male).

the differing slopes and intercepts of the regressions corresponded to low, moderate and high mobility/gene flow (Peterson 1996; Peterson & Denno 1998), giving support to the hypothesis of isolation by distance. However, exceptions to this approach have recently been found in brachypterous planthoppers, insects of assumed low mobility (Peterson *et al.* 2001). Assignment tests and Bayesian statistics are also being used to determine population genetic structure and dynamics (Waser & Strobeck 1998; Goldstein & Schlötterer 1999; Dawson & Belkhir 2001).

With the range of molecular markers now available, fascinating insights into insect population structure and dynamics have been gained that were largely, if not completely, impossible by other means. With mitochondrial markers, evidence exists for bottlenecks and founder events (e.g. Mun *et al.* 1999). Using nuclear genomic markers, there is evidence for founder events (Taberner *et al.* 1997), geographical invasions (Stone & Sunnucks 1993), small-scale movements, as illustrated in Fig. 2.6 (Loxdale *et al.* 2000), and large-scale displacements (Daly & Gregg 1985; Daly 1989), including perhaps movement of entire population demes (Loxdale &

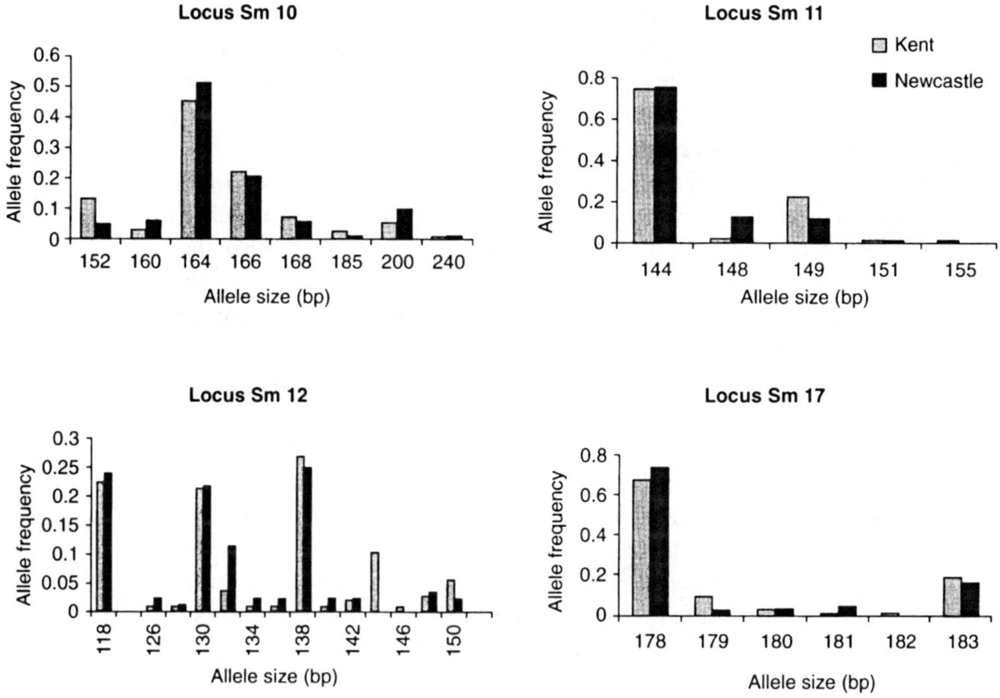

Figure 2.7 Histograms showing allele frequencies at four microsatellite loci (Sm 10, 11, 12 and 17) in samples of the grain aphid, *Sitobion avenae* (F.) collected from 12.2 m high suction traps at two sites. One trap was in the south (Wye, Kent), the other in the north (Newcastle), of England, some 470 km apart. At this large spatial scale, allele frequencies appear generally rather similar at both sites, leading to the inference that the insect is highly migratory (K.S. Llewellyn *et al.*, unpubl. data).

Lushai 1999) and even altitudinal movements related to habitat patchiness and persistence (Liebherr 1988). An example is shown in Fig. 2.7 of spatial and temporal microsatellite data derived from the grain aphid *Sitobion avenae* (F.) collected in the Rothamsted Insect Survey network of suction traps (Llewellyn 2000). The allele frequency data are suggestive of high levels of gene flow and/or migration over hundreds of kilometres.

Where individual genes or gene-like sequences (e.g. microsatellites) are concerned, statistical estimates denote degrees of relative movement, and definite statements can rarely be made concerning range, duration or direction of flight (Loxdale & Lushai 2001). But if, for example, a genotype which occurs at high frequency is found in a new area, then it can be inferred with some confidence that a definite displacement has taken place. Sometimes, geographical barriers appear to delimit population spread temporarily (Stone & Sunnucks 1993; Napolitano & Descimon 1994) or 'permanently', as with some butterflies (Thomas & Lewington 1991; Marchi *et al.* 1996), although other ecological parameters (temperature, host plant, parasites, etc.) could also be involved in such scenarios. The existence of fixed, unique alleles in different species or subspecies populations is usually good evidence for a total lack of gene flow (Marchi *et al.* 1996), whilst the discovery of heterozygotes may denote various degrees of introgression (Allegruci *et al.* 1997), usually in an allopatric sense but sometimes apparently sympatric (Sunnucks *et al.* 1997).

In summary, molecular markers are very useful tools for examining insect movement along with other ecological parameters. Usually, they are applied as indirect measures of population divergence (here in a population dynamic rather than phylogenetic sense) and not as direct measures, although the latter could be possible in some circumstances, e.g. movement of a novel insecticide-resistance genotype into a susceptible population (Foster *et al.* 1998). The most powerful of genetic approaches is ultimately derived from sequencing stretches of the genome and comparing across individuals within and between populations, to deduce spatial and temporal differences (e.g. Garnery *et al.* 1992). Whatever molecular markers are employed, consideration of sampling regime is essential for any study to be meaningful. When large spatial scales are considered, sampling will often tend to be relatively unstructured. At small spatial scales, more concise sampling is preferable, e.g. sampling at geometric distances concentrically from a fixed point (Byrne *et al.* 1996). Molecular markers are best used in combination with ecological approaches to reduce ambiguity in the interpretation of data (e.g. Lewis *et al.* 1997).

Conclusions

The features which allow insects to disperse so successfully over space and time also make study of the phenomenon most challenging. Researchers have risen to that challenge, developing a great variety of techniques flexible enough to study a multitude of species over a variety of landscapes and spatial scales. Recently, rapid advances have been made due to use of new technologies (Reynolds & Riley 2002). Technology has been used to develop specific methods, such as the radar systems and the molecular approaches discussed above, but it has also been used in more general terms. For example, the increased use of remote-sensing and satellite-collected data, and the availability of vast computational power, have dramatically changed the way in which insect movement can be monitored.

Direct methods of following insects will always be more precise than indirect methods. However, it is still very difficult to track insects directly over large distances, either individually or *en masse*. Direct methods, such as mark–recapture, also have the disadvantage that they often involve disturbance of the animal before or during dispersal. Molecular markers provide a valuable, usually indirect, approach. They have the advantage of being suitable for a variety of spatial and temporal scales, and the insect is not disturbed whilst moving (although destructive sampling is usually necessary). Indeed, without the application of molecular markers to the study of insect dispersal at various spatial and temporal scales, the understanding of many population displacements, and complex ecological interactions, is likely to remain incomplete. However, care must be taken with analysis and interpretation of evidence for dispersal based on molecular markers, particularly with regard to their level of neutrality (which is often assumed but is usually unknown). Common to all studies of insect dispersal is the need for rigorous analytical approaches, capable of identifying any bias in the data, and further work is needed in this area. A multidisciplinary approach, involving direct techniques (such as mark–recapture and tracking) in combination with indirect techniques (such as trapping networks and use of genetic markers) is essential in the future study of insect dispersal.

References

Allegruci, G., Minasi, M.G. & Sbordoni, V. (1997) Patterns of gene flow and genetic structure in cave-dwelling crickets of the Tuscan endemic, *Dolichopoda schiavazii* (Orthoptera, Rhaphidophoridae). *Heredity* **78**, 665–673.

Avise, J.C. (1994) *Molecular Markers, Natural History and Evolution.* Chapman & Hall, New York.

Bartlett, A. & Lingren, P. (1984) Monitoring pink bollworm (Lepidoptera: Gelechiidae) populations, using the genetic marker sooty. *Environmental Entomology* **13**, 543–550.

Byrne, D.N. (1999) Migration and dispersal by the sweet potato whitefly, *Bemisia tabaci. Agricultural and Forest Meteorology* **97**, 309–316.

Byrne, D.N., Rathman, R.J., Orum, T.V. & Palumbo, J.C. (1996) Localized migration and dispersal by the sweet potato whitefly, *Bemisia tabaci. Oecologia* **105**, 320–328.

Capaldi, E.A., Smith, A.D., Osborne, J.L. *et al.* (2000) Ontogeny of orientation flight in the honeybee revealed by harmonic radar. *Nature* **403**, 537–540.

Carreck, N.L., Osborne, J.L., Capaldi, E.A. & Riley, J.R. (1999) Tracking bees with radar. *Bee World* **80**, 124–131.

Chapco, W., Kelln, R.A. & McFadyen, D.A. (1994) Mitochondrial DNA variation in North American *Melanopline* grasshoppers. *Heredity* **72**, 1–9.

Chapman, J.W., Smith, A.D., Woiwod, I.P., Reynolds, D.R. & Riley, J.R. (2002) Development of vertical-looking radar technology for monitoring insect migration. *Computers and Electronics in Agriculture* (in press).

Claridge, M.F., Dawah, H.A. & Wilson, M.R. (eds) (1997) *Species: the Units of Biodiversity.* Chapman & Hall, London.

Cooter, R.J. (1989) Swarm flight behaviour in flies and locusts. In: *Insect Flight* (eds G.J. Goldsworthy & C.H. Wheeler), pp. 165–203. CRC Press, Boca Raton, Florida.

Cullen, M.J. (1974) The distribution of asynchronous muscle in insects with particular reference to the Hemiptera: an electron microscope study. *Journal of Entomology (A)* **49**, 17–41.

Daly, J.C. (1989) The use of electrophoretic data in a study of gene flow in the pest species *Heliothis armigera* (Hübner) and *H. punctigera* Wallengren (Lepidoptera: Noctuidae). In: *Electrophoretic Studies on Agricultural Pests* (eds H.D. Loxdale & J. den Hollander), pp. 115–141. Systematics Association Special Vol. 39. Oxford University Press, Oxford.

Daly, J.C. & Gregg, P. (1985) Genetic variation in *Heliothis* in Australia: species identification and gene flow in two pest species *H. armigera* (Hübner) and *H. punctigera* Wallengren (Lepidoptera: Noctuidae). *Bulletin of Entomological Research* **75**, 169–184.

Danthanarayana, W. (ed.) (1986) *Insect Flight: Dispersal and Migration.* Springer Verlag, Berlin.

Dawson, K.J. & Belkhir, K. (2001) A Bayesian approach to the identification of panmictic populations and the assignment of individuals. *Genetical Research* **78**, 59–77.

Day, R. & Knight, J. (1995) Operational aspects of forecasting migrant insect pests. In: *Insect Migration: Tracking Resources through Space and Time* (eds V.A. Drake & A.G. Gatehouse), pp. 323–334. Cambridge University Press, Cambridge.

De Barro, P.J., Sherratt, T.N., Brookes, C.P., David, O. & Maclean, N. (1995) Spatial and temporal variation in British field populations of the grain aphid *Sitobion avenae* (F.) (Hemiptera: Aphididae) studied using RAPD-PCR. *Proceedings of the Royal Society, Series B* **262**, 321–327.

Denholm, I., Sawicki, R. & Farnham, A. (1985) Factors affecting resistance to insecticides in houseflies, *Musca domestica* L. (Diptera: Muscidae) IV. The population biology of flies on animal farms in south-eastern England and its implications for the management of resistance. *Bulletin of Entomological Research* **75**, 143–158.

Dingle, H. (1989) The evolution and significance of migratory flight. In: *Insect Flight* (eds G.J. Goldsworthy & C.H. Wheeler), pp. 99–114. CRC Press, Boca Raton, Florida.

Dingle, H. (1996) *Migration: the Biology of Life on the Move.* Oxford University Press, Oxford.

Dingle, H. (2001) The evolution of migratory syndromes in insects. In: *Insect Movement: Mechanisms and Consequences* (eds I.P. Woiwod, D.R. Reynolds & C.D. Thomas), pp. 159–181. CAB International, Wallingford, UK.

Drake, V.A. (1993) Insect-monitoring radar: a new source of information for migration research and operational pest forecasting. In: *Pest Control and Sustainable Agriculture* (eds S.A. Corey, D.J. Dall & W.M. Milne), pp.452–455. CSIRO Publications, Melbourne.

Drake, V.A. & Gatehouse, A.G. (eds) (1995) *Insect Migration: Tracking Resources through Space and Time.* Cambridge University Press, Cambridge.

Drake, V.A., Gregg, P.C., Harman, I.T. *et al.* (2001) Characterizing insect migration systems in inland Australia with novel and traditional methodologies. In: *Insect Movement: Mechanisms and Consequences* (eds I.P. Woiwod, D.R. Reynolds & C.D. Thomas), pp. 207–233. CAB International, Wallingford, UK.

Foster, S.P., Denholm, I., Harling, Z.K., Moores, G.D. & Devonshire, A.L. (1998) Intensification of insecticide resistance in UK field populations of the peach-potato aphid, *Myzus persicae* (Hemiptera: Aphididae) in 1996. *Bulletin of Entomological Research* **88**, 127–130.

Free, J. & Williams, I. (1978) The responses of the pollen beetle, *Meligethes aeneus*, and the seed weevil, *Ceuthorhynchus assimilis*, to oil-seed rape, *Brassica napus*, and other plants. *Journal of Applied Ecology* **15**, 761–774.

Gangwere, S., Chavin, W. & Evans, F. (1964) Methods of marking insects, with especial reference to Orthoptera (Sens. Lat.). *Annals of the Entomological Society of America* **57**, 662–669.

Garnery, L., Cornuet, J.M. & Solignac, M. (1992) Evolutionary history of the honey bee *Apis mellifera* inferred from mitochondrial DNA analysis. *Molecular Ecology* **1**, 145–154.

Gary, N.E., Witherell, P.C., Lorenzen, K. & Marston, J.M. (1977) Area fidelity and intra-field distribution of honeybees during the pollination of onions. *Environmental Entomology* **6**, 303–310.

Gatehouse, A.G. (1989) Genes, environment, and insect flight. In: *Insect Flight* (eds G.J. Goldsworthy & C.H. Wheeler), pp. 115–138. CRC Press, Boca Raton, Florida.

Gibo, D. (1986) Flight strategies of migrating monarch butterflies (*Danaus plexippus* L.) in southern Ontario. In: *Insect Flight: Dispersal and Migration* (ed. W. Danthanarayana), pp. 172–184. Springer Verlag, Berlin.

Goldstein, D.B. & Schlötterer, C. (1999) *Microsatellites, Evolution and Applications.* Oxford University Press, Oxford.

Graham, H.M., Wolfenbarger, D., Nosky, J., Hernandez Jr, N., Llanes, J. & Tamayo, J. (1978) Use of rubidium to label corn earworm and fall armyworm for dispersal studies. *Environmental Entomology* **7**, 435–438.

Gregg, P. (1993) Pollen as a marker for migration of *Helicoverpa armigera* and *H. punctigera* (Lepidoptera: Noctuidae) from western Queensland. *Australian Journal of Ecology* **18**, 209–219.

Gregg, P., Fitt, G., Zalucki, M. & Murray, D. (1995) Insect migration in an arid continent. II. *Helicopvera* spp. in eastern Australia. In: *Insect Migration: Tracking Resources through Space and Time* (eds V.A. Drake & A.G. Gatehouse), pp. 151–172. Cambridge University Press, Cambridge.

Hawksworth, D.L. (ed.) (1994) *Identification and Characterization of Pest Organisms.* CAB International, Wallingford, UK.

Hedin, J. & Ranius, T. (2002) Using radio telemetry to study dispersal of the beetle *Osmoderma eremita*, an inhabitant of tree hollows. *Computers and Electronics in Agriculture* (in press).

Hobson, K., Wassenaar, L. & Taylor, O. (1999) Stable isotopes (delta D and delta C-13) are geographic indicators of natal origins of monarch butterflies in eastern North America. *Oecologia* **120**, 397–404.

Hoy, M.A. (1994) *Insect Molecular Genetics.* Academic Press, London.

Jackson, J.K., McElravy, E.P. & Resh, V.H. (1999) Long-term movements of self-marked caddisfly larvae (Trichoptera: Sericostomatidae) in a California coastal mountain stream. *Freshwater Biology* **42**, 525–536.

Jones, C.J., Isard, S.A. & Cortinas, M.R. (1999) Dispersal of synanthropic diptera: Lessons from the past and technology for the future. *Annals of the Entomological Society of America* **92**, 829–839.

Jones, R.E., Gilbert, N., Guppy, M. & Nealis, V. (1980) Long-distance movement of *Pieris rapae*. *Journal of Animal Ecology* **49**, 629–642.

Kennedy, J.S. (1985) Migration, behavioural and ecological. *Contributions in Marine Science* **27** (Suppl.), 5–26.

Kennedy, P.J. (1994) The distribution and movement of ground beetles in relation to set-aside arable land. In: *Carabid Beetles: Ecology and Evolution* (eds K. Desender, M. Dufrêne, M. Loreau, M.L. Luff & J.-P. Maelfait), pp. 439–444. Kluwer Academic Publishers, Dordrecht.

Kennedy, P.J. & Young, M.R. (1993) Radar tracking the movements of ground beetles in a farmland mosaic. *Aberdeen Lectures in Ecology* **6**, 18–19.

Krall, S., Peveling, R. & Ba Diallo, D. (eds) (1997) *New Strategies in Locust Control.* Birkhäuser Verlag, Basel.

Lewis, O.T., Thomas, C.D., Hill, J.K. *et al.* (1997) Three ways of assessing metapopulation structure in the butterfly *Plebejus argus*. *Ecological Entomology* **22**, 283–293.

Liebherr, J.K. (1988) Gene flow in ground beetles (Coleoptera: Carabidae) of differing habitat preference and flight-wing development. *Evolution* **42**, 129–137.

Liebhold, A.M., Rossi, R.E. & Kemp, W.P. (1993) Geostatistics and geographic information systems in applied insect ecology. *Annual Review of Entomology* **38**, 303–327.

Lingren, P., Bryant Jr, V., Raulston, J., Pendleton, M., Westbrook, J. & Jones, G. (1993) Adult feeding host range and migratory activities of corn earworm, cabbage looper, and celery looper (Lepidoptera: Noctuidae) moths as evidenced by attached pollen. *Journal of Economic Entomology* **86**, 1429–1439.

Llewellyn, K.S. (2000) *Genetic structure and dispersal of cereal aphid populations.* PhD thesis, University of Nottingham, Nottingham.

Loxdale, H.D. (1994) Isozyme and protein profiles of insects of agricultural and horticultural importance. In: *The Identification and Characterization of Pest Organisms* (ed. D.L. Hawksworth), pp. 337–375. CAB International, Wallingford, UK.

Loxdale, H.D. & Lushai, G. (1998) Molecular markers in entomology (review). *Bulletin of Entomological Research* **88**, 577–600.

Loxdale, H.D. & Lushai, G. (1999) Slaves of the environment: the movement of insects in relation to their ecology and genotype. *Philosophical Transactions of the Royal Society, Series B* **354**, 1479–1495.

Loxdale, H.D. & Lushai, G. (2001) Use of genetic diversity in movement studies of flying insects. In: *Insect Movement: Mechanisms and Consequences* (eds I.P. Woiwod, D.R. Reynolds & C.D. Thomas), pp. 361–386. CAB International, Wallingford, UK.

Loxdale, H.D., Tarr, I.J., Weber, C.P., Brookes, C.P., Digby, P.G.N. & Castañera, P. (1985) Electrophoretic study of enzymes from cereal aphid populations. III. Spatial and temporal genetic variation of populations of *Sitobion avenae* (F.) (Hemiptera: Aphididae). *Bulletin of Entomological Research* **75**, 121–141.

Loxdale, H.D., Brookes, C.P. & De Barro, P.J. (1996) Application of novel molecular markers (DNA) in agricultural entomology. In: *The Ecology of Agricultural Pests: Biochemical Approaches* (eds W.O.C. Symondson & J.E. Liddell), pp. 149–198. Systematics Association Special Volume No. 53. Chapman & Hall, London.

Loxdale, H.D., Brookes, C.P. & Powell, W. (2000) Movements of aphid parasitoids over different spatial scales studied using molecular markers (abstract for invited lecture). *XXI Congress of Entomology, Iguassu Falls, Brazil, 20–26 August 2000.* Abstract Book No. 1, p. 207. Embrapa Soja, Londrina, Brazil.

Lushai, G., Loxdale, H.D., Brookes, C.P., von Mende, N., Harrington, R. & Hardie, J. (1997) Genotypic variation among different phenotypes within aphid clones. *Proceedings of the Royal Society, Series B* **264**, 725–730.

Lushai, G., De Barro, P.J., David, O., Sherratt, T.N. & Maclean, N. (1998) Genetic variation within a parthenogenetic lineage. *Insect Molecular Biology* **7**, 337–344.

Macaulay, E.D.M., Tatchell, G.M. & Taylor, L.R. (1988) The Rothamsted Insect Survey '12-metre' suction trap. *Bulletin of Entomological Research* **78**, 121–129.

MacDonald, D. & Smith, H. (1990) Dispersal, dispersion and conservation in the agricultural ecosystem. In: *Species Dispersal in Agricultural Habitats* (eds R.G.H. Bunce & D.C. Howard), pp. 18–64. Belhaven Press, London.

Mallet, J. (2001) Gene flow. In: *Insect Movement: Mechanisms and Consequences* (eds I.P. Woiwod, D.R. Reynolds & C.D. Thomas), pp. 337–360. CAB International, Wallingford, UK.

Marchi, A., Addis, G., Hermosa, V.E. & Crnjar, R. (1996) Genetic divergence and evolution of *Polyommatus coridon gennargenti* (Lepidoptera: Lycaenidae) in Sardinia. *Heredity* **77**, 16–22.

Mascanzoni, D. & Wallin, H. (1986) The harmonic radar: a new method of tracing insects in the field. *Ecological Entomology* **11**, 387–390.

McManus, M.L. (1988) Weather, behaviour and insect dispersal. *Memoirs of the Entomological Society of Canada* **146**, 71–94.

Mun, J.H., Song, Y.H., Heong, K.L. & Roderick, G.K. (1999) Genetic variation among Asian populations of rice planthoppers, *Nilaparvata lugens* and *Sogatella furcifera* (Hemiptera: Delphacidae): mitochondrial DNA sequences. *Bulletin of Entomological Research* **89**, 245–253.

Napolitano, M. & Descimon, H. (1994) Genetic structure of French populations of the mountain butterfly *Parnassius mnemosyne* L. (Lepidoptera: Papilionidae). *Biological Journal of the Linnean Society* **53**, 325–341.

Osborne, J.L. & Williams, I.H. (2001) Site constancy of bumble bees in an experimentally patchy habitat. *Agriculture, Ecosystems and Environment* **83**, 129–141.

Osborne, J.L., Clark, S.J., Morris, R.J. *et al.* (1999) A landscape scale study of bumble bee foraging range and constancy, using harmonic radar. *Journal of Applied Ecology* **36**, 519–533.

Pedgley, D.E. (ed.) (1981) *Desert Locust Forecasting Manual*, Vols 1 and 2. Centre for Overseas Pest Research, London.

Peterson, M.A. (1996) Long-distance gene flow in the sedentary butterfly, *Euphilotes enoptes* (Lepidoptera: Lycaenidae). *Evolution* **50**, 1990–1999.

Peterson, M.A. & Denno, R.F. (1998) The influence of dispersal and diet breadth on patterns of genetic isolation by distance in phytophagous insects. *American Naturalist* **152**, 428–446.

Peterson, M.A., Denno, R.F. & Robinson, L. (2001) Apparent widespread gene flow in the predominantly flightless planthopper *Tumidagena minuta*. *Ecological Entomology* **26**, 629–637.

Reineke, A., Karlovsky, P. & Zebitz, C.P.W. (1999) Amplified fragment length polymorphism analysis of different geographic populations of the gypsy moth, *Lymantria dispar* (Lepidoptera: Lymantriidae). *Bulletin of Entomological Research* **89**, 79–88.

Reynolds, D.R. & Riley, J.R. (1997) *Flight Behaviour and Migration of Insect Pests: Radar Studies in Developing Countries.* NRI Bulletin No. 71. Natural Resources Institute, University of Greenwich, Chatham, UK.

Reynolds, D.R. & Riley, J.R. (2002) Remote-sensing, telemetric and computer-based technologies for investigating insect movement: a survey of existing and potential techniques. *Computers and Electronics in Agriculture* (in press).

Reynolds, D., Riley, J., Armes, N., Cooter, R., Tucker, M. & Colvin, J. (1997) Techniques for quantifying insect migration. In: *Methods in Ecological and Agricultural Entomology* (eds D. Dent & M. Walton), pp. 111–145. CAB International, Wallingford, UK.

Richardson, B.J., Baverstock, P.R. & Adams, M. (1986) *Allozyme Electrophoresis. A Handbook for Animal Systematics and Population Studies.* Academic Press, London.

Riley, J.R. & Osborne, J.L. (2001) Flight trajectories of foraging insects: observations using harmonic radar. In: *Insect Movement: Mechanisms and Consequences* (eds I.P. Woiwod, D.R. Reynolds & C.D. Thomas), pp. 129–157. CAB International, Wallingford, UK.

Riley, J.R., Smith, A.D., Reynolds, D.R. *et al.* (1996) Tracking bees with harmonic radar. *Nature* **379**, 29–30.

Riley, J.R., Valeur, P., Smith, A.D., Reynolds, D.R., Poppy, G.M. & Löfstedt, C. (1998) Harmonic radar as a means of tracking the pheromone-finding and pheromone-following flight in male moths. *Journal of Insect Behavior* **11**, 287–296.

Riley, J.R., Reynolds, D.R., Smith, A.D. *et al.* (1999) Compensation for the wind by foraging bumble bees. *Nature* **400**, 126.

Robinson, T. (1995) Geographical information systems and remotedly sensed data for determining the seasonal distribution of habitats for migrant insect pests. In: *Insect Migration: Tracking Resources through Space and Time* (eds V.A. Drake & A.G. Gatehouse), pp. 335–352. Cambridge University Press, Cambridge.

Rochester, W.A., Dillon, M.L., Fitt, G.P. & Zalucki, M.P. (1996) A simulation model of long-distance migration of *Helicoverpa* spp. moths. *Ecological Modelling* **86**, 151–156.

Roderick, G.K. (1996) Geographic structure of insect populations: gene flow, phylogeography, and their uses. *Annual Review of Entomology* **41**, 325–352.

Roderick, G. & Caldwell, R. (1992) An entomological perspective on animal dispersal. In: *Animal Dispersal: Small Mammals as a Model* (eds N. Stenseth & W. Lidicke), pp. 274–290. Chapman & Hall, London.

Roehrdanz, R.L. & Degrugillier, M.E. (1998) Long sections of mitochondrial DNA amplified from fourteen orders of insects using conserved polymerase chain reaction primers. *Annals of the Entomological Society of America* **91**, 771–778.

Roland, J., McKinnin, G., Backhouse, C. & Taylor, P.D. (1996) Even smaller radar tags on insects. *Nature* **381**, 120.

Rose, D.J.W., Dewhurst, C.F. & Page, W.W. (2000) *The African Armyworm Handbook: the Status, Biology, Ecology, Epidemiology and Management of Spodoptera exempta (Lepidoptera: Noctuidae)*, 2nd edn. Natural Resources Institute, Chatham, UK.

Rousset, F. (2001) Genetic approaches to the estimation of dispersal rates. In: *Dispersal* (eds J. Clobert, E. Danchin, A.A. Dhondt & J.D. Nichols), pp. 18–28. Oxford University Press, Oxford.

Schaefer, G.W. (1976) Radar observations of insect flight. In: *Insect Flight* (ed. R.C. Rainey), pp. 157–197. Symposia of the Royal Entomological Society Publication No. 7. Blackwell, Oxford.

Schmidt-Koenig, K. (1993) Orientation of autumn migration in the monarch butterfly. *Natural History Museum of Los Angeles County. Science Series* **38**, 275–283.

Schneider, J.C. (1999) Dispersal of a highly vagile insect in a heterogenous environment. *Ecology* **80**, 2740–2749.

Seeley, T.D. & Morse, R. (1978) Nest site selection by the honey bee, *Apis mellifera*. *Insectes Sociaux* **25**, 323–337.

Showers, W.B. (1997) Migratory ecology of the black cutworm. *Annual Review of Entomology* **42**, 393–425.

Slatkin, M. (1985) Gene flow in natural populations. *Annual Review of Ecology and Systematics* **16**, 393–430.

Slatkin, M. (1987) Gene flow and the geographic structure of natural populations. *Science* **236**, 787–792.

Smith, A.D., Riley, J.R. & Gregory, R.D. (1993) A method for routine monitoring of the aerial migration of insects by using a vertical-looking radar. *Philosophical Transactions of the Royal Society, Series B* **340**, 393–404.

Smith, A.D., Reynolds, D.R. & Riley, J.R. (2000) The use of vertical-looking radar to continuously monitor the insect fauna flying at altitude over southern England. *Bulletin of Entomological Research* **90**, 265–277.

Southwood, T.R.E. & Henderson, P. (2000) *Ecological Methods*, 3rd edn. Blackwell Science, Oxford.

Srygley, R.B. & Oliveira, E.G. (2001) Orientation mechanisms and migration strategies within the flight boundary layer. In: *Insect Movement: Mechanisms and Consequences* (eds I.P. Woiwod, D.R. Reynolds & C.D. Thomas), pp. 183–206. CAB International, Wallingford, UK.

Srygley, R.B., Oliveira, E.G. & Dudley, R. (1996) Wind drift compensation, flyways, and conservation of diurnal, migrant neotropical Lepidoptera. *Proceedings of the Royal Society of London, Series B* **263**, 1351–1357.

Stinner, R., Saks, M. & Dohse, L. (1986) Modelling of agricultural pest displacement. In: *Insect Flight: Dispersal and Migration* (ed. W. Danthanarayana), pp. 235–241. Springer Verlag, Berlin.

Stone, G.N. & Sunnucks, P. (1993) Genetic consequences of an invasion through a patchy environment—the cynipid gallwasp *Andricus quercuscalicis* (Hymenoptera: Cynipidae). *Molecular Ecology* **2**, 251–268.

Sunnucks, P., De Barro, P.J., Lushai, G., Maclean, N. & Hales, D. (1997) Genetic structure of an aphid studied using microsatellites: cyclic parthenogenesis, differentiated lineages, and host specialisation. *Molecular Ecology* **6**, 1059–1073.

Sunnucks, P., Wilson, A.C.C., Beheregaray, L.B., Zenger, K., French, J. & Taylor, A.C. (2000) SSCP is not so difficult: the application and utility of single-stranded conformation polymorphism in evolutionary biology and molecular ecology. *Molecular Ecology* **9**, 1699–1710.

Taberner, A., Dopazo, J. & Castañera, P. (1997) Genetic characterization of populations of a *de novo* arisen sugar beet pest, *Aubeonymus mariaefranciscae* (Coleoptera, Curculionidae), by RAPD analysis. *Journal of Molecular Evolution* **45**, 24–31.

Tatchell, G.M. (1991) Monitoring and forecasting aphid problems. In: *Aphid–Plant Interactions: Populations to Molecules* (eds D.C. Peters, J.A. Webster & C.S. Chloubor), pp. 215–231. Oklahoma Agricultural Experiment Station Miscellaneous Publication No. 132. Oklahoma Agricultural Experiment Station, Stillwater, OK.

Taylor, L.R. (1974) Insect migration, flight periodicity, and the boundary layer. *Journal of Animal Ecology* **43**, 225–238.

Thomas, J. & Lewington, R. (1991) *The Butterflies of Britain and Ireland.* Dorling Kindersley, London.

Tu, Z. (2001) Eight novel families of miniature inverted repeat transposable elements in the African malaria mosquito, *Anopheles gambiae. Proceedings of the National Academy of Sciences, USA* **98**, 1699–1704.

Tucker, M.R. & Holt, J. (1999) Decision tools for managing migrant insect pests. In: *Decision Tools for Sustainable Development* (eds I.F. Grant & C. Sear), pp. 97–128. Natural Resources Institute, Chatham, UK.

Turchin, P. (1998) *Quantitative Analysis of Movement: Measuring and Modeling Population Redistribution in Animals and Plants.* Sinauer Associates, Sunderland, MA.

Turchin, P. & Omland, K. (1999) Quantitative analysis of insect movement. In: *Ecological Entomology* (eds C. Huffaker & A. Gutierrez), pp. 463–502. Wiley, New York.

Turnock, W., Gerber, G. & Sabourin, D. (1980) An evaluation of the use of elytra and bodies in X-ray energy-dispersive spectroscopic studies of the red turnip beetle, *Entomoscelis americana* (Coleoptera: Chrysomelidae). *Canadian Entomologist* **112**, 609–614.

Urquhart, F. (1987) *The Monarch Butterfly: International Traveller.* Nelson-Hall, Chicago.

Urquhart, F. & Urquhart, N. (1979) Vernal migration of the monarch butterfly (*Danaus P. plexippus*, Lepidoptera: Danaidae) in North America from the overwintering site in the neo-volcanic plateau of Mexico. *Canadian Entomologist* **111**, 15–18.

Vaughn, T.T. & Antolin, M.F. (1998) Population genetics of an opportunistic parasitoid in an agricultural landscape. *Heredity* **80**, 152–162.

Vos, P., Hogers, R., Bleeker, M. *et al.* (1995) AFLP: a new technique for DNA fingerprinting. *Nucleic Acids Research* **23**, 4407–4414.

Wallin, H. & Ekbom, B.S. (1988) Movements of carabid beetles (Coleoptera: Carabidae) inhabiting cereal fields: a field tracing study. *Oecologia* **77**, 39–43.

Wallin, H. & Ekbom, B. (1994) Influence of hunger level and prey densities on movement patterns in 3 species of Pterostichus beetles (Coleoptera, Carabidae). *Environmental Entomology* **23**, 1171–1181.

Wallis, G.P. (1999) Do animal mitochondrial genomes recombine? *Trends in Ecology and Evolution* **14**, 209–210.

Waser, P.M. & Strobeck, C. (1998) Genetic signatures of interpopulation dispersal. *Trends in Ecology and Evolution* **13**, 43–44.

Wellington, W., Johnson, D. & Lactin, D. (1999) Quantitative analysis of insect movement. In: *Ecological Entomology* (eds C. Huffaker & A. Gutierrez), pp. 313–354. Wiley, New York.

Winston, M.L. (1987) *Biology of the Honey Bee.* Harvard University Press, Harvard.

Woiwod, I.P. & Harrington, R. (1994) Flying in the face of change: the Rothamsted Insect Survey. In: *Long-term Experiments in Agricultural and Ecological Sciences* (eds R.A. Leigh & A.E. Johnston), pp. 321–342. CAB International, Wallingford, UK.

Woiwod, I.P., Reynolds, D.R. & Thomas, C.D. (eds) (2001a) *Insect Movement: Mechanisms*

and Consequences. CAB International, Wallingford, UK.
Woiwod, I.P., Reynolds, D.R. & Thomas, C.D. (2001b) Introduction and overview. In: *Insect Movement: Mechanisms and Consequences* (eds I.P. Woiwod, D.R. Reynolds & C.D. Thomas), pp. 1–18. CAB International, Wallingford, UK.
Wratten, S.D. (1994) *Video Techniques in Animal Ecology and Behaviour.* Chapman & Hall, London.

Chapter 3

From marking to modelling: dispersal study techniques for land vertebrates

Robert E. Kenward, Steve P. Rushton, Chris M. Perrins, David W. Macdonald and Andy B. South

Introduction

Dispersal has been identified as a key process in many theoretical and applied studies. Concept models show, for example, how dispersal processes may crucially affect conservation in fragmenting landscapes (Hanski & Thomas 1994; Baillie *et al.* 2000). Dispersal data are also especially important for predicting the spread of genes, course of invasions or success of reintroductions (e.g. Rushton *et al.* 1997; South *et al.* 2000). Yet dispersal remains one of the least understood factors in conservation biology (Macdonald & Johnson 2001), in no small part because of difficulties in obtaining robust data. This chapter reviews existing and novel methods for the study of dispersal, concentrating on the land vertebrates that are the subjects for a quarter of the papers in this volume, but including some mention of fish and larval amphibia.

The choice of study methods is inevitably influenced by the question posed. However, the choice of method is also liable to influence the findings of a study, as explained later. In this paper, we consider four main categories of methods for studying dispersal in land vertebrates: namely markers, measurements, maps and models. We describe the different methods in each category, compare their ease of use and warn of their shortcomings. Attention is focused mainly on markers and measurements, because a subsequent chapter delves more deeply into the modelling of dispersal, especially when mapped into a spatially explicit context (South *et al.*, this volume).

We also note the particular issues of defining dispersal and of recording dispersal as it occurs. Much research on dispersal has avoided the challenge in these issues, by measurement after the event. This is convenient for an event that is typically hard to detect and rare in the life of an individual. However, although *post hoc* study can readily record the 'where' of dispersal, it is less good for recording the 'when' and very poor for recording 'how' animals moved in terms of their travel details, such as encounters with habitats, competitors and predators. Recording these aspects of dispersal is important for understanding how animals make decisions about: (i) whether to disperse, (ii) how and where to travel, and (iii) where to settle (see

Andreassen *et al.*, this volume). Knowledge of those decision mechanisms is essential for understanding dispersal, for modelling it and for putting the discoveries to practical use.

The advent of active markers, such as radio tags, has potentially made the 'when' of dispersal much easier to detect. However, radio tags also show that dispersal can be much more complicated than the simple start and end co-ordinates recorded by passive marking, and this complexity can make the 'when' hard to define. Moreover, researchers often cannot afford to wait for individuals to disperse. We therefore describe new techniques for objective and automated detection of dispersal.

Markers

Studies of dispersal need not involve tagging. For example, if observers record sightings of a species systematically as it colonizes new areas, simple presence/absence data can be used not merely to estimate invasion rates but also to infer mean dispersal in each generation (Hengeveld 1989; Okubo *et al.* 1989; Lensink 1997; Shigesada & Kawasaki 1997). Of course, this is likely to be most effective for conspicuous species, like the squirrels and raptors in these studies, and where there is a good network of competent observers with a recording service to collate the data.

Another alternative to tagging is the use of visible natural markers to identify individuals, either subjectively or using photographs (Pennycuik 1978; Scott 1978). Even markers that are not visible can be used. For example, individuals can be identified from DNA in faeces (Taberlet & Bouvet 1992; Morin & Woodruff 1996; Kohn & Wayne 1997). However, in either case, the collection of records over a wide enough area to investigate dispersal behaviour requires considerable resources. At population level, past dispersal patterns have sometimes been estimated using genetic markers (e.g. Slatkin 1985; Holder *et al.* 2000; Raybould *et al.*, this volume), albeit with some criticism (Whitlock & McCauley 1999).

Approaches of this type generally provide only extensive data, but these can be used to construct maps of spread that are of great value for testing predictions from sophisticated models. To construct those models, studies are usually based on artificial markings.

Artificial markings

Artificial markings can take the form of modifying the animal to enable individual recognition, typically by colouring or clipping it in some way; alternatively artificial markers can be attached as tags that transmit signals or as passive identifiers that are read in some way. These markings and tags can identify individuals at short, medium or long ranges, effectively 'in the hand', 'in view' and from signals that can be detected in 'line of sight' but may be diminished or blocked by vegetation or hills (Table 3.1).

The simplest form of artificial marking has been the clipping of fur, feathers, fins, ears or digits in combinations that serve to identify sites of origin, year classes or individuals. Generally, only the more permanent of these markings are suitable

Table 3.1 Types of marking and how they are used in dispersal studies.

Range of detection	Type of detection	Examples of marking types
Short (<2 m)	In the hand	Clipping, colouring, bands, coded wires, transponders
Medium (<2 km)	In view	Fur clipping, blanching, collars, colour bands, wing tags, streamers
Long (>2 km)	In line of sight	Radio tags

for dispersal studies, and often only if animals can be readily recaptured so that markings can be examined in the hand. The number of codes for individuals may also be quite limited. This approach is most advantageous if animals can only be caught when too small for other markings, for example when marking amphibian larvae to determine the proportion of each year class that disperses within a group of breeding ponds (Reading *et al.* 1991).

Passive marking for visual identification at medium distances is practical with paints or dyes, and the more permanent blanching of fur or feathers by freeze-marking. Markers based purely on artificial colouring of fur, feathers, scales or carapaces can identify quite large numbers of individuals if alphanumeric symbols or a distinct pattern can be seen readily. For example, the division of bird wings into multiple fields for freeze-marking or bleaching flight feathers can create up to 50 combinations that are highly visible on large, soaring birds (Frey 1999; H. Frey, pers. comm.). Fur clipping of large animals, such as badgers, *Meles meles*, can be used to monitor dispersal by siting video cameras at their dens (Stewart *et al.* 1997).

Passive tags

There are very many types of tags for identifying individual animals (see reviews in Stonehouse 1978; Bub & Oelke 1980; Parker *et al.* 1990; Calvo & Furness 1992; Nietfeld *et al.* 1994). Studies with passive, short-range tags have been especially widespread, dominated by the use of metal rings (bands) on the legs of birds and clips on the fins of fish. In these cases, there is widespread incidental detection of tags in the hand, by people finding corpses, and through professional or recreational bird trapping and fishing. Despite recovery rates of 1–5% or less, adequate data are obtained to estimate patterns of dispersal if thousands of individuals can be marked with these inexpensive tags. Alternatively, individuals can be injected with isotopes that are then shed in their faeces and can be found by searching appropriate areas (Kruuk *et al.* 1979; Jenkins 1980).

Microtransponders, also called passive integrated transponders (PITs), have been used as short-range tags in wildlife for more than a decade (Fagerstone & Johns 1987). These tags, now minimally 2–7 mm in size, emit a weak resonant signal for *c.* 150 mm in air (300 mm in water) if they are activated by a powerful nearby transmitter on the correct frequency and are programmed to return an individual code. Their short range and need of specialized detection equipment means that they have not been used widely in dispersal studies, but they are convenient individual markers for snakes and other herpetofauna (Reading & Davies 1996). For animals that must be marked individually when they are too small even for PITs, there is an alternative system of binary-coded wire tags (Crook & White 1995), which are inexpensive but again require special detection equipment and may be especially prone to be shed from live animals.

For identification of individuals that are in sight but not in the hand, many studies have used larger passive markers such as collars, rings (bands), earclips, wing tags and streamers. With coding as alphanumeric symbols on a variety of background colours, large numbers of individuals can be identified. They are at their best on conspicuous birds and mammals, in areas with an extensive network of competent observers.

Active tags

The main advantage of active tags for dispersal studies is for providing data in conditions of poor visibility or at great range, especially when relatively few individuals can be marked or where there is no observer network. Poor visibility at night motivated the development of medium-range markers that identify individuals by emitting coded light sequences (Batchelor & McMillan 1980; Wolcott 1980), but these seem not to have been used in dispersal studies. Active tags used in dispersal studies are primarily radio transmitters.

Radio tagging is a powerful and versatile study technique, but requires careful planning of equipment, attachment procedures and techniques for collection and analysis of data. Anyone contemplating this method is at risk of wasting resources and even creating animal welfare problems unless they prepare carefully. During the last two decades, there have been at least four books on radio-tagging techniques (Mech 1983; Kenward 1987, 2001; White & Garrott 1990) and other useful reviews (e.g. Macdonald & Amlaner 1980; Harris *et al.* 1990; Samuel & Fuller 1994).

Two main types of radio tag are available. The greatest range can be obtained from platform transmitter terminals (PTTs) that send ultra high-frequency signals to satellites. Current designs operate on the Franco-American ARGOS system; a Doppler shift in tag frequency is used to estimate vectors from satellites to tags. The tags cost about US$3000 to buy and operate for a year. With PTTs as small as 20 g, location resolution is typically in kilometres, although subkilometre resolution is possible if tag frequency is exceptionally stable and tag altitude is known (see also Boyd, this volume). Greater accuracy, which is important if visited habitats are to be assessed in detail, can be obtained if these tags estimate their position with miniature global positioning system (GPS) receivers and use the ARGOS system to relay the estimate (Seegar *et al.* 1996). Resolution better than 100 m is possible from GPS tags (Moen *et al.* 1997; Rempel & Rogers 1997) now that the military-motivated artificial inaccuracy of the NAVSTAR GPS system has been removed. Moreover, if the tags store a sequence of GPS locations for relaying by ARGOS, data on travel routes can be more detailed than the 1–5 locations normally recorded per day by the 2–3 satellites operated for ARGOS. However, at 80 g for the smallest GPS-enabled PTTs, these are only for work on animals of at least 2700 g (using 3% loading) and for projects with generous budgets: such tags cost more than US$5000 to purchase and operate for a year. Data from GPS-relay tags can be downloaded by methods other than PTTs, provided there is another way of knowing the location of the animal concerned.

Quantitative work on dispersal, involving tens or hundreds of radio-tagged animals, has mainly used much simpler tags with very high-frequency (VHF) transmissions. The smallest of these tags is about 0.5 g, but 1 g tags would be needed for a transmission of several weeks during dispersal. On animals that can support antennae of 250–350 mm, detection ranges of 40 km are possible from hills and 80 km from aircraft, but small animals with 100 mm antennae are not usually detectable much beyond 1 km. Optimal animal sizes for work on dispersal are at least 150 g, for carrying tags of 5 g that can last a year. Animals of 500–1000 g can manage tags of 20–30 g for long-range detection during 3–4 years. Projects can purchase and track 30–50 of these tags per year, for effective recording of dispersal to about 100 km, on an equipment budget of US$5000–10 000. This is the cost of two or three PTTs. However, the need to add labour and travel expenses for tracking VHF tags can make satellite telemetry more cost effective for studies that cover very large areas, especially where access is poor.

Measurements

What is measured as dispersal is a matter of definition. Howard (1960) defined dispersal as 'the permanent movement by an individual from birth place to place of reproduction', and this is similar to the definition of 'intergenerational movement' used for this volume. Howard's (1960) definition in which breeding sites were the origin and destination points was convenient at that time, because the young of birds, amphibia, some reptiles and mammals and a few fish could conveniently be marked at natal sites and later be captured at such sites as breeders. With the capture of many birds repeatedly at breeding sites, it was recognized that adults might move between sites, so Greenwood and Harvey (1982) followed Greenwood (1980) by separating natal dispersal, as 'the movement of an individual from the nest to the site of first reproduction', from breeding dispersal between breeding attempts.

Of course, animals do not exist only at breeding sites, but cover a much wider area used for eating, drinking, sleeping, displaying and other activities. Studies of marked and retrapped animals gave wide acceptance for a concept of home range as an area 'traversed by the individual in its normal activities of food gathering, mating and caring for young' (Burt 1943). Ideas of natal and breeding dispersal reconcile easily with this concept of home range. In principle, natal dispersal can be defined as the transition movement from a natal site or natal home range to a home range where an animal breeds, and breeding dispersal as a transition between separate nuptial sites or nuptial home ranges. In practice, however, if movements are recorded in detail by radio tracking an animal along its 'life path' (Baker 1978), several problems of definition emerge.

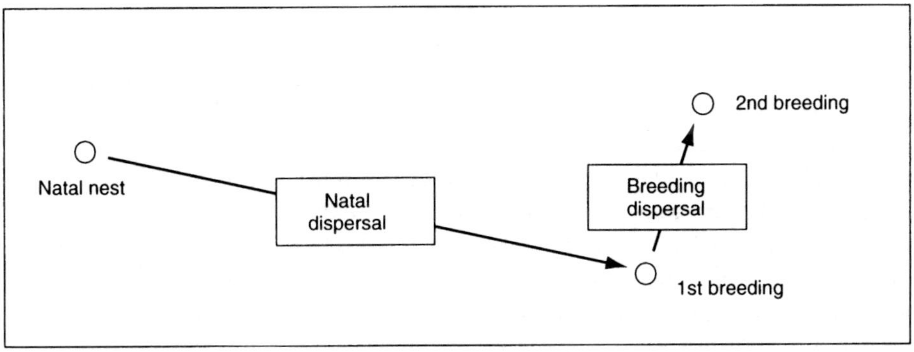

(a)

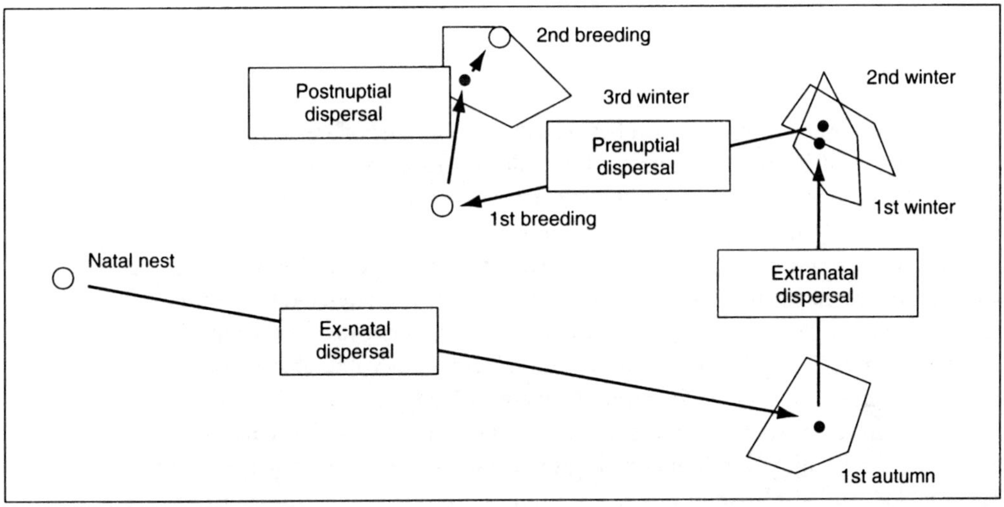

(b)

Figure 3.1 Differences in definition of lifetime movements available from: (a) banding–recapture, and (b) radio tracking. Open circles represent nest sites and polygons surrounding filled circles represent home ranges.

Definitions in dispersal

Some problems arise because movements between successive home ranges often do not fit into a simple framework of natal and breeding dispersal. Instead of a single transition movement, there may be several separate home ranges between the natal area and any breeding area (Fig. 3.1). Studies that recognized this problem have provided terms for the initial movement from the nest, of 'post-fledging dispersal' (Alonso *et al.* 1987; Eden 1987), 'juvenile dispersal' (Gonzalez *et al.* 1989) or 'winter dispersal' (Haig & Orring 1988; Warkentin & James 1990). However, dispersal is not confined to birds, nor is an initial transition necessarily in the juvenile first year or in winter, nor must an animal necessarily breed to have dispersed.

A simple way of dealing with these problems is to modify the terminology of Greenwood (1980), using Latin throughout (Kenward *et al.* 2001b). Thus, the transition 'out of' (Lat. *ex*) the natal area to the home range in which an individual first settled is an *ex-natal* movement, and subsequent transitions between home ranges 'outside' (Lat. *extra*) the natal area are *extranatal* movements. A spring movement that results in pairing or nesting can be termed *prenuptial*, and movements between nuptial sites (the breeding dispersal of Greenwood 1980) are *postnuptial*. This approach has the advantage of encompassing records of natal and breeding dispersal, for example from ringing (Fig. 3.1a), but adding terms for the processes recognized by radio tagging. Radio tracking too can measure natal dispersal, as a vector from site of birth to breeding site, but that vector may include separate components for ex-natal, extranatal and prenuptial movements (Fig. 3.1b).

Definitions in home ranges

Movement vectors are easy to measure between the co-ordinates of breeding sites, or to the point at which a marked animal is found dead, assuming the use of accurate maps, GPS equipment, etc. The process is less simple for movements between home ranges, because a home range is not a point in space. However, a focal point can be estimated from multiple locations of an animal in two main ways. One approach, which estimates the means or medians of the *x* and *y* co-ordinates, has been criticized because this centre of distribution can be in an area unvisited by the animal (White & Garrott 1990); also, the arithmetic mean is especially vulnerable to a skewing effect of outlying locations. These problems are avoided by the second approach, which estimates a centre of density at the location where the mean of a distance function to all other locations is minimal. This approach was introduced by Spencer and Barrett (1984) with the harmonic (inverse reciprocal) mean distance function. Lair (1987) showed that squirrels *Tamiasciurus hudsonicus* (as fairly typical central-place foragers) had nests closer to this focal point than to any other. Although the harmonic mean function is less robust mathematically than other kernel functions (Worton 1989), because of the infinite reciprocal of the distance between locations at the same point (= 1/0), it is also relatively insensitive to outliers that may occur in excursions preceding dispersal.

Another problem with defining start and end points for dispersal occurs when animals are not static. For example, young foxes *Vulpes vulpes* tend to 'drift' across the landscape (Storm *et al.* 1976) to such an extent that Doncaster and Macdonald (1991) provided the concept of a 'prevailing range' in which an animal was constantly adding new areas and ceasing to visit others. Gautestad and Mysterud (1995) used drift data from a wandering flock of domestic sheep, and observations from many studies that home ranges expand as animals excurse into new areas, to suggest that the concept of a definable home range is a 'ghost'. Others have noted the need to set a timeframe for home range definitions when animals move seasonally, and for the problem of deciding what is a 'normal' activity (Cooper 1978; White & Garrott 1990).

These problems are resolved in principle if home ranges are defined simply as 'areas repeatedly traversed within a life path' (Kenward *et al.* 2001b). This approach is especially convenient because it enables unidirectional dispersal to be separated analytically from a home range (with a seasonal or other temporal filter to separate dispersal from migratory movements between home ranges). A 'repeated traverse' concept includes excursions, which are there-and-back movements, within home ranges. It also accommodates the prevailing range of Doncaster and Macdonald (1991) which omits areas no longer traversed; indeed the 'window' aspect of a prevailing range may help separate drift from dispersal (or migration).

After separating home ranges from transition movements along a life path, further analyses can define excursion-exclusive home range cores, if these are required for studies of sociality and resource use (Kenward *et al.* 2001a). Alternatively, relationships between dispersal and habitats or social effects can be investigated in circles of sizes appropriate for core, excursive and visible areas (Kenward *et al.* 2001b). In this case, it is important that the range centre chosen as the origin for such circles is relatively unaffected by delayed detection of dispersal or any 'ghostly growth' that adds outlying locations. If the circles are based on the harmonic mean focal point, the inverse distances to remote locations have tiny values and therefore minimal influence on the location of the range centre.

Dispersal detection

With animals that can in principle be tracked continuously along a life path, detection of timing as well as extent of dispersal becomes important. Departure times can be recorded by logging systems. However, despite the value of data from tracking a disperser in detail, to discover how settlement may depend on areas visited and other animals encountered, researchers usually cannot wait in the field for a rare dispersal event to occur. A solution would be real-time analysis of tag signals, to detect dispersal when it is initiated and then to contact biologists by mobile phone with an event–alert (EVAL) system (Kenward 2001).

The repeated-traverse concept of home range provides the basis for *post hoc* detection of dispersal in spatial data. So far, studies have typically used subjective rules. For example, plotting goshawk *Accipiter gentilis* movements showed that they did not return after travelling more than 1 km from nests, so travel beyond 1 km was used to define dispersal (Kenward *et al.* 1993). Sometimes, however, dispersal criteria become complex. Buzzards *Buteo buteo* made excursions up to 25 km from the nest before dispersing, so in their case the rule was 'travelling more than 1 km and not returning within a month' (Walls & Kenward 1998).

Ranges V software (Kenward & Hodder 1996) provides a first vectored dispersal detector, for objective assessment of when a dispersal or migration movement occurs along a life path. The principle is that if sampled locations conform to a particular distribution before dispersal, consecutive daily locations after dispersal will: (i) fall outside that distribution, and (ii) occur in one direction (Fig. 3.2). The daily

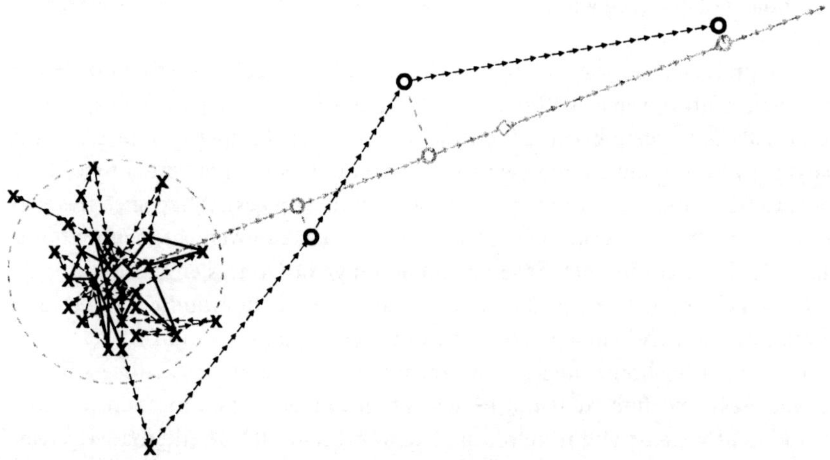

Figure 3.2 Dispersal detection along a vector (grey line) through the mean (grey diamond) of the last three locations (open circles) from the mean (black diamond) of the previous locations (black crosses). Dispersal has occurred because the perpendicular distances of the three locations along the vector are all beyond the 95% confidence limit of a circular normal distribution for the previous locations.

sampling and direction (vector) criteria are important to avoid triggering by short-term excursion movements. The Ranges V implementation uses $n = 3$ consecutive new daily locations and starts with $N = 3$ previous locations, with a circular normal distribution for N and a vector linking the arithmetic means of N and n. Dispersal is flagged if the orthogonal distances of the n locations along this vector are beyond the distribution of the first N, at a chosen level of significance.

Refinements in this simple detector permit a minimum distance for dispersal to be flagged, to reserve triggering for when animals leave a natal area and not when they merely leave a nest, and for triggering with $n < 3$ if subsequent readings were absent because an animal disappeared. However, the process of rejecting dispersal when an animal makes consecutive excursions in different directions leads to delayed or missed detection if the animal makes an initial movement in one direction and then leaves in the opposite direction. Improvements based on a contoured distribution for N and modified vector estimation might be worthwhile.

The prevalence of problems with drift has yet to be investigated, but may be least in birds because these can travel so rapidly. Raptors tend to disperse abruptly: as in goshawks, ex-natal dispersal was abrupt for all but one of 77 buzzards, *Buteo buteo*, tracked through their first year (Walls & Kenward 1995). For species that drift extensively when young, or even remain nomadic (Glover 1952; Baker 1978; Lidicker & Stenseth 1992), there may be little scope for analyses of dispersal categories beyond natal and nuptial. Moreover, even the best objective detector may not cover all occasions without added subjective rules, for example to account for social animals that

Table 3.2 How the type of sampling combines with the type of marker to place constraints on dispersal studies. Local sampling intensity is inversely related to the spatial extent of sampling. Constraints shown in italics are usually insurmountable.

	Local sampling intensity		
Detectability	High (researchers)	Medium (volunteer network)	Low (public)
High (signals)	Animal size	*Expensive*	*Impractical*
Medium (sightings)	Local only	Training	*Unreliable*
Low (in the hand)	Site only	Mass marking	Mass marking

first make excursions to neighbouring social groups and eventually change group completely (Woodroffe *et al.* 1995).

Sampling

The ways in which measurements are collected depend on research goals. Good advice on this aspect of sampling can be found in manuals for radio tagging, banding and marking in general (e.g. Spencer 1984; Bookhout 1994; Kenward 2001). However, it is also important to understand how detectability and cost, which depend on the type of marker, can constrain sampling (Table 3.2).

Active tags can provide a detection probability that approaches 1, which enables location-on-demand of individuals to record details of dispersal. Active tags permit continuous tracking during dispersal; they also enable repeated sampling at regular intervals, to investigate timing effects that can indicate mechanisms. Details of timing showed that ex-natal dispersal is earliest for buzzards with the most siblings, with a 'first-goes-farthest' effect on distances travelled (Kenward *et al.* 2001b). However, active tags are relatively expensive and require animals large enough to carry them (Table 3.2, top left).

Tags with low detectability, like metal bird bands, can also give adequate sample sizes to investigate dispersal at population level, for example to investigate seasonal and regional variation in the pattern of movements (e.g. Thomson *et al.* 1997). However, unless the animal movements are constrained to a few sites (Table 3.2, bottom left), such as amphibian breeding ponds or rivers with PIT detectors on fish ladders, detailed dispersal data depend on marking huge numbers and collating reports from an interested public (Table 3.2, bottom right).

An interesting situation arises with the use of a competent volunteer network, such as the members of the British Trust for Ornithology, and visual markers. Although reports of visual tags by the public may be relatively infrequent, because of a need to identify the species as well as to read the tag (requiring binoculars or telescopes), volunteers can be trained to report effectively and encouraged to visit appropriate sites. Useful data for dispersal analyses have been obtained in this way during British re-introductions of sea eagles (*Haliaeetus albicilla*) and red kites (*Milvus milvus*) (Green *et al.* 1996; Evans *et al.* 1999).

Bias

If detection probability approaches 1, it can be shown that there is little chance of collection bias for survival data (Bunck *et al.* 1995). The same is probably true for dispersal, provided rules are followed to avoid, for example, reduced sampling of animals that have travelled furthest. However, survival data are vulnerable to capture bias, such that marked animals are unrepresentative of the population, and to tag bias if the marking reduces survival. Again, the same is probably true of dispersal studies. Bias from loss of tags, on the other hand, is more likely in estimates of survival than of dispersal.

Capture bias can probably be avoided in dispersal studies if animals are easy to capture at natal sites. The impact of tags on survival, foraging and grooming are assessed relatively frequently for radio tags compared with other types of marking (Murray & Fuller 2000), but studies of tag impact on dispersal are rare. Such studies should assess survival and dispersal separately, and not just record persistence of animals at sites: Smith (1980) showed that tagging could cause little apparent effect on local persistence if a decrease in survival compensated for a decrease in dispersal. Another consideration is the difficulty of testing for tag impacts on dispersal unless one compares two types of marker, one of which is assumed not to cause bias. This approach was used to show that dispersal distances of radio-tagged buzzards were comparable with those recorded by banding (Walls & Kenward 1998).

The worst problem for dispersal measurements is likely to be biased data collection arising from spatial variation in recording efficiency for mass markers. Mean dispersal distances are likely to be underestimated if detectability decreases with distance from a study site (Fig. 3.3a). This would explain why Village (1982) found home ranges of kestrels *Falco tinnunculus* to be smaller if recorded by wing tags than by radio tracking, with the difference greatest for the largest ranges. Records from the public or a volunteer network could also overestimate dispersal and bias its direction if, say, the bulk of the human population lives south of a northern marking area or if resighting is especially likely at distant study sites. The 'fat tails' observed on some dispersal distributions (Turchin 1998) could sometimes result from this bias (Fig. 3.3b).

Maps

Maps are needed in dispersal studies for three main reasons. Firstly, distances travelled by animals are measured between co-ordinates taken from maps; even when the GPS receivers are used, mapping calculations based on the shape of the earth are used to estimate the distances. Second, maps provide data on habitats, human developments and other co-variates that may be important for investigating dispersal mechanisms; for example, was there a significant relationship between local habitat and whether an animal dispersed? Finally, spatially explicit modelling requires map data.

Sophisticated analyses and modelling can be constrained by the type and scale of map data and the speed and memory of computers. The use of spatial data in

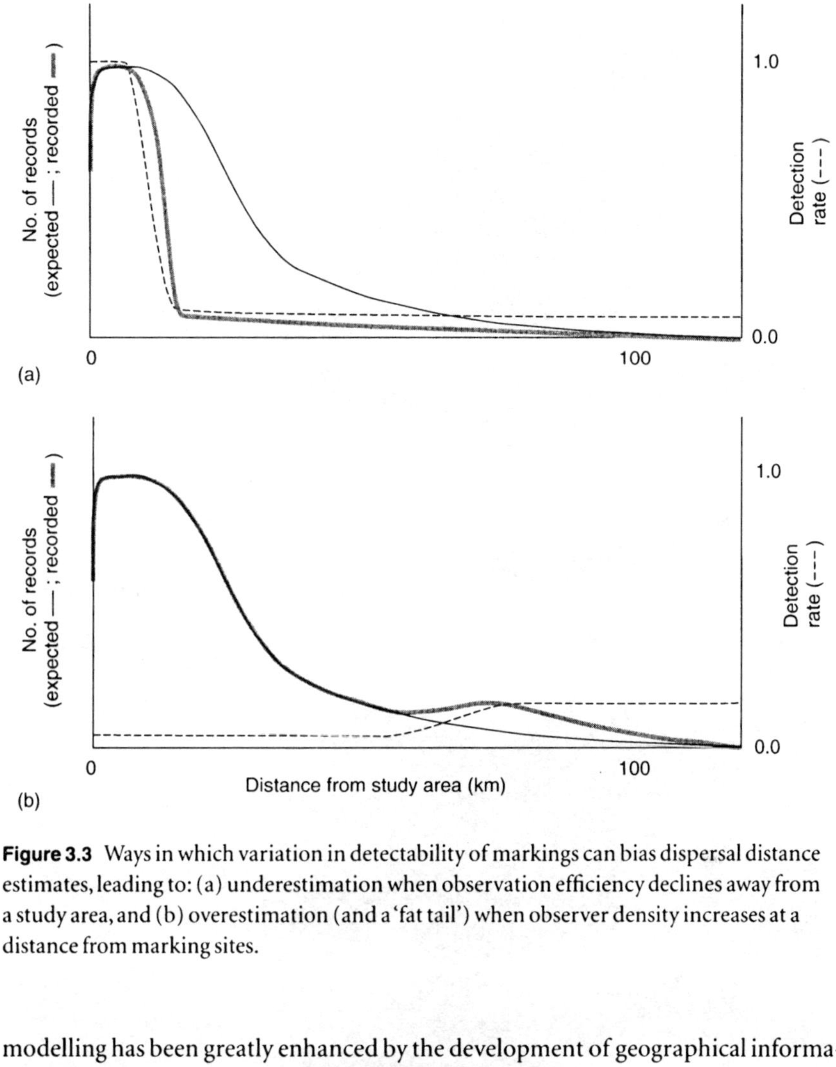

Figure 3.3 Ways in which variation in detectability of markings can bias dispersal distance estimates, leading to: (a) underestimation when observation efficiency declines away from a study area, and (b) overestimation (and a 'fat tail') when observer density increases at a distance from marking sites.

modelling has been greatly enhanced by the development of geographical information systems (GIS), which provide efficient systems for storing information and algorithms for analysing their spatial structure. Maps are held on computers in two formats, vector and raster. Vector maps are composed of lines or curves drawn between points, to create shapes that may or may not be filled (Fig. 3.4a). The lines have no width, so shapes can be defined in detail at any scale, and it is relatively simple to change the attributes of fields as crops are changed, or of forest compartments as trees age. In contrast, a raster map is an array of identical shapes (typically square) with a defined size (Fig. 3.4b). It cannot depict detail below that size and local features cannot easily be edited once the map is completed. Vector maps typically require less storage space than raster maps for a given degree of detail, but also require much computation to fill, which can make complex vector maps slower than raster maps in analyses.

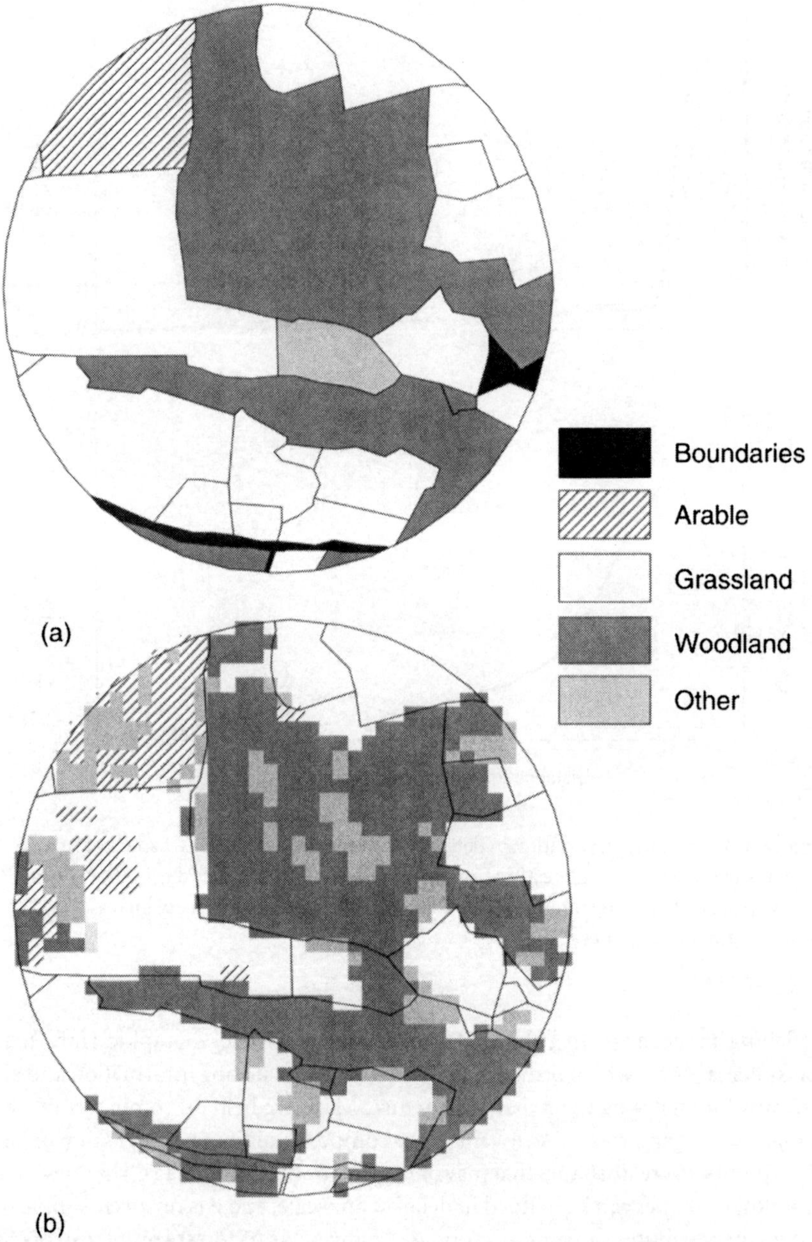

Figure 3.4 Two maps of a 500 m buffer around a buzzard nest, using: (a) vector outlines digitized by hand and filled by ground survey, and (b) 25 m rasters from the Landsat-based Land Cover Map of Great Britain. Land cover has been reduced to five categories for clarity. (Adapted from Hodder 2001.)

Several recent studies of dispersal and colonization models (e.g. Rushton *et al.* 1997; Walls *et al.* 1999; Kenward *et al.* 2001b) have used the raster Land Cover Map of Great Britain (LCMGB) (Fuller *et al.* 1994). The LCMGB was prepared from Landsat images, and depicts 25 landcover types at 25 m resolution. Its great strength is that it covers the whole of Great Britain. Although a vector map of boundaries surveyed by the Ordinance Survey could be prepared, surveying the land use within those boundaries is not currently practical, except from remote images. A comparison of the LCMGB with a local boundary-based map shows that the LCMGB provides fine detail, especially in the categorization of undeveloped areas of heathland, scrub, marshland, etc. that sometimes explains use by buzzards better than the vector map (Hodder 2001). In large, unmapped areas only a remote-imaged raster map is practical.

The weakness of a remote-imaged raster map is that it must be repeated annually if annual changes in field use are to be represented. Rotation of grass to arable land is a particular problem. In a remote map, classification is also constrained by imaging and analysis capabilities, typically preventing distinctions between ages of tree stands, or many types of crop, or a natural plant that is an important food resource (Rushton *et al.* 2000). Moreover, it assesses only the outer layer of vegetation: where there are trees, it registers the canopy but not the shrub and herb layers beneath. This was not a problem for the population-level study of squirrel *Sciurus carolinensis* colonization (Rushton *et al.* 1997), for which the map usefully separated areas of deciduous trees and conifers, or for work on buzzards at home range level (Walls *et al.* 1999) because each buzzard range typically enclosed >100 rasters. However, a dispersal study of individual rural blackbirds *Turdus merula* and thrushes *Turdus philomelos* (Hill 1998) had to use a vector map, not only because the home ranges of these species encompassed <10 rasters but also to enable separate analyses for canopy, shrub and herb layers.

Studies of dispersal at the individual level require software that can combine animal locations and map data. Analyses can be within a general purpose GIS, perhaps with special tools to handle the animal data (Hooge & Eichenlaub 1997), or can use software dedicated to analysing animal location data, with maps imported from another GIS if required. The advantage of the latter approach, if using Ranges V, is the availability of dispersal detectors and other tools for working on life paths; Ranges V is also designed for map-based analyses to produce spreadsheet-ready data from large numbers of animals. A general purpose GIS is typically designed to conduct one analysis at a time and requires editing of output data into rows for further analysis. There is additional discussion and advice on these issues in Kenward (2001).

Models and experiments

Models used in dispersal studies are discussed in detail elsewhere in this symposium, but are covered here briefly for the sake of completeness. Of two model types, one seeks to model dispersal mechanisms, such that observed distributions of dispersal distances emerge from the model. An early example is the cellular occupancy approach of Waser (1985), in which modelled individuals moved outwards from a starting point until they encountered empty living space or died.

The second type seeks to model population processes, such as metapopulation persistence or colonization. Four main approaches have developed (Fig. 3.5), varying in assumptions about movement mechanisms and effects of landscape. Diffusion models (e.g. Okubo 1980; Okubo *et al.* 1989; Shigesada & Kawasaki 1997) represent the spread of a species as the sum of random movements of the individuals making up that population. Spatial contact models (e.g. Hengeveld 1989; Lensink 1997) represent dispersal using observed distributions of dispersal distances. These two approaches are currently unable to incorporate effects of habitat heterogeneity. Therefore, they provide a method for predicting colonizations where landscapes are effectively homogenous for animals, due to generalist behaviour or continental-scale movements, but are unable to predict the effects of landscape on spread rates of habitat specialists at smaller scales. Metapopulation models use simple measures of landscape, as habitable patches and dispersal matrix between, in their representation of a series of subpopulations linked by dispersal (e.g. Levins 1969; Hanski & Thomas 1994). However, a complete representation of variation in landscape and temporal effects, for predicting responses of populations to changing climate and land use at all levels of scale, is practical only with more detailed spatially explicit population models (Dunning *et al.* 1995; South *et al.*, this volume).

Spatially explicit population models incorporate assumptions about the mechanisms that initiate dispersal and eventual settling, but these mechanisms are not well understood. In the field, there have been investigations of what initiates dispersal (e.g. Holleback 1974, Davies 1976; Macdonald 1983; Alonso *et al.* 1987; Bustamente & Hiraldo 1990; Lidicker & Stenseth 1992; Larsen & Boutin 1994; Andreassen *et al.*, this volume; Sutherland *et al.*, this volume). Importantly, since mechanistic modelling based on correlations is inevitably suspect, there have also been experiments to test causality. These have looked at the initiation roles of nutrition and parental behaviour in birds (e.g. Davies 1978; Nilsson 1990; Kenward *et al.* 1993), sociality in mammals (e.g. Ims 1990) and recently include some elegant studies of prenatal conditions in lizards (Massot & Clobert 1995, 2000; Lèna *et al.* 1998). Experimentation ('meddling') is therefore an appropriate addition to the dispersal study methods of 'marking', 'measurement', 'maps' and 'models'. However, the social and environmental factors affecting settlement decisions are hard to measure at individual level (Turchin 1998), let alone to meddle with. Experimental work to elucidate settlement mechanisms remains a major challenge for students of dispersal.

Conclusions

The distance at which markers of any type can be detected reduces with animal size, in a stepped fashion (Fig. 3.6). Only the larger species can carry active tags (PTTs, VHF radios) for distant detection, and their range reduces with antenna dimensions. This means that only animals weighing several hundred grams can be equipped with tags that transmit far and long enough to collect detailed data on

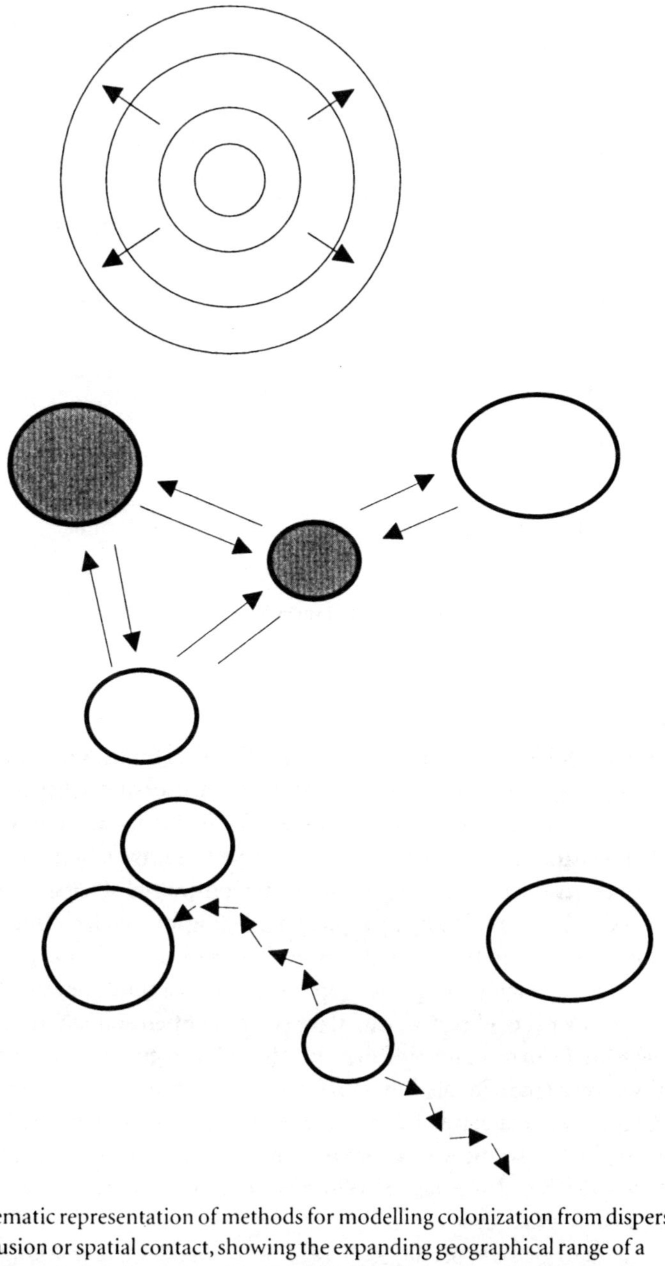

(a)

(b)

(c)

Figure 3.5 Schematic representation of methods for modelling colonization from dispersal data by: (a) diffusion or spatial contact, showing the expanding geographical range of a species, (b) metapopulation dynamics with occupied (filled) or unoccupied (empty) patches, and (c) individual-based, spatially explicit techniques, in which ellipses represent areas of habitat that may be occupied by populations or individuals; dispersal between them is dependent on their position in space, and may be represented by explicitly simulating dispersal paths through the matrix. Dispersal movements are represented by arrows.

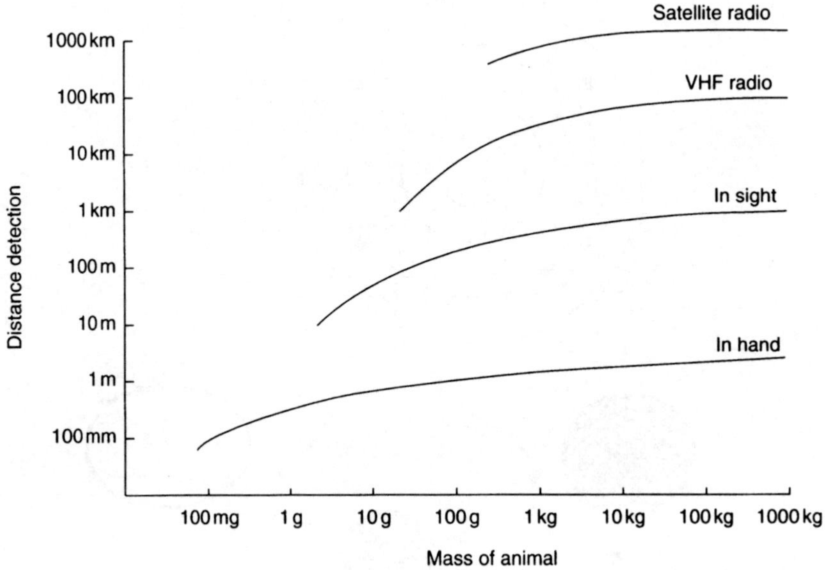

Figure 3.6 The approximate detection distances for different types of marking as a function of animal mass.

dispersal of individuals, over long and short distances for the several years of life in which dispersal may occur. However, short-distance, ex-natal dispersal of animals weighing tens of grams can be recorded if individuals are monitored closely. Another constraint on the study of small species is the availability of map data at a fine enough scale over large enough areas. The constraints of tag size and map detail hinder individual-based spatially explicit modelling for small, mobile vertebrates.

Visual markers can be detected and read on large animals that are hundreds of metres away. For highly conspicuous species like large raptors or water birds visiting specific sites, a net of observers can make sightings often enough to gather an occasional record from most marked individuals. This can give data on dispersal timing and distances of individuals, but with less precision and more risk of bias than with active tags. For a majority of vertebrate species, that are either small or have relatively cryptic behaviour, visual markers are only noticeable and readable at short distances. Only a dense net of observers may then provide worthwhile data, locally, at individual level.

For markers that are usually only read in the hand, mass marking of common species can provide data on seasonal, annual and regional variation in dispersal distances at population level, from reports by the general public. However, rare species may provide too few data to study variation in dispersal distances. Moreover, a low probability of detection provides scope for bias when recording data.

All methods of measuring dispersal are vulnerable to bias. The high detectability of active tags reduces the risk of biased records, although the size of the tags themselves may have an impact on dispersal behaviour. Ideally, distance data from active tags should be checked for tag bias with data from mass marking, in an area where recording bias is minimized by an even distribution of observers. Detailed data from individuals with active tags can then be analysed with increased confidence to investigate mechanisms.

Software based on the 'repeated traverse' concept of home range can now separate dispersal movements from home ranges objectively. Other computer routines can then estimate range centres, and areas around these centres in which mapped habitats and social pressures may explain an individual's tendency to die, or to disperse a recorded distance from that centre and then perhaps to breed. These are important building blocks for spatially explicit models based on the behaviour of individuals. Now that there are sophisticated systems to collect life-path data automatically, for large species at least, we could automate modelling of their movements and demography. It is amazing to consider that only 20 years ago, animals with simple radio tags were tracked mostly on foot, acetate overlays were used to show their movements across maps, and home ranges were still estimated by weighing paper polygons.

References

Alonso, J.C., Gonzalez, L.M., Heredia, B. & Gonzalez, L. (1987) Parental care and the transition to independence of Spanish imperial eagles *Aquila heliaca* in Donana National Park, southwest Spain. *Ibis* **129**, 212–224.

Baillie, S.R., Sutherland, W.J., Freeman, S.N., Gregory, R.D. & Paradis, E. (2000) Consequences of large-scale processes for the conservation of bird populations. *Journal of Applied Ecology* **37**, 88–102.

Baker, R.R. (1978) *The Evolutionary Ecology of Animal Migration.* Hodder & Stoughton, London.

Batchelor, T.A. & McMillan, J.R. (1980) A visual marking system for nocturnal animals. *Journal of Wildlife Management* **44**, 497–499.

Bookhout, T.A. (1994) *Research and Management Techniques for Wildlife and Habitats,* 5th edn. The Wildlife Society, Bethesda, MD.

Bub, H. & Oelke, H. (1980) *Markierungsmethoden für Vögel.* Die Neue Brehm-Bücherei, Wittenberg-Lutherstadt.

Burt, W.H. (1943) Territoriality and home range concepts as applied to mammals. *Journal of Mammalogy* **24**, 346–352.

Bunck, C.M., Chen, C.-L. & Pollock, K.H. (1995) Robustness of survival estimates from radiotelemetry studies with uncertain relocation of animals. *Journal of Wildlife Management* **59**, 790–793.

Bustamente, J. & Hiraldo, F. (1990) Factors influencing family rupture and parent–offspring conflict in the black kite *Milvus migrans. Ibis* **132**, 58–67.

Calvo, B. & Furness, R.W. (1992) A review of the use and the effects of marks and devices on birds. *Ringing and Migration* **13**, 129–151.

Cooper, W.E. (1978) Home range size and population dynamics. *Journal of Theoretical Biology* **75**, 327–337.

Crook, D.A. & White, R.W.G. (1995) Evaluation of subcutaneously implanted visual implant tags and coded wire tags for marking and benign recovery in a small scaleless fish, *Galaxias truttaceus* (Pisces: Galaxiidae). *Marine and Freshwater Research* **46**, 943–946.

Davies, N.B. (1976) Parental care and the transition to independent feeding in the young spotted fly-catcher (*Muscicapa striata*). *Behaviour* **59**, 280–295.

Davies, N.B. (1978) Parental meaness and offspring independence: an experiment with hand reared great tits *Parus major. Ibis* **120**, 509–514.

Doncaster, C.P. & Macdonald, D.W. (1991) Drifting territoriality in the red fox *Vulpes vulpes. Journal of Animal Ecology* **60**, 423–439.

Dunning, J.B., Stewart, D.J., Danielson, B.J. *et al.* (1995) Spatially explicit population models: current forms and future uses. *Ecological Applications* **5**, 3–11.

Eden, S.F. (1987) Natal philopatry of the magpie *Pica pica*. *Ibis* **129**, 477–490.

Evans, I.M., Summers, R.W., O'Toole, L. *et al.* (1999) Evaluating the success of translocating red kites *Milvus milvus* in the UK. *Bird Study* **46**, 129–144.

Fagerstone, K.A. & Johns, B.E. (1987) Transponders as permanent identification markers for domestic ferrets, black-footed ferrets and other wildlife. *Journal of Wildlife Management* **51**, 294–297.

Frey, H. (1999) Report on releases in 1999. In: *Bearded Vulture Reintroduction in the Alps, Annual Report 1999* (eds H. Frey, G. Schaden & M. Bijleveld van Lexmond), pp. 21–27. Foundation for the Conservation of the Bearded Vulture, Veterinary Medical University of Vienna, Vienna.

Fuller, R.M., Groom, G.B. & Jones, A.R. (1994a) The Land Cover Map of Great Britain: an automated classification of Landsat Thematic Mapper data. *Photogrammatic Engineering and Remote Sensing* **60**, 553–562.

Gautestad, A.O. & Mysterud, I. (1995) The home range ghost. *Oikos* **74**, 195–204.

Glover, B. (1952) Movements of birds in South Australia. *South Australia Ornithology* **20**, 82–91.

Gonzalez, L.M., Heredia, B., Gonzalez, J.L. & Alonso, J.C. (1989) Juvenile dispersal of Spanish imperial eagles. *Journal of Field Ornithology* **60**, 369–379.

Green, R.E., Pienkowski, M.W. & Love J.A. (1996) Long-term viability of the re-introduced population of the white-tailed eagle. *Haliaeetus albicilla* in Scotland. *Journal of Applied Ecology* **33**, 357–368.

Greenwood, P.J. (1980) Mating systems, philopatry and dispersal in birds and mammals. *Animal Behaviour* **28**, 1140–1162.

Greenwood, P.J. & Harvey, P.H. (1982) The natal and breeding dispersal of birds. *Annual Review of Ecology and Systematics* **13**, 1–21.

Haig, S.M. & Orring, L.W. (1988) Distribution and dispersal in the piping plover. *Auk* **105**, 630–638.

Hanski, I. & Thomas, C.D. (1994) Metapopulation dynamics and conservation: a spatially explicit model applied to butterflies. *Biological Conservation* **68**, 167–180.

Harris, S., Cresswell, W.J., Forde, P.G., Trewella, W.J., Woollard, T. & Wray, S. (1990) Home-range analysis using radio-tracking data—a review of problems and techniques particularly as applied to the study of mammals. *Mammal Review* **20**, 97–123.

Hengeveld, R. (1989) *Dynamics of Biological Invasions*. Chapman & Hall, London.

Hill, I.F. (1998) *Post-nesting mortality and dispersal in blackbirds and song thrushes*. DPhil thesis, University of Oxford, Oxford.

Hodder, K.H. (2001) *The common buzzard in lowland UK: relationships between food availability, habitat use and demography*. PhD thesis, University of Southampton, Southampton.

Holder, K., Montgomerie, R. & Friesen, V.L. (2000) Glacial vicariance and historical biogeography of rock ptarmigan (*Lagopus mutus*) in the Bering region. *Molecular Ecology* **9**, 1265–1278.

Holleback, M. (1974) Behavioural interactions and the dispersal of the family in black-capped chikadees. *Wilson Bulletin* **86**, 466–468.

Hooge, P.N. & Eichenlaub, B. (1997) *Animal movements extension to ArcView version 1.1*. Alaska Biological Science Center, US Geological Survey, Anchorage.

Howard, W.E. (1960) Innate and environmental dispersal of individual vertebrates. *American Midland Naturalist* **63**, 152–161.

Ims, R.A. (1990) Determinants of natal dispersal and space use in grey-sided voles, *Clethrionomys rufocanus*: a combined field and laboratory experiment. *Oikos* **57**, 106–113.

Jenkins, D. (1980) Ecology of otters in northern Scotland. I. Otter (*Lutra lutra*) breeding and dispersion in mid-Deeside, Aberdeenshire in 1974–79. *Journal of Animal Ecology* **49**, 713–735.

Kenward, R.E. (1987) *Wildlife Radio Tagging—Equipment, Field Techniques and Data Analysis*. Academic Press, London.

Kenward, R.E. (2001) *A Manual for Wildlife Radio Tagging*. Academic Press, London.

Kenward, R.E. & Hodder, K.H. (1996) *Ranges V. An Analysis System for Biological Location Data*. Institute of Terrestrial Ecology, Wareham.

Kenward, R.E., Marcström, V. & Karlbom, M. (1993) Post-nestling behaviour in goshawks, *Accipiter gentilis*: I. The causes of dispersal. *Animal Behaviour* **46**, 365–370.

Kenward, R.E., Clarke, R.T., Hodder, K.H. & Walls, S.S. (2001a) Density and linkage estimators of home range: nearest-neighbor clustering defines multinuclear cores. *Ecology* **82**, 1905–1920.

Kenward, R.E., Walls, S.S. & Hodder, K.H. (2001b). Life path analysis: scaling indicates priming effects of social and habitat factors on dispersal distances. *Journal of Animal Ecology* **70**, 1–13.

Kohn, M.H. & Wayne, R.K. (1997) Facts from faeces revisited. *Trends in Ecology and Evolution* **12**, 223–227.

Kruuk, H., Gorman, M. & Parish, T. (1979) The use of ^{65}Zn for estimating populations of carnivores. *Oikos* **34**, 206–208.

Lair, H. (1987) Estimating the location of the focal center in red squirrel home ranges. *Ecology* **68**, 1092–1101.

Larsen, K.W. & Boutin, S. (1994) Movements, survival and settlement of red squirrels (*Tamiasciurus hudsonicus*) offspring. *Ecology* **75**, 214–223.

Lèna, J.-P., Clobert, J., de Fraipont, M., Lecompte, J. & Guyot, G. (1998) The relative influence of density and kinship on dispersal in the common lizard. *Behavioural Ecology* **9**, 500–507.

Lensink, R. (1997) Range expansion of raptors in Britain and the Netherlands since the 1960s: testing an individual-based diffusion model. *Journal of Animal Ecology* **66**, 811–826.

Levins, R. (1969) Some demographic and genetic consequences of environmental heterogeneity for biological control. *Bulletin of the Entomological Society of America* **15**, 237–240.

Lidicker, W.Z. & Stenseth, N.C. (1992) To disperse or not to disperse: who does it and why? In: *Animal Dispersal: Small Mammals as a Model* (eds N.C. Stenseth & W.Z. Lidicker), pp. 33–45. Chapman & Hall, London.

Macdonald, D.W. (1983) The ecology of carnivore social behaviour. *Nature* **301**, 379–384.

Macdonald, D.W. & Amlaner, C.J. (1980) A practical guide to radio tracking. In: *A Handbook on Biotelemetry and Radio Tracking* (eds C.J. Amlaner & D.W. Macdonald), pp. 143–159. Pergamon Press, Oxford.

Macdonald, D.W. & Johnson, D.D.P. (2001) Dispersal in theory and practice: consequences for conservation biology. In: *Causes, Consequences and Mechanisms of Dispersal at the Individual, Population and Community Level* (eds J. Clobert, J.D. Nichols, E. Danchin & A. Dhondt), pp. 361–374. Oxford University Press, Oxford.

Massot, M. & Clobert, J. (1995) Influence of maternal food availability on offspring dispersal. *Behavioural Ecology and Sociobiology* **37**, 413–418.

Massot, M. & Clobert, J. (2000) Processes at the origin of similarities in dispersal behaviour among siblings. *Journal of Evolutionary Biology* **13**, 707–719.

Mech, L.D. (1983) *Handbook of Animal Radio-Tracking*. University of Minnesota, St Paul, MN.

Moen, R., Pastor, J. & Cohen, Y. (1997) Accuracy of GPS telemetry collar locations with differential correction. *Journal of Wildlife Management* **61**, 530–539.

Morin, P.A. & Woodruff, D.S. (1996) Noninvasive genotyping for vertebrate conservation. In: *Molecular Genetic Approaches in Conservation* (eds R.K. Wayne & T.B. Smith), pp. 298–313. Oxford University Press, Oxford.

Murray, D.L. & Fuller, M.R. (2000) Effects of marking on the life history patterns of vertebrates. In: *Research Techniques in Ethology and Animal Ecology* (eds L. Boitani & T. Fuller), pp. 15–64. Columbia University Press, New York.

Nietfeld, M.T., Barrett, M.W. & Silvy, N. (1994) Wildlife marking techniques. In: *Research and Management Techniques for Wildlife Habitats*, 5th edn (ed. T.A. Bookhout), pp. 140–168. The Wildlife Society, Bethesda, MD.

Nilsson, J.Å. (1990) Family flock break-up: spontaneous dispersal or parental aggression. *Animal Behaviour* **40**, 1001–1003.

Okubo, A. (1980) Diffusion and ecological problems: mathematical models. Springer Verlag, Berlin.

Okubo, A., Maini, P.K., Williamson, M.H. & Murray, J.D. (1989) On the spatial spread of the grey squirrel in Britain. *Proceedings of the Royal Society, Series B* **238**, 113–125.

Parker, N.C., Giorgi, A.E., Heidinger, R.C., Jester, D.B., Prince, E.D. & Winans, G.A. (1990) *Fish-Marking Techniques.* American Fisheries Society, Bethesda, MD.

Pennycuick, C.J. (1978) Identification using natural markings. In: *Animal Marking: Recognition Marking of Animals in Research* (ed. B. Stonehouse), pp. 147–159. University Park Press, Baltimore, MD.

Reading, C.J. & Davies, J.L. (1996) Predation by grass snakes (*Natrix natrix* L.) at a site in southern England. *Journal of Zoology* **239**, 73–82.

Reading, C.J., Loman, J. & Madsen, T. (1991) Breeding pond fidelity in the common toad, *Bufo bufo*. *Journal of Zoology* **225**, 201–211.

Rempel, R.S. & Rodgers, A.R. (1997) Effects of differential correction on accuracy of a GPS location system. *Journal of Wildlife Management* **61**, 525–530.

Rushton, S.P., Lurz, P.W.W., Fuller, R.M. & Garson P.J. (1997) Modelling the distribution of the red and grey squirrel at the landscape scale: a combined GIS and population dynamics approach. *Journal of Applied Ecology* **34**, 1137–1154.

Rushton, S.P., Barreto, G.W., Cormack, R.M., Macdonald, D.W. & Fuller, R.M. (2000) Modelling the effects of mink and habitat fragmentation on the water vole. *Journal of Applied Ecology* **37**, 475–490.

Samuel, M.D. & Fuller, M.R. (1994) Wildlife radiotelemetry. In: *Research and Management Techniques for Wildlife Habitats*, 5th edn (ed. T.A. Bookhout), pp. 370–418. The Wildlife Society, Bethesda, MD.

Scott, D.K. (1978) Identification of individual Bewick's swans by bill patterns. In: *Animal Marking: Recognition Marking of Animals in Research* (ed. B. Stonehouse), pp. 160–168. University Park Press, Baltimore, MD.

Seegar, W.S., Cutchis, P.N., Fuller, M.R., Suter, J.J., Bhatnagar, V. & Wall, J.S. (1996) Fifteen years of satellite tracking development and application to wildlife research and conservation. *Johns Hopkins Advanced Physics Laboratory Technical Digest* **17**, 305–315.

Shigesada, N. & Kawasaki, K. (1997) *Biological Invasions: Theory and Practice*. Oxford University Press, Oxford.

Slatkin, M. (1985) Gene flow in natural populations. *Annual Review of Ecology and Systematics* **16**, 393–430.

Smith, H.R. (1980) Growth, reproduction and survival in *Peromyscus leucopus* carrying intraperitoneally implanted transmitters. In: *A Handbook on Biotelemetry and Radio Tracking* (eds C.J. Amlaner & D.W. Macdonald), pp. 367–374. Pergamon Press, Oxford.

South, A.B., Rushton, S.P. & Macdonald, D.W. (2000). Simulating the proposed reintroduction of the European beaver (*Castor fiber*) to Scotland. *Biological Conservation* **93**, 103–116.

Spencer, R. (1984) *The Ringer's Manual*, 3rd edn. British Trust for Ornithology, Thetford.

Spencer, W.D. & Barrett, R.H. (1984) An evaluation of the harmonic mean measure for defining carnivore activity areas. *Acta Zoologica Fennica* **171**, 255–259.

Stewart, P.D., Ellwood, S.E. & Macdonald, D.W. (1997) Remote video surveillance of wildlife; our experience with the European badger *Meles meles*. *Mammal Review* **27**, 185–204.

Stonehouse, B. (1978) *Animal Marking: Recognition Marking of Animals in Research*. Macmillan, London.

Storm, G.L., Andrews, R.D., Phillips, R.L., Bishop, R.A., Siniff, D.B. & Tester, J.R. (1976) Morphology, reproduction, dispersal and mortality of midwestern red fox populations. *Wildlife Monographs* **49**, 1–82.

Taberlet, P. & Bouvet, J. (1992) Bear conservation genetics. *Nature* **358**, 197.

Thomson, D.L., Ballie, S.R. & Peach W.J. (1997) The demography and age-specific annual survival of song thrushes during periods of population stability and decline. *Journal of Animal Ecology* **66**, 414–424.

Turchin, P. (1998) *Quantitative Analysis of Movement*. Sinauer Associates, Sunderland, MA.

Village, A. (1982) The home range and density of kestrels in relation to vole abundance. *Journal of Animal Ecology* **51**, 413–428.

Walls, S.S. & Kenward, R.E. (1995) Movements of radio-tagged buzzards *Buteo buteo* in their first year. *Ibis* **137**, 177–182.

Walls, S.S. & Kenward, R.E. (1998) Movements of radio-tagged buzzards *Buteo buteo* in early life. *Ibis* **140**, 561–568.

Walls, S.S. Mañosa, S., Fuller, R.M., Hodder, K.H. & Kenward, R.E. (1999) Is early dispersal enterprise or exile? Evidence from radio-tagged buzzards. *Journal of Avian Biology* **30**, 407–415.

Warkentin, I.G. & James, P.C. (1990) Dispersal terminology: changing definitions in midflight? *Condor* **92**, 802–803.

Waser, P.M. (1985) Does competition drive dispersal? *Ecology* **66**, 1170–1175.

White, G.C. & Garrott, R.A. (1990) *Analysis of Wildlife Radio-tracking Data.* Academic Press, New York.

Whitlock, M.C. & McCauley, D.E. (1999) Indirect measures of gene flow and migration: $F_{ST} \neq 1/(4Nm+1)$. *Heredity* **82**, 117–125.

Wolcott, T.G. (1980) Optical and radio optical techniques for tracking nocturnal animals. In: *A Handbook on Biotelemetry and Radio Tracking* (eds C.J. Amlaner & D.W. Macdonald), pp. 333–338. Pergamon Press, Oxford.

Woodroffe, R., Macdonald, D.W. & Da Silva, J. (1995) Dispersal and philopatry in the European badger, *Meles meles. Journal of Zoology (London)* **237**, 227–239.

Worton, B.J. (1989) Kernel methods for estimating the utilization distribution in home-range studies. *Ecology* **70**, 164–168.

Chapter 4

The measurement of dispersal by seabirds and seals: implications for understanding their ecology

Ian L. Boyd

Introduction

We usually view endothermic predators in marine systems, such as seals and seabirds, as long-lived species that have high fidelity to breeding sites. From the smallest of seabirds to the largest of seals (note that *seals* in the context of the present paper is equivalent to *pinnipeds* and includes all true seals, fur seals and sea lions), the common feature that unites all of these groups is the separation of the foraging site from the breeding site. Thus, the site where energy is gained may be from tens to thousands of kilometres from the breeding site where energy is turned into offspring, thereby contributing to fitness.

The marine environments are characterized by high spatial and temporal variability when compared with terrestrial environments (Steele 1985). Combining the necessity to return to land to breed with this variability means that seabirds and seals are restricted to breeding at locations that can be at highly variable distances from the location where food can be found. In order to cope with this unpredictability, these predators have highly conservative breeding strategies. These can be defined as strategies which favour extended parental longevity, a relatively low rate of investment in offspring and extended periods of sexual immaturity. The extended period of sexual immaturity is most probably associated with the time it takes to develop the skills or body size (Laws 1956; Ricklefs 1968; Nelson 1980) needed to forage with sufficient success to earn an energy surplus that can be invested into reproduction.

Understanding this ecological framework for predation by marine species that retain a terrestrial component to the breeding cycle is critical to understanding dispersal in seabirds and seals. Several predictions about dispersal arise from such a framework. First, we would predict that species that have small foraging ranges, especially during the breeding season, would tend to disperse more readily than those that have the capability of long-range feeding movements. This is because those species that feed over a long range have the ability to integrate spatial variability over much larger spatial scales than those that feed over short ranges. On this basis, dispersal would occur less in the long-range foragers because it would be less likely to result in access to new food resources.

The second prediction is that, amongst species like seabirds and seals, in which the location of food is unpredictable, colonial breeding is favoured because: (i) suitable land breeding sites in close proximity to food are limited, and (ii) individuals may gain information from conspecifics at the breeding site about the local abundance or even the actual location of food. On this basis, social behaviour may provide the ultimate mechanisms that drive dispersal between colonies.

I do not believe that the strict definitions of dispersal as they apply to the casting of a new generation of seeds, the permanent movement away from the parent of larvae or even the permanent movement of song birds away from the parental nest, are easy to apply to long-lived iteroparous animals like seabirds and seals. Evidence from the literature also suggests that I am not alone in seeing this as a difficulty (e.g. Kooyman *et al.* 1996; Veit & Prince 1997). There are, of course, shades in the spectrum between classic dispersal, defined in the preface as the intergenerational movement of individuals, and what I am referring to as dispersal in seabirds and seals. Seabirds and seals are also not unique in this respect. The problem I am investigating could apply to many of the higher vertebrates. I think we have to accept that dispersal is not always a one-off process in the lives of individual organisms; that it can take place at any time in the life of an animal; that it may happen on several occasions for a multitude of complex reasons; and that the mechanisms which drive dispersal can be closely tied in with the ecology of individuals, including their movements between dispersal events (see also Kenward *et al.*, this volume). We also have to accept that there may be a close evolutionary link between the rates of dispersal that are characteristic of a species (or a group of species) and the dynamics of food dispersion, which may itself be linked to the foraging movements of individuals.

In this paper I examine dispersal in and between populations of seabirds and seals. I ask four fundamental questions: (1) how often do these animals disperse; (2) at what stages in the life cycle does dispersal happen; (3) what are the consequences of dispersal processes for the dynamics, stability and resilience of these populations; and (4) what ecological and behavioural processes stimulate and control the rate of dispersal? Seabirds and seals present special challenges when studying dispersal processes. Individuals tend to have particularly high reproductive values so that the dispersal of specific individuals within populations tends to be important. In practice, these species also have convenient features mainly involving the relative ease with which it is possible to observe individuals, over years or even decades, at distinct and easily defined breeding colonies or sites. Conversely, long generation times and difficulties with following marked individuals throughout the whole of their life cycle are disadvantages to the study of dispersal in these species.

Measurement of dispersal in seabirds and seals

A fundamental requirement in the measurement of dispersal is to be able to measure the movements of individuals, either directly or by statistical inference, over sufficiently long periods to be able to distinguish true dispersal from other patterns of movement. The techniques used for measuring dispersal in seabirds and seals fall into three categories. These involve sequential population survey, mark–recapture and genetic analysis. There are two main methods involved in mark–recapture: the tagging or ringing (also known as banding) of animals, and the use of radio or satellite tags attached to animals to gain multiple 'recaptures' using terrestrial, aerial or earth-orbiting receiving stations.

Population survey

Sometimes the most appropriate way to examine the dispersal of individuals is to consider the dynamics of dispersion at the population level. Thus, the dispersion of pupping in grey seals (*Halichoerus grypus*) in Orkney has changed through time but because of what we know about the breeding biology of this species, we can interpret this in terms of dispersal (Fig. 4.1). The pattern of change in the grey seal population in Orkney shows that, as the population increased, seals moved to new sites or islands to breed. Grey seals are believed to show high degrees of fidelity to particular pupping sites (Pomeroy *et al.* 1994, 2000). The pattern illustrated in Fig. 4.1 suggests that there was a shift in the balance of attractiveness for new recruits to settle in old versus new sites as the densities of seals on the colonized islands increased (see also Sutherland *et al.*, this volume). This led to progressive overspill from one site to another. Population counts in different regions were also used to infer dispersal in harbour seals (*Phoca vitulina*) during recovery from an epizootic (Ries *et al.* 1999). Numerous examples exist to show this in seabirds too. One of the clearest long-term examples is the progressive colonization of coastal regions in the North Atlantic by fulmars (*Fulmarus glacialis*). In this species the breeding range has expanded over the past 200 years through a process of dispersal to new breeding sites as a consequence of population increase (Stenhouse & Montevecchi 1999).

Mark–recapture

Seasonal and interannual movements, rather than intergeneration movements, have received most attention in seabirds and seals. This reflects the relative difficulty of testing hypotheses over short (months to years) versus long (years to decades) time periods. Dispersal, as a unit of measurement, concerns the rate of change in the location of individuals relative to their location of birth. Ringing, tagging and branding have all been used successfully to investigate dispersal. New technological solutions have assisted with understanding seasonal movements but have, as yet, had

Figure 4.1 (*opposite*) Changes in the number of grey seal (*Halichoerus grypus*) pups born on selected islands in Orkney, UK. Numbers were estimated annually for aerial surveys. The islands were (a) Muckle Greenholm, (b) Holm of Spurness, (c) Sweynholm and Gairsay, (d) Lingaholm, (e) Stroma, and (f) Calf of Eday. The bold lines in (e) show the expected rate of increase from the initial colonization of Stroma assuming a maximum potential rate of increase of 16% per annum or the mean population rate of increase in Britain of 6% per annum. (Data are from annual aerial surveys conducted on behalf of the UK Natural Environment Research Council by the Sea Mammal Research Unit.)

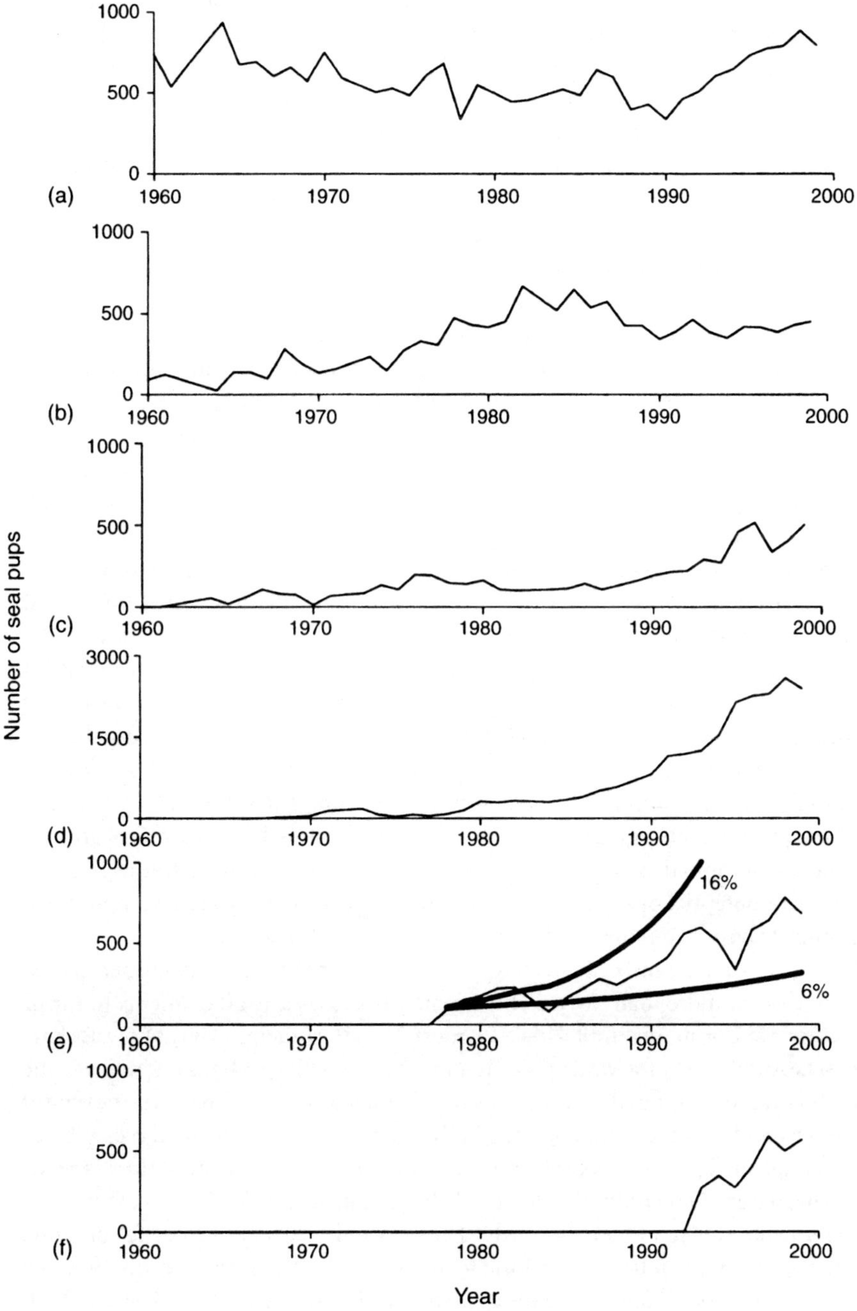
1000
500
0
(a)
1960
1970
1980
1990
2000
1000
500
0
(b)
1960
1970
1980
1990
2000
1000
500
0
(c)
1960
1970
1980
1990
2000
3000
1500
0
(d)
1960
1970
1980
1990
2000
1000
500
0
(e)
16%
6%
1960
1970
1980
1990
2000
1000
500
0
(f)
1960
1970
1980
1990
2000
Number of seal pups
Year

little impact upon our knowledge of dispersal. Nevertheless, these technological solutions have provided insights into the foraging ranges of seabirds and seals that contribute to our understanding of dispersal processes. The very large ranges and the practical difficulties of tracking individuals over the oceans has meant that there are few alternatives to using electronic instrumentation.

Satellite tagging is an especially powerful tool for measuring seasonal movements. At present, this uses the ARGOS system (CLS, 18 Avenue Edouard Belin, 31401 Toulouse, Cedex, France). It has proved to be as useful in both marine and terrestrial environments. In the best circumstances, this system will provide locations of marine animals to an accuracy of 0.5–1.5 km (Boyd *et al.* 1998) but 5–12 km may be more normal (Burns & Castellini 1998) and the accuracy can vary with the behaviour of the animal (Plotkin 1998). Nevertheless, the accuracy is normally sufficient to address most current questions about the patterns of dispersion in seabirds and seals because seasonal movements usually occur on scales of tens to thousands of kilometres. The main limitation is that the animal has to carry a sizeable externally mounted tag (also known technically as a platform transmitter terminal or PTT). There is a trade-off between carrying a battery large enough to power regular, long-term transmissions to a satellite in the earth's orbit and the effect of the size of the instrument on the animal's behaviour (Obrecht *et al.* 1988; Bannasch *et al.* 1994; Croll *et al.* 1996; Watson & Granger 1998). Remarkable progress has been made over the past 10 years in the design and capability of these devices. For example, the first PTTs to be deployed on albatrosses weighed 180 g and could only be used on the largest species (Jouventin & Weimerskirch 1990). Now, PTTs with equivalent capabilities weigh 15 g. Equivalent technological strides have been made with PTTs being deployed on diving animals (McConnell *et al.* 1999; Pütz *et al.* 2000). Although only a little larger than those now being used on flying seabirds (where weight is of greatest concern), some of these tags can now withstand pressures of up to 200 atm (20 MPa) and have potential operational duration of over a year. They are also now small enough to be used on some of the diving seabirds (Pütz *et al.* 2000).

In general, these instruments have not yet been used to examine dispersal over timescales of more than 1 year. This is partly because early attachments to fur or feathers were prone to moult. However, most PTT attachments to birds now use harnesses. Another way forward may be to mount the tracking systems internally in the body of the animal, but this also brings with it other difficulties involving the type of surgical methods that could be used in these circumstances, the materials used to coat implanted instruments and, of course, the issue of whether or not this is an ethical procedure. One study (Hatch *et al.* 2000) examined the effectiveness of implanting satellite transmitters in seabirds. This was only partially successful because, although the transmitters were found to be effective for tracking the birds, they had a short operational lifetime and the implanted birds experienced unacceptably high mortality. In the case of the larger seabirds, we may not be far from having PTTs that are small enough to mount on a leg ring and the latest technological developments, including the use of solar power to reduce the need for on-board battery power, may ultimately make this possible.

An intriguing and still developmental system of measuring dispersal that has mainly been applied to marine predators—although it could also be a useful tool for examining migration patterns in the terrestrial system—is the use of light loggers (De Long 1992; Tuck *et al.* 1999). These microelectronic instruments are extremely simple, small and inexpensive (about £30 per instrument to build). They record light levels at preset intervals and, through the *post hoc* application of analytical algorithms, it is possible to estimate the position of an individual. The principle is that the time between dusk and dawn provides an estimate of latitude and the time of local apparent noon provides an estimate of longitude. The main problems with this system are that the errors in the estimated location tend to be very large (hundreds of kilometres) and the only way to recover the data is by recapturing the animal. Consequently, the system is only applicable to species that move over distances of the scale of ocean basins. For example, it is possible to say if a black-browed albatross has been located over the shelf off Patagonia or in the Benguela Current off South Africa, but it may not provide a way of distinguishing between foraging within or beyond the Falkland Islands Fisheries Zone (Prince *et al.* 1997; Gremillet *et al.* 2000). The problem of having to recover the data is reduced by the tendency of individuals to return to the same breeding location over consecutive breeding attempts. Although the accuracy of this method of examining dispersal (Gremillet *et al.* 2000) is poor, the inaccuracies can be minimized by using long-term averages and by framing the question being addressed around the capabilities of the instrument. The greatest strength of the method is that it could be used to examine questions about interannual dispersal because the instrument can be deployed on a leg ring, at least on the larger seabirds. However, marine fouling can also cause attenuation of the light signal and this may be a problem for the long-term deployment of these instruments on diving species.

The power of the global positioning system (GPS) remains to be harnessed in marine wildlife tracking but it seems likely that useful operational systems will be in place during the next few years. One exploratory study has been carried out on monk seals (Sisak 1998). However, GPS may not provide much more information about dispersal than is already available from satellite tracking and light-level geolocators, although it could be cheaper to operate. This is because GPS still has the dual problems of size and durability that are present in satellite tags. Until recently, the utility of GPS has been limited by the need for relatively large amounts of battery power, mainly to support on-board data processing. However, new data processing algorithms and circuitry requiring less power is beginning to reduce these difficulties. GPS is likely to provide at least an order of magnitude greater accuracy of location than current satellite tracking and so may be useful for answering specific questions about the movement patterns and the detailed locations of individuals relative to breeding sites or food supplies, such as trawlers that tend to attract feeding seabirds. Unless the GPS data produced are linked to a satellite PTT or very high-frequencey (VHF) system for transmission, GPS has the additional restriction that the device normally has to be retrieved from the animal in order to gain access to the data.

The most effective systems for measuring dispersal have been substantially less sophisticated than PTTs, GPS and other data storage tags, which can be very costly both in terms of the costs of the instruments and their operation (e.g. paying for satellite time with the ARGOS system). This can be a major obstacle to examining long-term movements in a sample size large enough to provide statistical power sufficient to examine dispersal.

As with many other bird species, ringing in seabirds has been the most effective long-term system of marking and for studying dispersal (Coulson & Demevergnies 1992; Chastel *et al.* 1993; Prince *et al.* 1994; Pendelow *et al.* 1995; Harris *et al.* 1996; Annett & Pierotti 1999; Frederiksen & Petersen 2000). In pinnipeds, plastic or metal tags have been used extensively (Testa 1987; Oosthuizen 1991; Reiter & Le Boeuf 1991; Baker *et al.* 1995; Boyd *et al.* 1995) but tag loss rates, especially when tags are placed on young juveniles, tend to be high (Testa 1987; Boyd *et al.* 1995) and this has resulted in the method having quite limited utility. Branding of seals has proved to be one of the most effective ways of marking these animals (Ingham 1967), but animal welfare issues now mean that this type of system cannot be maintained in all circumstances. The use of passive integrated transponder (PIT) tags provides a potentially permanent form of individual identification. These small electronic transponders are embedded in glass and can be injected under the skin through a needle. They have no battery power of their own but, when powered by a low-frequency signal from a 'wand' held close to the surface of the animal at the location of the tag, they transmit a unique digital code. This system is particularly popular for monitoring movements of freshwater fish (e.g. Lucas *et al.* 1999) but, while it is in common use on seals and seabirds, there is little evidence from the literature that this system of marking is as successful as it is in fish. Moreover, the utility of PIT tags is usually limited by the need to know in advance that an animal is carrying an implant; so, even with PIT tags, there is usually a need for external marking.

Photographic identification of individual seals using pelage patterns has been used successfully to track the movements of grey seals (A. R. Hiby, pers. comm.). It is possible that this non-invasive method of tracking individuals could be used to examine dispersal in the longer term but its potential remains to be fully exploited.

The difficulties of tracking individuals through long time periods (on occasions for decades) and across large areas result in profound difficulties with the measurement of dispersal in marine predators. Consequently, virtually no study has been able to provide a comprehensive view of dispersal. However, through the methods described above some generalizations are possible.

Genetic analysis

Ideally, we might wish to measure dispersal in terms of genes rather than individuals, although we are still some way from doing this with seabirds and seals. As defined above, dispersal can be measured directly by the movements of marked individuals. Alternatively, it can be inferred indirectly from the spatial distribution of alleles. Direct methods, such as mark–recapture, tend to underestimate long-distance dispersal and they are also costly and time consuming. Several genetic techniques have examined the degree of philopatry amongst seabirds. These include the use of allozymes, the mitochondrial cytochrome *b* gene (Austin 1996; Friesen *et al.* 1996) and restriction analysis of mitochondrial DNA (Ovenden *et al.* 1991); although mitochondrial techniques are generally not good for defining population structures. Similar methods have been used to examine the origins and introgression of seal populations and species (Hoelzel *et al.* 1999). Amongst grey seals, genetic evidence from polymorphic microsatellite loci have confirmed that there is limited dispersal of grey seals between some colonies around the coast of the UK (Allen *et al.* 1995) and a similar result has been obtained for harbour seals (Goodman 1998).

Although many of these methods have provided insights into the history of populations, especially in terms of their recovery from genetic bottlenecks (Hoelzel 1999), it may be premature to infer dispersal mechanisms or processes from them until there is a more complete understanding of avian and pinniped genetic variation (Crochet 1996; Amos & Harwood 1998).

Dispersal and population dynamics

Even though it may not be possible always to detect the dispersal patterns of individuals, we can often detect the presence of dispersal processes from knowledge of the population dynamics of the species concerned. The progressive change in the dispersion of grey seals illustrated in Fig. 4.1 suggests a dispersal process, but what is not clear from the figure is how this has taken place. We do not know whether a single founder event has established each new colony or if this is the result of a protracted process of immigration. From our knowledge of grey seal population dynamics, we can recognize two phases of colonization (Fig. 4.1): a phase when the population is increasing and a phase when it is relatively stable. The rate of increase in the population during the first phase could be fuelled by dispersal from other colonies. During the second phase it is possible that all recruitment is derived from offspring born at the colony or that emigration is balanced by immigration. In this example, the likelihood is that continued immigration has an important role. Figure 4.1e shows two scenarios: one in which the population increases at 16% per annum (close to the theoretical maximum without immigration) after the initial founding event (in this case involving 90 pups) and another in which the population increases at 6%. A 6% increase was close to the measured rate of increase for the UK grey seal population as a whole and this rate probably reflects the biologically realistic rate of increase for a self-sustaining population in which there is no immigration. The difference between the 6% scenario and the observed population change suggests that the locally observed population increase in this species was likely to have been augmented by a continuous recruitment of individuals that originated from other breeding sites.

Another example of a dispersal process illustrated by sequential population survey comes from Audouin's gull (*Larus audouinii* Payraudeau) colonies in the Mediterranean (Oro & Ruxton 2001). In this case, colony growth at one particular site where there had been an apparent improvement in the quality of habitat for Audouin's gull could only be explained by large-scale immigration of individuals into the colony. This conclusion was based upon stochastic modelling of the dynamics of the population from probability distributions of the vital rates.

This method of sequential population assessment, together with an analysis of the dynamics of the population, often does not provide us with any clue, however, as to which section of the population is dispersing. Although age structures can be influenced by many different factors leading to complex interpretations, in general if the age structure of a new colony also has a preponderance of young individuals then we can probably assume that dispersal is mainly fuelled by the emigration of juveniles. Nevertheless, age structures are impossible to measure for most marine predators (especially seabirds).

The example in Fig. 4.1 indicates that dispersal may be important in the dynamics of a seal population. It is likely that dispersal has equal importance for seabirds. Metapopulation structure, represented as geographical fragmentation, can influence the effective population size (Amos & Harwood 1998) but it may also increase the resilience of populations to environmental variability and reduce susceptibility to disease. This may operate by isolating individuals into groups, thus reducing the probability of disease transmission and by acting as a mechanism that could introduce genetic diversity into the population. Therefore, although there is little direct evidence to support the view that dispersal mechanisms will have evolved in these species in order to reduce the risk to fitness from disease or environmental catastrophe, circumstantial evidence would support the hypothesis that there may be a dynamic equilibrium between the forces mitigating for and against dispersal. At an evolutionary scale, the assessments made by individuals will determine the fitness of the individual strategies. The decision mechanisms actively involved in this assessment process are considered in the next section.

Patterns and proximate causes of dispersal in seabirds and seals

The relatively high level of interest in seabirds amongst the research community means that most information about patterns of dispersal come from this group, whereas there is relatively little information from pinnipeds. It is known for many species of both seabirds and seals that, once they are established at a breeding site, adults tend to remain there to breed. However, there is less information about the factors that cause adults to emigrate. In effect these may be the same set of factors that modulate the degree of philopatry amongst new recruits, although they may have different threshold levels at which they become effective in adults compared with new recruits.

For many years it was believed that there were very high levels of philopatry amongst colonial seabirds and this is still assumed to exist in some species (e.g. Prince *et al.* 1994; Bretagnolle & Genevois 1997). However, there is growing awareness of the importance of intergenerational dispersal for seabird population dynamics (Coulson & Demevergnies 1992; Frederiksen & Petersen 2000). This awareness may derive from a more statistically stratified search effort for marked birds. As more information becomes available about the resighting of ringed birds, the analysis of this information becomes more sophisticated. It is probable that previous views of high philopatry resulted from information biased towards re-sightings centred on the location or colony at which marking took place (Coulson & Demevergnies 1992; Harris *et al.* 1996).

Genetic evidence tends to support the view that philopatry occurs at a lower rate than previously expected. Despite evidence of philopatry from capture and resighting studies, there is little genetic differentiation between North Atlantic colonies of Brünnich's guillemot (*Uria lomvia*) (Friesen *et al.* 1996). Similar results were obtained for the fairy prion (*Pachyptila turtur*) (Ovenden *et al.* 1991), Galapagos cormorants (*Compsohalieus harrisi*) (Valle 1995) and sooty terns (*Sterna fuscata*) (Avise *et al.* 2000). Although not always recognized in the literature, the concept of 'strict philopatry' (Austin *et al.* 1994) in seabirds is not compatible with processes of immigration that lead to the founding of new colonies or the rapid recovery of colonies from reduction. In contrast, the genetic evidence does support morphological differentiation between Atlantic colonies of the common guillemot (*Uria aalge*) (Friesen *et al.* 1996), but the genetic techniques are not yet sufficiently developed to be able to estimate rates of dispersal between colonies. In addition, the degree of genetic differentiation between colonies can be influenced by biogeographic history, founder effects and rates of population expansion and may not always be a good indicator of current rates of dispersal (Congden *et al.* 2000).

Actual estimates of rates of philopatry include 36% in kittiwakes (*Rissa tridactyla*) (Coulson & Demevergnies 1992), up to 50% in black guillemots (*Cepphus grylle*) (Frederiksen & Petersen 2000), 42% in the common guillemot (*Uria aalge*) (Harris *et al.* 1996) and only 13% in snow petrels (*Pagodroma nivea*) (Chastel *et al.* 1993). There is less information about how far birds tend to disperse from their natal colony although studies of kittiwakes suggest that dispersal takes place over distances of hundreds of kilometres (Coulson & Demevergnies 1992).

The degree of philopatry shown by pinnipeds is less well documented, mainly because of the difficulties of marking individuals over a sufficiently long period of time. However, from anecdotal accounts, there appears to be a relatively high degree of philopatry amongst both male and female pinnipeds and this is supported by genetic evidence (Allen *et al.* 1995; Stanley *et al.* 1996; Goodman 1998; Slade *et al.* 1998). Amongst northern fur seals (*Callorhinus ursinus*) philopatry is greater amongst females than amongst males (Baker *et al.* 1995) but in this case, and in the closely related Antarctic fur seal, there is a degree of fidelity not just to the breeding colony but to sites within a few metres of the natal area (Lunn & Boyd 1991; Baker *et al.* 1995). In contrast, philopatry in harbour seals operates at scales of 300–500 km (Goodman 1998). Based on this contrast in scale with fur seals, there may be some difficulty with the definition of what constitutes philopatry in these species.

Studies of terns and gulls have provided the most complete information about the mechanisms underlying dispersal. Tern colonies are renowned for being ephemeral. A study of the locations occupied by tern colonies in coastal Virginia showed that there was no serial correlation in the sites occupied in successive years (Erwin *et al.* 1998). In these species, site occupancy was also unrelated to reproductive success in

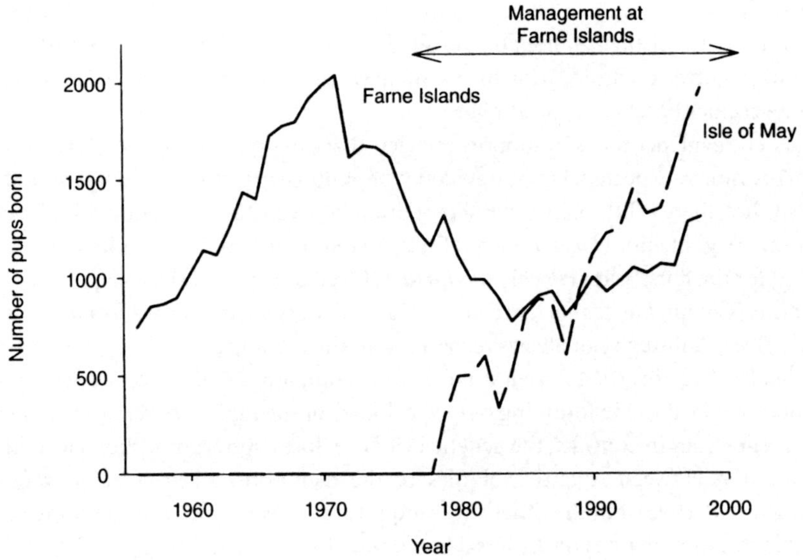

Figure 4.2 Changes in the number of grey seal pups born on the Farne Islands, Northumberland, UK, compared with the number on the neighbouring colony at the Isle of May, Firth of Forth. Management measures were introduced at the Farne Islands in 1979 to keep seals off some islands in order to protect the soils from erosion by the seals. This led to dispersal of seals to the Isle of May, including some emigration of adults. (Data are from annual aerial surveys conducted on behalf of the UK Natural Environment Research Council by the Sea Mammal Research Unit and also from ground counts carried out by the National Trust.)

the previous year. Although gulls are assumed to be mainly philopatric they are also known to disperse from their natal colony. Several studies have shown that this can be related to breeding failure both amongst established breeders and new recruits (Forbes & Kaiser 1994; Oro *et al.* 1999).

Amongst pinnipeds, there is evidence from grey seals that disturbance can lead to emigration. At the Farne Islands, which lie off the northeast coast of England, there has been a management policy in place since the late 1970s which has disturbed any female grey seals attempting to breed on islands where the covering of soil is fragile and important to breeding seabirds. This has resulted in emigration of seals, or at least changes in the recruitment patterns of seals, which resulted in an increase in the population at the Isle of May, the nearest alternative breeding site (Fig. 4.2). Therefore, the mechanisms of dispersal in colonially breeding seals may be similar to those tested experimentally in seabirds, and there is evidence of social mechanisms to support this process (Pomeroy *et al.* 2001).

Factors cited as leading to breeding dispersal in seabirds include predation (Bried & Jouventin 1999; Oro *et al.* 1999), parasitism (Boulinier & Lemel 1996), environmental perturbation (Veit *et al.* 1996; Erwin *et al.* 1998) and poor breeding success

(Coulson & Thomas 1985). The proximate physiological cause of many of these factors may be elevated levels of stress indicated by high corticosterone levels (Wingfield & Ramenofsky 1997). In addition, some of the early ideas developed by Darling (1938) and Wynne-Edwards (1959) about the functions of social interactions in colonial birds, but particularly seabirds, are relevant to more recent explanations of dispersal processes, albeit within the context of evolution at the level of individuals. This involves the hypothesis that individuals either recruiting into a colony, or that are already breeding in a colony, glean information from conspecifics and use this to help determine whether to remain in the colony or to disperse.

This information transfer hypothesis can potentially affect dispersal both within and between colonies. Amongst kittiwakes (*Rissa tridactyla*), which are probably the best studied species, it is known that breeding success is influenced by the apparent quality of individuals, a feature which is also represented by their position within the colony (Coulson & Thomas 1985). A similar situation may also exist in seals (Pomeroy *et al.* 2001). In several species, including kittiwakes, it has been shown that the reproductive success of current breeders in a colony is predictive of the reproductive success of new recruits (Forbes & Kaiser 1994; Danchin *et al.* 1998; Schjorring *et al.* 1999). Prospecting by young birds in years up to breeding may also result in them choosing breeding sites close to those which are most productive (Forbes & Kaiser 1994; Schjorring *et al.* 1999). In addition, reproductive success of locally breeding conspecifics may be sufficient to over-ride an individual's own breeding experience when deciding whether to emigrate (Danchin *et al.* 1998).

These recent studies have emphasized the strong role that social behaviour has in determining rates of dispersal (see also Andreassen *et al.*, this volume). Recruits to a colony have to assess the risks associated with breeding at the colony compared with emigrating. The costs of staying in terms of increased competition for space and resources must be weighed up against both the risks of emigrating and the indicators given by current inhabitants of a colony as to the local abundance of food or safe breeding sites. Therefore, there is a complex set of interacting environmental and social factors that could influence the rates of immigration and emigration that drives dispersal; these factors may be highly influential in determining the cohesiveness of colonial breeding systems.

Modulation of dispersal

At the beginning of this chapter, I predicted that there was a linkage between the foraging movements of individuals within a species during the breeding season and the rate of dispersal. Within the context of the limited information that exists about dispersal in these species it may be possible to come to some broad conclusions about this prediction. Amongst the seabirds, the albatrosses are perhaps those that forage over the greatest ranges during breeding and, consequently, they have the capability to integrate food availability across large spatial scales (Brothers *et al.* 1997, 1998; Weimerskirch *et al.* 1997; Stahl & Sagar 2000; Waugh *et al.* 2000; Wood *et al.* 2000). In contrast, terns and gulls tend to feed close to the breeding sites

Table 4.1 A generalized view of the differing strategies used by seabirds and seals to cope with the high degree of spatiotemporal variability in the marine environment. There may be a trade-off between dispersal and non-dispersal depending upon the species. Those that have small foraging ranges (especially during breeding) are less able to cope with variability in food availability and are, therefore, more likely to disperse, whereas those with large foraging ranges can integrate variability over larger scales and tend to have lower rates of dispersal.

Species group			Foraging ranges (in support of	
Seabirds	Seals	Critical habitats	breeding)	Dispersal
Terns	Sea lions	Coastal/shelf zone	Small	Disperser
Cormorants	Fur seals	↓	↓	↓
Gulls	Small phocids			
Auks	Large phocids			
Petrels				
Albatrosses		Ocean ecosystems	Large	Non-disperser

(Garthe 1997; Oro *et al.* 1997; Suryan *et al.* 2000) and these species have a greater tendency to disperse than albatrosses. These are clearly extremes on a spectrum that may also be represented as a difference between coastally distributed and oceanic species. This spectrum may also be representative of species adaptations to different scales of spatiotemporal heterogeneity of food. There may be a trade-off between dispersal and the capacity to forage over large distances amongst these species (Table 4.1).

Conclusions

In this chapter, I have shown that seabirds and seals exhibit greater rates of dispersal than had previously been considered to be the case. Although the methodological difficulties of measuring dispersal in these species are gradually being overcome (mainly through the application of advanced technology), the long generation times and the potential for dispersal over distances of hundreds to thousands of kilometres has meant that there are relatively few comprehensive studies of dispersal in these species. However, in general, even though most dispersal appears to occur amongst juveniles, with adults showing relatively high degrees of breeding site fidelity, environmental perturbation or breeding failure can lead to dispersal of all age and sex groups. It is also clear that in many of the land-based marine predators, dispersal may have important implications for the stability and resilience of populations. Finally, there is strong evidence of complex social as well as environmental modulation of dispersal and, at an evolutionary scale, species in which foraging occurs amongst individuals over most of the available habitat may show relatively lower levels of dispersal.

References

Allen, P.J., Amos, W., Pomeroy, P.P. & Twiss, S.D. (1995) Microsatellite variations in grey seals (*Halichoerus grypus*) shows evidence of genetic differentiation between two British breeding colonies. *Molecular Ecology* **4**, 653–662.

Amos, W. & Harwood, J. (1998) Factors affecting levels of genetic diversity in natural populations. *Philosophical Transactions of the Royal Society of London, Series B* **353**, 177–186.

Annett, C.A. & Pierotti, R. (1999) Long-term reproductive output in western gulls: consequences of alternate tactics in diet choice. *Ecology* **80**, 288–297.

Austin, J.J. (1996) Molecular phylogenetics of *Puffinus* shearwaters: preliminary evidence from mitochondrial cytochrome *b* gene sequences. *Molecular Phylogenetics and Evolution* **6**, 77–88.

Austin, J.J., White, R.W.G. & Ovenden, J.R. (1994) Population genetic structure of a philopatric colonially nesting seabird, the short-tailed shearwater (*Puffinus tenuirostris*). *Auk* **111**, 70–79.

Avise, J.C., Nelson, W.S., Bowen, B.W. & Walker, D. (2000) Phylogeography of colonially nesting seabirds, with special reference to global matrilineal patterns in the sooty tern (*Sterna fuscata*). *Molecular Ecology* **9**, 1783–1792.

Baker, J.D., Antonelis, G.A., Fowler, C. & York, A.E. (1995) Natal site fidelity in northern fur seals, *Callorhinus ursinus*. *Animal Behaviour* **50**, 237–247.

Bannasch, R., Wilson, R.P. & Culik, B. (1994) Hydrodynamic aspects of design and attachment of a back-mounted device in penguins. *Journal of Experimental Biology* **194**, 83–96.

Boulinier, T. & Lemel, J.Y. (1996) Spatial and temporal variations of factors affecting breeding habitat quality in colonial birds: some consequences for dispersal and habitat selection. *Acta Oecologica* **17**, 531–552.

Boyd, I.L., Croxall, J.P., Lunn, N.J. & Reid, K. (1995) Population demography of Antarctic fur seals: the costs of reproduction and implications for life-histories. *Journal of Animal Ecology* **64**, 505–518.

Boyd, I.L., McCafferty, D.J., Reid, K., Taylor, R. & Walker, T.R. (1998) Dispersal of male and female Antarctic fur seals. *Canadian Journal of Fisheries and Aquatic Science* **55**, 845–852.

Bretagnolle, V. & Genevois, F. (1997) Geographic variation in the call of the blue petrel: effects of sex and geographical scale. *Condor* **99**, 985–989.

Bried, J. & Jouventin, P. (1999) Influence of breeding success on fidelity in long-lived birds: an experimental study. *Journal of Avian Biology* **30**, 392–398.

Brothers, N., Reid, T.A. & Gales, R.P. (1997) At-sea distribution of shy albatrosses *Diomedea cauta cauta* derived from records of band recoveries and colour-marked birds. *Emu* **97**, 231–239.

Brothers, N., Gales, R., Hedd, A. & Roberston, G. (1998) Foraging movements of the shy albatross *Diomedea cauta* breeding in Australia; implications for interactions with longline fisheries. *Ibis* **140**, 446–457.

Burns, J.M. & Castellini, M.A. (1998) Dive data from satellite tags and time-depth recorders: a comparison in Weddell seal pups. *Marine Mammal Science* **14**, 750–764.

Chastel, O., Weimerskirch, H. & Jouventin, P. (1993) High annual variability in reproductive success and survival of an Antarctic seabird, the snow petrel *Pagodroma nivea*—a 27-year study. *Oecologia* **94**, 278–285.

Congden, B.C., Piatt, J.F., Martin, K. & Friesen, V.L. (2000) Mechanisms of population differentiation in marbled murrelets: historical versus contemporary processes. *Evolution* **54**, 974–986.

Coulson, J.C. & Demevergnies, G.N. (1992) Where do young kittiwakes *Rissa tridactyla* breed, philopatry or dispersal? *Ardea* **80**, 186–197.

Coulson, J.C. & Thomas, C. (1985) Differences in the breeding performance of individual kittiwake gulls, *Rissa tridactyla* (L.). In: *Behavioural Ecology* (eds R.M. Sibly & R.H. Smith), pp. 489–503. Blackwell Scientific Publications, Oxford.

Crochet, P.A. (1996) Can measures of gene flow help to evaluate bird dispersal? *Acta Oecologica* **17**, 459–474.

Croll, D.A., Jansen, J.K., Goebel, M.E., Boveng, P.L. & Bengtson, J.L. (1996) Foraging behavior and reproductive success in chinstrap penguins: the effects of transmitter attachment. *Journal of Field Ornithology* **67**, 1–9.

Danchin, E., Boulinier, T. & Massot, M. (1998) Conspecific reproductive success and breeding

habitat selection: implications for the study of coloniality. *Ecology* **79**, 2415–2428.
Darling, F.F. (1938) *Bird Flocks and the Breeding Cycle.* Cambridge University Press, Cambridge.
De Long, R.L. (1992) Documenting migrations of northern elephant seals using day length. *Marine Mammal Science* **8**, 155–159.
Erwin, R.M., Nichols, J.D., Eyler, T.B., Stotts, D.B. & Truitt, B.R. (1998) Modeling colony site dynamics: a case study of gull-billed terns (*Sterna nilotica*) in coastal Virginia. *Auk* **115**, 970–978.
Forbes, L.S. & Kaiser, G.W. (1994) Habitat choice in breeding seabirds: when to cross the information barrier. *Oikos* **70**, 377–384.
Frederiksen, M. & Petersen, A. (2000) The importance of natal dispersal in a colonial seabird, the black guillemot *Cepphus grylle. Ibis* **142**, 48–57.
Friesen, V.L., Montevecchi, W.A., Baker, A.J., Barrett, R.T. & Davidson, W.S. (1996) Population differentiation and evolution in the common guillemot *Uria aalge. Molecular Ecology* **5**, 793–805.
Garthe, S. (1997) Influence of hydrography, fishing activity, and colony location on summer seabird distribution in the south-eastern North Sea. *ICES Journal of Marine Science* **54**, 566–577.
Goodman, S.J. (1998) Patterns of extensive genetic differentiation and variation among European harbor seals (*Phoca vitulina vitulina*) revealed using microsatellite DNA polymorphisms. *Molecular Biology and Evolution* **15**, 104–118.
Gremillet, D., Wilson, R.P., Wanless, S. & Chater, T. (2000) Black-browed albatrosses, international fisheries and the Patagonian Shelf. *Marine Ecology Progress Series* **195**, 269–280.
Harris, M.P., Halley, D.J. & Wanless, S. (1996) Philopatry in the common guillemot *Uria aalge. Bird Study* **43**, 134–137.
Hatch, S.A., Meyers, P.M., Mulcahy, D.M. & Douglas, D.C. (2000) Performance of implantable satellite transmitters in diving seabirds. *Waterbirds* **23**, 84–94.
Hoelzel, A.R. (1999) Impact of population bottlenecks on genetic variation and the importance of life-history: a case study of the northern elephant seal. *Biological Journal of the Linnean Society* **68**, 23–39.
Hoelzel, A.R., Stephens, J.C. & O'Brien, S.J. (1999) Molecular genetic diversity and evolution at the MHC DQB locus in four species of pinnipeds. *Molecular Biology and Evolution* **16**, 611–618.
Ingham, S.E. (1967) Branding elephant seals for life-history studies. *Polar Record* **13**, 447–449.
Jouventin, P. & Weimerskirch, H. (1990) Satellite tracking of wandering albatrosses. *Nature* **343**, 746–748.
Kooyman, G.L., Kooyman, T.G., Horning, M. & Kooyman, C.A. (1996) Penguin dispersal after fledging. *Nature* **383**, 397.
Laws, R.M. (1956) Growth and sexual maturity in aquatic mammals. *Nature* **178**, 193–194.
Lucas, M.C., Mercer, T., Armstrong, J.D., McGinty, S. & Rycroft, P. (1999) Use of a flat-bed passive integrated transponder antenna array to study the migration and behaviour of lowland river fishes at a fish pass. *Fisheries Research* **44**, 183–191.
Lunn, N.J. & Boyd, I.L. (1991) Pupping site fidelity of Antarctic fur seals at Bird Island, South Georgia. *Journal of Mammalogy* **72**, 202–206.
McConnell, B.J., Fedak, M.A., Lovell, P. & Hammond, P.S. (1999) Movements and foraging areas of grey seals in the North Sea. *Journal of Applied Ecology* **36**, 573–590.
Nelson, J.B. (1980) *Seabirds, their Biology and Ecology.* Hamlyn, London.
Obrecht, H.H., Pennycuick, C.J. & Fuller, M.R. (1988) Wind-tunnel experiments to assess the effect of back-mounted radio transmitters on bird body drag. *Journal of Experimental Biology* **135**, 265–273.
Oosthuizen, W.H. (1991) General movements of South African (Cape) fur seals *Arctocephalus pusillus pusillus* from analysis of recoveries of tagged animals. *South African Journal of Marine Science* **11**, 21–29.
Oro, D. & Ruxton, G.D. (2001) The formation and growth of seabird colonies: Audouin's gull as a case study. *Journal of Animal Ecology* **70**, 527–535.
Oro, D., Ruiz, X., Jover, L., Pedrocchi, V. & Gonzalez-Solis, J. (1997) Diet and adult time budgets of Audouin's gull *Larus audouinii* in response to changes in commercial fisheries. *Ibis* **139**, 631–637.
Oro, D., Pradel, R. & Lebreton, J.D. (1999) Food availability and nest predation influence life history traits in Audouin's gull, *Larus audouinii. Oecologia* **118**, 438–445.
Ovenden, J.R., Wuststaucy, A., Bywater, R., Brothers, N. & White, R.W.G. (1991) Genetic evidence for philopatry in a collonially nesting seabird, the fairy prion (*Pachyptila turtur*). *Auk* **108**, 688–694.

Pendelow, J.A., Nichols, J.D., Nisbet, I.C.T. *et al.* (1995) Estimating annual survival and movement rates of adults within a metapopulation of roseate terns. *Ecology* **76**, 2415–2428.

Plotkin, P.T. (1998) Interaction between behavior of marine organisms and the performance of satellite transmitters: a marine turtle case study. *Marine Technology Society Journal* **32**, 5–10.

Pomeroy, P.P., Anderson, S.S., Twiss, S.D. & McConnell, B.J. (1994) Dispersion and site fidelity of breeding female grey seals (*Halichoerus grypus*) on North Rona, Scotland. *Journal of Zoology* **233**, 429–447.

Pomeroy, P.P., Twiss, S.D. & Redman, P. (2000) Philopatry, site fidelity and local kin associations within grey seal breeding colonies. *Ethology* **106**, 899–919.

Pomeroy, P.P., Worthington Wilmer, J., Amos, W. & Twiss, S.D. (2001) Reproductive performance links to fine scale spatial patterns of female grey seal relatedness. *Proceedings of the Royal Society, Series B* **268**, 711–717.

Prince, P.A., Rothery, P., Croxall, J.P. & Wood, A.G. (1994) Population dynamics of black-browed and grey-headed albatrosses *Diomedia melanophris* and *D. chrysostoma* at Bird Island, South Georgia. *Ibis* **136**, 50–71.

Prince, P.A., Corxall, J.P., Trathan, P.N. & Wood, A.G. (1997) The pelagic distribution of South Georgia albatrosses and their relationships with fisheries. In: *Albatross Biology and Conservation* (eds G. Roberston & R. Gales), pp. 137–167. Surrey Beatty & Sons, Chipping Norton, UK.

Pütz, K., Ingham, R.J. & Smith, J.G. (2000) Satellite tracking of the winter migration of Magellanic penguins *Spheniscus magellanicus* breeding in the Falkland Islands. *Ibis* **142**, 614–622.

Reiter, J. & Le Boeuf, B.J. (1991) Life history consequences of variation in age at primiparity in northern elephant seals. *Behavioral Ecology and Sociobiology* **28**, 153–160.

Ricklefs, R.E. (1968) Patterns of growth in birds. *Ibis* **110**, 419–451.

Ries, E.H., Traut, I.M., Brinkman, A.G. & Reijnders, P.J.H. (1999) Net dispersal of harbour seals within the Wadden Sea before and after the 1988 epizootic. *Journal of Sea Research* **41**, 233–244.

Schjorring, S., Gregersen, J. & Bregnalle, T. (1999) Prospecting enhances breeding success of first-time breeders in the great cormorant, *Phalacrocorax carbo sinensis*. *Animal Behaviour* **57**, 647–654.

Sisak, M.M. (1998) Animal-borne GPS and the deployment of a GPS based archiving datalogger on Hawaiian monk seal (*Monachus schauinslandi*). *Marine Technology Society Journal* **32**, 30–36.

Slade, R.W., Moritz, C., Hoelzel, A.R. & Burton, H.R. (1998) Molecular population genetics of the southern elephant seal *Mirounga leonina*. *Genetics* **149**, 1945–1957.

Stahl, J.C. & Sagar, P.M. (2000) Foraging strategies of southern Buller's albatrosses *Diomedea b. bulleri* breeding on The Snares, New Zealand. *Journal of the Royal Society of New Zealand* **30**, 299–318.

Stanley, H.F., Casey, S., Carnahan, J.M., Gooman, S., Harwood, J. & Wayne, R.K. (1996) Worldwide patterns of mitochondrial DNA differentiation in the harbor seal (*Phoca vitulina*). *Molecular Biology and Evolution* **13**, 368–382.

Steele, J.H. (1985) A comparison of terrestrial and marine ecological systems. *Nature* **313**, 355–358.

Stenhouse, I.J. & Montevecchi, W.A. (1999) Increasing and expanding populations of breeding northen fulmars in Atlantic Canada. *Waterbirds* **22**, 382–391.

Suryan, R.M., Irons, D.B. & Benson, J. (2000) Prey switching and variable foraging strategies of black-legged kittiwakes and the effect on reproductive success. *Condor* **102**, 374–384.

Testa, J.W. (1987) Long-term reproductive patterns and sighting bias in Weddell seals (*Leptonychotes weddelli*). *Canadian Journal of Zoology* **65**, 1091–1099.

Tuck, G.N., Polacheck, T., Croxall, J.P., Weimerskirch, H., Prince, P.A. & Wotherspoon, S. (1999) The potential of archival tags to provide long-term movement and behaviour data for seabirds: first results from wandering albatross *Diomedea exulans* off South Georgia and the Crozet Islands. *Emu* **99**, 60–68.

Valle, C.A. (1995) Effective population size and demography of the rare flightless Galapagos cormorant. *Ecological Applications* **5**, 601–617.

Veit, R.R. & Prince, P.A. (1997) Individual and population level dispersal of black-browed albatrosses *Diomedea melanophris* and grey-headed albatrosses *D. chrysostoma* in response to Antarctic krill. *Ardea* **85**, 129–134.

Veit, R.R., Pyle, P. & McGowan, J.A. (1996) Ocean warming and long-term change in pelagic bird abundance within the California current system. *Marine Ecology, Progress Series* **139**, 11–18.

Watson, K.P. & Granger, R.A. (1998) Hydrodynamic effect of a satellite transmitter on a juvenile green turtle (*Chelonia mydas*). *Journal of Experimental Biology* **201**, 2497–2505.

Waugh, S.M., Weimerskirch, H., Cherel, Y. & Prince, P.A. (2000) Contrasting strategies of provisioning and chick growth in two sympatrically breeding albatrosses at Campbell Island, New Zealand. *Condor* **102**, 804–813.

Weimerskirch, H., Cherel, Y., Cuenot-Chaillet, F. & Ridoux, V. (1997) Alternative foraging strategies and resource allocation by male and female wandering albatrosses. *Ecology* **78**, 2051–2063.

Wingfield, J.C. & Ramenofsky, M. (1997) Corticosterone and facultative dispersal in response to unpredictable events. *Ardea* **85**, 155–166.

Wood, A.G., Naef-Daenzer, B., Prince, P.A. & Croxall, J.P. (2000) Quantifying habitat use in satellite-tracked pelagic seabirds: application of kernel estimation to albatross locations. *Journal of Avian Biology* **31**, 278–286.

Wynn-Edwards, V.C. (1959) The control of population density through social behaviour: a hypothesis. *Ibis* **101**, 436–441.

Chapter 5

Inferring patterns of dispersal from allele frequency data

Alan F. Raybould, Ralph T. Clarke, Joanna M. Bond, Ruth E. Welters and Chris J. Gliddon

Introduction

Genetic data can be used to detect dispersal directly, or to make indirect estimates of dispersal from spatial and/or temporal variation of allele frequencies. In most cases, genetic methods are used to make estimates of dispersal that results in breeding success. This process is usually referred to as gene flow, and 'dispersal' is used in this sense in this chapter. However, powerful genetic fingerprinting techniques can be used to infer annual dispersal if the same multilocus genotype is detected in different places at different times (genetic tagging; Palsbøll 1999). It may not even be necessary to capture an individual as its presence at a site can be established by genotyping sloughed skin, feathers or even cells contained in faeces.

A very simple way of using genetic markers to estimate dispersal is to introduce individuals with novel markers into a population and to follow the spread of the marker. For example, this method has been used to measure the dispersal of adult and juvenile freshwater shrimps (Hancock & Hughes 1999) and pollen dispersal from pine trees (Latta *et al.* 1998). Numerous studies of pollen dispersal from genetically modified crops have also adopted this approach (e.g. Llewellyn & Fitt 1996; Hokanson *et al.* 1997).

A more sophisticated direct method is parentage analysis. Here highly polymorphic markers, such as microsatellites, are used to identify with high likelihood the parentage of individuals. If there is prior knowledge of the spatial distribution of potential parents, then dispersal can be inferred directly from the data. The method is particularly useful for paternity analysis of plants to infer pollen dispersal (e.g. Dow & Ashley 1998), but it can also be used to estimate seed dispersal (Ennos 2001) and the movement of individuals in animal populations (e.g. Prodöhl *et al.* 1998).

Of course, in many instances the direct methods estimate actual gene flow from previous generations rather than dispersal. For example, in plants not all pollen will effect fertilization; therefore the pollen distribution estimated from the spread of a genetic marker will probably give a lower absolute measure of dispersal compared with capturing pollen in traps. In addition, the shapes of the dispersal curves may differ. The fitness of progeny can be dependent on the distance between parents. If a species suffers outbreeding depression, the shape of the curve estimated from markers may be more leptokurtic than that estimated from trapping pollen (e.g. Waser *et al.* 2000), whereas inbreeding depression may result in a more platykurtic curve (e.g. Byers 1998). Distance from the maternal parent may also be important; larger seeds may disperse shorter distances but be more likely to survive and produce offspring; seed may have to disperse a certain distance away from the mother plant to avoid shading, and so on (see Greene & Calogeropoulos, this volume).

Direct methods provide estimates of short-term or 'contemporary' dispersal and are analogous to mark–recapture studies (Cain *et al.* 2000). These approaches are ideal for measuring dispersal of particular individuals in a particular site. Although direct methods can often give virtually unambiguous indications of dispersal, they can require large amounts of data to derive estimates of migration rates, especially for the longer dispersal distances, which are often of most ecological interest. Also they are 'snapshots'. Unless studies are made at several sites and times, which may be difficult for methods that need large sample sizes, they give limited information about temporal or spatial variation and may not detect rare long-distance dispersal (e.g. Slatkin 1985; Cain *et al.* 2000).

There is a variety of methods for indirectly inferring dispersal from distributions of allele frequency data. An advantage of most of these methods is that they are sensitive to rare events and capture an average picture of 'successful' dispersal over many generations. In general, sampling can be less intensive than for direct methods, and therefore temporal and spatial variation can be investigated even when research budgets are limited. A major disadvantage is that estimates of dispersal from allele frequency data are very sensitive to assumptions about demographic processes in the populations under study.

There are several excellent reviews of procedures to derive estimates of dispersal from allele frequency data (e.g. Slatkin 1985; Neigel 1997; Ouborg *et al.* 1999; Cain *et al.* 2000). Therefore our purpose here is not to provide an overview of genetic methods and the assumptions on which they are based, but to examine in some detail the set of methods based on Wright's *F*-statistics (Wright 1921, 1943, 1951, 1965, 1978). In simple terms, F_{ST} is a parameter that describes average difference in allele frequency between two or more populations. F_{ST} is zero when populations have identical allele frequencies, and equal to one when they are fixed for different alleles. Because migration reduces variation in allele frequencies between populations, the rate of migration per generation is inversely proportional to F_{ST}.

Estimation of F_{ST} is probably the most common means of analysing genetic structure of populations and thereby making inferences about migration. However, in spite of the development of user-friendly software to estimate and test the significance of F_{ST}, there is still confusion over the meaning of this term, related parameters and their estimators. Also, the use of transformations of F_{ST} to give estimates of the numbers of migrants per generation between populations ($N_e m$) has been criticized on a number of grounds. Indeed, some authors suggest that the whole F_{ST}-based approach has little power to infer patterns of migration (e.g. Ennos 2001). In this paper we highlight some often overlooked features of F_{ST} and its relationship to migration rates, and we suggest ways in which it can be informative in studies of dispersal.

F_{ST}, related parameters and their estimation

Derivation of F_{ST}

To the non-specialist in population genetics, the literature on *F*-statistics can appear vast, confusing and even intimidating. There is misunderstanding over whether certain quantities are parameters, estimators or estimates. A parameter is a quantitative attribute of a distribution or model; an estimator is strictly the rule, method or algorithm to derive an estimate of a parameter from a set of data. Unfortunately, the general term statistic is ambiguous, as it has variously been used to imply estimate, estimator and parameter. Also, even among experts, there is lack of consensus over the most appropriate estimators of *F*-statistics (see Slatkin 1985; Chakraborty & Danker-Hopfe 1991; Neigel 1997). A comprehensive discussion of these issues is outside the scope of this paper, however clarification of some key points may be useful as a guide to the literature.

Wright (1921) defined a correlation, *f* (later designated *F*) between the presence or absence of an allele in uniting gametes compared with the expected correlation if pairs of gametes were drawn randomly from the population. He also (e.g. Wright 1922) designated *F* as the inbreeding coefficient. As Slatkin (1985) pointed out, there is confusion between *F* the correlation coefficient and *F* the inbreeding coefficient (the expected correlation between uniting gametes). The actual correlation is an estimator of the inbreeding coefficient (parameter), given certain assumptions such as neutrality of alleles.

Wright (1951) defined a set of correlations for subdivided populations. In the simplest case of a population divided into a series of subpopulations*, three quantities were defined: (i) F_{IS} (the correlation between uniting gametes within a subpopulation compared with the correlation between gametes selected randomly from within that subpopulation); (ii) F_{IT} (the correlation between uniting gametes in the whole population compared with the correlation between gametes selected randomly from the whole population); and (iii) F_{ST} (the correlation between gametes selected randomly within subpopulations compared with the correlation between gametes selected randomly from the whole population). These correlations are related by the formula $(1 - F_{IT}) = (1 - F_{IS}) \times (1 - F_{ST})$. F_{IT} measures the extent of non-random mating with the whole population, F_{IS} measures the extent of non-random mating within subpopulations and F_{ST} the extent to which alleles occur in the same subpopulation. Again, the correlations are estimators of inbreeding coefficients (now called fixation indices), although the estimators and the parameters they are estimating were given the same notation (Slatkin 1985).

This lack of distinction between the *F*-statistics as estimators and parameters is at the heart of the plethora of definitions of fixation indices, inbreeding coefficients and their estimators. Chakraborty and Danker-Hopfe (1991) have covered the relationship between various parameters and estimators in detail, but it is worth men-

* *In the literature on the theory of F-statistics, it is usual to describe spatially discrete samples as 'subpopulations'. However, in papers describing empirical work, the samples from separate locations are usually referred to as 'populations'.*

tioning two quantities, G_{ST} and θ, as they are commonly used interchangeably with F_{ST}.

G_{ST} was originally devised as a method to derive F_{ST} explicitly for multiple alleles and multiple loci (Nei 1973)—Wright had developed *F*-statistics for two alleles and pooled all but the commonest allele. G_{ST} is the textbook formulation of F_{ST} in terms of heterozygote deficits:

$$G_{ST} = 1 - \frac{\overline{H}_S}{H_T} \tag{5.1}$$

$\overline{H}_S$ is the average over subpopulations of the proportion of heterozygotes in a subpopulation assuming random mating and H_T is the expected proportion of heterozygotes if the subpopulations are pooled and mated randomly. $\overline{H}_S$ and H_T are averaged over loci to give the multilocus G_{ST}. In this form, G_{ST} is an *estimator* of F_{ST} (Weir & Cockerham 1984). G_{ST} has also been defined as a *parameter* by Crow and Aoki (1984). The idea is similar to the approach of Wright (1951), except that probabilities rather than correlations are used:

$$G_{ST} = \frac{(F_0 - \overline{F})}{(1 - \overline{F})} \tag{5.2}$$

F_0 is the probability that two identical alleles will be picked at random from a subpopulation and $\overline{F}$ is the probability that two identical alleles will be picked from the entire population. This formulation is identical to Wright's F_{ST} in the case of a single locus with two alleles (Crow & Aoki 1984). Cockerham and Weir (1993) give a useful discussion of the relationships between the various quantities defined as G_{ST}.

θ is a parameter called co-ancestry, the correlation between alleles in different individuals within a subpopulation, and is analogous to an intraclass correlation coefficient. Cockerham (1969, 1973) discusses the properties of this parameter and its relationship to Wright's F_{ST}. The important difference is that F_{ST} (and the other *F*-statistics) consider alleles in gametes, whereas θ is concerned with alleles within individuals. In other words, for individuals A and B, *F*-statistics are concerned with the correlation between alleles in a randomly selected gamete produced by A and a randomly selected gamete produced by B. In the co-ancestry approach, the correlations are between a randomly selected allele from A and a randomly selected allele from B. If meiosis is regular, the correlations are identical. Differences between F_{ST} and θ can arise under particular systems of mating in groups of individuals (Cockerham 1973), but estimates of θ can be considered estimates of F_{ST} (Weir & Cockerham 1984). This is important for what follows.

Estimation and testing of F_{ST}

The most commonly used method of estimating F_{ST} is that of Weir and Cockerham (1984). They actually derive an estimator, $\hat{\theta}$, of θ but, for reasons outlined above, they consider that their formula also serves as an estimator of F_{ST} (it should be noted that $\hat{\theta}$, not θ itself, is the estimator of F_{ST}). Weir and Cockerham use this particular notation, because they consider θ to be a parameter unambiguously, whereas F_{ST} has been defined both as a parameter and its estimator.

Weir and Cockerham (1984) used an analysis of variance approach to derive an expression for $\hat{\theta}$. Consider a single locus with two alleles A and a, with overall frequencies p and $(1-p)$, respectively. Arbitrarily assign values of $X=1$ for A and $X=0$ for a. If X_{ijk} denotes the value (0 or 1) for the ith allele in the jth individual in the kth population, then the total variance of X can be partitioned as follows:

$$X_{ijk} = p + a_k + b_{jk} + w_{ijk} \tag{5.3}$$

where a_k represents the difference (from overall p) in the frequency of A in population k (with variance of the set of a_k equal to σ_a^2), b_{jk} denotes differences between individuals within populations (variance σ_b^2), and w_{ijk} represents differences between alleles within individuals (variance σ_w^2). The total variance of $X = p(1-p) = \sigma_T^2 = \sigma_a^2 + \sigma_b^2 + \sigma_w^2$. Let $\hat{\sigma}_a^2$, $\hat{\sigma}_b^2$ and $\hat{\sigma}_w^2$ (with $\hat{\sigma}_T^2 = \hat{\sigma}_a^2 + \hat{\sigma}_b^2 + \hat{\sigma}_w^2$) denote the analysis of variance sample estimates of the variance components. The expected value of $\hat{\sigma}_a^2$ equals $p(1-p)\theta$; so θ (and hence F_{ST}) can be estimated by

$$\hat{\theta} = \hat{\sigma}_a^2 / \hat{\sigma}_T^2 \tag{5.4}$$

which is an estimator of the proportion of the total variance that is due to differences in the mean allele frequency between populations.

Weir and Cockerham take equation (5.4) and modify it so that the estimates of the variance components are corrected for the (potentially) small number of subpopulations sampled and the (potentially) small and unequal number of individuals sampled within each subpopulation.

There are different ways of treating loci with more than two alleles and for combining data from several loci. Weir and Cockerham's (1984) suggested approach, which is widely used, is to calculate the overall $\hat{\theta}$ as:

$$\hat{\theta} = \sum_r \sum_s \hat{\sigma}_{a(rs)}^2 \Big/ \sum_r \sum_s \hat{\sigma}_{T(rs)}^2 \tag{5.5}$$

which is effectively an average of the individual $\hat{\theta}$ values ($\hat{\theta}_{(rs)} = \hat{\sigma}_{a(rs)}^2 / \hat{\sigma}_{T(rs)}^2$) for each allele s of each locus r, weighted by respective total variances ($\hat{\sigma}_{T(rs)}^2$). The simple, uncorrected form of the equation for $\hat{\theta}$ is used for clarity. A different approach to multiple alleles and loci has been proposed by Long (1986), but most standard population genetics computer programs (e.g. Goudet 1995; Raymond & Rousset 1995a) use Weir and Cockerham's approach.

Several methods have been proposed for testing the significance of $\hat{\theta}$. Weir and Cockerham (1984) suggested bootstrapping over loci when data are from multiple loci, and jackknifing over populations for single or small numbers of loci. The bootstrap approach has been criticized by Raymond and Rousset (1995b) and Van Dongen (1995), largely because most data sets have insufficient independent (i.e.

unlinked) loci. Van Dongen (1995) suggests that more than 20 independent loci are necessary in order that the bootstrap distribution is more or less continuous and does not exhibit 'peculiar' properties because of the limited number of possible values the resampled estimate can take.

An alternative to bootstrapping is a one-sided randomization test. Here, alleles are allocated randomly to subpopulations, preserving the original sample sizes, and $\hat{\theta}$ is calculated from the randomized data. Multiple iterations of this procedure generate a distribution under the null hypothesis that $\theta = 0$. If the observed $\hat{\theta}$ is greater than 95% of the $\hat{\theta}$ values generated from randomized data, the null hypothesis ($\theta = 0$) is rejected. If the estimate of F_{IS} is significant, alleles within individuals are not independent and the correct units of permutation are genotypes (Goudet 1995). The testing of $\hat{\theta}$ is discussed in detail by Raymond and Rousset (1995b) and Goudet *et al.* (1996).

Relationship between F_{ST} and the number of migrants ($N_e m$)

Early papers on *F*-statistics are based on a very simple demographic model of population structure called the infinite island model, which assumes an infinite number of subpopulations with equal but finite effective population size (N_e). Mating is random within subpopulations, and in each generation a proportion (m) of individuals in each subpopulation are immigrants from other subpopulations. As there is an infinite number of islands, the allele frequency (q_t) of the population as a whole does not change (i.e. there is no drift overall), but the frequency within each population (q) does change due to drift and migration. Wright (1943) showed that at equilibrium the variance of the subpopulation allele frequencies ($\sigma_a^2 = \sigma_q^2$ in Wright 1943) is given by:

$$\sigma_a^2 = q_t(1-q_t)\big/\left[2N_e - (2N_e - 1)(1-m)^2\right] \tag{5.6}$$

If the proportion of migrants in each generation is small, the terms containing m^2 can be ignored, which gives:

$$\sigma_a^2 = q_t(1-q_t)/(4N_e m - 2m + 1) \approx q_t(1-q_t)/(4N_e m + 1) \tag{5.7}$$

Rearranging this expression gives:

$$\sigma_a^2/q_t(1-q_t) = F_{ST} \approx 1/(4N_e m + 1) \tag{5.8}$$

Because computer programs make the estimation of F_{ST} from allele frequency data using Weir and Cockerham's formula relatively straightforward, the average number of migrants per generation from/to each subpopulation ($N_e m$) is easily estimated from genetic data, using this model.

Clearly the infinite island model is an idealized structure (for recent reviews see Bossart & Prowell 1998; Whitlock & McCauley 1999) and its assumptions may be violated in real populations (Table 5.1). One prerequisite of the model that is

Table 5.1 The effects on estimates of migration (numbers of individuals migrating into and out of islands per generation) when assumptions of the infinite island model (see text) are violated.

Assumption	Violation	Effect on $\hat{N}_e m$
No selection	Underdominance	Reduction
	Overdominance	Increase
	Local adaptation	Reduction
	Balancing selection	Increase
	Frequency dependent selection	Increase (single equilibrium)
	Frequency dependent selection	Reduction (multiple equilibria)
	Inbreeding depression	Increase
	Outbreeding depression	Reduction
No mutation	High rates of stepwise mutation	Increase
Equal and stable population sizes	Source-sink populations	Increase
	Extinction/recolonization	Reduction
No spatial structure within populations	Non-random mating	Increase
No spatial structure among populations	Stepping stone model	Reduction
Equilibrium between gene flow and drift	Recent range expansion	Increase
	Recent contact	Reduction

almost always overlooked when using estimates of F_{ST} to infer $N_e m$ is that there should be random mating within subpopulations. Empirically, it is easy to see that disregarding the requirement for random mating can lead to peculiar outcomes. For example consider a species with the distribution illustrated in Fig. 5.1; such a distribution might occur in a plant with a high selfing rate and restricted seed dispersal. If the four subpopulations, each of eight individuals, on either side of the mountain range are collected separately the overall subpopulation $F_{ST} = 1$, and the reasonable assumption that there is no gene flow is made. However, if the populations on either side of the mountains had been collected as a single sample and then analysed by F-statistics, the correct estimate of $F_{ST} = 0.2$ between the two 'populations' would be obtained, but the subsequent inference that $N_e m = 1$ from equation (5.8) is clearly wrong.

The reason for this discrepancy is that F_{ST} is not a measure of the absolute amount of differentiation among subpopulations, rather it measures the extent to which the process of fixation (the presence of a single allele within a subpopulation) has gone to completion (Wright 1978, p. 82). Clearly, if there are more alleles than subpopulations, complete fixation cannot be achieved, which sets an upper limit for F_{ST} for hypervariable loci (Hedrick 1999). Pooling populations that do not have identical allele frequencies inevitably means that the estimate of gene flow among the pooled groups is greater (F_{ST} is lower) than that when they are treated separately. The analysis of differentiation among non-randomly mating subpopulations is discussed by Wright (e.g. 1946), but a theoretical treatment of migration among non-randomly mating subpopulations, even in the simple infinite island model, appears lacking.

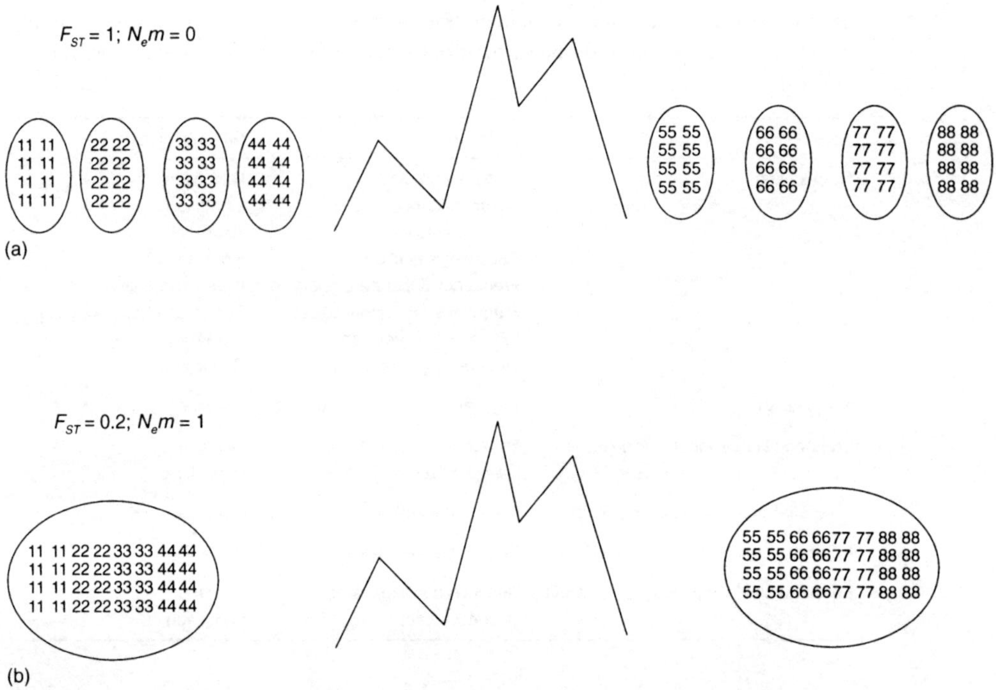

Figure 5.1 Illustration of the effect of failure to consider non-random mating within groups on estimates of N_em (see text for details). (a) Samples are treated as spatially discrete populations; whereas in (b), samples on either side of the barrier are treated as single populations. This example gives an extreme case of the Wahlund effect; N_em is also estimated incorrectly if non-random mating within samples is the result of inbreeding.

A practical result of this observation is that, if feasible, it makes sense to record the position of every individual sampled. If this is done, and one finds that estimates of F_{IS} in sampled 'populations' are significantly positive, the samples can be subdivided on the basis of spatial information to identify randomly mating groups within the populations. Estimates of N_em among the non-randomly mating populations can be derived from the average F_{ST} between pairs of randomly mating groups in different populations (Raybould *et al.* 1997).

Detecting restricted dispersal from pairwise F_{ST} estimates

Many of the criticisms of the infinite island model for estimating N_em assume two things. First, that the average number of migrants per generation among all subpopulations is actually the parameter of interest; and second, that estimates of it obtained by transformation of F_{ST} values (eqn 5.8) are taken at face value. Nevertheless, although the absolute values of average N_em derived from an over all population estimate of F_{ST} are likely to be, at best, imprecise, it is still possible to use F_{ST} estimates to infer information about migration.

One of the unrealistic assumptions about the infinite island model is that migration is equally restricted between all pairs of subpopulations. Wright referred to the process of differentiation in this model as 'isolation by distance'. More realistic models have been proposed such as one- and two-dimensional stepping stone models (Kimura & Weiss 1964), in which migration occurs between neighbouring subpopulations but not (directly) between non-neighbours. In these models, the expectation is that genetic differentiation among subpopulations increases with their separation (for a brief review and relevant references, see Slatkin 1993). Isolation by distance has now come to represent the phenomenon of positive correlation between genetic differentiation and geographic distance, rather than simply the spatial subdivision of a population into genetically differentiated groups.

Slatkin (1993) used computer simulations to show that under a variety of stepping stone models there is, in effect, a positive correlation between the F_{ST} and distance (number of steps) apart of pairs of populations. Specifically, he converted F_{ST} to a measure of gene flow, $\hat{M}$, which is the value of $N_e m$ for a pair of populations derived from the pair's F_{ST} estimate using the relationship for an island model (i.e. eqn 5.8.). In stepping stone models, where in each generation direct migration only occurs between neighbouring populations, $\hat{M}$ does not estimate the direct per generation rate of movement between a pair of populations, but does represent the effective rate of gene flow (in the infinite island model gene flow and per generation movement are the same thing). Slatkin showed there is a negative correlation between $\log_{10} \hat{M}$ and $\log_{10}$ distance between all pairs of populations. Statistically significant correlations could also occur even though the system may not be in migration–drift equilibrium. Therefore, in principle, restricted dispersal (isolation by distance) in real populations can be detected by estimating values of pairwise F_{ST} and distance among spatially discrete samples. The robustness of the test in the face of variation in demographic models and non-equilibrium conditions suggests that it should be widely applicable, but of course it also means that in its simplest form it has limited power to discriminate between alternative models of restricted dispersal.

The pairwise nature of the data required for Slatkin's test of isolation by distance creates some statistical problems. The usual test of significance of a Pearson product–moment correlation is not applicable. If N subpopulations have been sampled, there will be $N(N - 1)/2$ pairwise observations used in the calculation of the correlation coefficient, but only N independent sampling units. The solution is to use a randomization test developed by Mantel (1967) to detect spatial clustering of diseases. Mantel's test randomizes the N populations to generate a new matrix of $\log_{10} \hat{M}$ values, while keeping the $\log_{10}$ distance matrix constant. The correlation coefficient between the randomized data is calculated. The randomization process is repeated a large number (say 10 000) of times to generate the probability distribution of the correlation coefficient under the null hypothesis of no correlation between $\log_{10} \hat{M}$ and $\log_{10}$ distance. As there is an *a priori* expectation that $\hat{M}$ will decline with distance, a one-sided test of significance is usually used.

As Slatkin (1993) and others (e.g. Whitlock & McCauley 1999) have pointed out, $\hat{M}$ is not intended as an estimate of the number of migrants per generation, but as a measure of differentiation; the relationship between differentiation and distance, not the average amount of migration among populations, provides us with information about dispersal. We should therefore ask whether $\hat{M}$ is the best measure of differentiation. When F_{ST} estimates are low, say < 0.2, small changes in $\hat{F}_{ST}$ lead to large changes in $N_e m$ (using eqn 5.8); however $\log_{10} N_e m$ is much less sensitive to small changes in $\hat{F}_{ST}$. Rousset (1997) suggested that a correlation of $\hat{F}_{ST}/(1-\hat{F}_{ST})$ and distance was the most powerful test of isolation by distance in a linear set of populations, and $\hat{F}_{ST}/(1-\hat{F}_{ST})$ and $\log_{10}$ distance best for two-dimensional sets of populations. We suggest that a rank correlation of $\hat{F}_{ST}$ and distance may be a useful way of generalizing the test for isolation by distance using pairwise data as the rank correlation is unaffected by any such monotonic transformations of $\hat{F}_{ST}$ or geographic distance.

Using some of our data, we examined variation among transformations of $\hat{F}_{ST}$ and distance in detecting isolation by distance (Table 5.2). In general, Slatkin's method of correlating $\log_{10} \hat{M}$ and $\log_{10}$ distance gives similar results to other transformations of $\hat{F}_{ST}$ and distance. The rank correlation of $\hat{F}_{ST}$ and distance also appears to work well and makes fewest assumptions. It would be interesting to use computer simulations to examine the power of the rank correlation to detect isolation by distance under a variety of dispersal models.

In the following sections we illustrate how F_{ST}-based approaches have the power to test hypotheses about dispersal patterns. Unless otherwise stated, from now on 'F_{ST}' refers to sample estimates of F_{ST} (i.e. $\hat{F}_{ST}$).

Detecting differences in dispersal patterns—brown trout in the River Severn

The data in Table 5.2 show that it is possible to detect isolation by distance in natural populations. However, does this provide us with useful information about dispersal? We tested for isolation by distance among populations of brown trout *Salmo trutta* in two tributaries of the River Severn on the England–Wales border (Fig. 5.2). In the River Tanat there are potential barriers to dispersal of fish, such as waterfalls and acid run-off from conifer plantations (Crisp & Beaumont 1996). In the River Teme, however, there appear to be no barriers to dispersal other than long distances. If the putative barriers in the Tanat do prevent dispersal of fish, then the expectation is that neighbouring populations will be no more similar genetically than populations further apart. In the Teme, on the other hand, if there are no barriers to dispersal, genetic differentiation will be correlated with distance (unless fish are so mobile that they disperse randomly around the system).

We used seven microsatellite loci to estimate F_{ST} between all pairs of populations of brown trout (J. M. Bond, unpubl. data). Plots of F_{ST} against distance (Fig. 5.3) show a clear difference between the rivers. In the Tanat, F_{ST} values are relatively high (>0.1) and do not show a statistically significant relationship with distance. In the

Table 5.2 Effects of various transformations of pairwise F_{ST} estimates and geographic distance on tests for isolation by distance.

	No. of populations	$\log_{10} \hat{M}$ and $\log_{10}$ dist.		F_{ST} and $\log_{10}$ dist.		$F_{ST}/(1-F_{ST})$ and $\log_{10}$ dist.		$F_{ST}/(1-F_{ST})$ and dist.		Rank F_{ST} and rank dist.		References
		r	*P*	*r*	*P*	*r*	*P*	*r*	*P*	*r*	*P*	
Sea beet *Beta vulgaris* spp. (RFLP)	10	−0.631	0.0005	0.517	0.0014	0.488	0.0023	0.457	0.0047	0.494	0.0038	Raybould *et al.* 1996b
Sea beet *Beta vulgaris* spp. (isozymes)	10	0.003	0.5161	−0.024	0.5703	−0.029	0.5990	−0.127	0.8419	−0.070	0.6991	Raybould *et al.* 1996b
Wild cabbage *Brassica oleracea* (microsatellites)	7	−0.539	0.0113	0.560	0.0106	0.549	0.0122	0.554	0.0249	0.555	0.0126	Raybould *et al.* 1999
Wild cabbage *Brassica oleracea* (isozymes)	7	−0.441	0.0385	0.438	0.0502	0.433	0.0530	0.461	0.0817	0.495	0.0275	Raybould *et al.* 1999
Agrostis curtisii	30	−0.170	0.0000	0.210	0.0005	0.209	0.0002	0.265	0.0008	0.254	0.0015	Warren *et al.* 1998
Lolium perenne	27	0.069	NT	0.004	0.4110	0.003	0.4561	−0.131	0.9832	−0.069	0.8141	Warren *et al.* 1998
Brown trout *Salmo trutta* (River Teme)	7	−0.593	0.0127	0.579	0.0138	0.570	0.0167	0.574	0.0267	0.632	0.0097	J. M. Bond, unpubl. data
Brown trout *Salmo trutta* (River Tanat)	7	−0.173	0.2551	0.173	0.2568	0.179	0.2471	0.185	0.3180	0.157	0.3204	J. M. Bond, unpubl. data

dist., distance; NT, not tested.
P values from one-sided Mantel tests.

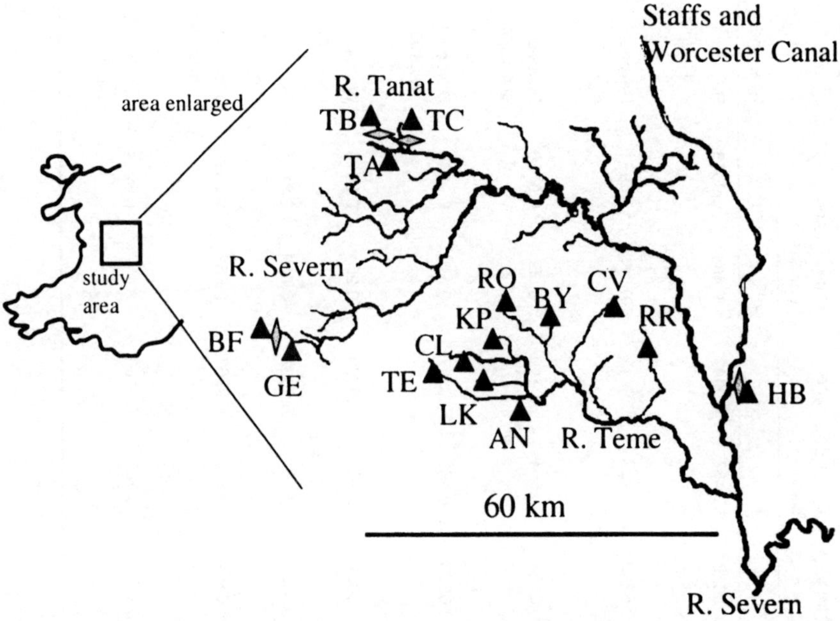

Figure 5.2 Sampling sites for the study of brown trout dispersal barriers.

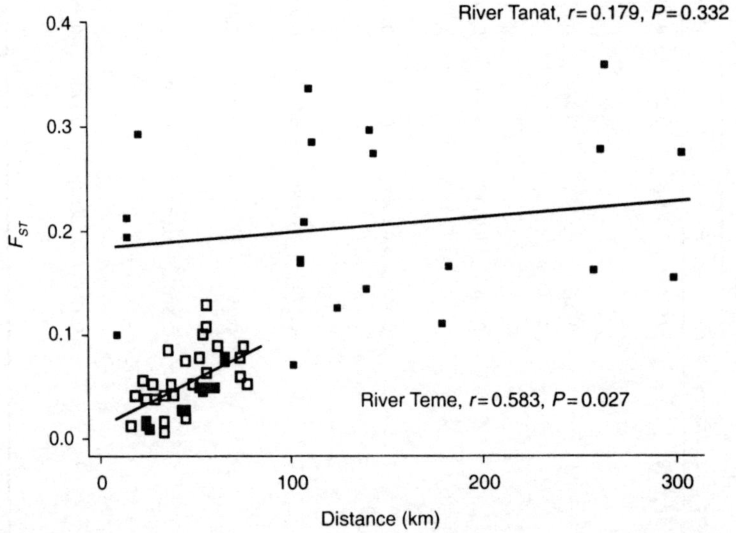

Figure 5.3 Positive correlation between F_{ST} and distance among brown trout populations in the River Teme (no apparent barriers to dispersal except distance) compared with no correlation in the River Tanat where putative barriers (waterfalls, acid run-off) appear to operate to prevent dispersal.

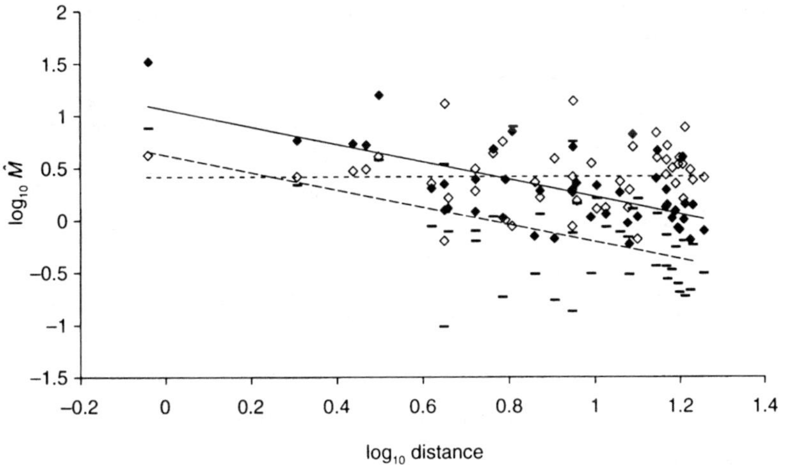

Figure 5.4 Regressions of $\log_{10} \hat{M}$ on $\log_{10}$ distance for sea beet (Raybould *et al.* 1996b). Solid diamonds and solid line are based on RFLP data (one-sided Mantel test, $P = 0.0007$); open diamonds and short dashed line are based on allozyme data (one-sided Mantel test, $P = 0.4910$); dashes and long dashed line show the difference in $\hat{M}$ ($\log_{10} \hat{M}_{\mathrm{RFLPs}} - \log_{10} \hat{M}_{\mathrm{allozymes}}$) (two-sided Mantel test, $P = 0.0064$).

Teme, F_{ST} values are generally less than 0.1 and show a significant positive relationship with distance. Thus the genetic data support the view that there are barriers to dispersal in the Tanat, but that isolation by distance operates in the Teme.

Testing the difference between patterns of dispersal

The patterns in the trout data give information about contrasting amounts of dispersal in two river systems. However, it is not straightforward to demonstrate that the patterns are different from each other statistically. The problem is that conventional tests of significance (such as a *t*-test on regression coefficients) are not appropriate because of the pairwise nature of the data.

Raybould and Clarke (1999) proposed a solution for the special case when different data are sampled from the same sites. Raybould *et al.* (1996b) estimated F_{ST} among 10 populations of sea beet *Beta vulgaris* ssp. *maritima* in Dorset (UK) using allele frequency estimates at restriction fragment length polymorphism (RFLP) and allozyme loci. A regression of pairwise $\log_{10} \hat{M}$ and $\log_{10}$ distance was significant when $\hat{M}$ was derived from RFLP data but not for $\hat{M}$ using allozyme data (Fig. 5.4). However, Raybould *et al.* (1996b) were unable to test the significance of the difference between the slopes.

Raybould and Clarke (1999) pointed out that if the RFLP and allozyme slopes are not significantly different (i.e. they are parallel), the slope of the regression of the difference between the two $\log_{10} \hat{M}$ values and $\log_{10}$ distance will be zero. Therefore, an

Table 5.3 Estimates of the distance at which $\hat{M} = 1$ in the RFLP data set from sea beet (Raybould *et al.* 1996b).

Estimation method $d(\hat{M} = 1)$	Critical distance (km)	Lower 95% confidence limit (km)	Upper 95% confidence limit (km)
Independent points	13.9	10.3	24.1
MLPE	14.5	6.7	39.3
Jackknife over populations	13.9	6.8	279.1

MLPE, maximum likelihood population effects model (R.T. Clarke, unpubl. data).

appropriate test of the difference between the slopes is a regression of $\log_{10}$ $(\hat{M}_{RFLPs}/\hat{M}_{allozymes})$ and $\log_{10}$ distance, with the significance of the regression coefficient obtained with a two-sided Mantel test. For the sea beet example, the test of the difference was significant, showing that the RFLP and allozyme slopes were indeed different from each other (Fig. 5.4).

This test can be applied to any situation where multiple matrices of pairwise data are obtained from the same sites. One useful application of the method would be to compare seed dispersal and pollen dispersal. Ennos (1994) showed that a comparison of F_{ST} from nuclear markers and chloroplast markers gives a test of whether the average amounts of gene flow through seed and pollen are different (in species in which chloroplasts are inherited through one sex only). The test described by Raybould and Clarke (1999) would allow a test of whether the spatial patterns are different, based on regressions or correlations of $(F_{ST(nuclear)}/F_{ST(chloroplast)})$ and distance (or suitable transformations of these variables).

A general test of the difference between two regressions (or correlations) based on pairwise data is not yet available. It is, however, possible to estimate confidence intervals on regression coefficients between two matrices (e.g. Manly 1997) and confidence intervals on point estimates from such regressions (Clarke *et al.*, in press). For example in the sea beet example described above, we can estimate, say, the average distance at which $\hat{M} = 1$. Assuming the pairwise points are independent gives confidence intervals that are too narrow, whereas jackknifing over populations gives confidence intervals that are very conservative. Estimates based on a 'maximum likelihood population effects' (MLPE) model are probably the best available (Table 5.3). Estimates of the distance at which $\hat{M}$ (or F_{ST}) equals a particular value may be useful in comparing spatial scales and patterns of migration across species. Comparisons of F_{ST} estimates across a large number of species (e.g. Hamrick *et al.* 1979; Loveless & Hamrick 1984; Hamrick & Godt 1996; Bohonak 1999) have provided useful generalizations, but are not standardized for spatial scale.

Vicariance—sea beet on Furzey Island

A criticism of the method proposed by Slatkin (1993) is that if a positive correlation

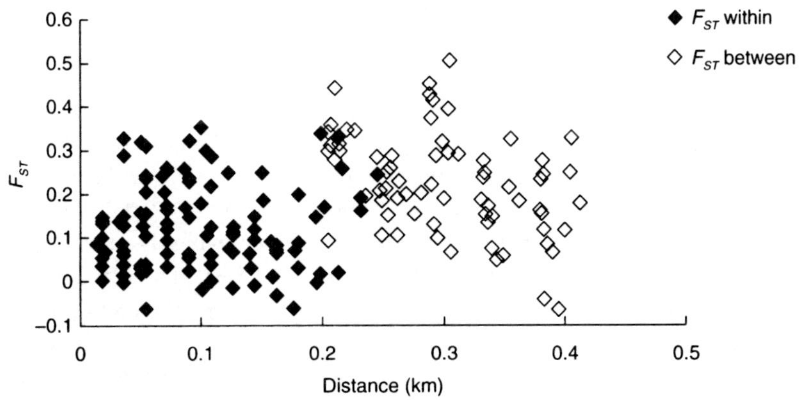

Figure 5.5 The problem of distinguishing between isolation by distance and vicariance: this graph shows the relationship between pairwise F_{ST} estimates and distance among randomly mating patches of sea beet in two transects on Furzey Island (Raybould *et al.* 1996a). The different symbols for within and between transect estimates illustrate that most long distances are between transect comparisons and hence vicariance, rather than isolation by distance, and could be the cause of the positive correlation between F_{ST} and distance (r= 0.328, P=0.0043).

between F_{ST} and distance is detected, it is not possible without other independent information to say whether the relationship is due to isolation by distance or vicariance (Bossart & Prowell 1998).

Vicariance is 'the subdivision of a population into distinct but related species etc. by the appearance of a geographical barrier' (*Shorter Oxford English Dictionary*). A geographical barrier may completely prevent gene flow between populations on either side, leading to genetic differentiation between the two sets of populations. Suppose we sample several populations on either side of the barrier and estimate F_{ST} between all pairs of populations and find a positive correlation between these values and distance. How should we interpret this result? It is likely that all of the long distances in the correlation are between pairs of populations on different sides of the barrier, whereas the short distances are between populations on the same sides of the barrier. It is possible, therefore, that the positive correlation may not be the result of isolation by distance, but the result of the barrier.

The problem can be illustrated in a data set obtained by Raybould *et al.* (1996a). They sampled every adult sea beet plant found along two transects, one on the north shore of Furzey Island in Poole Harbour (UK) and one on the southwest shore. Using the spatial co-ordinates of each plant, they identified randomly mating groups within each transect and estimated F_{ST} between all pairs of groups. In the whole data set there was a significant correlation between F_{ST} and distance between pairs of groups (r = 0.328; P = 0.0043). However, because most of the long distances are between groups in different transects (Fig. 5.5), the effect may be due to allele frequency differences between transects (vicariance), rather than because of isolation by distance.

It is possible to test whether isolation by distance or vicariance was the source of the correlation by creating a variable to denote transect membership. A pair was coded '1' if the groups were in different transects and '0' if the groups were in the same transect. A partial regression (Smouse *et al.* 1986) of F_{ST} versus distance, removing the effect of transect membership, was not significant ($P > 0.05$), whereas a partial regression of F_{ST} versus transect membership, removing the effect of distance, was highly significant ($P < 0.001$) (Raybould *et al.* 1996a). This demonstrates that partial regressions have the power to distinguish between isolation by distance and vicariance in data sets in which there is an overall effect of distance, but where populations are clustered into two or more regional groups. Further uses of partial Mantel tests to control for correlations between ecological variables and distance are illustrated in Thorpe and Baez (1993).

Mutation mechanism

A crucial usual assumption of F_{ST}-based approaches to studies of dispersal is that alleles that are identical in state are also identical by descent. This means that in addition to assuming that mutation rates at marker loci are low compared with migration, it is also assumed that mutation follows the infinite alleles or *k*-alleles model (IAM and KAM, respectively) (e.g. Slatkin 1995). In the IAM, when an allele mutates it changes to one of an infinite number of alleles that do not already exist; the KAM is similar except that mutation is to one of $k-1$ possible different states. In the IAM, alleles that are identical in state are by definition also identical by descent; in the KAM, the probability of being both identical in state and by descent decreases as *k* becomes smaller and as the mutation rate increases.

Microsatellite markers have become very popular markers in population genetic studies because of their generally high amounts of polymorphism (e.g. Jarne & Lagoda 1996). However, they may violate the assumptions of IAM and/or KAM and low mutation rates. Microsatellites are tandemly repeated short DNA sequences; precise definitions in terms of repeat length and number vary, but six base pairs (bp) has been considered the maximum length of repeat and about 100 bp the maximum length of the whole unit to qualify as a microsatellite (Chambers & MacAvoy 2000). It is not possible here to give a comprehensive review of microsatellite biology, but for discussions of the applicability of F_{ST} two points are salient: mutation tends to occur by addition or subtraction of single repeat units (stepwise mutation) and at rates that can be of the same order as migration. These factors mean that for a particular microsatellite allele (number of repeats), identity in state does not necessarily imply identity by descent, which can affect the performance of F_{ST} as a measure of differentiation.

An alternative to F_{ST} is a parameter called Φ_{ST}, which is similar to F_{ST} except that it is defined as the correlation between random haplotypes (not alleles) within subpopulations, relative to the correlation between haplotypes in the whole population

(Excoffier *et al.* 1992). Estimators of Φ_{ST} for microsatellites use the same analysis of variance (ANOVA) approach as Weir and Cockerham (1984), except that the dependent variable is allele length rather than frequency (Michalakis & Excoffier 1996).

A number of ANOVA estimators are available for Φ_{ST}, which differ in the weights given to samples of different size (Rousset 1996). For microsatellites, Michalakis and Excoffier (1996) suggest:

$$\hat{\Phi}_{ST} = \frac{S_B - S_W}{S_B} \tag{5.9}$$

where S_W and S_B are, respectively, the average squared difference in allele size (number of repeats) between alleles drawn randomly from within subpopulations and from different subpopulations. Slatkin (1995) defined a similar quantity, R_{ST}:

$$R_{ST} = \frac{\bar{S} - S_W}{\bar{S}} \tag{5.10}$$

where $\bar{S}$ is the average squared difference in allele size between alleles drawn randomly from a set of d subpopulations. Both Rousset (1996) and Feldman *et al.*, (1999) consider the relationship between $\hat{\Phi}_{ST}$ and R_{ST}. If N diploid individuals are sampled from each of d subpopulations, both authors define the relationship as:

$$R_{ST} = \frac{(1-c)\hat{\Phi}_{ST}}{1-c\hat{\Phi}_{ST}} \tag{5.11}$$

where $c = (2N-1)/(2Nd-1)$. The relationship between R_{ST} and $\hat{\Phi}_{ST}$ is of the same form as that between G_{ST} and $\hat{\theta}$ (Slatkin 1995; Rousset 1996; Feldman *et al.* 1999).*

As yet there is no consensus on whether F_{ST}- or Φ_{ST}-based approaches are better for handling microsatellite data. Both have disadvantages: Φ_{ST} estimators overestimate differentiation if microsatellites frequently mutate by large insertions or deletions (see, for example, Di Rienzo *et al.* 1994); on the other hand, F_{ST} estimators underestimate differentiation if mutation is largely stepwise (Slatkin 1995).

One implication of these results is that for some microsatellite data sets $\hat{\theta}$ might be a better estimator of differentiation between populations with high gene flow (e.g. short distances apart), while R_{ST} might be better for populations with low gene flow (e.g. long distances apart) (Slatkin 1995; Rousset 1996; Raybould *et al.* 1998). For example, Rousset (1996) investigated the properties of F_{ST} and Φ_{ST} in a circular stepping stone model with 80 subpopulations, each with 1000 individuals and with pure stepwise mutation at a rate of 10^{-4} per generation. At any given distance (steps be-

* *There are some differences in the treatment of R_{ST} among authors. Michalakis and Excoffier (1996) describe a relationship between $\hat{\Phi}_{ST}$ and $\hat{R}_{ST}$ (their eqn 9 which is actually incorrect; Feldman et al. 1999), implying that they regard R_{ST} as a parameter. The implication of Rousset's approach is that he regards R_{ST} as an estimator of Φ in the same way the Nei's G_{ST} is an estimator of F_{ST}. Cockerham and Weir's (1993) discussion of the difference between Nei's and Crow and Aoki's formulations of G_{ST} is useful here.*

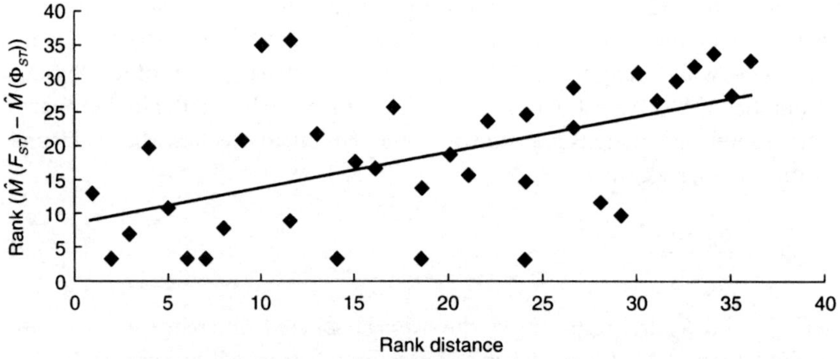

Figure 5.6 Plot of the rank of $\hat{M}$ estimated from F_{ST} minus $\hat{M}$ estimated from Φ_{ST} against rank distance for the River Teme trout data. The significant positive correlation ($r=0.532$, $P=0.0283$) indicates that the difference between $\hat{M}$ (F_{ST}) and $\hat{M}$ (Φ_{ST}) increases with distance, as predicted by Rousset's (1996) simulation of stepwise mutation in a circular stepping stone model.

tween subpopulations), M estimated from F_{ST} was highter than that from Φ_{ST} and the difference increased with distance. We detected this effect in the trout data from the Teme (Fig. 5.6), even though these populations are likely to be more like a one-dimensional stepping stone model. The simulations and sample data suggest that the choice of best estimator of genetic differentiation when analysing microsatellite data is not trivial, and extensive simulation studies may be needed to clarify the issue.

Temporal variation and artefacts of sampling

The above discussions of F_{ST} estimates have shown ways in which they can be used to infer spatial patterns of dispersal averaged over time. However, it is possible to use F_{ST} to examine temporal processes explicitly. As part of a study of the population dynamics of Atlantic salmon *Salmo salar* in the River Frome (Dorset, UK) we obtained estimates of allele frequencies at three microsatellite loci in three consecutive years from 11 sites (R. E. Welters, unpubl. data). Our objective was to discover whether adult salmon returned randomly to spawning sites or whether there was genetic differentiation among sites. However, due to restrictions on sampling adults at spawning time, we had to infer allele frequencies in the adults from their young progeny (parr).

In each of 3 years, we found that F_{ST} estimates over all populations were small (between 0.013 and 0.053), but statistically significant ($P < 0.001$ in all cases). One explanation of the data is that adult fish return non-randomly. However, an alternative is that fish return randomly, and the significant F_{ST} estimates result from the non-random sampling of families of parr (e.g. Allendorf & Phelps 1981; Hansen *et al.* 1997).

To attempt to distinguish between these alternatives we reasoned that one possible outcome of non-random returning is a positive correlation between allele frequencies across years within sites, if there is a tendency for fish of the same genotype to return to the same part of the river ('homing'). There is no reason to expect such a correlation if the significant F_{ST} estimates are the result of sampling effects. To test for within-site correlations in consecutive years we estimated F_{ST} within each site between years (F_{ST}w) and between all pairs of sites between years (F_{ST}b). We excluded pairwise estimates of F_{ST} between sites within years to control for possible systematic changes in allele frequency between years. We ranked the F_{ST} estimates (because some estimates were negative) and calculated the mean rank of F_{ST}w $(= Q_1)$ and F_{ST}b $(= Q_2)$. The test statistic, Q, is given by:

$$Q = \frac{Q_1 - Q_2}{(n \times n/2)} \tag{5.12}$$

where n is the number of populations sampled in each year. Q can vary between -1 and $+1$, with negative values indicating that allele frequencies within sites between years are on average more similar than between sites between years. The significance of Q is obtained by randomizing the populations in one of the years to obtain the distribution of Q under the null hypothesis that $Q = 0$.

Our values of Q were -0.019 ($P > 0.05$) for 1998–1999 and -0.109 ($P \approx 0.09$) for 1999–2000. Therefore, we cannot distinguish between sampling effects and non-random returning (with or without 'homing') in our study, although more loci may give us better resolution. Nevertheless, the test does have the power to detect correlations within sites over years. Data on Atlantic salmon populations in the Sainte-Marguerite River in Canada published by Garant *et al.* (2000) gave a Q value of -0.259 ($P = 0.02$), suggesting at least some of the observed differentiation in allele frequency is due to 'homing'. The test is applicable to any situation where one is interested in testing the temporal stability of patterns of differentiation.

Conclusions

It is now clear that an estimate of the average migration among a set of populations obtained by transformation of the single over all populations estimate of F_{ST} is likely to be unreliable in virtually all cases. However, demolishing the straw man $F_{ST} = 1/(4N_e m + 1)$ does not mean, as some authors have implied, that F_{ST} estimates tell us nothing about dispersal patterns. We have tried to show in this chapter that the pairwise F_{ST} approach introduced by Slatkin (1993) can be useful and that there are solutions to commonly voiced problems, such as the inability of F_{ST}-based approaches to distinguish between alternative dispersal models. For example, partial regressions can discriminate between vicariance and isolation by distance (and can also control for correlations between ecological variables and distance); and randomization procedures can be devised to test whether differentiation is due to non-random migration patterns or sampling effects. We believe, therefore, that F_{ST}-based methods still have much to offer ecologists interested in dispersal.

References

Allendorf, F.W. & Phelps, S.R. (1981) Use of allele frequencies to describe population structure. *Canadian Journal of Fisheries and Aquatic Sciences* **38**, 1507–1514.

Bohonak, A.J. (1999) Dispersal, gene flow and population structure. *Quarterly Review of Biology* **74**, 21–45.

Bossart, J.L. & Prowell, D.P. (1998) Genetic estimates of population structure and gene flow: limitations, lessons and new directions. *Trends in Ecology and Evolution* **13**, 202–206.

Byers, D.L. (1998) Effect of cross proximity on progeny fitness in a rare and common species of *Eupatorium* (Asteraceae). *American Journal of Botany* **85**, 644–653.

Cain, M.L., Milligan, B.G. & Strand, A.E. (2000) Long-distance seed dispersal in plants. *American Journal of Botany* **87**, 1217–1227.

Chakraborty, R. & Danker-Hopfe, H. (1991) Analysis of population structure: a comparative study of different estimators of Wright's fixation indices. In: *Handbook of Statistics. Vol. 8. Statistical Methods in Biological and Medical Sciences* (eds C.R. Rao & R. Chakraborty), pp. 203–254. Elsevier Science Publishers, Amsterdam.

Chambers, G.K. & MacAvoy, E.S. (2000) Microsatellites: consensus and controversy. *Comparative Biochemistry and Physiology, Part B* **126**, 455–476.

Clarke, R.T., Rothery, P. & Raybould, A.F. (in press) Confidence limits for regression relationships between distance matrices: estimating gene flow with distance. *Journal of Agricultural, Environmental and Biological Statistics* (in press).

Cockerham, C.C. (1969) Variance of gene frequencies. *Evolution* **23**, 72–84.

Cockerham, C.C. (1973) Analyses of gene frequencies. *Genetics* **74**, 679–700.

Cockerham, C.C. & Weir, B.S. (1993) Estimation of gene flow from *F*-statistics. *Evolution* **47**, 855–863.

Crisp, D.T. & Beaumont, W.R.C. (1996) The trout (*Salmo trutta* L) populations of the headwaters of the rivers Severn and Wye, mid-Wales, UK. *Science of the Total Environment* **177**, 113–123.

Crow, J.F. & Aoki, K. (1984) Group selection for a polygenic behavioural trait: estimating the degree of population subdivision. *Proceedings of the National Academy of Sciences of the USA* **81**, 6073–6077.

Di Rienzo, A., Peterson, A.C., Garza, J.C., Valdes, A.M., Slatkin, M. & Freimer, N.B. (1994) Mutational processes at simple-sequence repeat loci in human populations. *Proceedings of the National Academy of Sciences of the USA* **91**, 3166–3170.

Dow, B.D. & Ashley, M.V. (1998) High levels of gene flow in burr oak revealed by paternity analysis using microsatellites. *Journal of Heredity* **89**, 62–70.

Ennos, R.A. (1994) Estimating the relative rates of pollen and seed migration among plant populations. *Heredity* **72**, 250–259.

Ennos, R.A. (2001) Inferences about spatial processes in plant populations from the analysis of molecular markers. In: *Integrating Ecology and Evolution in a Spatial Context* (eds J. Silvertown & J. Antonovics), pp. 45–71. Blackwell Science, Oxford.

Excoffier, L., Smouse, P.E. & Quattro, J.M. (1992) Analysis of molecular variance inferred from metric distances among DNA haplotypes: application to human mitochondrial DNA restriction data. *Genetics* **131**, 479–491.

Feldman, M.W., Kumm, J. & Pritchard, J. (1999) Mutation and migration models of microsatellite evolution. In: *Microsatellites: Evolution and Applications* (eds D.B. Goldstein & C. Schlötterer), pp. 98–115. Oxford University Press, Oxford.

Garant, D., Dodson, D.D. & Bernatchez, L. (2000) Ecological determinants and temporal stability of the within river population structure in Atlantic salmon (*Salmo salar* L.). *Molecular Ecology* **9**, 615–628.

Goudet, J. (1995) FSTAT (version 1.2): a computer program to calculate *F*-statistics. *Journal of Heredity* **86**, 485–486.

Goudet, J., Raymond, M., de Meeüs, T. & Rousset, F. (1996) Testing differentiation in diploid populations. *Genetics* **144**, 1933–1940.

Hamrick, J.L. & Godt, M.J.W. (1996) Effects of life history traits on genetic diversity in plant species. *Philosophical Transactions of the Royal Society, Series B* **351**, 1291–1298.

Hamrick, J.L., Linhart, Y.B. & Mitton, J.B. (1979) Relationships between life history characteristics and electrophoretically detectable variation in plants. *Annual Review of Ecology and Systematics* **10**, 173–200.

Hancock, M.A. & Hughes, J.M. (1999) Direct measures of instream movement in a freshwater shrimp using a genetic marker. *Hydrobiologia* **416**, 23–32.

Hansen, M.M., Nielsen, E.E. & Mensberg, K.L.D. (1997) The problem of sampling families rather than populations: relatedness among individuals in samples of juvenile brown trout *Salmo trutta* L. *Molecular Ecology* **6**, 469–474.

Hedrick, P.W. (1999) Highly variable loci and their interpretation in evolution and conservation. *Evolution* **53**, 313–318.

Hokanson, S.C., Hancock, J.F. & Grumet, R. (1997) Direct comparison of pollen-mediated movement of native and engineered genes. *Euphytica* **96**, 397–403.

Jarne, P. & Lagoda, J.L. (1996) Microsatellites, from molecules to populations and back. *Trends in Ecology and Evolution* **11**, 424–429.

Kimura, M. & Weiss, G.H. (1964) Stepping stone model of population structure and the decrease of genetic correlation with distance. *Genetics* **49**, 561–576.

Latta, R.G., Linhart, Y.B., Fleck, D. & Elliot, M. (1998) Direct and indirect estimates of seed versus pollen movement within a population of ponderosa pine. *Evolution* **52**, 61–67.

Llewellyn, D. & Fitt, G. (1996) Pollen dispersal from two field trials of transgenic cotton in the Namoi Valley, Australia. *Molecular Breeding* **2**, 157–166.

Long, J.C. (1986) The allelic correlation structure of Gainj- and Kalam-speaking people. I. The estimation and interpretation of Wright's *F*-statistics. *Genetics* **112**, 629–647.

Loveless, M.D. & Hamrick, J.L. (1984) Ecological determinants of genetic structure of plant populations. *Annual Review of Ecology and Systematics* **15**, 65–95.

Manly, B.F.J. (1997) *Randomisation, Bootstrap and Monte Carlo Methods in Biology*, 2nd edn. Chapman & Hall, London.

Mantel, N. (1967) The detection of disease clustering and a generalised regression approach. *Cancer Research* **27**, 209–220.

Michalakis, Y. & Excoffier, L. (1996) A generic estimation of population structure using distances between alleles with special reference for microsatellite loci. *Genetics* **142**, 1061–1064.

Nei, M. (1973) Analysis of gene diversity in subdivided populations. *Proceedings of the National Academy of Sciences of the USA* **70**, 3321–3323.

Neigel, J.E. (1997) A comparison of alternative strategies for estimating gene flow from genetic markers. *Annual Review of Ecology and Systematics* **28**, 105–128.

Ouborg, N.J., Piquot, Y. & van Groenendael, J.M. (1999) Population genetics, molecular markers and the study of dispersal in plants. *Journal of Ecology* **87**, 551–568.

Palsbøll, P.J. (1999) Genetic tagging: contemporary molecular ecology. *Biological Journal of the Linnean Society* **68**, 3–22.

Prodöhl, P.A., Loughry, W.J., McDonough, C.M., Nelson, W.S., Thompson, E.A. & Avise, J.C. (1998) Genetic maternity and paternity in a local population of armadillos assessed by microsatellite DNA markers and field data. *American Naturalist* **151**, 7–19.

Raybould, A.F. & Clarke, R.T. (1999) Defining and measuring gene flow. In: *Gene Flow and Agriculture—Relevance for Transgenic Crops*, pp. 41–48. British Crop Protection Council, Farnham, UK.

Raybould, A.F., Goudet, J., Mogg, R.J., Gliddon, C.J. & Gray, A.J. (1996a) Genetic structure of a linear population of *Beta vulgaris* ssp. *maritima* (sea beet) revealed by isozyme and RFLP analysis. *Heredity* **76**, 111–117.

Raybould, A.F., Mogg, R.J. & Clarke, R.T. (1996b) The genetic structure of *Beta vulgaris* ssp. *maritima* (sea beet) populations: RFLPs and isozymes show different patterns of gene flow. *Heredity* **77**, 245–250.

Raybould, A.F., Mogg, R.J. & Gliddon, C.J. (1997) The genetic structure of *Beta vulgaris* ssp. *maritima* (sea beet) populations. II. Differences in gene flow estimated from RFLP and isozyme loci are habitat-specific. *Heredity* **78**, 532–538.

Raybould, A.F., Mogg, R.J., Aldham, C., Gliddon, C.J., Thorpe, R.S. & Clarke, R.T. (1998) The genetic structure of *Beta vulgaris* ssp. *maritima* (sea beet) populations. III. Detection of isolation by distance at microsatellite loci. *Heredity* **80**, 127–132.

Raybould, A.F., Mogg, R.J., Clarke, R.T., Gliddon, C.J. & Gray, A.J. (1999) Variation and population structure at microsatellite and isozyme loci in wild cabbage (*Brassica oleracea* L.) in Dorset (UK). *Genetic Resources and Crop Evolution* **46**, 351–360.

Raymond, M. & Rousset, F. (1995a) GENEPOP (version 1.2): a population genetics software for exact test and ecumenism. *Journal of Heredity* **86**, 248–249.

Raymond, M. & Rousset, F. (1995b) An exact test for population differentiation. *Evolution* **49**, 1280–1283.

Rousset, F. (1996) Equilibrium values of measures of population subdivision for stepwise mutation processes. *Genetics* **142**, 1357–1362.

Rousset, F. (1997) Genetic differentiation and estimation of gene flow from *F*-statistics under isolation by distance. *Genetics* **145**, 1219–1228.

Slatkin, M. (1985) Gene flow in natural populations. *Annual Review of Ecology and Systematics* **16**, 393–430.

Slatkin, M. (1993) Isolation by distance in equilibrium and non-equilibrium populations. *Evolution* **47**, 264–279.

Slatkin, M. (1995) A measure of population subdivision based on microsatellite allele frequencies. *Genetics* **139**, 457–462.

Smouse, P.E., Long, J.C. & Sokal, R.R. (1986) Multiple regression and correlation extensions of the Mantel test of matrix correspondence. *Systematic Zoology* **35**, 627–632.

Thorpe, R.S. & Baez, M. (1993) Geographic variation in scalation of the lizard *Gallotia stehlini* within the island of Gran Canaria. *Biological Journal of the Linnean Society* **48**, 75–87.

Van Dongen, S. (1995) How should we bootstrap allozyme data? *Heredity* **74**, 445–447.

Warren, J.M., Raybould, A.F., Ball, T., Gray, A.J. & Hayward, M.D. (1998) Genetic structure in the perennial grasses *Lolium perenne* and *Agrostis curtisii*. *Heredity* **81**, 556–562.

Waser, N.M., Price, M.V. & Shaw, R.G. (2000) Outbreeding depression varies among cohorts of *Ipomopsis aggregata* planted in nature. *Evolution* **54**, 485–491.

Weir, B.S. & Cockerham, C.C. (1984) Estimating *F*-statistics for the analysis of population structure. *Evolution* **38**, 1358–1370.

Whitlock, M.C. & McCauley, D.E. (1999) Indirect measures of gene flow and migration: F_{ST} (1/(4*Nm* + 1). *Heredity* **82**, 117–125.

Wright, S. (1921) Systems of mating. *Genetics* **6**, 111–178.

Wright, S. (1922) Coefficients of inbreeding and relationship. *American Naturalist* **56**, 330–338.

Wright, S. (1943) Isolation by distance. *Genetics* **28**, 114–138.

Wright, S. (1946) Isolation by distance under diverse systems of mating. *Genetics* **31**, 39–59.

Wright, S. (1951) The genetical structure of populations. *Annals of Eugenics* **15**, 323–354.

Wright, S. (1965) The interpretation of population structure by *F*-statistics with special regard to systems of mating. *Evolution* **19**, 395–420.

Wright, S. (1978) *Evolution and the Genetics of Populations. Vol. 4. Variability Within and Among Natural Populations.* University of Chicago Press, Chicago.

Part 2

Dispersal in behavioural and evolutionary ecology

Chapter 6

Sailing with the wind: dispersal by small flying insects

Steve G. Compton

Introduction

Dispersal is central to the population dynamics of species (Hassell 2000). It also determines the distribution of species at a variety of spatial scales, has a major influence on the ability of species to respond to environmental change and their ability to invade new environments (Hengeveld 1989; Kareiva 1996), and influences the genetic structure of populations and their likelihood of differentiation and speciation (Reaka & Manning 1987; Mopper & Strauss 1998). The movements of each species also have broader, even community-wide effects, shaping diverse ecological interactions such as predation, seed dispersal and pollination (Crawley 1997).

Many, perhaps most, insects are associated with resources that are highly patchy in distribution. The scale of this resource patchiness is highly variable and its significance for insect population dynamics depends as much on the dispersal ability of a species as on the absolute physical distances between patches. Furthermore, resource patches are also ephemeral or are only suitable for limited periods each year, forcing insects to disperse from one patch of resource to another at some stage of their life cycle, as well as to quite different resources such as overwintering sites. Over longer time periods the location of resources also changes at the landscape scale, due mainly to ecological succession. As a consequence, herbivorous insects utilizing weedy annual plants must themselves adopt a more transitory strategy than species associated with long-lived perennial hosts (Price 1992).

This review examines how small insects respond to the challenge of long-distance dispersal between host plants. It concentrates on two well-studied, but superficially disparate groups of small, slow-flying insects: aphids (Homoptera: Aphididae) and fig wasps (Hymenoptera: Agaonidae). They have contrasting ecological and environmental backgrounds, but both groups are shown to exhibit a similar two-phase sequence of flight behaviour that enables them to combine long-distance, undirected dispersal with directed flights to suitable hosts. The cues used during directed flights none the less differ in importance and may reflect the differing nature of the relationship between the insects and their host plants.

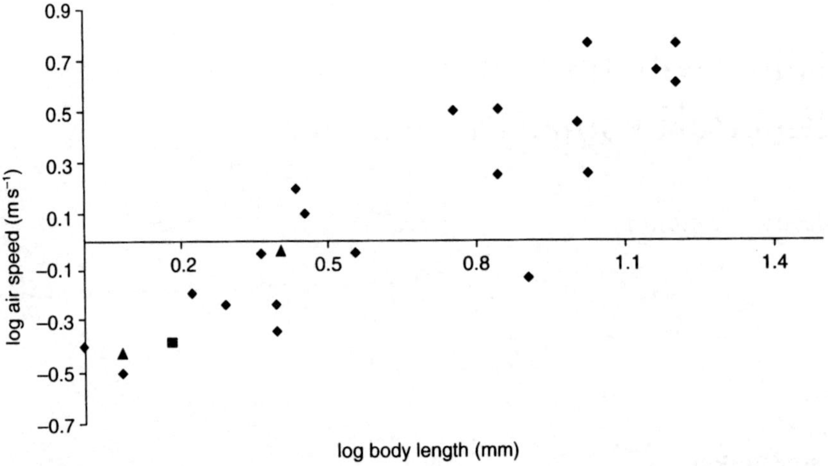

Figure 6.1 Flight speeds of insects in relation to body length. The recorded flight speeds of aphids (triangles) and fig wasps (square) are typical for insects of their small size. Diamonds represent insects from a range of other taxa. (Adapted from Dudley 2000.)

It is hard to get from 'a' to 'b' when you are small

Although almost all pterygote insects can walk (or swim), their dispersion over long distances generally involves flight (but not always; Moran *et al.* 1982) and is therefore necessarily a feature of the adult stage. Flight clearly has many advantages for rapid dispersal, but insects that take to the wing cannot guarantee to move in an optimal direction. This is because the air is rarely totally still, and the direction that an insect moves over the ground is related not only to the heading of its flight, but also the direction in which the wind is blowing. A major consequence of this is that once an insect enters an air column where the air is moving faster than its own maximum airspeed it will inevitably be carried down wind and its angle of flight can only marginally influence its vector (Gatehouse 1997). This is not a large problem for fast-flying insects because they will rarely encounter wind velocities greater than their own speed of flight, but it is routinely the situation for those insects with slower flight. Insect airspeeds are closely correlated with body size (Fig. 6.1), which means that controlling direction of flight in moving air is much more problematical for small insects than large ones (Dudley 2000).

Wind speeds are, of course, highly variable from place to place and time to time, but do vary reasonably consistently with respect to altitude (the higher it is the windier it is), with time of day (it is often windier during the day than at night) and with cover (wind speeds close to and within vegetation are lower than outside it). Even in high winds a narrow boundary layer of essentially still air can form close to physical surfaces (Taylor 1974). Small insects can therefore maximize their chances of travelling in the direction they would prefer by flying at those times when wind speeds are reduced, or by staying out of the general air column and restricting them-

Table 6.1 The dilemma of distance versus direction for small flying insects.

Choice	Costs	Benefits
Disperse low down among vegetation	Limited dispersal distances	Allows control of flight direction
'Sail with the wind'	No control of direction	Makes long-distance dispersal possible

selves to 'trivial' flights close to the ground or within vegetation. There is a trade-off with such behaviour, however, in that the distances that small (and often short-lived) insects can travel are necessarily restricted by their slower flight speeds (Table 6.1). Moving air, therefore, provides small insects with the opportunity for rapid dispersal over long distances, but dictates its direction. Small insects must forfeit control of their flight direction, at least temporarily, if they are to make use of transport by the wind, but this may be their only option if their life style requires long-distance dispersal.

Insects may end their flight as a result of either endogenous changes in their behaviour (linked perhaps to the time they have been flying), in response to general changes in their environment such as sun set, or in response to visual or olfactory information from the ground (Vinson 1985). If these external stimuli are indicative of potential host plants, or are otherwise attractive, then the insect can respond by attempting to reach their source. Many phytophagous insects are attracted by the greens and yellows that are indicative of vegetation (though other wavelengths can also be attractive; Finch 1992; Kostal & Finch 1996) aided by the visual contrast with the plants' immediate surroundings (Kostal & Finch 1994). In so far as potential host plants can be perceived from some distance ahead, even a small flying insect can potentially respond to visual stimuli before being carried past by the wind. Truly host plant-specific responses to the colour of vegetation are unlikely, but colour contrasts may allow some differentiation between vegetation types, and trees to be distinguished from low-lying vegetation (Kennedy *et al.* 1959; Finch & Collier 2000).

In contrast to visual cues, volatile compounds released from plants have the potential to provide highly specific information to flying insects, indicative not only of a plant's identity, but also its physiological condition and the presence of other herbivores (Bernasconi *et al.* 1998; Quiroz & Niemayer 1998). Although the likelihood of detecting volatiles declines with distance from a source, clear-cut odour gradients are prevented by air turbulence (Fig. 6.2). Volatile attractants are necessarily detected downwind of their source, which means that insects have to travel against the prevailing wind direction to reach their source (positive anemotaxis). For fast-flying insects this may not be a problem, and some moths can follow pheromone plumes for a kilometre or more. For small insects in air columns moving faster than their flight speeds, directed upwind movement towards a source of attractant volatiles is not an option, as all this will achieve is a reduction in the rate that they are being carried downwind. Their alternative is to fly down to an area where the air is

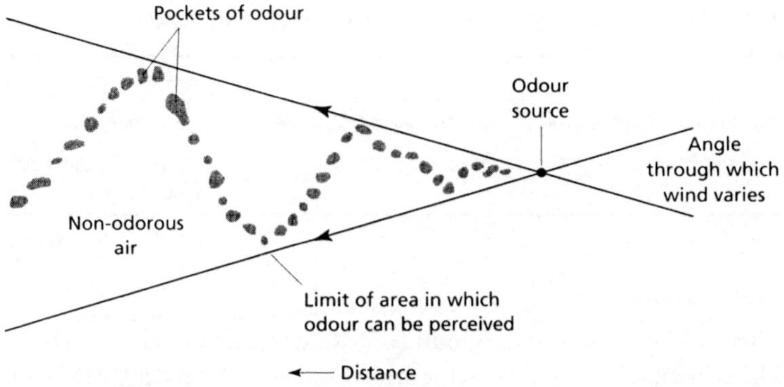

Figure 6.2 Diagrammatic representation of the dispersal of volatiles away from a source such as a potential host plant. Atmospheric instability prevents the formation of concentration gradients and results in the volatiles being distributed within discrete filaments (pockets) separated by odour-free air. Consequently, insects responding to the volatiles only encounter and detect them intermittently. The frequency of encounter will increase towards the source. (Adapted from Bernays & Chapman 1994.)

moving sufficiently slowly for them to move upwind (Gatehouse 1997). The ease with which insects can leave the air column (and the heights that they are flying at before they attempt to) will depend on the extent to which airflow is laminar and without eddies that can carry the insects upwards, whatever their direction of flight. Airflow is only laminar up to about $1.5\,m\,s^{-1}$ ($5\,km\,h^{-1}$) and so will not be the norm for insects travelling away from vegetation. Once in areas with reduced wind speeds they can then potentially travel towards the source of volatiles, either by tracking the volatiles directly or flying or walking upwind. Odour sources from host plants tend to form relatively large and diffuse odour plumes, and this is considered to enhance the viability of anemotactic strategies, without the need for the side-to-side casting typical of insects seeking out more discrete pheromone plumes emanating from point sources (Cardé 1996).

Aphids and fig wasps: shared problems?

Aphids (Hemiptera, Aphididae) include many plant pest species, particularly at temperate latitudes. In contrast, fig trees (*Ficus* spp., Moraceae) are largely tropical and depend for pollination on fig wasps (Hymenoptera, Chalcidoidea, Agaonidae) (Wiebes 1979). The diverse non-pollinating fig wasps are not considered here. Fig trees are often described as 'keystone' species in tropical forests because of the large numbers of birds and mammals that feed on their ripe fruits (Janzen 1979; Terborgh 1986; Lambert & Marshall 1992). The dispersal abilities of aphids and fig wasps are therefore of particular interest to two disparate branches of applied biology—to those concerned with predicting the arrival and abundance of aphids on field crops (Taylor 1974) and those interested in how fragmentation of tropical forests may re-

Table 6.2 Aphids and fig wasps: similarities and differences in characteristics related to dispersal.

Aphids (Aphididae)	Fig wasps (Agaonidae)
Host plant specificity variable	Absolute host specificity
Small size (1.0–10.0 mm)	Even smaller (0.8–3.0 mm)
Adults feed	Adults do not feed
Adult lifespan variable, exceptionally to 48 weeks	Adult lifespan only 1 or 2 days
Host plants aggregated or dispersed	Often highly dispersed host plants
Mainly temperate–relatively simple landscapes	Mainly tropical—often more diverse landscapes
Host plant suitability often prolonged	Host suitability often very brief
Harms hosts	Benefits hosts

duce fruiting of a keystone resource, and its consequences for tropical biodiversity (Mawdsley *et al.* 1998).

Aphids utilize host plants from many different families and may be monophagous, polyphagous or display host alternation (Dixon 1998). Most species are highly polymorphic and reproduction is also diverse, with both parthenogenesis and sexual reproduction. Alates (winged adults of one or both sexes) are responsible for long-distance dispersion in most, but not all, species. Fig wasps are much more uniform, with extreme host plant specificity, sexual reproduction, females that are always fully winged and flightless males that never leave their natal host plants.

Aphids and fig wasps share the problems with aerial dispersal faced by small, delicate insects in general, but there are differences between them (Table 6.2). They are all small insects, with correspondingly slow fight speeds, but fig wasps tend to be smaller. Aphids can achieve flight speeds up to about 0.9 m s^{-1} (Lewis & Taylor 1967), while fig wasps manage about 0.4 m s^{-1} (Fig. 6.1) (Ware & Compton 1994a). Reflecting their predominantly temperate distribution (Mackenzie *et al.* 1994), aphids often display diapause or aestivation. Fig wasps, in contrast, reproduce continuously, although the rate of development of temperate species can be greatly slowed during the winter period (Bronstein & Patel 1992). Adult aphids continue to feed throughout their life, and can potentially live for several days or weeks (Dixon 1998). In contrast, adult fig wasps do not feed (Compton *et al.* 1994a) and even under optimal conditions have never been kept alive for more than 48 h (Kjellberg *et al.* 1988). Most are likely to die the day they emerge unless they manage to find a fig quickly. This race against time is shortened in many species by the presence of predatory nematodes that begin feeding on the wasps as they emerge (Herre 1993).

Both aphids and fig wasps utilize hosts of varying growth forms and apparency (although there are no non-woody fig trees, some are small bushes). On the plants, the sites sought by the insects range from the canopy to below ground (in the case of root-feeding aphids or those fig wasps that pollinate subterranean figs). The distances that have to be travelled to reach new host plants are largely determined by the host specificity of the insects and the extent to which suitable hosts are aggregated, and as such are likely to be highly variable between species. None the less, average distances between hosts are likely to be higher for fig wasps, given that many fig trees are found at low densities in tropical rainforests (Nason *et al.* 1998). The greater botanical diversity of tropical, relative to temperate, habitats also means that fig wasps must often detect host-associated stimuli from a particularly complex background of volatile compounds and visual stimuli. The rarity of suitable host plants for fig wasps is greatly increased by the small proportion of the trees that are suitable at any one time. Many fig trees have highly synchronized fruit development on individual trees, but fruiting is asynchronous at the population level (Compton *et al.* 1994a). This means that emerging fig wasps always have access to trees with figs at a suitable stage (so long as the host population is large enough), but such trees make up only a small proportion of the overall population. In general, then, the difficulties in successfully finding a suitable host plant faced by fig wasps are at least as great as those faced by aphids. Fig wasps none the less may have one potential advantage in that natural selection will have favoured those plants that are most conspicuous to fig wasps—they 'want' to be found.

Aphids and fig wasps: shared solutions?

A sequence of events between take-off by aphids and their final settlement on a new host after a long-distance flight was described by Moericke (1955) and has formed the basis for subsequent studies. Dispersal is seen as comprising a single long-distance flight, with descent followed by one or a series of directed ('attack') flights to and between host plants, with the stages determined by changes in the behaviour ('moods') of the insects (Fig. 6.3a). Ware and Compton (1994a, 1994b) suggested a rather similar sequence of events for dispersing fig wasps, with the major difference being that termination of long-distance flight was seen as a direct response to stimuli from suitable host plants (Fig. 6.3b).

Take-off

Take-off is an important component of the dispersal of small insects because it determines the conditions under which subsequent long-distance flight is attempted. Timing of take-off from natal host plants is particularly significant. Due to warming effects, day-flying species are more likely to be carried high above the ground by ascending air, and subsequent horizontal movement increases with height. Diurnal species are therefore likely to be carried further than night-flying species (Gatehouse 1997).

On emergence, female fig wasps immediately fly away from their natal figs. This reduces the impact of specialist predators such as phorid flies and generalist predators, particularly ants, that concentrate on the figs as the wasps emerge (Compton & Robertson 1988; Compton & Disney 1991). Long-distance flight is initiated immediately, unless there are figs suitable for entry within the vicinity, in which case the female fig wasps will fly directly to them. There are both night- and day-flying species. Take-off times vary accordingly and can be in the morning, early afternoon or evening (S. G. Compton, unpubl. data). In areas with seasonal climates,

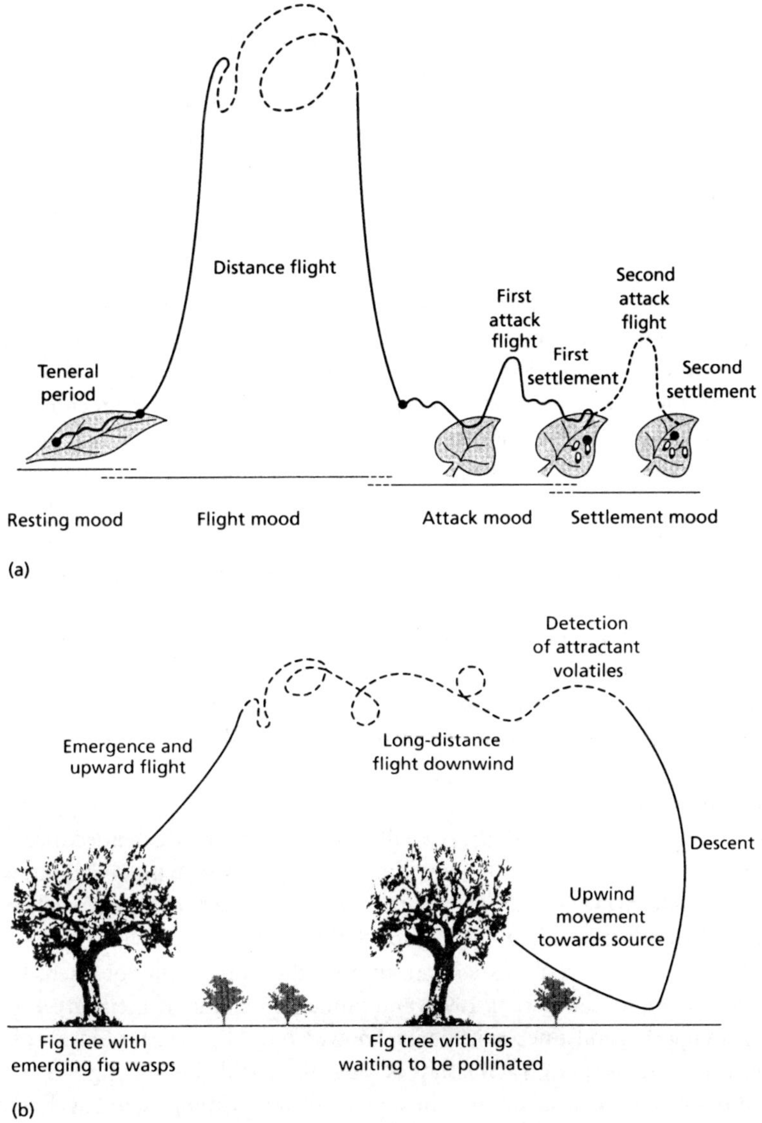

Figure 6.3 (a) The putative sequence of long-distance flight followed by host finding in aphids seeking herbaceous plants. The direction of the extended flight depends on the direction of the wind. After descent, subsequent short-distance flights occur under conditions where directed flight can be achieved. (From Moericke 1955.) (b) The putative sequence of long-distance flight and host finding in fig wasps seeking suitable fig trees in a savanna habitat is broadly similar to that recorded for aphids (Ware & Compton 1994a, 1994b).

it also varies according to temperature. *Elisabethiella baijnathi* in South Africa emerge shortly after dawn in the summer, but later in the day during the winter thereby ensuring that temperatures are above the thresholds required for flight (Ware & Compton 1994a). There is no evidence that take-off is influenced by ambient wind speeds or direction, and night-flying fig wasps do not display lunar phobia (Sutton 1989).

Take-off by aphids does not necessarily occur immediately after the teneral period is complete, and not all winged individuals undertake even short-distance flights. Alate aphids routinely feed before take-off and some also reproduce before departure (Shaw 1970). Take-off can also be influenced by ambient conditions, with species varying in their minimum temperature requirements and their willingness to take-off when wind speeds exceed flight speeds. In most aphids, take-off followed by upwards flight occurs during the day, once temperatures are warm enough for them to do so, but there is considerable variation between and within species. In *Drepanosiphum platanoidis* first-generation individuals take-off in mid-afternoon, when wind speeds are likely to be high, whereas second-generation individuals largely avoid long-distance flights by taking off mainly at dusk, when wind speeds tend to be lowest (Dixon & Mercer 1983). In an analagous situation, dispersing individuals of the whitefly *Bemisia tabaci* may either stay close to the ground (and thereby concentrate their host plant search locally) or fly upwards and achieve long-distance dispersal (Byrne *et al.* 1996). More generally, Haine (1955) concluded that migratory aphids take-off in higher wind speeds than non-migrants, reflecting their greater willingness to be carried long distances by the wind.

Sailing with the wind

Long-distance flights in aphids are typically diurnal, but can be extended throughout the night (Loxdale *et al.* 1993). How long fig wasps can stay in the air is unknown. Visual images from the ground are important for flying insects as they allow them to orientate their direction of flight in relation to the direction of the wind. By flying at an angle to the wind, small insects can improve their probability of intercepting attractant volatiles (Zanen *et al.* 1994) and influence marginally their direction of travel, though they must necessarily travel downwind. Migrating aphids display a temporary refractory period at the start of long-distance flights, during which they are not responsive to visual or other host plant-related stimuli (Kennedy & Booth 1963; Nottingham & Hardie 1989). There is no evidence of a similar refractory period in fig wasps, as they are willing and able to enter suitable figs immediately after emergence, even if they are on the same host plant.

Flying aphids and fig wasps can travel great distances, reflecting the speed and direction of the winds that carry them. They need to remain flying to stay aloft, but if they are in ascending air currents then even this requirement is reduced, at least temporarily, as eddies can take up to an hour to circulate (Thomes *et al.* 1977). The presence of both groups above the Pacific Ocean (Gressitt & Yoshimoto 1963) or the famous landings of aphids on the icy wastes of Spitzbergen (Elton 1925) illustrate the chance nature of where they are carried.

Johnson (1969) quotes exceptional dispersal distances in excess of 1000 km for aphids, while an isolated fig tree in the desert of Namibia showed that numerous fig wasps can find a host tree that is over 60 km from its nearest conspecific. This only provides a minimum estimate of how far the fig wasps had travelled, and they may have come from much further afield (S. Ahmed & S. G. Compton, unpubl. data). Extreme dispersal distances have important consequences for gene flow and colonization events, but the routine distances that are achieved are just as interesting. Taylor *et al.* (1979) showed that the dispersal by hop aphids (*Phorodon humuli*) away from two localized centres of hop production in England was not directional. Females had a median dispersal distance of over 15 km, while the 95% confidence limits extended to over 100 km. The density gradient was that which could be predicted assuming that the insects were transported by turbulent air currents. Cammell *et al.* (1989) similarly monitored the spring movements of another host-alternating species, *Aphis fabae*, away from its primary host plant (spindle), which has a mainly southern distribution in England. The subsequent density gradient of alates in the spring broadly reflected overwintering egg densities, but showed that effectively all of England, Scotland and Wales was being colonized by migrants from the south (Fig. 6.4).

The pollen carried by fig wasps has provided an indirect means of assessing their movements. Seeds pollinated by the wasps can be assessed for their paternity, allowing estimates to be made of the number of sources of fig wasps arriving at a particular tree. When this is combined with known densities of trees releasing wasps in an area, dispersal distances can be calculated. In the rainforest of Panama, breeding units for each species of fig tree extended over hundreds of square kilometres, with routine dispersal distances of 6–14 km (Nason *et al.* 1998). Routine dispersal distances of this length clearly increase the likelihood that even isolated plants will be pollinated.

Fig trees were among the early colonizers of the new volcanic island of Anak Krakatau, situated between Java and Sumatra. Collections made before any fig trees were fruiting showed that small numbers of pollinators were already there, arriving from the surrounding islands 2 or 3 km away, or from further afield (Compton *et al.* 1988). A few years later, when several species of fig trees had begun to fruit on Anak Krakatau, the rarest species were adequately pollinated, whereas more common species were strongly pollinator-limited. It was hypothesized that the 'continuous rain' (Johnson 1969) of fig wasps onto the island was more like a fine drizzle, providing adequate numbers of pollinators for the rarer *Ficus* species, but insufficient numbers for the more common species (Compton *et al.* 1994b). This limitation was predicted to last only until tree populations on Anak Krakatau built up to the point where the island could maintain independent pollinator populations. It will be some time before this can be tested, because recent volcanic activity has 'set the clock back' on Anak Krakatau, by destroying many of the trees.

Wind speeds increase with distance from the ground. In a savanna habitat with small, scattered fig trees a few metres high, the upward flight of emerging fig wasps immediately exposes them to an air column that is faster than their flight speed, and they are carried downwind. Reflecting this, aerial densities of fig wasps in this habi-

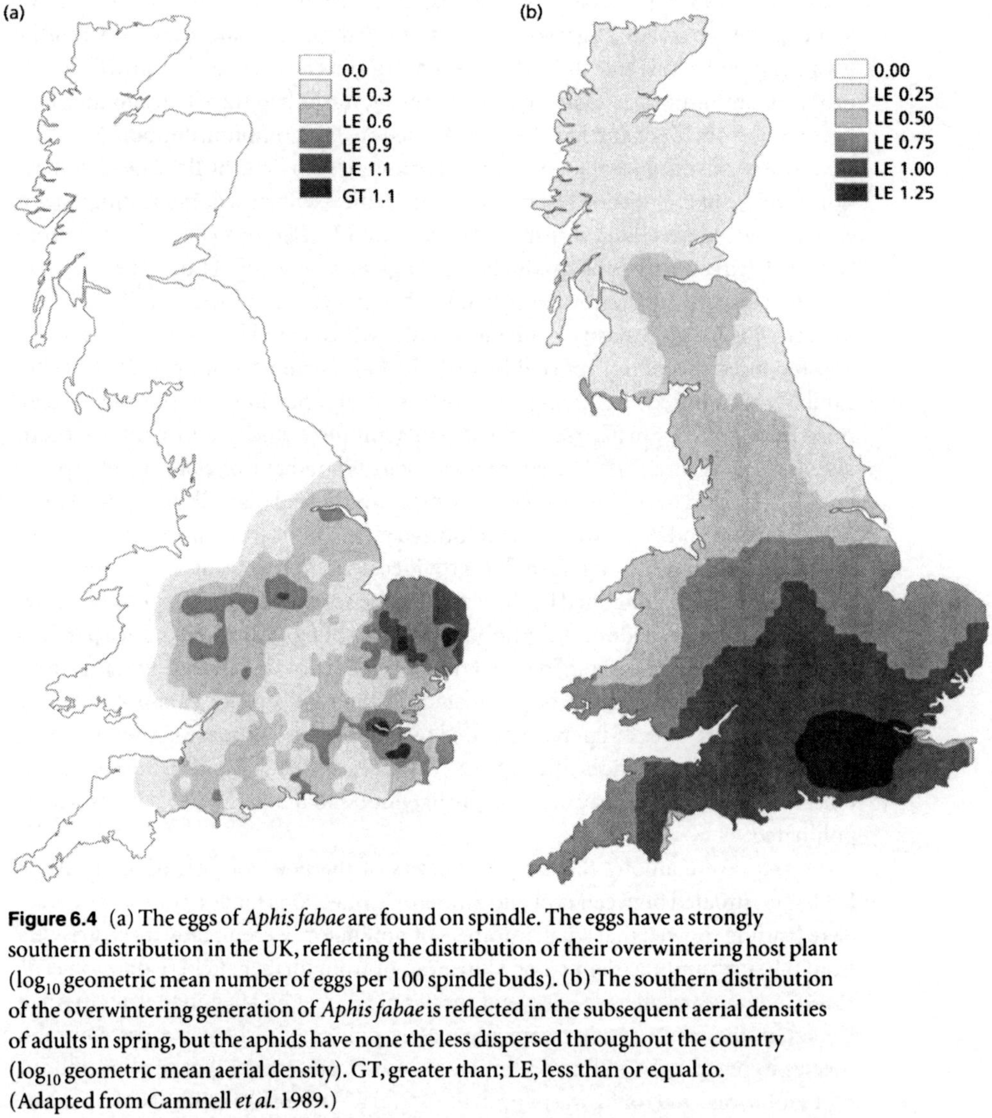

Figure 6.4 (a) The eggs of *Aphis fabae* are found on spindle. The eggs have a strongly southern distribution in the UK, reflecting the distribution of their overwintering host plant ($\log_{10}$ geometric mean number of eggs per 100 spindle buds). (b) The southern distribution of the overwintering generation of *Aphis fabae* is reflected in the subsequent aerial densities of adults in spring, but the aphids have none the less dispersed throughout the country ($\log_{10}$ geometric mean aerial density). GT, greater than; LE, less than or equal to. (Adapted from Cammell *et al.* 1989.)

tat (trapped away from any fruiting trees and corrected for wind speed) were greater at 4 m than closer to the ground (Ware & Compton 1994b). For rainforest species, the habitat is much more vertically structured, and fig wasps may be emerging at ground level, in the understorey, or in the canopy (Shanahan & Compton 2001). Wind speeds within the forest are typically very low, providing small insects with the opportunity for directed flight, but at their own flight speeds (Fig. 6.5) (Baynton *et al.* 1964). In practice, in primary rainforest most opt to fly upwards and expose themselves to the much faster winds that are blowing above the canopy, thereby allowing themselves to be transported downwind for much greater distances than would have been possible unaided (Fig. 6.6) (Compton *et al.* 2000). Fig wasp densi-

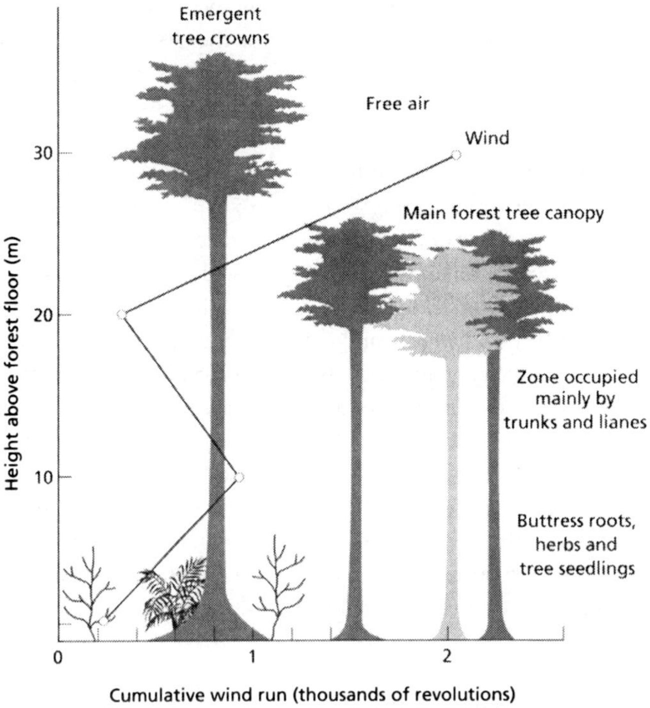

Figure 6.5 The relationship between wind speed and height in an area of rainforest at Morowali, Sulawesi. Little air movement was recorded within and beneath the canopy, but it greatly increased in the more open conditions of the overstorey. (Adapted from Sutton 1989.)

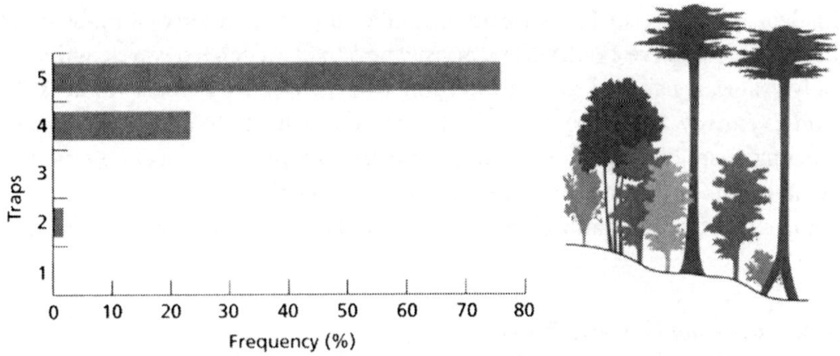

Figure 6.6 The relationship between height above the ground, tree heights and the numbers of fig wasps captured on sticky traps in rainforest at Danum Valley Field Centre, Sabah. The five traps were placed at regular intervals from ground level to 37 m (S.G. Compton, unpubl. data), but most of the fig wasps were collected above the canopy. Note that the numbers of trapped wasps are given, not their aerial densities.

ties appear to be highest just above the general canopy (the overstorey) and decline with further height. This location should favour the detection and response to volatiles released from the vegetation below. Dioecious fig trees grow mainly beneath the forest canopy, but some at least of their pollinators also utilize the overstorey.

High mortality rates are inevitably associated with long-distance dispersal (Loxdale & Lushai 1999), and among aphids Ward *et al.* (1998) estimate that only 0.6% of the autumn migrants of *Rhopalosiphum padi* reach new host plants. Female fig wasps are pro-ovigenic, emerging from the figs with their lifetime egg load. The number of eggs they carry provides an indication of the maximum mortality rates they can sustain, and ranges from around 70 in *Elisabethiella baijnathi* to around 150 in the larger *Eupristina belagaumensis* (Nefdt & Compton 1996; Kathuria *et al.* 1999). With its strongly female-biased sex ratio, *Elisabethiella baijnathi* could maintain its populations if about one in 60 (1.7%) dispersing females could produce the full 70 adult progeny. In practice, some females that reach a fig die trying to penetrate the narrow fig ostiole (the bract-lined entrance), and even if they pass this hurdle they often fail to lay all their eggs due to competition for oviposition sites (Kathuria *et al.* 1999). Developing larvae are also routinely killed by parasitoids (Compton *et al.* 1994a) implying that considerably more than one in 60 females disperse successfully.

Termination of long-distance flight

The period of flight during which small insects are carried largely passively by the wind provides them with the opportunity to be carried for much greater distances than they would be able to fly on their own. This increases the likelihood that they will pass over or near to suitable host plants, but will be of little value in itself if those plants cannot be detected and subsequently reached. Alate aphids generally have larger eyes than apterae, and generally retain the ocelli that have been lost by other aphid morphs (Anderson & Bromley 1987). Fig wasps show loss of ocelli to varying extents and this may be related to the size of the eyes, which are noticeably bigger in night-flying species. Visual cues are important for the termination of long-distance flights in aphids, where a renewed responsiveness to plant-related wavelengths, especially yellow, is believed to be a forerunner to their active descent towards the ground (Kennedy & Fosbrooke 1973; David & Hardie 1988). It is not known whether fig wasps display similar changes in behaviour prior to descent. At moderate wind speeds airflow is laminar, which would allow them to have some control over their flight heights and perhaps optimize detection and responses to host stimuli.

Directed movement towards hosts

It is believed that long-distance flights by aphids are not resumed after a descent. Once away from fast-moving air they can potentially control their direction of flight. Broadbent (1948), for example, collected more aphids on the windward side of traps positioned more than 1 m above a potato field, but found no effect of wind direction on captures from lower traps, where the insects had control of flight direction. Movements towards hosts are typically upwind (Kennedy & Fosbrooke 1973) and at a low level—over 70% of the *Phorodon humuli* (hop aphids) making short flights towards their hosts were beneath 1.5 m (Born 1968).

Early studies of host detection in aphids suggested that arrival at a suitable host plant individual is largely fortuitous (Kennedy *et al.* 1959). Based on visual differences, flying *Myzus persicae* and *Aphis fabae* were none the less shown to favour landing on plants in general over other objects, and distinguished trees from low-growing vegetation, but were no more likely to land on suitable than unsuitable herbaceous plants (Kennedy *et al.* 1961). Host plant selection was therefore envisaged as being carried out largely after landing had taken place, with the aphids resuming short flights again if they were on an unsuitable host (or often even if they had landed on what appeared to be a suitable host, though this may be more of a feature of species with broad host ranges; Bernays & Funk 1999). In contrast, colour preferences in the reed-feeding *Hyalopterus pruni* allow it to distinguish its host from most other plants (Moericke 1969) and may be typical of less polyphagous aphid species.

Colour seems to play little part in the attraction of fig wasps towards individual fig trees (Gibernau *et al.* 1998). Sticky traps placed in a glasshouse environment, although effective in capturing sciarid flies, captured only small numbers of the fig wasp *Liporrhopalum tentacularis*, and yellow traps were no more likely to capture the wasps than white or blue (S. G. Compton, unpubl. data). At the time when they are visited by pollinators, figs are typically green or yellow, more or less blending in with the plant's foliage and do not appear to offer conspicuous visual targets. Vision may none the less be significant for their orientation towards host silhouettes.

All plants release a range of 'green leaf volatiles': aldehydes and alcohols produced as a consequence of their general metabolism (Bernays & Chapman 1994). In addition to these almost universal components, plants also release volatiles characteristic of particular taxa—the familiar smell of isothiocyanates produced by cabbages and other Brassicaceae is an example. Several tens of compounds contribute to the overall bouquet of each plant, but as smaller molecular weight compounds are the most volatile, they are more likely to contribute to long-distance attraction of insects.

Insects detect volatile chemicals using their antennal sensilla (Keil 1999). Apparently uniquely among chalcid wasps, elongate hair-like sensilla have evolved independently several times in female fig wasps (Ware & Compton 1992). The greater surface area of these sensilla is likely to be correlated with increased chemosensory ability, reflecting the importance of detecting volatiles among female fig wasps. Interestingly, among other chalcid wasps it is the males that have repeatedly evolved such elongate sensilla, presumably in order to facilitate the detection of sex pheromones. Few fig wasps have evolved alternative means of increasing the surface area of the sensilla, through having elongate or branched antennae, perhaps because of the problems associated with exit and entry into figs. Again reflecting the need for enhanced olfactory abilities among dispersing individuals, alate aphids have longer antennae than apterae, and their placoid sensilla (rhinaria) are more numerous and

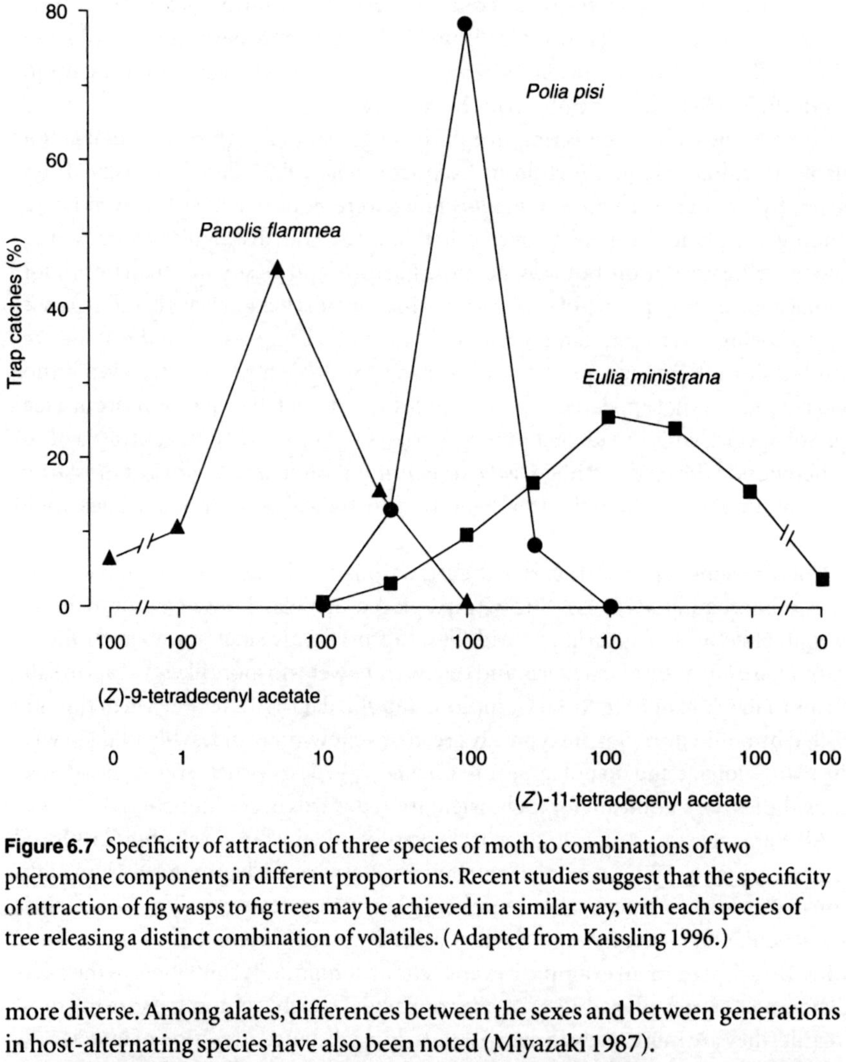

Figure 6.7 Specificity of attraction of three species of moth to combinations of two pheromone components in different proportions. Recent studies suggest that the specificity of attraction of fig wasps to fig trees may be achieved in a similar way, with each species of tree releasing a distinct combination of volatiles. (Adapted from Kaissling 1996.)

more diverse. Among alates, differences between the sexes and between generations in host-alternating species have also been noted (Miyazaki 1987).

The sensilla of aphids are capable of responding differentially to a wide range of green leaf volatiles, as well as to more taxonomically restricted plant volatiles (Nottingham *et al.* 1991; Campbell *et al.* 1993). Electroantennograms from the antennae of the fig wasp *Liporrhopalum tentacularis* showed that the sensilla of fig wasps, like those of aphids, are responsive to a wide range of green leaf volatiles (M. Wakefield & J. Chambers, pers. comm.). This suggests that both aphids and fig wasps have the sensory equipment to detect and the ability to respond to volatile signals emanating from plants in general, and that different combinations of volatiles could provide highly specific information on their source, as has been found with blends of moth pheromones (Fig. 6.7) (Kaissling 1996).

Combinations of compounds as an attractant might be expected to have an inherent disadvantage over simpler pheromone-type signals in that they necessarily require several different molecules, with differing volatilities, to be detected simultaneously for them to be effective. If simple concentration gradients were established away from source plants, then their compositions would necessarily change with distance. However, the turbulent diffusion of odour plumes carried downwind of a source are more complex (Baker & Haynes 1989; Murlis *et al.* 1992), generating pulses of intact volatile blends ('filaments') many metres away, although the frequency at which the bursts are encountered declines with distance (Fig. 6.2) (Cardé 1996).

For small insects, detection of attractant volatiles is only the first step towards reaching a host plant, because of the physical difficulties associated with movement upwind. The distances covered by aphids when carrying out directed flights is unclear, as is how visual and olfactory stimuli are interrelated in different species. At least some aphids can use host volatiles as cues, though these are not necessarily plant specific (Campbell & Ridout 2001). They can also be deterred from landing by other volatiles, including aphid alarm pheromones (Chapman *et al.* 1981; Wohlers 1981; Hardie *et al.* 1994; Bernasconi *et al.* 1998; Quiroz & Niemayer 1998). *Rhopalosiphum padi* are attracted by volatiles of *Prunus* (Pettersson 1970) and Anderson and Bromley (1987) concluded that the terpenes characteristic of trees and shrubs may be the major cue used by host-alternating species when returning to their woody hosts in autumn, but doubted whether they could be detected at long range.

Cabbage root flies and onion flies (*Delia* spp.) display broadly similar behaviour to aphids when seeking their host plants, with short upwind flights and frequent landings (Hawkes & Coaker 1979). Finch and Collier (2000) examined the movements towards host plants in a range of crucifer-feeding insects (including cabbage aphid) and produced a general model for their dispersal and host finding. Their model differs from the classic Moericke (1955) scheme for dispersal in aphids in that termination of long-distance flights is seen as being in response to volatiles emanating from host plants, and if confirmed for aphids in general would remove one of the apparent differences in dispersal behaviour between aphids and fig wasps (Fig. 6.3). They suggest, however, that the volatiles are generally at too low a density to provide directional information and that subsequent local flights lead to landings on green plants, but are otherwise non-specific. Eventual host plant acceptance involves a combination of short-distance flights in combination with cues obtained from plant surfaces.

Volatile compounds appear to be much more important than visual cues for host finding in fig wasps. Experiments involving placing figs in muslin bags in field situations demonstrated that fig wasps are specifically attracted to their host figs, and not other plant parts. They also showed that the figs are only attractive during the period when they are ready to be pollinated, that the figs of other species are not attractive, and that when the ostiole is blocked the figs are no longer attractive (van Noort *et al.* 1989). Attraction to pentane extracts of figs has confirmed the role of volatiles in long-range attraction (Hossaert-McKey *et al.* 1994; Gibernau *et al.* 1997).

The volatile profiles of figs have been shown to differ between species, and to change at the time when they become attractive (Ware *et al.* 1993; Gibernau *et al.*

1997). Identification of the volatiles shows that they are mainly terpenoids and comprise both widespread and uncommon floral compounds (Grison *et al.* 1999). Closely related species often have rather similar blends, pointing to a strong phylogenetic determinant of their composition and a probable role of the volatiles in speciation of the plants and their pollinators (Grison *et al.* 1999). Fig wasps would benefit greatly from being able to distinguish between the sexes in dioecious fig species, because they produce no progeny if they are attracted to a female tree. They do not appear to be able to tell the sexes apart, however, and mutual mimicry among the two sexes of the plants has been postulated (Grafen & Godfray 1991). It has therefore come as something of a surprise that there are often differences in the volatile blends of male and female figs of the same species (Grison *et al.* 1999). The differences are presumably among volatiles that are not significant for attraction of the pollinators, but clearly the mimicry between the sexes is not complete.

Fig wasps making host-seeking local flights in open situations travel upwind, close to the ground. More *Elisabethiella baijnathi*, the pollinator of a small tree (*Ficus burtt-davyi*), were trapped downwind than upwind of attractive plants and their average flight heights downwind were lower than upwind of the trees, or when away from attractive trees (Ware & Compton 1994b). Another species of fig wasp, not associated with this fig tree, did not show the same pattern of distribution, showing that it was not a consequence of the physical presence of the trees alone. Studies in India have shown that under light wind conditions, fig wasps were able to travel upwind for 100 m across a grassy parkland to reach attractive figs placed in muslin bags (P. Kathuria-Gupta & S. G. Compton, unpubl. data). The extent of the attractant radius from a fig tree that may have hundreds of thousands of attractant figs must be greater than this, though the likelihood that the wasps will successfully fly upwind to such sources must decline rapidly with distance.

The greater the quantity of attractant volatiles that are released by a plant, the more likely it is to be detected. The quantities of volatiles emitted by rainforest strangler figs, with synchronously developing crops of hundreds of thousands, borne high in the canopy, may be the lighthouses of the genus, drawing in their pollinators from a wide radius. At the other extreme, this must be quite different from the attractant radii of some small, understorey, usually dioecious fig trees, where small crop sizes and asynchronous fruiting mean that the potential for attractant release is far smaller (Harrison 2000).

It is not known whether fig wasps ever deliberately fly on again after encountering trees with figs that are suitable for entry, but it seems unlikely. In laboratory trials *Liporrhopalum tentacularis* generally enter the first figs that they land on, so long as the figs are at the correct stage to be pollinated (J. Moore & S. G. Compton, unpubl. data). Some other species may be more discriminating (Gibernau *et al.* 1998). Given the generally isolated nature of their host plants and their brief longevity, rejection of host trees with attractive figs would clearly be risky. Once they have entered a fig the option of moving on is not normally available to fig wasps as even if they re-emerge their wings have been removed by the plant.

Aphids and fig wasps: models or misfits?

Aphids and fig wasps achieve long-distance dispersal in the same way. By first exposing themselves to fast-moving air they travel much longer distances than would be possible unaided, and then by flying within more sheltered conditions they are capable of directed flight towards their hosts. In this way the dilemma of direction versus distance outlined in Table 6.1 is avoided. In aphids, the decision whether or not to undergo long-distance flight is flexible and responsive to local conditions, while in fig wasps it is typically the default, and only circumvented if suitable hosts are detected at the time of emergence.

There are also differences in the ways that suitable host plants are detected, with fig wasps having a much greater reliance on olfactory cues. The host-finding behaviour of fig wasps may prove to be exceptional, due to the nature of the relationship with their host plants. Unlike aphids and most other plant-feeding insects, fig wasps have a major potential advantage in that natural selection will have favoured those fig trees that are most conspicuous to fig wasps—they 'want' to be found and so may be exceptionally 'apparent'. More direct data than is currently available is required to determine whether fig wasps are unusually successful at finding hosts, compared with aphids and other small insects. The distances that fig wasps can apparently track upwind to host plants certainly contrasts sharply with values for others insects orientating to plant volatiles, where maximum values of just a few metres have been recorded (Finch 1980), and more closely resembles those recorded for insects responding to pheromones. The casting behaviour displayed by fig wasps downwind of attractive figs (van Noort *et al.* 1989) is also typical of insects seeking pheromone plumes, and is in contrast to the searching behaviour outlined for aphids and insects in general by Finch and Collier (2000). This implies that the quantities of volatile attractants released from figs may be exceptional. Quantitative estimates of the amounts of attractant are available from, for example, crucifers (Finch 1980), and similar data are needed from fig trees to test this idea.

Aphids and fig wasps are typical insects in that they are small, but how typical of small insects is their willingness to undertake long-distance dispersal? For example, the cyclical parthenogenesis and host alternation typical of many aphids both allows them to spread the risks associated with dispersal within clones and also necessitates migrations between winter and summer hosts. Similarly, other groups of rainforest chalcidoids do not show as marked a preferance for flying in the overstorey as fig wasps (Compton *et al.* 2000). Mymarid wasps, which are even smaller than fig wasps, appear to have a much more localized dispersal strategy, largely restricting their flights to the subcanopy. It may turn out that the resource patches to which aphids and fig wasps are responding are exceptionally diffuse, compared with those of many other small insects. Studies of the relationship between the dispersal of small insects and their 'habitat template' (Southwood 1977) are needed if we are to achieve a more predictive overview of dispersal behaviour in small insects.

References

Anderson, M. & Bromley, A.K. (1987) Sensory system. In: *Aphids, their Biology, Natural Enemies and Control. World Crop Pests 2a* (eds A.K. Minks & P. Harrewijn), pp. 153–162. Elsevier, Amsterdam.

Baker, T.C. & Haynes, K.F. (1989) Field and laboratory electroantennographic measurements of pheromone plume structure correlated with oriental fruit moth behaviour. *Physiological Entomology* **14**, 1–12.

Baynton, H.W., Biggs, W.G., Hamilton, H.L., Sherr, P.E. & Worth, J.J.B. (1964) Wind structure in and above a tropical forest. *Journal of Applied Meteorology* **4**, 670–675.

Bernasconi, M.L., Turlings, T.C.J, Ambrosetti, L. & Bassetti, P. (1998) Herbivore-induced emissions of maize volatiles repel the corn leaf aphid *Rhopalosiphum maidis. Entomologioa Experimentalis et Applicata* **87**, 133–142.

Bernays, E.A. & Chapman, R.F. (1994) *Host-plant Selection by Phytophagous Insects.* Chapman & Hall, New York.

Bernays, E.A. & Funk, D.J. (1999) Specialists make faster decisions than generalists: experiments with aphids. *Proceedings of the Royal Society of London, Series B* **266**, 151–156.

Born, M. (1968) Beirage zur bionomie von *Phorodon humuli* (Schrank, 1801). *Archiv für Planzenschutz* **4**, 37–52.

Broadbent, L. (1948) Aphis migration and the efficiency of the trapping method. *Annals of Applied Biology* **35**, 379–394.

Bronstein, J.L. & Patel, A. (1992) Temperature-sensitive development: consequences for local persistence of two subtropical fig wasp species. *American Midland Naturalist* **18**, 397–403.

Byrne, D.N., Athman, R.J., Orum, T.V. & Palumbo, J.C. (1996) Localised migration and dispersal by the sweet potato whitefly, *Bemisia tabaci. Oecologia* **105**, 320–328.

Cammell, M.E., Tatchell, G.M. & Woiwod, I.P. (1989) Spatial pattern of abundance of the black bean aphid, *Aphis fabae*, in Britain. *Journal of Applied Ecology* **26**, 463–472.

Campbell, C.A.M. & Ridout, M.S. (2001) Effects of plant spacing and interplanting with oilseed rape on colonisation of dwarf hops by the damson-hop aphid, *Phorodon humuli. Entomologia Experimentalis et Applicata* **99**, 211–216.

Campbell, C.A.M., Pettersson, J., Pickett, J.A., Wadhams L.J. & Woodcock, C.M. (1993) Spring migration of damson-hop aphid, *Phorodon humuli* (Homoptera, Aphididae), and summer host plant-derived semiochemicals released on feeding. *Journal of Chemical Ecology* **19**, 1569–1576.

Cardé, R.T. (1996) Odour plumes and odour-mediated flight in insects. In: *Olfaction in Mosquito–Host Interactions* (eds G. Bock & G. Cardew), pp. 54–70. Wiley, Chichester.

Chapman, R.F., Bernays, E.A. & Simpson, S.J. (1981) Attraction and repulsion of the aphid *Cavariella aegopodii*, by plant odours. *Journal of Chemical Ecology* **7**, 881–888.

Compton, S.G. & Disney, R.H.L. (1991) New species of *Megaselia* (Diptera: Phoridae) whose larvae live in figs (Urticales: Moraceae), and adults prey on fig wasps (Hymenoptera: Agaonidae). *Journal of Natural History* **25**, 203–219.

Compton, S.G. & Robertson, H.G. (1988) Complex interactions between mutualisms: ants tending homopterans protect fig seeds and pollinators. *Ecology* **69**, 1302–1305.

Compton, S.G., Thornton, I.W.B., New, T.R. & Underhill, L. (1988) Colonisation of the Krakatau Islands by fig wasps and other chalcids. *Philosophical Transactions of the Royal Society of London, Series B* **322**, 459–470.

Compton, S.G., Rasplus, J.-Y. & Ware, A.B. (1994a) African fig wasp parasitoid communities. In: *Parasitoid Community Ecology* (eds B.A. Hawkins & W. Sheehan), pp. 243–370. Oxford University Press, Oxford.

Compton, S.G., Ross, S.J. & Thornton, I.W.B. (1994b) Pollinator limitation of fig tree reproduction on the island of Anak Krakatau (Indonesia). *Biotropica* **26**, 180–186.

Compton, S.G., Ellwood, M.D.F., Davis, A.J. & Welch, K. (2000) The flight heights of chalcid wasps (Hymenoptera: Chalcidoidea) in a lowland Bornean rainforest: fig wasps are the high fliers. *Biotropica* **32**, 515–522.

Crawley, M.J. (ed.) (1997) *Plant Ecology.* Blackwell Science, Oxford.

David, C.T. & Hardie, J. (1988) The visual responses of free-flying summer and autumn forms of the black bean aphid, *Aphis fabae*, in an automated flight chamber. *Physiological Entomology* **13**, 277–284.

Dixon, A.F.G. (1998) *Aphid Ecology*, 2nd edn. Chapman & Hall, London.

Dixon, A.F.G. & Mercer, D.R. (1983) Flight behaviour in the sycamore aphid: factors affecting take-off. *Entomologia Experimentalis et Applicata* **33**, 43–49.

Dudley, R. (2000) *The Biomechanics of Insect Flight*. Princeton University Press, New Jersey.

Elton, C.S. (1925) The dispersal of insects to Spitzbergen. *Transactions of the Royal Entomological Society of London* **1925**, 289–299.

Finch, S. (1980) Chemical attractants of plant-feeding insects to plants. In: *Applied Biology V* (ed T.H. Coaker), pp. 67–143. Academic Press, London.

Finch, S. (1992) Improving the selectivity of water traps for monitoring populations of the cabbage root fly. *Annals of Applied Biology* **120**, 1–7.

Finch, S. & Collier, R.H. (2000) Host-plant selection by insects—a theory based on 'appropriate/inappropriate' landings by pest insects of cruciferous crops. *Entomolgia Experimentalis et Applicata* **96**, 91–102.

Gatehouse, A.G. (1997) Behavior and ecological genetics of wind-borne migration by insects. *Annual Review of Entomology* **42**, 475–502.

Gibernau, M., Buser, H.R., Frey, J.E. & Hossaert-McKey, M. (1997) Volatile compounds from extracts of figs of *Ficus carica*. *Phytochemistry* **46**, 241–244.

Gibernau, M., Hossaert-McKey, M., Frey, J. & Kjellberg, F. (1998) Are olfactory signals sufficient to attract fig pollinators? *Ecoscience* **5**, 306–311.

Grafen, A. & Godfray, H.C.J. (1991) Vicarious selection explains some paradoxes in dioecious fig pollinator systems. *Proceedings of the Royal Society of London, Series B* **245**, 73–76.

Gressit, J.L. & Yoshimoto, C.M. (1963) Dispersal of animals in the Pacific. In: *Pacific Basin Biogeography: a Symposium* (ed. J.L. Gressitt), pp. 283–292. Bishop Museum Press, Honolulu.

Grison, L., Edwards, A.A. & Hossaert-McKey, M. (1999) Interspecies variation in floral fragrances emitted by tropical *Ficus* species. *Phytochemistry* **52**, 1293–1299.

Haine, E. (1955) Aphid take-off in controlled wind speeds. *Nature* **175**, 474–475.

Hardie, J., Storer, J.R., Nottingham, S.F. *et al.* (1994) The interaction of sex pheromone and plant volatiles for field attraction of male bird-cherry aphid *Rhopalosiphum pisum*. In: *Brighton Crop Protection Conference: Pests and Diseases. Proceedings of an International Conference Organised by the British Crop Protection Council 21–24 November 1994* (ed. British Crop Protection Council), pp. 1223–1230. British Crop Protection Council, Farnham, UK.

Harrison, R.D. (2000) Repercussions of el Nino: drought causes extinction and the breakdown of mutualism in Borneo. *Proceedings of the Royal Society of London, Series B* **267**, 911–915.

Hassell, M.P. (2000) *The Spatial and Temporal Dynamics of Host–Parasitoids Interactions*. Oxford University Press, Oxford.

Hawkes, C. & Coaker, T.H. (1979) Factors affecting the behavioural responses of the adult cabbage root fly, *Delia brassicae*, to host plant odour. *Entomologia Experimentalis et Applicata* **25**, 45–58.

Hengeveld, R. (1989) *Dynamics of Biological Invasions*. Chapman & Hall, London.

Herre, E.A. (1993) Population structure and the evolution of virulence in nematode parasites of fig wasps. *Science* **259**, 1442–1445.

Hossaert-McKey, M., Gibernau, M., & Frey, J.E. (1994) Chemosensory attraction of fig-wasps to substances produced by receptive figs. *Entomologia Experimentalis et Applicata* **70**, 185–191.

Janzen, D.H. (1979) How to be a fig. *Annual Review of Ecology and Systematics* **10**, 13–51.

Johnson, C.G. (1969) *Migration and Dispersal of Insects by Flight*. Methuen, London.

Kaissling, K.-E. (1996) Peripheral mechanisms of pheromone reception in moths. *Chemical Senses* **20**, 257–268.

Kareiva, P. (1996) Developing a predictive ecology for non-indigenous species and ecological invasions. *Ecology* **77**, 1651–1652.

Kathuria, P., Greeff, J.M., Compton, S.G. & Geneshaiah, K.N. (1999) What fig wasp sex ratios may or may not tell us about sex allocation strategies. *Oikos* **87**, 520–530.

Keil, T.A. (1999) Morphology and development of the peripheral olfactory organs. In: *Insect Olfaction* (ed. B.S. Hansson), pp. 5–47. Springer Verlag, Berlin.

Kennedy, J.S. & Booth, C.O. (1963) Free flight of aphids in the laboratory. *Journal of Experimental Biology* **40**, 67–85.

Kennedy, J.S. & Fosbrooke, I.H.M. (1973) The plant in the life of an aphid. In: *Insect/Plant Relation-*

ships (ed. H.F. van Emden), pp. 129–140. Symposia of the Royal Entomological Society of London No. 6. Blackwell Science, London.

Kennedy, J.S., Booth, C.O. & Kershaw, W.J.S. (1959) Host finding by aphids in the field. I. Gynoparae of *Myzus persicae* (Sulzer). *Annals of Applied Biology* **47**, 410–423.

Kennedy, J.S., Booth, C.O. & Kershaw, W.J.S. (1961) Host finding by aphids in the field III. Visual attraction. *Annals of Applied Biology* **49**, 1–21.

Kjellberg, F., Doumesche, B. & Bronstein, J.L. (1988) Longevity of a fig wasp (*Blastophaga psenes*). *Proceedings of the Koninklijke Nederlandse Akademie van Wetenschappen C* **91**, 117–122.

Kostal, V. & Finch, S. (1994). Influence of background on host-plant selection and subsequent oviposition by the cabbage root fly (*Delia radicum*). *Entomologia Experimentalis et Applicata* **70**, 153–163.

Kostal, V. & Finch, S. (1996) Preference of the cabbage root fly *Delia radicum* (L) for coloured traps: influence of sex and physiological status of the flies, trap background and experimental design. *Physiological Entomology* **21**, 123–130.

Lambert, F.R. & Marshall, A.G. (1992) Keystone characteristics of bird-dispersed *Ficus* in a Malaysian lowland rain forest. *Journal of Ecology* **79**, 793–809.

Lewis, T. & Taylor, L.R. (1967) *Introduction to Experimental Ecology*. Academic Press, London.

Loxdale, H.D. & Lushai, G. (1999) Slaves of the environment: the movement of herbivorous insects in relation to their ecology and genotype. *Philosophical Transactions of the Royal Society of London, Series B* **354**, 1479–1495.

Loxdale, H.D., Hardie, J., Halberts, S., Foottit R., Kidd, N.A.C. & Carter, C.I. (1993) The relative importance of short- and long-range movement of flying aphids. *Biological Reviews* **68**, 291–311.

Mackenzie, A., Dixon, A.F.G. & Kindlmann, P. (1994) The relationship between the regional number of aphid species and plant-species diversity. *European Journal of Entomology* **91**, 135–138.

Mawdsley, N.A., Compton, S.G. & Whittaker, R.J. (1998) Population persistence, pollination mutualisms, and figs in fragmented tropical landscapes. *Conservation Biology* **12**, 1416–1420.

Miyazaki, M. (1987) Forms and morphs of aphids. In: *Aphids, their Biology, Natural Enemies and Control. World Crop Pests 2a* (eds A.K. Minks & P. Harrewijn), pp. 27–50. Elsevier, Amsterdam.

Moericke, V. (1955) Uber die lebensgewohnheiten der geflugelten blattlause (Aphidina) unter besonderer berucksichtigung des verhaltens beim landen. *Zeitscrift fur Angewalde Entomologie* **37**, 29–91.

Moericke, V. (1969) Hostplant specific colour behaviour by *Hyalopterus pruni* (Aphididae). *Entomologia Experimentalis et Applicata* **1**, 524–534.

Mopper, S. & Strauss, S.Y. (eds) (1998) *Genetic Structure and Local Adaptation in Natural Insect Populations: Effects of Ecology, Life History and Behavior*. Chapman & Hall, New York.

Moran, V.C., Gunn, B.H. & Walter, G.H. (1982). Wind dispersal and settling of first-instar crawlers of the cochineal insect *Dactylopius austrinus* (Homoptera: Coccoidea: Dactylopiidae). *Ecological Entomology* **7**, 409–419.

Murlis, J., Elkington, J.S. & Cardé, J.T. (1992) Odor plumes and how insects use them. *Annual Review of Entomology* **37**, 505–532.

Nason, J.D., Herre, E.A. & Hamrick, J.L. (1998) The breeding structure of a tropical keystone plant resource. *Nature* **391**, 685–687.

Nefdt, R.J.C. & Compton, S.G. (1996) Regulation of seed and pollinator production in the fig–fig wasp mutualism. *Journal of Animal Ecology* **65**, 170–182.

Nottingham, S.F. & Hardie, J. (1989) Migration and targeted flight in seasonal forms of the black bean aphid *Aphis fabae*. *Physiological Entomology* **14**, 451–458.

Nottingham, S.F., Hardie, J., Dawson, G.W. *et al.* (1991) Behavioral and electrophysiological responses of aphids to host and nonhost plant volatiles. *Journal of Chemical Ecology* **17**, 1231–1242.

Pettersson, J. (1970) Studies on *Rhopalosiphum padi* (L.) I. Laboratory studies on olfactometric responses to winter host, *Prunus padus*. *Lantbruskshogskolans Annaler* **36**, 381–399.

Price, P.W. (1992) Plant resources as the mechanistic basis for insect herbivore population dynamics. In: *Effects of Resource Distribution on Animal–Plant Interactions* (eds M.D. Hunter, T. Ohgushi & P.W. Price), pp. 139–173. Academic Press, San Diego.

Quiroz, A. & Niemayer, H.M. (1998) Olfactomer-assessed responses of aphid *Rhopalosiphum padi* to wheat and oat volatiles. *Journal of Chemical Ecology* **24**, 113–124.

Reaka, M.L. & Manning, R.B. (1987) *The Significance of Body Size, Dispersal Potential, and Habitat for Rates of Morphological Evolution in Stomatopod Crustacea.* Smithsonian Press, Washington, DC.

Shanahan, M. & Compton, S.G. (2001) Vertical stratification of figs and fig-eaters in a Bornean lowland rainforest: how is the canopy different? *Plant Ecology* **153**, 121–132.

Shaw, M.J.P. (1970) Effects of population density on alienicolae of *Aphis fabae* Scop. II. The effects of crowding on the expression of migratory urge among alatae in the laboratory. *Annals of Applied Biology* **65**, 197–203.

Southwood, T.R.E. (1977) Habitat, the template for ecological studies? *Journal of Animal Ecology* **46**, 337–365.

Sutton, S.L. (1989) The spatial distribution of flying insects. In: *Ecosystems of the World 14b. Tropical Rain Forest Ecosystems* (eds H. Lieth & M.J.A. Werger), pp. 427–436. Elsevier, Amsterdam.

Taylor, L.R. (1974) Insect migration, flight periodicity and the boundary layer. *Journal of Animal Ecology* **43**, 225–238.

Taylor, L.R., Woiwod, I.P. & Taylor, R.A.J. (1979) The migratory ambit of the hop aphid and its significance in aphid population dynamics. *Journal of Animal Ecology* **48**, 955–972.

Terborgh, J. (1986) Keystone plant resources in the tropical forest. In: *Conservation Biology: the Science of Scarcity and Diversity* (ed. M.E. Soulé), pp. 330–344. Sinauer Associates, Sunderland, MA.

Thomes, A.A.G., Ludlow, A.R. & Kennedy, J.S. (1977) Sinking speeds of falling and flying aphids. *Ecological Entomology* **2**, 315–326.

van Noort, S., Ware, A.B. & Compton, S.G. (1989) Release of pollinator-specific volatile attractants from the figs of *Ficus burtt-davyi. South African Journal of Science* **85**, 323–324.

Vinson, S.B. (1985) The behavior of parasitoids. In: *Comprehensive Insect Physiology, Biochemistry and Pharmacology*, Vol. 9 (eds G.A. Kerkut & L.I. Gilbert). Pergamon Press, Oxford.

Ward, S.A., Leather, S.R., Pickup, J. & Harrington, R. (1998) Mortality during dispersal and the cost of host-specificity in parasites: how many aphids find hosts? *Journal of Animal Ecology* **67**, 763–773.

Ware, A.B. & Compton, S.G. (1992) Repeated evolution of elongate sensilla in female fig wasps (Hymenoptera: Agaonidae: Agaoninae). *Proceedings of the Koninklijke Nederlandse Akademie van Wetenschappen C* **95**, 275–292.

Ware, A.B. & Compton, S.G. (1994a) Dispersal of adult female fig wasps, I: arrivals and departures. *Entomologia Experimentalis et Applicata* **73**, 221–230.

Ware, A.B. & Compton, S.G. (1994b) Dispersal of adult female fig wasps, II: movements between trees. *Entomologia Experimentalis et Applicata* **73**, 231–238.

Ware, A.B., Kaye, P.T., Compton, S.G. & van Noort, S. (1993) Fig volatiles: their role in attracting pollinators and maintaining pollinator specificity. *Plant Systematics and Evolution* **186**, 147–156.

Wiebes, J.T. (1979) Co-evolution of figs and their insect pollinators. *Annual Review of Ecology and Systematics* **10**, 1–12.

Wohlers, P. (1981) Aphid avoidance of plants contaminated with alarm pheromone (E)-β-farnesene. *Zeitsschrift fur Angewandte Entomologie* **93**, 102–108.

Zanen, P.O., Sabelis, M.W., Buonaccorsi, J.P. & Cardé, R.T. (1994) Search strategies of fruit flies in steady and shifting wind in the absence of food odors. *Physiological Entomology* **19**, 335–341.

Chapter 7

Density-dependent dispersal in animals: concepts, evidence, mechanisms and consequences

William J. Sutherland, Jenny A. Gill and Ken Norris

Introduction

A major cause of dispersal in animals is competition for resources resulting from changes in population density. Although density-dependent dispersal is probably a widespread process, it has been the subject of relatively few empirical studies, possibly because wide ranges of population densities are rarely encountered over the relatively short duration of many intensive field studies. However, density-dependent dispersal has been the subject of many conceptual and experimental studies. Density-dependent dispersal can be a critical determinant of patterns of use of sites differing in quality, and hence has important implications for individual fitness and for population ecology. In this chapter we describe the conceptual approaches to density-dependent dispersal in animals and the mechanisms by which density dependence can act within sites. We also review the evidence for density-dependent dispersal and discuss the consequences for population ecology.

Theoretical concepts

There are four major concepts that describe density-dependent dispersal: the ideal free distribution, ideal despotic distribution, sources and sinks and the buffer effect. Although each of these is associated with particular conditions, they are not mutually exclusive. Rather, the different concepts provide different levels of explanation of density-dependent dispersal patterns. For example, the extent to which a good (source) and a poor (sink) area are used may be explained in terms of the ideal free distribution (or ideal despotic distribution) but may also be compatible with the buffer effect.

Ideal free distribution

The ideal free distribution (Fretwell & Lucas 1970) is probably the most commonly known concept of density-dependent dispersal (Fig. 7.1a). It describes the distribution of animals between patches when individuals have perfect knowledge of all patches ('ideal') and are capable of unconstrained movement ('free') to the area with

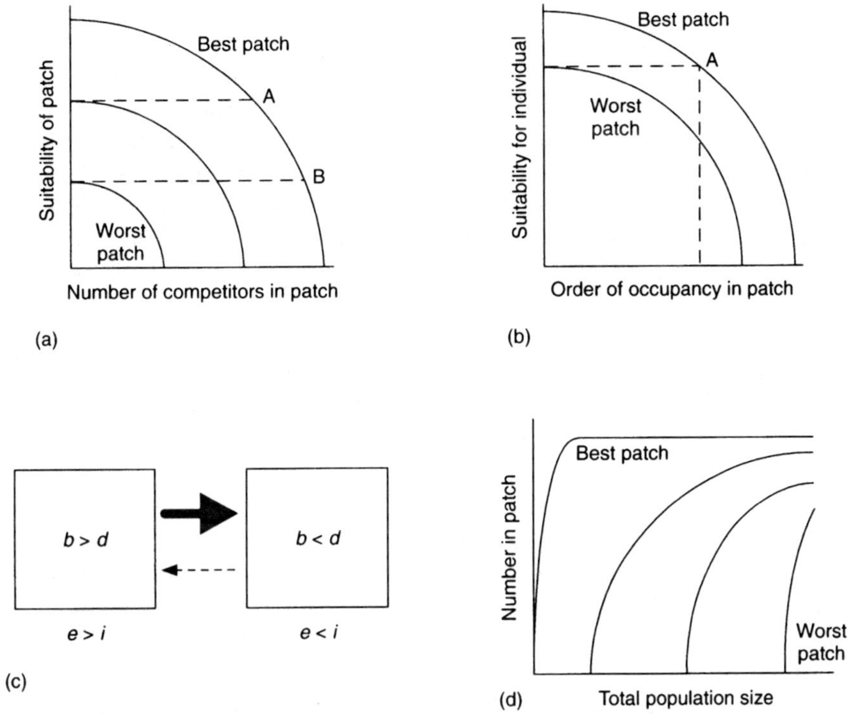

Figure 7.1 Four main mechanisms for density-dependent dispersal. (a) The ideal free distribution: individuals initially occupy the best patch until the increase in density has reduced the suitability of the patch to the point A where it is equal to the next best patch, which individuals will then occupy. (b) The ideal despotic distribution: individuals again initially occupy the best patch but exclude others from it, so that the second patch is occupied when the suitability for the individual is equal in both patches, even though the average reward is higher in the first patch. (c) Sources and sinks: in source locations births (*b*) outnumber deaths (*d*) and emigration (*e*) exceeds immigration (*i*), whereas the reverse is true in sink locations. (d) The buffer effect: individuals initially occupy the best patch but as population size increases, numbers in poorer-quality sites increase disproportionately.

the highest suitability (such as food density or expected reproductive success). Thus, at low population sizes only the best areas are occupied. As the population increases, the increasing density reduces the suitability of the good patches and individuals begin to occupy poorer-quality sites. Consequently, the theoretical expectation is that the suitability of all occupied patches will be equal. Numerous modifications have been made to the ideal free distribution including, for example, incorporating the consequences of individuals differing in competitive ability (Parker & Sutherland 1986; Sutherland & Parker 1992), the consequences of sampling errors and the costs of travelling (Bernstein *et al.* 1991).

Ideal despotic distribution

The ideal despotic distribution (Fretwell & Lucas 1970) is similar to the ideal free distribution except that individuals can defend resources (Fig. 7.1b), for example through territorial behaviour. Thus, at low population density individuals will occupy the best territories in the best patch but will then exclude others from these territories. Further individuals will have a choice of occupying a low-ranking territory on a good patch or the best territory on a poor patch. The presence of further individuals does not affect the success of those that obtained the best quality territories and, as a result, average suitability is higher in the better patch.

Sources and sinks

The initial concept of sources and sinks (Pulliam 1988) considers the case of areas in which population growth (λ) exceeds 1, which thus act as sources exporting individuals at high density, while other sites have negative growth ($\lambda < 1$) and hence act as sinks. The emigration of individuals from sources to sinks maintains populations in each (Fig. 7.1c). For example, in an experimental study of root vole *Microtus oeconomus* populations, Gundersen *et al.* (2001; see also Andreassen *et al.*, this volume) demonstrated that in the presence of sink patches (from which individuals were removed), density-dependent dispersal from source patches to sinks resulted in sink populations increasing, whereas the source populations did not grow. Thus, the presence of sink patches had a marked effect on the dynamics of source populations.

Watkinson and Sutherland (1995) pointed out that it is exceedingly difficult to identify sinks. This is because reasonable quality locations, capable of sustaining a population, may still receive immigrants from adjacent better quality sites. The increase in density within the site would reduce the growth rate as a result of density dependence. Such a site would then appear as a sink as deaths would exceed births ($\lambda < 1$) and the number of immigrants would exceed the number of emigrants. Such areas have been termed pseudosinks.

Buffer effect

The buffer effect arises when sites vary in quality and animals emigrate from preferred to less preferred sites at high density. Consequently, fluctuations in population size result in large changes in numbers in less preferred sites but small changes in number in preferred sites (Fig. 7.1d). The buffer effect was first described by Kluyver and Tinbergen (1953) and, since then, numerous studies of a range of taxa have described buffer effect patterns. Unlike the other three concepts in this section, the buffer effect is largely empirically determined and has not had a clear theoretical basis. Nevertheless, the buffer effect can be generated by theoretical models based on the ideal free distribution. To illustrate this we used a model of the ideal free distribution with unequal competitors (Sutherland 1983, 1996; Sutherland & Parker 1985; Parker & Sutherland 1986; Bernstein *et al.* 1988, 1991; Sutherland & Dolman 1994). In this model, consumers occupy an area of habitat that consists of 15 distinct patches of equal size that vary in initial resource density. The consumer population occupies this habitat during a particular part of its life cycle for 40 time units (referred to hereafter as the 'season'). During each time unit, the model calculates the optimal distribution pattern of the consumers between the patches (i.e. so that each individual consumer maximizes its consumption rate) and the number of resource items removed by the consumers. Consumers in this habitat experience two types of competition: depletion and interference. Depletion occurs because we assumed there is no growth or reproduction in the resource population during the season, so consumers reduce resource densities over the season. Interference occurs because the consumption rate of an individual is reduced when exploiting resources in a patch with one or more conspecifics. We assumed that each consumer would maximize its fitness by maximizing its consumption rate, so consumers have to balance the conflicting demands of occupying the patch with the highest resource density and avoiding intense interference competition due to other individuals trying to do the same. We used a game theoretic approach to determine the optimal solution to this problem for a range of model conditions (see also Sutherland & Parker 1985; Parker & Sutherland 1986; Sutherland & Dolman 1994).

The consumption rate of a consumer in patch i was described by Holling's disk equation (Holling 1959):

$$\frac{N_i}{T} = \frac{a_i' d_i}{1 + a_i' d_i T_h} \tag{7.1}$$

where N_i is the number of resource items exploited by a consumer during time unit T, a_i' is the searching efficiency of a consumer in patch i, d_i is the density of resources in patch i, and T_h is the handling time of a single resource item. We assume that T_h is constant in different patches.

If a consumer is searching for resources in patch i with other conspecifics then its searching efficiency is reduced in relation to consumer density according to the expression (see Hassell & Varley 1969):

$$a_i' = QP_i^{-m} \tag{7.2}$$

where Q is the quest constant (i.e. the value of a_i' that would be achieved in patch i by a solitary individual), P_i is the density of consumers and m is the interference coefficient. During certain simulations, we assumed that individuals varied in their susceptibility to interference. That is, the decline in searching efficiency in relation to an increase in consumer density varied between individuals. This was incorporated as:

$$a_{S,i}' = QP_i^{-mR_{S,i}} \tag{7.3}$$

where $a_{S,i}'$ is the searching efficiency of phenotype S in patch i and $R_{S,i}$ is the relative competitive ability of phenotype S in patch i, expressed as the mean competitive ability of all individuals in patch i divided by the competitive ability of an individual of phenotype S (Sutherland & Parker 1985).

During each time unit we calculated the optimal distribution pattern, so that each individual occupied the patch that maximized its consumption rate. At the start of the time unit, individuals of each competitive ability phenotype were distributed evenly between patches. Note that if the simulation involved equal competitors then the entire consumer population was initially distributed evenly between patches. Each phenotype was then allowed to replicate in each patch in proportion to its consumption rate in each patch (i.e. replication was fastest in the patch providing the highest consumption rate). This iterative process was repeated up to 6000 times when a stable distribution had been reached (see also Parker & Sutherland 1986). At this point, resources were depleted from the patch by the consumers. This entire process was repeated for the 40 time units during the season.

Using this model, we ran a range of model simulations that included consumer populations of differing size in each simulation (range 10–200 individuals). Each simulation began with the same initial resource density in each of the 15 patches. These values were drawn at random from a uniform distribution that ranged from 0 to 40 resource items per patch. The actual number of resource items in each patch at the start of the season were: 0, 7, 11, 13, 14, 15, 20, 21, 24, 25, 31, 33, 37, 37, 39. The simplest version of the model included no interference and so is analogous to the spatial depletion model of Sutherland and Anderson (1993). We also examined how interference (with unequal competitors) affected the buffer effect with reference to these depletion-only effects. For simulations with unequal competitors, each consumer was assigned to a competitive ability phenotype by randomly choosing a value from a normal distribution of values (mean = 20 and SD = 2).

Figure 7.2a shows the results of simulations including depletion but no interference. At low population sizes, the animals occupy the best-quality patch(es), but as population size increases progressively more animals are forced to occupy poorer-quality patches as the season proceeds due to resource depletion. As a result, the abundance of animals in the best-quality patch(es) is a decelerating function of overall population size, whereas abundance in the poorest quality patch only starts to increase after a threshold population size is reached. Figure 7.2b shows the results of the model including resource depletion, interference ($m = 0.3$) and unequal competitive abilities between animals in the populations (e.g. Sutherland & Dolman 1994). In this case, animal abundance in the best-quality patch is again a decelerating function of total population size, indicating a buffer effect. However, the distribution of animals between patches is much more even than in the spatial depletion model because the poorest competitors are displaced from the best-quality patches at relatively low population sizes due to the strength of interference competition they experience.

Figure 7.2 (*opposite*) The buffer effect in theoretical models based on: (a) spatial resource depletion, and (b) an ideal free model incorporating resource depletion, interference and unequal competitors. Each graph shows total population size as total bird days in an area of habitat containing the patches plotted against bird days in each particular patch.

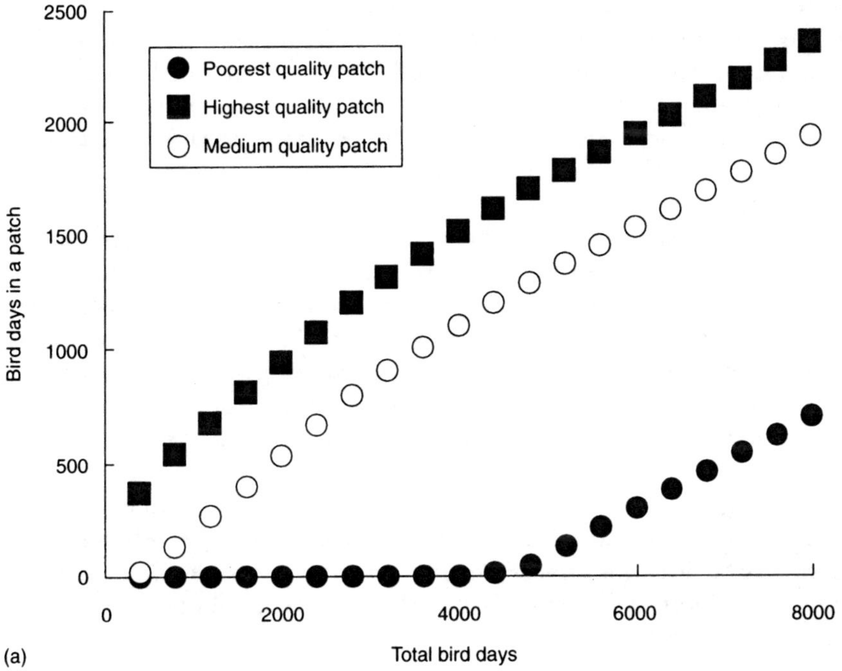
2500
2000
1500
1000
500
0
Poorest quality patch
Highest quality patch
Medium quality patch
Bird days in a patch
0
2000
4000
6000
8000
Total bird days

(a)

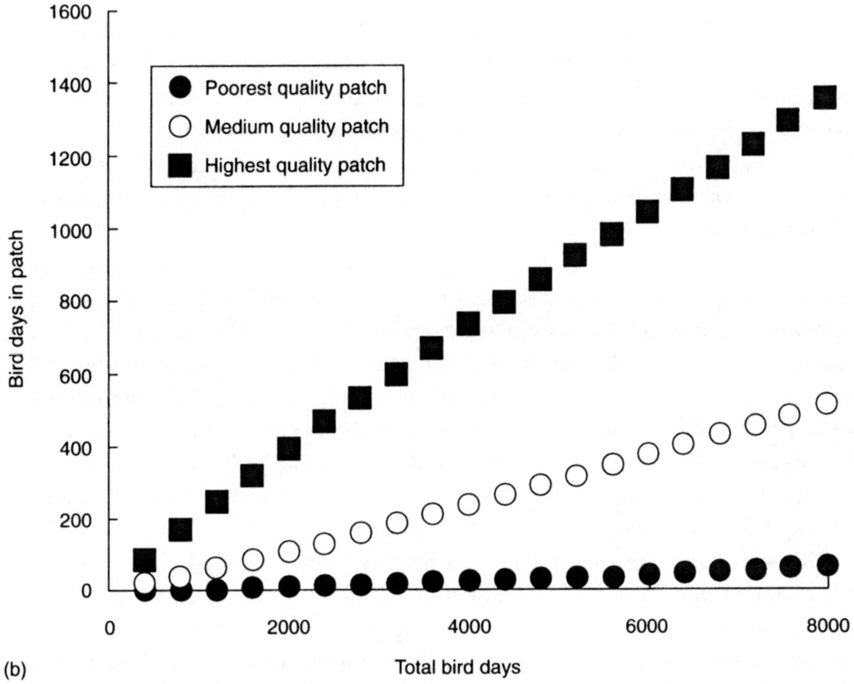
1600
1400
1200
1000
800
600
400
200
0
Poorest quality patch
Medium quality patch
Highest quality patch
Bird days in patch
0
2000
4000
6000
8000
Total bird days

(b)

Density-dependent factors within sites

Density dependence within sites is fundamental to driving density-dependent dispersal. In the absence of density dependence the best sites would be selected regardless of population density. In this section we consider the main density-dependent mechanisms for animals and how they might lead to density-dependent processes.

Interference

Interference is the short-term reversible decline in food intake due to the presence of others (Hassell & Varley 1969; Goss-Custard 1980). It has a number of possible mechanisms including fighting, intraspecific kleptoparasitism (Norris & Johnstone 1998; Triplet *et al.* 1999), prey disturbance (Yates *et al.* 2000), attraction of interspecific kleptoparasites (Carbone *et al.* 1997) and avoiding aggressive interactions (Goss-Custard 1970; Cresswell 1997). Interference can be incorporated within the ideal free distribution (Sutherland 1983) to explain the distribution for a given level of interference. Subsequent studies (Parker & Sutherland 1986; Sutherland & Parker 1992) have shown that it is also important to consider the variation in interference experienced by individuals of differing competitive ability, a consequence of which may be that poor competitors leave. Interference can therefore result in predator densities that are lower than that which could otherwise be supported by the resource density.

Depletion

Depletion is the removal of resources that could be used by others. Sutherland and Anderson (1993) provided an analytical version of Royama's (1971) graphical model in order to explain the pattern of distribution of predators in relation to prey depletion. The Sutherland and Anderson model has been used successfully to explain the pattern of prey depletion and distribution of black-tailed godwits *Limosa limosa* across scales varying from small patches of mudflat up to adjacent estuaries (Gill *et al.* 2001b).

Depletion models incorporating natural prey mortality have also been used to explain the distribution of bean geese *Anser fabalis* and wigeon *Anas penelope* and the competition between the two on pasture fields (Sutherland & Allport 1994). Similarly, the use of intertidal foraging habitats by dark-bellied brent geese *Branta b. bernicla* (Rowcliffe *et al.* 2001) and light-bellied brent geese *B. b. hrota* (Percival *et al.* 1996, 1998) could be explained by the interaction between depletion of food supply by the geese and natural mortality through storms, and Vickery *et al.* (1995) showed that the depletion of algae and saltmarsh by dark-bellied brent geese explained the timing of the switch of birds to inland foraging sites.

Although interference and depletion models have been used almost entirely to explain the distribution of individuals in the non-breeding season, they are also appropriate for studies of dispersal patterns in the breeding season. This would be particularly important for colonial species in which depletion of food around the colony or interference in acquiring that food can play a major role (Birkhead & Furness 1985).

Territoriality

Many species defend territories and exclude others. This usually applies in the breeding season but some species defend separate territories during the non-breeding season. Density dependence can result both from individuals acquiring poor-quality territories with lower expected reproductive output and through an increasing proportion of individuals deciding to defer breeding and thus reducing mean reproductive output. Kokko and Sutherland (1998) and Kokko *et al.* (2001) have determined the population density at which individuals should decide not to breed but to wait for a better site to be vacated.

Liley (1999) used the ideal despotic distribution and territorial behaviour to explain the use of different sections of beaches by breeding ringed plovers *Charadrius hiaticula*. Broader sections of beach were preferred as these had lower predation rates. The broader beaches were selected first and were more likely to be re-used if the initial nesting attempt failed. This behaviour can result in density-dependent fecundity, as selection of the best locations at low population size results in high reproductive success; but as the population increases, a wider range of habitats are used and mean reproductive output declines.

Similarly Ens *et al.* (1992) examined the behaviour of breeding oystercatchers *Haematopus ostralegus* on the island of Schiermonnikoog in the Netherlands. The oystercatchers had two possible breeding strategies. Residents acquired territories along the boundary between the mudflat and the saltmarsh and could thus take their chicks down on the shore to feed. Leapfrogs bred in the saltmarsh set back from the shore and individuals therefore had to fly over the residents to get food and then return to feed the chicks. The leapfrog strategy is clearly less efficient and reproductive success of leapfrog strategy was approximately one-third of the success of residents. Leapfrogs very rarely move to a resident territory so once they decide to become a leapfrog they keep that strategy for the rest of their lives. The question then was how did these strategies coexist? The answer seems to be that competition to acquire resident territories is very severe; individuals usually have to defer breeding for several years in order to acquire a resident territory and many individuals waiting to occupy resident territories die before breeding. The average expected reproductive success is then equal for the two strategies. From this it is possible to determine the expected density-dependent reproductive success (Sutherland 1996); as the population increases more individuals become leapfrogs or queue to become a resident and thus mean reproductive success declines.

Rank

Dominance hierarchies may also affect survival probabilities and expected reproductive output. There is little theory addressing this issue so far and the consequences for density-dependent mortality and fecundity are poorly understood. For example, this may apply to breeding systems such as olive baboons *Papio anubis* in which males move from troops with high numbers of males to female-biased troops (Packer 1979).

Predators and parasites

Survival and breeding success may be affected by predation and parasitism and the effects of these may increase with density. For example, Nams (1997) showed that skunks *Spinachia spinachia* improved their abilities to detect prey in a manner that would create density-dependent predation. Potts (1986) showed that productivity of partridges *Perdix perdix* was density dependent, which was thought to be as a result of individual carrion crows *Corvus corone* learning to detect partridge nests and then specializing upon them. At low densities few individual crows specialized but at high densities, crow predation increased greatly. Parasitism is also often likely to be related to density, for example, red grouse *Lagopus lagopus* are more likely to be infected with parasites at high grouse densities than at low densities. As parasitism level affects fecundity this may drive population cycles of the species (Hudson *et al.* 1998).

Evidence for density-dependent dispersal

The evidence for density-dependent dispersal has resulted both from direct observations and has been assumed from variation in population size. The published results can be divided into six broad categories.

1 *Density-dependent emigration.* Several studies have shown that emigration is more likely to occur at high population densities. For example, Fonseca and Hart (1996) showed that dispersal rates of newly hatched black fly *Simulium vittatum* larvae were strongly density dependent (Fig. 7.3a), possibly in response to reduced feeding rates at high densities. Similarly, goldenrod leaf beetles *Trihabda virgata* dispersed at significantly higher rates from dense populations than from uncrowded conditions (Herzig 1995). As the destination of the emigrants is not known, such studies do not quantify the extent to which this behaviour results in redistribution.

2 *Density-dependent dispersal observed between sites.* Studies of marked individuals can show which individuals move and where to. For example, individual marking of barnacle geese *Branta leucopsis* at breeding colonies on islands in the Baltic Sea showed that an increasing proportion of juvenile males dispersed to non-natal breeding locations as population size increased (Fig. 7.3b) (van der Jeugd 1999).

3 *Local buffer effects within seasons.* Merkt (1981) showed that after experimental removal of deer mice *Peromyscus maniculatus* from an area of shrub and grassland, newly arriving individuals inhabited shrub areas preferentially and only inhabited grassland when densities in shrub areas were high. Goss-Custard (1977) also showed that, on the Wash Estuary in England, both knot *Calidris canutus* and oystercatchers *Haematopus ostralegus* concentrated within their preferred feeding areas at the start of the autumn, but as numbers increased through the autumn they increasingly used other feeding areas (Fig. 7.3c). Interpreting such patterns is not straightforward as changes in population size and total resource depletion may be correlated. Thus the changing distribution pattern of oystercatchers and knot could also be caused by prey depletion within the preferred sites causing the birds to use a wider range of sites, which would happen irrespective of increases in total population size.

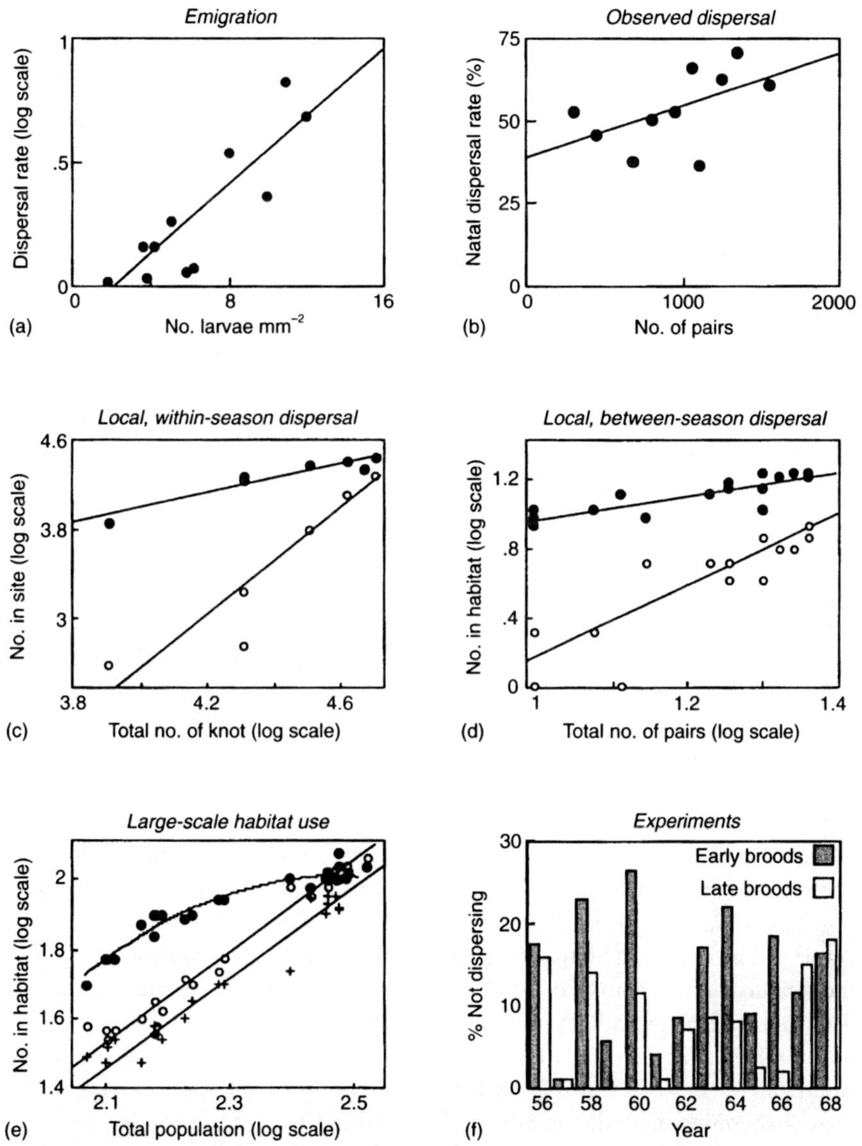

Figure 7.3 Examples of categories of evidence for density-dependent dispersal. (a) Dispersal rates of fly larvae increased with density (data from Fonseca & Hart 1996). (b) The natal dispersal rate of male barnacle geese *Branta leucopsis* increased with density (data from van der Jeugd 1999). (c) Increasing numbers of knot *Calidris canutus* resulted in an increasing use of offshore (open circle) rather than inshore (filled circle) locations (data from Goss-Custard 1977). (d) The numbers of breeding blue tits *Parus caeruleus* increased more in pine (open circle) than in mixed woodland (filled circle) in years with higher population size (data from Kluyver & Tinbergen 1953). (e) As the population size of British-breeding skylarks *Alauda arvensis* declined, the proportion breeding in arable (cross) and mixed (open circle) farming habitats declined faster than in grazed farmland (filled circle) (data from Browne 1996). (f) The removal of the first broods of great tits *Parus major* in 1967 and 1968 resulted in a greater proportion of second brood birds not dispersing than in previous years (data from Kluyver 1971).

4 *Local buffer effects between seasons.* The classic evidence for the buffer effect applies to sites of differing quality which are adjacent or very near one another. Numbers using high-quality sites are consistently high and vary little whereas the poorer-quality sites only attract large numbers when overall population size is high. One of the earliest examples was Kluyver and Tinbergen's (1953) study of great tits *Parus major* and blue tits *P. caeruleus* in strips of mixed woodland within a pine forest. Over a 10-year period, numbers of both species varied little in the mixed woodland but varied positively with overall population size in the pine forest (Fig. 7.3d). In the same study area, Glas (1960) showed that numbers of chaffinches *Fringilla coelebs* over 10 years also varied greatly in pine woodland but not in mixed woodland. Similarly, over a 12-year period, territory occupancy by nuthatches *Sitta europea* in areas of spruce forest containing patches of deciduous woodland varied little in good-quality territories but fluctuated with total population size in poorer territories (Nilsson 1987). Dhondt *et al.* (1992) also showed that blue tit use of good-quality territories (those with large clutch sizes) did not vary with population size, but that use of poor-quality territories (with small clutch sizes) increased in years of high population density.

In addition to these small-scale studies, buffer effects have also been demonstrated at larger spatial scales, including countrywide patterns. For example, Moser (1988) showed that numbers of grey plovers *Pluvialis squatarola* wintering in Britain increased at a higher rate on estuaries on which numbers had initially been scarce.

5 *Large-scale habitat use.* The relative use of different habitat types can also be affected by overall population sizes. Thus, across Britain numbers of great tits using farmland increased much more rapidly with overall population size than the number using woodland (O'Connor 1980). Similarly, as numbers of skylarks *Alauda arvensis* in Britain have declined, the rate of decline has been significantly greater in arable and mixed farmland than in grazed farmland (Fig. 7.3e) (Browne 1996). Although such patterns can be detected on large spatial scales, the dispersal processes determining them may still be occurring on a local scale.

6 *Experiments.* Krebs (1971) demonstrated that the removal of great tits from woodland territories resulted in birds from nearby hedgerow territories immediately abandoning these sites in favour of the vacant woodland territories. Nesting success was significantly higher in woodland than hedgerow territories and a significantly greater proportion of birds in hedgerow territories were first-year birds. In this system there was therefore clear density-dependent dispersal between sites of differing quality. Kluyver (1971) experimentally removed 90% of the first broods of great tits and showed that the dispersal distance of the second brood was greatly reduced (Fig. 7.3f).

How is density-dependent dispersal achieved?

We suggest that density-dependent dispersal occurs primarily due to four processes: exclusion, settlement in relation to habitat quality, movement in relation to success and movement in relation to condition. Some of these processes involve assessment of the quality of a range of sites. Sampling is, therefore, likely to be an important constraint and is likely to be a greater constraint over large distances. For example, over adjacent mudflats and estuaries, the distribution of black-tailed godwits *Limosa limosa* reflects that of the food supply (Gill *et al.* 2001b); however at a countrywide scale there are differences in survival rates between wintering locations (Gill *et al.* 2001a).

Exclusion

Emigration may take place regardless of the actual surrounding site quality, for example because individuals are excluded or unable to survive if they stay. This may well apply to single territories. For larger sites there is often likely to be the opportunity to either become a floater or to occupy a poor-quality area within the site. If this is the case then exclusion will be rare and dispersal is likely to be the consequence of active decisions of individuals to leave rather than being forced out.

Differential settlement

This occurs when one section of a population tends to disperse further than other sections. In many cases, it is animals in their first year of life which disperse and settle in new areas, although there is often also a sex bias in dispersal distances (e.g. Favre *et al.* 1997; Verhulst *et al.* 1997; Lindberg *et al.* 1998). In an analysis of 69 species of British breeding birds, Paradis *et al.* (1998) found that in 61 of those species, natal dispersal distances were greater than dispersal distances between breeding attempts. Thus, juveniles appeared to be dispersing and selecting a breeding location to which they were subsequently highly philopatric.

Dispersal of juvenile or immature individuals in the non-breeding season also frequently appears to be greater than adult dispersal. For example, Sutherland (1982) examined the consequences of an enormous spat-fall of cockles *Cerastoderma edule* on the Ribble Estuary, UK. The number of oystercatchers *Haematopus ostralegus* feeding on the cockles increased by over four times and the proportion of juveniles and immatures in the population increased considerably. Similarly, under deteriorating conditions adults often appear to remain philopatric whereas juveniles are more likely to move elsewhere. As the cockle population on the Ribble subsequently crashed, the proportion of juveniles and immatures declined markedly. This implies that much of the redistribution in relation to changes in habitat quality may occur due to differential settlement of juveniles, with adults tending to stay in the sites selected as a juvenile.

Success-mediated dispersal

A common phenomenon in breeding bird studies is that pairs often return to the same location to breed if they have previously bred successfully there, but are more likely to switch breeding locations if they were previously unsuccessful (e.g. Jackson 1994; Flyn *et al.* 1999; Forero *et al.* 1999). Similarly, pairs may stay together if successful but divorce if breeding is not successful. For example, experimental manipulation of brood size in great tits resulted in higher divorce rates for birds with experimentally reduced clutches and lower divorce rates for experimentally enlarged clutches (Linden 1991). Similarly, studies of red squirrels *Sciurus vulgaris* have shown that dispersal rates are often linked to habitat quality (Wauters *et al.* 1995; Lurz *et al.* 1997). The consequence of dispersing in response to failure is that emigration will tend to occur from unsuccessful sites but not from successful sites. This can clearly be a mechanism for density-dependent dispersal but one that is likely to apply largely on local scales.

Condition-dependent dispersal

Dispersal as a consequence of poor individual condition can also be an example of density-dependent dispersal. For example, aphids provided with low-quality food or kept at high densities are more likely to give birth to offspring with wings (Dixon 1985). Ims and Hjermann (2001) review the literature on this subject and show that responses are inconsistent across species. Natal dispersal in crested tits *Parus cristatus* is later for birds from poor-quality habitat and these individuals have a lower probability of settling in good-quality habitat (Lens & Dhondt 1994), whereas experimental food provisioning of young Belding's ground squirrel *Spermophilus beldingi* resulted in higher emigration of females but had no effect on male dispersal (Nunes *et al.* 1997). Ims and Kjermann point out that optimal dispersal responses are likely to depend upon the scale at which population fluctuations take place (i.e. there is little point in dispersing to avoid high densities if densities are currently high everywhere). The time delay is also important. As conditions often depend upon the mother's nutritional state there will be a considerable delay between the dispersal event and the time at which condition is determined. This may result in individuals dispersing from sites that were once at high density but no longer are.

Consequences for population ecology

Density-dependent dispersal has a range of consequences for population ecology. The best known is the concept of sources and sinks (Pulliam 1988) in which, as a result of continuing immigration, populations are able to persist in areas in which they could otherwise not survive. Identifying sources and sinks will therefore be important for conservationists when identifying which populations are most in need of protection.

For sensible logistical reasons biologists generally pick areas of high population densities as study areas. The dynamics of such populations may be atypical and have higher growth rates and emigration rates than is usual. For example, density-dependent dispersal may have played an important role in the collapse of Atlantic cod *Gadus morhua* in the northwest Atlantic. Hutchings (1996) described how random trawl surveys identified a decline in low- and medium-density aggregations of cod but the sites with high-density aggregations were maintained until there were no low- or medium-density aggregations left, at which point the whole population collapsed. Hutchings' interpretation was that as population size declined, cod moved to the preferred high-density areas thus maintaining population sizes at these sites. As trawlers preferentially fished in the high-density areas, the population decline was not identified until the collapse occurred. The discrepancy between the numbers of cod fished and the trawl survey data hindered resolution of the biological evidence until the total population had almost collapsed and the Canadian government decided that the only realistic option was to shut the fishery, which resulted in enormous social and economic problems.

The density-dependent response of an entire population will depend upon the interaction between the within-site density dependence and the movement that occurs between sites. Under ideal free or ideal despotic distributions, as population size increases a wider range of sites will be used. This increased allocation of individuals across sites must reduce the density dependence. However, if the poorer sites have lower suitability but are used in contravention of the ideal free distribution then this will result in increased density dependence. The latter seems to be the case for black-tailed godwits (Gill *et al.* 2001a), where areas with lower survival and delayed arrival on the breeding grounds were increasingly used at high total population sizes.

Site quality may not only be related to food availability but may also be related to other factors such as hunting intensity, pollution, climatic conditions or disturbance (Gill *et al.* 1996). Thus, as populations increase a higher proportion may use less suitable sites and may then be prone to increased mortality resulting from hunting, pollution or severe weather.

A lack of density-dependent dispersal can also cause problems. Many species show strong natal philopatry (e.g. Thompson *et al.* 1994; Wheelwright & Mauck 1998). Kokko and Sutherland (2001) show that in the absence of other information or in the presence of misleading information, natal philopatry is an excellent rule for enabling individuals to respond to changes in habitat quality. This is because sites which produce surviving individuals are likely to be of reasonable quality. However, philopatry can potentially be a costly rule if populations increase rapidly. An extreme example is the lesser snow goose *Anser caerulescens* in which there is considerable habitat deterioration due to overgrazing within the centre of the colony resulting in reduced breeding success and growth of goslings (Ganter & Cooke 1998). Similarly, van der Jeugd (1999) showed that in the Baltic breeding grounds, the average breeding success of Barnacle geese *Branta leucopsis* was markedly lower in larger colonies than in small new colonies but that females in particular showed a considerable degree of philopatry. Spear *et al.* (1998) also showed that philopatric western gulls *Larus occidentalis* had a lower breeding success than did less philopatric individuals.

Conclusions

There are four major concepts describing density-dependent dispersal: ideal free distribution, ideal despotic distribution, sources and sinks and buffer effects. Each of these relies on density dependence operating within sites causing animals to disperse, through processes such as interference, competition, resource depletion and territorial exclusion. Evidence for density-dependent dispersal comes from a wide range of taxa at a range of spatial scales. Dispersal can result from exclusion, differential settlement of sections of a population, or movement in response to success or individual condition. The consequence of density-dependent dispersal can be use of sites of varying quality, hence understanding patterns of distribution and dispersal in response to density can have important implications for population ecology and conservation.

References

Bernstein, C., Kacelnik, A. & Krebs, J.R. (1988) Individual decisions and the distribution of predators in a patchy environment. *Journal of Animal Ecology* **57**, 1007–1026.

Bernstein, C., Kacelnik, A. & Krebs, J.R. (1991) Individual decisions and the distribution of predators in a patchy environment. II The influence of travel costs and the structure of the environment. *Journal of Animal Ecology* **60**, 205–225.

Birkhead, T.R. & Furness, R.W. (1985) Regulation of seabird populations. In: *Behavioural Ecology: the Ecological Consequences of Adaptive Behaviour* (eds R.H. Smith & R.M. Sibly), pp. 145–167. Blackwell Scientific Publications, Oxford.

Browne, S.J. (1996) Breeding skylarks survey. *BTO News* **207**, 11.

Carbone, C., DuToit, J.T. & Gordon, I.J. (1997) Feeding success in African wild dogs: does kleptoparasitism by spotted hyenas influence hunting group size? *Journal of Animal Ecology* **66**, 318–326.

Cresswell, W. (1997) Interference competition at low competitor densities in blackbirds *Turdus merula*. *Journal of Animal Ecology* **66**, 461–471.

Dhondt, A.A., Kempenaers, B. & Adriansen, F. (1992) Density-dependent clutch size caused by habitat heterogeneity. *Journal of Animal Ecology* **61**, 643–648.

Dixon, A.F.G (1985) *Aphid Ecology*. Blackie, Glasgow.

Ens, B.J., Weissing, F.J. & Drent, R.H. (1995) The despotic distribution and deferred maturity: two sides of the same coin. *American Naturalist* **146**, 625–650.

Favre, L., Balloux, F., Goudet, J. & Perrin, N. (1997) Female-biased dispersal in the monogamous mammal *Crocidura russula*: evidence from field data and microsatellite patterns. *Proceedings of the Royal Society of London, Series B* **264**, 127–132.

Flyn, L., Nol, E. & Zharikov, Y. (1999) Philopatry, nest-site tenacity and mate-fidelity of semipalmated plovers. *Journal of Avian Biology* **30**, 47–55.

Fonseca, D.M. & Hart, D.D. (1996) Density-dependent dispersal of black fly neonates is mediated by flow. *Oikos* **75**, 49–58.

Forero, M.G., Donazar, J.A., Blas, J. & Hiraldo, F. (1999) Causes and consequences of territory change and breeding dispersal distance in the black kite. *Ecology* **80**, 1298–1310.

Fretwell, S.D & Lucas Jr, H.J. (1970) On territorial behaviour and other factors influencing habitat selection in birds. *Acta Biotheoretica* **19**, 16–36.

Ganter, B. & Cooke, F. (1998) Colonial nesters in a deteriorating habitat: site fidelity and colony dynamics of lesser snow geese. *Auk* **115**, 642–652.

Gill, J.A., Sutherland, W.J. & Watkinson, A.R. (1996) A method to quantify the effects of human disturbance on animal populations. *Journal of Applied Ecology* **33**, 786–792.

Gill, J.A., Norris, K., Potts, P. *et al.* (2001b) The buffer effect and large-scale population regulation in migratory birds. *Nature* **412**, 436–438.

Gill, J.A., Sutherland, W.J. & Norris, K. (2001b) Depletion models can predict shorebird distribution at different spatial scales. *Proceedings of the Royal Society of London, Series B* **268**, 369–376.

Glas, P. (1960) Factors governing density in chaffinch in different types of wood. *Archives Néerlandaises de Zoologie* **13**, 466–472.

Goss-Custard, J.D. (1970) Feeding dispersion in some overwintering wading birds. In: *Social Behaviour in Birds and Mammals* (ed. J.H. Crook), pp. 3–35. Academic Press, London.

Goss-Custard, J.D. (1977) The ecology of the Wash. III. Density-related behaviour and the possible effects of a loss of feeding grounds on wading birds (Charadrii). *Journal of Applied Ecology* **14**, 721–739.

Goss-Custard, J.D. (1980) Competition for food and interference among waders. *Ardea* **68**, 31–52.

Gunderson, G., Johannesen, E., Andreassen, H.P. & Ims, R.A. (2001) Source–sink dynamics; how sinks affect demography of sources. *Ecology Letters* **4**, 14–21.

Hassell, M.P. & Varley, G.C. (1969) New inductive population model for insect parasites and its bearing on biological control. *Nature* **223**, 1133–1136.

Herzig, A.L. (1995) Effects of population density on long-distance dispersal in the goldenrod beetle *Trirhabda virgata*. *Ecology* **76**, 2044–2054.

Holling, C.S. (1959) Some characteristics of simple types of predation and parasitism. *Canadian Entomologist* **91**, 385–398.

Hudson, P.J., Dobson, A.P. & Newborn, D. (1998) Prevention of population cycles by parasites. *Science* **282**, 2256–2258.

Hutchings, J.A. (1996) Spatial and temporal variation in the density of northern cod and a review of hypotheses for the stock's collapse. *Canadian Journal of Fisheries and Aquatic Sciences* **53**, 943–962.

Ims, R.A. & Hjermann, D.Ø. (2001) Condition-dependent dispersal. In: *Dispersal* (eds J. Clobert, E. Danchin, A.A. Dhondt & J.D. Nichols), pp. 203–216. Oxford University Press, Oxford.

Jackson, D.B. (1994) Breeding dispersal and site-fidelity in 3 monogamous wader species in the Western Isles, UK. *Ibis* **136**, 463–473.

Kluyver, H.N. (1971) Regulation of numbers in populations of great tits (*Parus m. major*). In: *Dynamics of Numbers in Populations* (eds P.J. den Boer & G.R. Gradwell), pp. 507–523. Centre for Agricultural Publishing and Documentation, Wageningen.

Kluyver, H.N. & Tinbergen, L. (1953) Territory and the regulation of density in titmice. *Archives Neerlandaises de Zoologie* **10**, 265–289.

Kokko, H. & Sutherland, W.J. (1998) Optimal floating strategies: consequences for density dependence and habitat loss. *American Naturalist* **152**, 354–366.

Kokko, H. & Sutherland, W.J. (2001) Ecological traps in changing environments: ecological and evolutionary consequences of a behaviourally mediated Allee effect. *Evolutionary Ecology Research* **3**, 537–551.

Kokko, H., Sutherland, W.J. & Johnstone, R.A. (2001) The logic of territory choice: implications for conservation and source–sink dynamics. *American Naturalist* **157**, 459–463

Krebs, J.R. (1971) Territory and breeding density in the great tit, *Parus major* L. *Ecology* **52**, 2–22.

Lens, L. & Dhont, A.A. (1994) Effects of habitat fragmentation on the timing of crested tit *Parus cristatus* natal dispersal. *Ibis* **136**, 147–152.

Liley, D. (1999) *Predicting the consequences of human disturbance, predation and sea -level rise for ringed plover population size.* PhD thesis, University of East Anglia, Norwich.

Lindberg, M.S., Sedinger, J.S., Derksen, D.V. & Rockwell, R.F. (1998) Natal and breeding philopatry in a black, *Brant Branta bernicla nigricans*, metapopulation. *Ecology* **79**, 1893–1904.

Linden, M. (1991) Divorce in great tits—chance or choice—an experimental approach. *American Naturalist* **138**, 1039–1048.

Lurz, P.W.W., Garson, P.J. & Wauters, L.A. (1997) Effects of temporal and spatial variation in habitat quality on red squirrel dispersal behaviour. *Animal Behaviour* **54**, 427–435.

Merkt, J.R. (1981) An experimental study of habitat selection by the deer mouse, *Peromyscus maniculatus*, on Mandarte Island, BC. *Canadian Journal of Zoology* **59**, 589–597.

Moser, M.E. (1988) Limits to the numbers of grey plover *Pluvialis squatarola* wintering on British estuaries. *Journal of Applied Ecology* **25**, 473–486.

Nams, V.O. (1997) Density-dependent predation by skunks using olfactory search images. *Oecologia* **110**, 440–448.

Nilsson, S.G. (1987) Limitation and regulation of population density in the nuthatch (*Sitta europea*) breeding in natural cavities. *Journal of Animal Ecology* **56**, 921–937.

Norris, K. & Johnstone, I. (1998) Interference competition and the functional response of oystercatchers searching for cockles by touch. *Animal Behaviour* **56**, 639–650.

Nunes, S., Zugger, P.A., Engh, A.L., Reinhart, K.O. & Holekamp, K.E. (1997) Why do female Belding's ground squirrels disperse away from food resources? *Behavioural Ecology and Sociobiology* **40**, 199–207.

O'Connor, R.J. (1980) Pattern and process in great tit (*Parus major*) populations in Britain. *Ardea* **68**, 165–183.

Packer, C. (1979) Inter-troop transfer and inbreeding avoidance in *Papio anubis*. *Animal Behaviour* **27**, 1–36.

Paradis, E., Baillie, S.R., Sutherland, W.J. & Gregory, R.D. (1998) Patterns of natal and breeding dispersal in birds. *Journal of Animal Ecology* **67**, 518–536.

Parker, G.A. & Sutherland, W.J. (1986) Ideal free distribution when individuals differ in competitive ability: phenotype-limited ideal free models. *Animal Behaviour* **34**, 1222–1242.

Percival, S.M., Sutherland, W.J. & Evans, P.R. (1996) A spatial depletion model of the responses of grazing wildfowl to changes in availability of intertidal vegetation during the autumn and winter. *Journal of Applied Ecology* **33**, 979–993.

Percival, S.M., Sutherland, W.J. & Evans, P.R. (1998) Intertidal habitat loss and wildfowl numbers: applications of a spatial depletion model. *Journal of Applied Ecology* **35**, 57–63.

Potts, G.R. (1986) *The Partridge: Pesticides, Predation and Conservation.* Collins, London.

Pulliam, H.R. (1988) Sources, sinks and population regulation. *American Naturalist* **132**, 652–661.

Rowcliffe, J.M., Watkinson, A.R. Sutherland, W.J. & Vickery, J.A. (2001) The depletion of algal beds by geese: a predictive model and test. *Oecologia* **127** (3), 361–371.

Royama, T. (1971) Evolutionary significance of a predator's response to local differences in prey density. A theoretical study. In: *Dynamics of Populations* (eds P.J. den Boer & G.R. Gradwell), pp. 344–357. Centre for Agricultural Publishing and Documentation, Wageningen.

Spear, L.B., Pyle, P. & Nur, N. (1998) Natal dispersal in the western gull: proximal factors and fitness consequences. *Journal of Animal Ecology* **67**, 165–179.

Sutherland, W.J. (1982) Food supply and dispersal in the determination of wintering population levels of oystercatchers *Haematopus ostralegus*. *Estuarine, Coastal and Shelf Science* **14**, 223–229.

Sutherland, W.J. (1983) Aggregation and the ideal free distribution. *Journal of Animal Ecology* **52**, 821–828.

Sutherland, W.J. (1996) *From Individual Behaviour to Population Ecology.* Oxford University Press, Oxford.

Sutherland, W.J. & Allport, G.A. (1994) A spatial depletion model of the interaction between bean geese and wigeon with the consequences for habitat management. *Journal of Animal Ecology* **63**, 51–59.

Sutherland, W.J. & Anderson, C.W. (1993) Predicting the distribution of individuals and the consequences of habitat loss: the role of prey depletion. *Journal of Theoretical Biology* **160**, 223–230.

Sutherland, W.J. & Dolman, P.M. (1994) Combining behaviour and population dynamics with applications for predicting consequences of habitat loss. *Proceedings of the Royal Society of London, Series B* **25**, 133–138.

Sutherland, W.J. & Parker, G.A. (1985) The distribution of unequal competitors. In: *Behavioural Ecology: the Ecological Consequences of Adaptive Behaviour* (eds R.H. Smith & R.M. Sibly), pp. 255–278. Blackwell Scientific Publications, Oxford.

Sutherland, W.J. & Parker, G.A. (1992) The relationship between continuous input and interference models of ideal free distributions with unequal competitors. *Animal Behaviour* **44**, 345–355.

Thompson, P.S., Baines, D., Coulson, J.C. & Longrigg, G. (1994) Age at 1st breeding, philopatry and breeding site-fidelity in the lapwing *Vanellus vanellus*. *Ibis* **136**, 474–484.

Triplet, P., Stillman, R.A. & Goss-Custard, J.D. (1999) Prey abundance and the strength of interference in a foraging shorebird. *Journal of Animal Ecology* **68**, 254–265.

Van der Jeugd, H. (1999) *Life history decisions in a changing environment: a long-term study of a temperate barnacle goose population.* PhD thesis, Uppsala University, Uppsala.

Verhulst, S., Perrins, C.M. & Riddington, R. (1997) Natal dispersal of great tits in a patchy environment. *Ecology* **78**, 864–872.

Vickery, J.A., Sutherland, W.J., Watkinson, A.R., Lane, S.J. & Rowcliffe, M.R. (1995) Habitat switching by dark-bellied brent geese *Branta b. bernicla* (L.) in relation to food depletion. *Oecologia* **103**, 499–508.

Watkinson, A.R. & Sutherland, W.J. (1995) Sources, sinks and pseudosinks. *Journal of Animal Ecology* **64**, 126–130.

Wauters, L.A., Lens, L. & Dhondt, A.A. (1995) Variation in territory fidelity and territory shifts among red squirrel, *Sciurus vulgaris*, females. *Animal Behaviour* **49**, 187–193.

Wheelwright, N.T. & Mauck, R.A. (1998) Philopatry, natal dispersal and inbreeding avoidance in an island population of savannah sparrows. *Ecology* **79**, 755–767.

Yates, M.G., Stillman, R.A. & Goss-Custard, J.D. (2000) Contrasting interference and foraging dispersion in two species of shorebird (Charadrii). *Journal of Animal Ecology* **69**, 314–322.

Chapter 8
Seed dispersal: the search for trade-offs

Ken Thompson, Louise C. Rickard, Dunmail J. Hodkinson and Mark Rees

Introduction

Plants and animals cannot be good at everything, and resources allocated to one function cannot be allocated to another, meaning plants and animals show specialization in character expression. Crawley (1997) provides a useful list of trade-offs. Trade-offs may arise from a variety of sources, including the difficulty of combining two traits. For example, if allocation of resources to reproduction is roughly constant, then seed number trades off against seed size (Shipley & Dion 1992; Turnbull *et al.* 1999; Jakobsson & Eriksson 2000). A large number of large seeds might be advantageous, but biophysical constraints forbid this. Evolution of other traits may owe more to shared evolutionary functions. For example, if seed dispersal and seed persistence in the soil both reduce the perception of environmental variability, the existence of one trait may reduce the adaptive value of the other (Venable & Brown 1988).

The approach adopted here is to seek evidence for trade-offs that, on theoretical grounds, one would expect to find. We consider three suggested trade-offs involving seed dispersal, each with a theoretical rationale, but each with limited and often contradictory empirical evidence. Two concern regenerative traits only, while the third connects dispersal with mature plant traits.

Dispersal and colonization abilities

Here we face the difficulty of quantifying dispersal ability, or in the case of the third trade-off, colonization ability. A simple solution is to treat dispersules as either effectively dispersed (with morphological adaptations for dispersal) or not (lacking such adaptations; e.g. Willson *et al.* 1990). This approach allows different dispersal modes to be treated on an equal basis, since all seeds with dispersal adaptations are considered to be equally effectively dispersed. In reality, of course, they are not, but it is futile (in practice, even if not in theory) to determine whether a particular wind-dispersal adaptation is more effective than another animal-dispersal adaptation. A disadvantage is that morphology is not always a guide to dispersal mode. For example, some large tree seeds are effectively dispersed by birds and rodents, yet they lack any visible evidence of this dispersal mechanism. Fischer *et al.* (1996) have shown that seed morphology is a poor predictor of which species are dispersed on the fleece of sheep. Therefore, apart from the reanalysis of Lokesha *et al.* (1992)—see below—for which we employed binary dispersal data, we considered only wind dispersal. Empirical and theoretical studies support the assumption that terminal velocity is a good predictor of wind-dispersal effectiveness (Sheldon & Burrows 1973; Greene & Johnson 1989, 1996; Andersen 1992; Jongejans & Schippers 1999). Although dispersal is also influenced by height of release, this variable cannot be quantified for individual species since it is essentially a feature of the relative height of the target plant and of the surrounding vegetation (Grace 1981). In addition, the relative importance of terminal velocity and height of release varies with dispersal distance and plant size (Tackenberg 2001). Note also that we have not attempted to rank species by mean (or modal, median or maximum) dispersal distance, since these rankings would change with environmental conditions. For example, median dispersal distance is proportional to wind speed, but 99th percentile distances increase exponentially with wind speed (van Dorp *et al.* 1996; Jongejans & Schippers 1999). Accordingly, seed terminal velocity in still air, measured as described in Askew *et al.* (1997), was employed as a measure of wind dispersal capacity. Terminal velocity data are available for *c.* 280 species, mostly temperate European herbs and a few trees. We excluded: (i) species whose primary mode of dispersal is zoochory, and (ii) water plants lacking adaptations for wind dispersal, which we assume are dispersed primarily by water.

Colonization ability is less straightforward. It is not obvious which plant traits are the best predictors of the ability to reach unoccupied gaps. Tilman (1994) did not define colonization ability, but clearly assumed that it was correlated with allocation of resources to reproduction. Effective wind dispersal has been strongly linked to colonization ability in a number of studies (Grime 1986; Dwzonko & Loster 1992; del Moral & Wood 1993; Nakashizuka *et al.* 1993; Stöcklin & Bäumler 1996). Nevertheless, achieving greater dispersal distances does not necessarily mean reaching more unoccupied gaps. Producing more seeds (ignoring dispersal differences) must translate directly into reaching more unoccupied gaps. Seed production data are unavailable for most species, but seed production is predicted reasonably well by seed size. Therefore, we conducted separate analyses of the competition–colonization trade-off, using terminal velocity and seed size as predictors of colonization ability.

Trade-off 1: seed dispersal and chemical composition

Light wind-dispersed seeds are better dispersed than heavy seeds of similar morphology (e.g. Meyer & Carlson 2001), although the generality of this relationship across all dispersal modes is questionable (Hughes *et al.* 1994). Lighter seeds can be produced by reducing seed size or by storing fats rather than proteins or carbohydrates. Fats yield about twice the energy per unit mass of carbohydrates. Therefore, a plant could make seeds only half as heavy by replacing stored carbohydrate with fats, with no sacrifice of energy storage, although in practice the potential weight savings are less than this. The advantages for wind dispersal are obvious and many plant species rely almost exclusively on fats for energy storage in their seeds (Harwood 1980). However, there are costs: lipid synthesis is more energetically demanding than either protein or carbohydrate production (Slack & Browse 1985). It has also been suggested (Lokesha *et al.* 1992), although with little supporting evidence, that vulnerability to lipid peroxidation might reduce longevity of fatty seeds.

Lokesha *et al.* (1992) therefore suggested that seeds of species with active dispersal mechanisms should store more fat than seeds of species with no apparent dispersal mechanism. Their analysis of the large data set in Earle and Jones (1962) and Jones and Earle (1966) supported this prediction—on average, passively dispersed seeds contained about 10% fat, while wind-dispersed seeds contained about 25% fat. However, this analysis failed to account for the possible confounding effect of relatedness on species characters (Harvey & Pagel 1991). In this case the results may arise from relatively few evolutionary divergences between major plant groups. For example, much the two largest families in the Earle and Jones data set were Fabaceae (mostly low in fat and rarely wind dispersed) and Asteraceae (normally high in fat and frequently wind dispersed). Treating species as independent data points under these circumstances generates extraneous degrees of freedom and increases the probability of wrongly rejecting the null hypothesis.

Here we first reanalyse the data of Lokesha *et al.* (1992) using phylogenetically independent contrasts (PICs) (Purvis & Rambaut 1995). We are particularly interested in the hypothesis that the wind-dispersal syndrome and seed fat content are positively related, but we also examine the relationship between fat content and other dispersal modes. Second, we conduct a new analysis to test the association between wind-dispersal effectiveness and fat content, again using PICs and accounting for the effects of seed mass and life history. This new analysis combines published data and new data on wind dispersal.

Seed chemistry, phylogeny and life history

Seed chemical compositions (percentage lipid content) and seed masses were taken from Earle and Jones (1962) and Jones and Earle (1966), part of a very large screening programme of the US Department of Agriculture. The two papers report seed mass, fat and protein content for 1010 samples, representing 103 families and 759 species. The data set is taxonomically highly unbalanced, with Fabaceae, Asteraceae, Brassicaceae and Poaceae much the largest families. Thirty-two families were represented by a single sample. Overall, the data set is biased towards temperate herbs, but trees and tropical families are also represented. Where there was more than one value for a species, we used the mean. Seed lipid contents for several additional species were measured by the methods described in Hendry and Thorpe (1993). Air-dry seed masses for additional species were measured from fresh collections.

The phylogeny used was a modification of the Chase *et al.* (1993) series B phylogeny, incorporating additional genera and species at the family and genus nodes according to Woodland (1990), Cronquist (1988) and Stace (1997). In addition, the Asteraceae have been further resolved by tribes according to the Dollo majority rule consensus tree (Jansen *et al.* 1990). The entire phylogeny contains many multiple nodes but covers most species from the Earle and Jones data set.

Species belonging to tribes of the Asteraceae not represented in Jansen *et al.* (1990) were excluded from the analyses, rather than creating polytomies at the tribe level.

A phylogeny is more satisfactory than a taxonomy for two reasons. Firstly, the phylogeny is more likely to reflect accurately the evolutionary history of each species and, secondly, a phylogeny contains many more nodes, each of which represents an evolutionary divergence where the hypothesis can be tested. The availability of a phylogeny for the Asteraceae is particularly important as many genera within this family are wind dispersed and the phylogeny adds valuable detail. On the basis of published sources (Hitchcock 1971; Hitchcock & Cronquist 1973; McDougall 1973; Willis 1985; Clapham *et al.* 1987; Fernald 1987; Grime *et al.* 1988; Stace 1997), we divided our species into either four or five life history classes.

Data analysis

The data were analysed using PICs and also (in the case of the new analysis) congeneric comparisons. PICs estimate the relationship between character states at nodes within the phylogeny. The null hypothesis is that there is no correlation between changes in traits at the nodes. The contrasts were produced using the CAIC (comparative analysis by independent contrasts) package (Purvis & Rambaut 1995). We assumed that all branch lengths were the same, but analyses using estimated branch lengths produced qualitatively similar results. The evolutionary model underlying the independent contrast method assumes that lineages are equally likely to make the same proportional change in a continuous character. To meet this assumption seed masses were logarithmically transformed. Other data were untransformed. In the reanalysis of Lokesha *et al.* (1992), wind, water, animal and passive dispersal were included in the analysis as dummy variables. In almost all analyses, the contrasts were calculated using the CRUNCH algorithm from CAIC; this standardizes contrasts to produce homogeneity of variance. In analyses using a single dichotomous variable (for example, wind-dispersed or not) the BRUNCH algorithm was used.

In addition, congeneric contrasts were calculated from the same data sets using the same predictor variables as the PICs. These contrasts were not standardized. Whereas PICs allow assessment of relationships through evolutionary time, congeneric contrasts focus on the most recent speciation events, but have the advantage that no values are estimated.

The hypothesis being tested predicts negative relationships between seed fat content and terminal velocity. All contrasts involving continuous or dummy variables were analysed by regression forced through the origin, and where appropriate partial correlations were calculated for each predictor in the equation. These test for the presence of correlated evolution independently of other variables in the analysis. The outputs from the BRUNCH algorithm were analysed by sign tests.

Results: reanalysis of Lokesha *et al.*

An initial descriptive analysis, using species as independent data points, confirmed that wind-dispersed seeds contain a significantly higher proportion of fat than passively dispersed seeds (results not shown). The relationship between dispersal mode

Table 8.1 Single variable model analysing the effect of a transition to a given dispersal mode at nodes in the phylogeny on seed fat content.

n	Dispersal mode	Sign test result
58	Wind	ns
72	Animal	ns
32	Water	ns
45	Passive	ns

ns, not significant.

Table 8.2 Full model where dispersal mode is fitted as a dummy variable (see text for further details) and all the terms are included simultaneously, thus showing the effects of dispersal mode on seed fat content, over and above the effects of adult longevity and seed mass. All regressions were forced through the origin.

Log seed mass	Longevity	Dispersal mode	r^2 and *n* for full model
ns	ns	Wind ns	$r^2=0.016$ $n=233$
ns	ns	Animal ns	$r^2=0.0078$ $n=233$
ns	ns	Water ns	$r^2=0.014$ $n=233$
ns	ns	Passive (–) **	$r^2=0.039$ $n=233$

$^{**}P<0.01$; ns, not significant.

and fat content contrasts was examined in a single variable model, including only dispersal mode (Table 8.1), and in a full model, showing the effects of dispersal mode over and above the effects of adult longevity and seed mass (Table 8.2). In the single variable model, no dispersal mode was significantly associated with fat content. In the full model, passive dispersal was significantly associated with a reduction in fat content. However, the relationship explained only a very small proportion of the variance, and was strongly dependent on a small number of influential outliers. Since the number of contrasts in these analyses was 233, a relationship that hinges on only a few outliers must be treated with caution.

Results: new analysis

Seventy-four contrasts were calculated from 112 species for seed fat content, terminal velocity, seed mass and adult longevity, using seed fat content as the predictor variable. These contrasts were analysed using multiple regression of the equation

$$\text{Seed fat content} = \beta_1 * \text{Terminal velocity} + \beta_2 * \text{Seed mass} + \beta_3 * \text{Adult longevity} \qquad (8.1)$$

This regression failed to account for a significant proportion of the variation in fat content ($r^2 = 0.012$, $F_{(3,71)} = 0.294$, $P = 0.829$). Multiple regression of 12 congeneric contrasts calculated from this data set using the same model was not significant ($r^2 = 0.22$, $F_{(3,9)} = 0.84$, $P = 0.51$).

Discussion: data sources

The large data set employed by Lokesha *et al.* (1992) allows the calculation of a large number of contrasts. Nevertheless, its power is limited by the differentiation of species into a limited number of dispersal modes. In contrast, the new terminal velocity data allow us to make distinctions within the wind-dispersed category (Askew *et al.* 1997).

Discussion: energy storage in wind-dispersed seeds

The reanalysis implies strongly that the relationship found by Lokesha *et al.* (1992) arises from relatively few evolutionary divergences at the family level. For example, Lokesha *et al.* listed 240 species as wind dispersed, with a mean fat content of 25%, while the data set contains 144 species of Asteraceae, with a mean fat content of 29%. Other families with wind-dispersed members may have high (Onagraceae, 28%), moderate (Betulaceae, 16%) or low (Aceraceae, 4%) fat contents. There is very little evidence for any relationship between fat content and wind dispersal below the family level. For example, the fat content of seeds of Asteraceae tribes in which wind dispersal is absent (e.g. Anthemidae) is indistinguishable from that of tribes in which it is universal (e.g. Lactuceae, results not shown). Thus, although oily—and hence relatively light—seeds *may* have favoured the widespread adoption of wind dispersal shown by Asteraceae, the comparative approach cannot deal with such rare events.

This conclusion is supported by the new analysis. There is no evidence to support the hypothesis that evolution of greater wind-dispersal ability is associated with increased seed fat content, after allowing for the effects of seed size and adult longevity. Given the clear weight advantages of fat storage, why is there no evidence of any close link between wind dispersal and increased fat content? One possibility is that terminal velocity is more sensitive to quite small changes in size or structure of the pappus (e.g. pappus/achene diameter or variation in the solidity or 'openness' of the pappus; Sheldon & Burrows 1973) than to seed mass. Seed fat content may also be responding to other selective pressures; for example, germination in shaded or anaerobic situations may select for increased (Levin 1974) or decreased (Crawford 1989) seed fat content, respectively.

Trade-off 2: seed dispersal and persistence in soil

Theory predicts that plant traits that reduce the impact of environmental variation will show negative correlations (Venable & Brown 1988; Rees 1993, 1996), for example the hypothesized trade-off between seed persistence in the soil and dispersal effectiveness (Rees 1993, 1996). Here we conduct a new analysis to test the association between wind-dispersal effectiveness and longevity in soil, again using PICs and accounting for the effects of seed mass and life history. This analysis uses the phylogeny and wind-dispersal data described above, plus seed persistence data from Thompson *et al.* (1997).

Seed persistence

Thompson *et al.* (1997) classified every available species record for the northwest European flora as either type 1 (transient, persistence <1 year), type 2 (short-term persistent, persistence >1 year but <5 years) and type 3 (persistence >4 years). To calculate a single longevity index for each species, we recognized only two classes of record: transient (type 1) and persistent (types 2 and 3). Our longevity index is defined as:

$$\frac{\sum(\text{type } 2 + \text{type } 3)}{\sum(\text{type } 1 + \text{type } 2 + \text{type } 3)} \qquad (8.2)$$

which can take any value from 0 (no persistent records) to 1 (all records persistent). In order to reduce the problems caused by the notorious variability of seed persistence data, only species with at least 10 persistence estimates were used (Thompson *et al.* 1998). Although seed persistence is placed initially into one of three discrete categories, the effect of using the mean of at least 10 values is to create a continuous variable.

Results: seed longevity versus terminal velocity

A preliminary analysis, using species as independent data points, revealed a significant negative correlation between seed longevity and terminal velocity ($r = -0.16$, $n = 185$, $P = 0.03$). However, this relationship explained very little of the variation in the data ($r^2 = 0.027$).

Seventy-five contrasts were calculated from 116 species for seed longevity, terminal velocity, seed mass and adult longevity, using seed longevity as the predictor variable. These contrasts were analysed using multiple regression of the equation:

$$\text{Seed longevity} = \beta_1 * \text{Terminal velocity} + \beta_2 * \text{Seed mass} + \beta_3 * \text{Adult longevity} \qquad (8.3)$$

The complete model showed a significant relationship ($r^2 = 0.404$, $F_{(3,72)} = 16.256$, $P < 0.001$). Partial correlation showed significant negative correlations between seed longevity and both seed mass ($\beta_2 = -0.497$, $t_{(72)} = -4.76$, $P < 0.001$) and adult longevity ($\beta_3 = -0.455$, $t_{(72)} = -4.94$, $P < 0.001$) and no correlation between seed longevity and terminal velocity ($\beta_1 = 0.136$, $t_{(72)} = 1.29$, $P = 0.201$). Multiple regression of 16 congeneric contrasts calculated from this data set using the same model was not significant ($r^2=0.360$, $F_{(3,13)}=2.435$, $P=0.11$). Consequently, we conclude that these results provide little evidence of a trade-off between dispersal in time and dispersal in space. On the other hand, our results are consistent with the frequently reported association between seed persistence, small seed size and short life histories (Rees 1993, 1996; Thompson *et al.* 1998).

Discussion: trade-offs between dispersal in time and space

This analysis provides very little evidence for a trade-off between dispersal in time through accumulation of a persistent seed bank and dispersal in space by wind. This contradicts a previous analysis, which found that species with morphological adaptations for dispersal in space had less persistent seeds (Rees 1993). Several factors may contribute to the contrasting results of the two analyses. First, as previously noted, treating dispersal capacity as a binary variable produces few contrasts when using the PIC method (Rees 1996), seriously reducing the power of the analysis.

Second, it seems clear that particular circumstances favour either effective dispersal and a persistent seed bank, or neither. For example, exploitation of forest gaps by shade-intolerant fugitives seems to be improved by both persistence and dispersal (Marks 1983; Murray 1988). The dominant grasses of Mediterranean grasslands, which experience spatially and temporally predictable seasonal droughts, seem to require neither persistence nor dispersal (Russi *et al.* 1992; Peco *et al.* 1998).

Third, both persistence in soil, at least in cool temperate floras (Rees 1993, 1996; Thompson *et al.* 1993; Funes *et al.* 1999), and wind dispersal (Willson *et al.* 1990; Hughes *et al.* 1994; but see Rees 1996) are strongly linked to small seed size. Thus, selection for small seeds (for whatever reason) may increase capacity for both wind dispersal and persistence in soil.

Fourth, trade-offs depend on the existence of costs and benefits. Seed persistence in soil has clear costs (Venable & Brown 1988), but seeds may enter the seed bank only if conditions are unsuitable for seedling establishment. Seeds are remarkably good at assessing habitat favourability by responding to rainfall (Went 1949; Gutterman 1993), fire (Gimingham 1972; Lamont *et al.* 1991; Thanos & Rundel 1995; Keeley & Fotheringham 1998), intensity, duration and wavelength of light (Gorski *et al.* 1977; Cresswell & Grime 1981; Vàzquez-Yanes & Orozco-Segovia 1990; Milberg & Andersson 1997), temperature alternations (Thompson & Grime 1983) and soil nitrate concentration (Pons 1989). If these mechanisms are sufficiently discriminatory, and if the probability of death for seeds that germinate under unfavourable conditions is high, then the cost of persistence may be low. There may be costs associated with the capacity to persist in the soil, for example slower germination under favourable conditions and higher investment in defence. We tested the former by comparing germination rates of fresh seeds of species with transient, short-term and long-term persistent seeds. There was no significant difference between germination rates in the three groups, whether the data were analysed with species as independent data points or as PICs (Fig. 8.1).

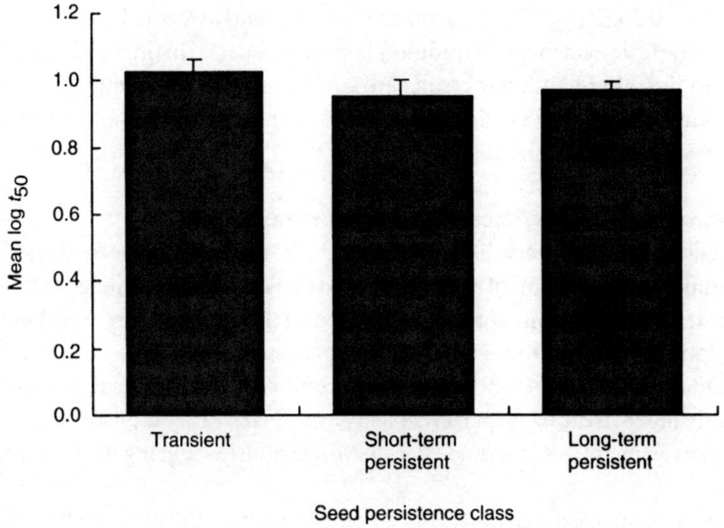

Figure 8.1 Mean log t_{50} (days to 50% of maximum germination) of fresh seeds in three persistence classes. Bars are standard errors. Classes are not significantly different ($F_{(2,238)} = 0.89, P = 0.41$). When analysed as PICs, with persistence classified into just two classes, the same data yielded 37 contrasts, with a mean t_{50} contrast not significantly different from 0 (mean ± sd = 0.00015 ± 0.149). (Based on Grime *et al.* 1981; Hodgson *et al.* 1995.)

Small seed size and burial may afford some protection against predation (Thompson 1987), but persistent seeds are still at risk from pathogens. Chemical defences of seeds against pathogens are not well understood, but phenols may be employed by many species (Hendry *et al.* 1994). Since phenols do not contain any 'costly' nutrients (e.g. nitrogen), it seems reasonable to assume that their cost to the plant is low. We therefore suspect that although seed persistence itself entails large costs, the capacity to persist in the soil may be relatively inexpensive.

Finally, the 'efficient dispersal' category of Rees (1993) included all modes of dispersal, while we explicitly considered only wind dispersal, which may reduce the range of seed dispersal and persistence and make it difficult to detect a significant relationship. In addition, structures that reduce terminal velocity (usually some form of parachute in herbs) are often relatively caducous and therefore may not impede burial. Hooks and barbs, in contrast, are usually tougher and probably do interfere significantly with burial. Furthermore, dispersal by animals (externally or internally) is compatible with larger seeds than wind dispersal (Willson *et al.* 1990; Hughes *et al.* 1994), and large seeds are rarely persistent (Rees 1993, 1996; Thompson *et al.* 1993). Therefore, it may be unrealistic to expect a universal trade-off between dispersal ability and persistence; the strength (or even existence) of such a trade-off may depend upon phylogeny, dispersal mode and habitat variables in a complex manner.

Trade-off 3: competition and colonization

Fundamental to the concept of reproductive allocation and much thinking on succession is a trade-off between vegetative and reproductive functions. If competitive ability depends on substantial allocation to stems, leaves and roots, good competitors will necessarily allocate proportionally fewer resources to flowers and seeds. This apparently inevitable trade-off will also be driven by natural selection; poor competitors that cannot coexist with competitive dominants, by way of shade tolerance or phenology, will be compelled to escape by evolving better dispersal or faster growth (Pacala & Rees 1998). Such a trade-off is also one of the few potential explanations for the existence of species-rich plant communities, although the coexistence of large numbers of species may require unreasonable assumptions about the nature of the trade-off, including instantaneous exclusion of the inferior competitor, even if the difference in competitive ability is slight (Adler & Mosquera 2000). Spatial subdivision of a homogeneous habitat into cells, each large enough to accommodate a single adult, can allow the coexistence of an unlimited number of competing species, as long as each has the appropriate competition–colonization trade-off (Tilman 1994). A widespread competition–colonization trade-off also has major implications for species extinctions; if the commonest species are the best competitors but poorest colonists, then they, rather than the rare species, might be the first to be driven to extinction by habitat destruction (Tilman *et al.* 1997).

How good is the evidence for a competition–colonization trade-off? Adult and regenerative traits are not independent; for example, there is a clear relationship between seed persistence and life history and between seed size and plant size (Salisbury 1942; Leishman *et al.* 1995). But do these broad statistical relationships imply a ubiquitous trade-off between traits related to competitive ability and those associated with colonization ability? The evidence that regenerative traits in general are tightly constrained by vegetative traits is poor. In 25 Canadian emergent macrophytes (Shipley *et al.* 1989), 300 species from Australian semiarid woodlands (Leishman & Westoby 1992), 100 species from a steep climatic gradient in Argentina (Díaz & Cabido 1997) and 273 British herbaceous species (Grime *et al.* 1987), coherent groups of species based on vegetative traits were not closely related to those based on seed traits. Nevertheless, unambiguous trade-offs can be seen, for example, during the course of secondary succession. Gleeson and Tilman (1990), in a study of 35 old fields in Minnesota, found a clear distinction between young fields (<15 years) and older fields (>15 years) (Fig. 8.2). Their data provided evidence of the trade-off between plant longevity and reproductive allocation: young fields, dominated by annuals, had high levels of reproductive allocation, while older fields, dominated by perennials, had much lower reproductive allocation. Ecologists have been aware of such successional trends in resource allocation since Clements' time, but they have no definite bearing on whether a competition–colonization trade-off plays any part in the organization of relatively stable, non-successional communities. Tilman (1994) suggested two tests for such a role. First, the component species of a community should show the appropriate trade-offs between competition and dispersal. Second, species abundances should often be limited by dispersal, and this limitation

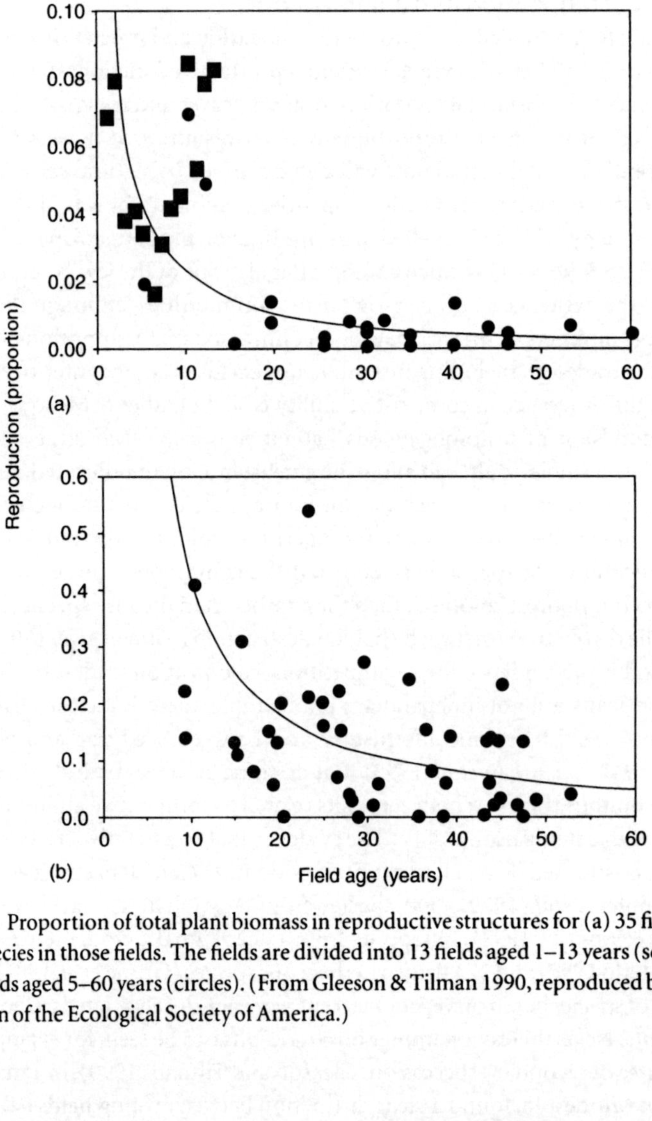

Figure 8.2 Proportion of total plant biomass in reproductive structures for (a) 35 fields and (b) the species in those fields. The fields are divided into 13 fields aged 1–13 years (squares) and 22 fields aged 5–60 years (circles). (From Gleeson & Tilman 1990, reproduced by permission of the Ecological Society of America.)

should be greater for the better competitors. That is, good competitors should show the largest increases in abundance when saturating densities of propagules are experimentally added (but see Bolker & Pacala 1999, pp. 589–590, for a discussion of the problems with this approach). Here we investigate these tests, both by reviewing the literature and by analysis of new data.

Competition and colonization

There is frequent disagreement about how both competition and colonization should be defined and measured. We adopted here a simple and conservative approach. First, we looked at two plant traits that should contribute to colonization

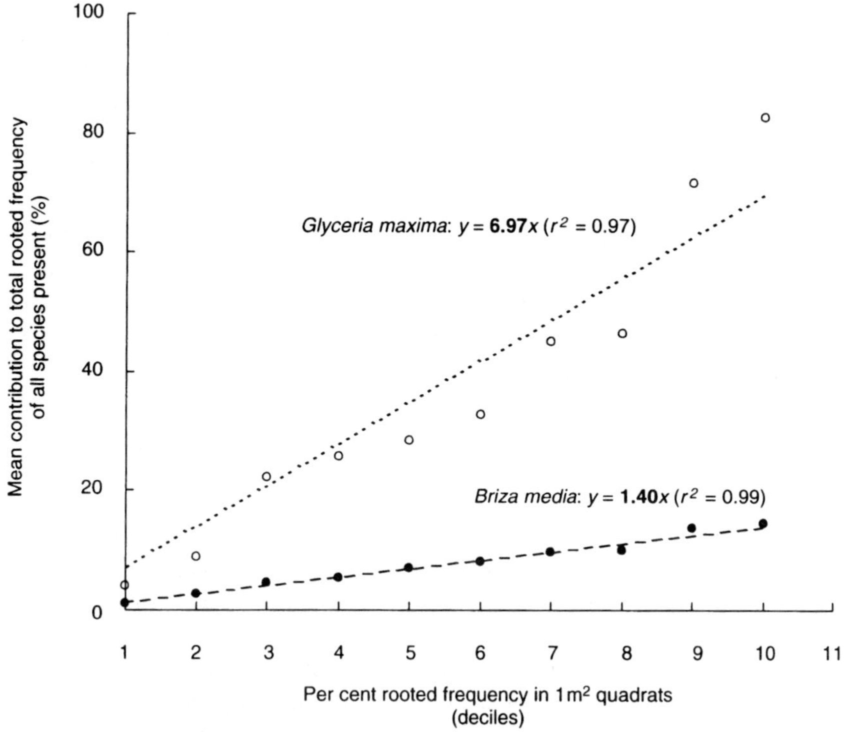

Figure 8.3 The basis of the method of estimating competitive ability: the steeper the slope, the greater the competitive ability of the species. See Hodgson *et al.* (1999) for assumptions and fuller details. (From Hodgson *et al.* 1999, reproduced by permission of Munksgaard International Publishers.)

ability—seed mass (related to seed production) and dispersal ability. As discussed above, we considered only wind dispersal.

Second, we adopted an operational definition of competitive ability: for a given species, this is the observed ability to suppress and exclude other species. Using a large database of field survey data (Grime *et al.* 1988), we plotted the total rooted frequency of all species in a quadrat against the proportion of rooted frequency accounted for by the target species, using mean values from many quadrats. The slope of the relation is a measure of the capacity of the target species to exclude competitors. For example, even when the weakly dominant grass *Briza media* occupies virtually all parts in a 1 m^2 quadrat, it commonly accounts for only one-tenth of the total abundance of all species present. The dominant grass *Glyceria maxima*, on the other hand, makes up about three-quarters of the total abundance in situations in which it appears throughout the quadrat (Fig. 8.3). This exercise was performed for 117 species. This operationally observed competitive ability was extended to a larger number of species by regressing it against a number of plant traits that were expected to be good predictors of competitive ability, including canopy height, lateral vegetative spread, specific leaf area (SLA) and leaf size. A regression equation including these predictor variables could then be used as a surrogate for competitive ability for any species. Finally, all species were placed in one of five categories of competitive ability (Hodgson *et al.* 1999).

Mean terminal velocity and mean log seed masses were calculated for the five classes of competitive ability, for all species for which data were available. Species with adaptations for zoochory and water plants lacking adaptations for wind dispersal were excluded from the terminal velocity analysis, but not from the seed mass analysis. There was a weak, non-significant tendency for good competitors to have lower terminal velocities, but a very strong positive relationship between seed mass and competitive ability (Fig. 8.4a, d). Note that analysis of PICs is not appropriate here—we are interested not in whether there is any evolutionary relationship between competitive and colonizing abilities, but in whether such a trade-off is responsible for structuring modern plant communities. Consequently, we repeated the analysis for two contrasted plant communities: a stable, largely perennial community (limestone pasture), and a community that is frequently destroyed and reassembled and contains many short-lived species (fallow arable). The results for both communities were very similar: there was no significant relationship between terminal velocity and competitive ability, but there was a significant tendency for seed mass to rise and then fall with increasing competitive ability (Fig. 8.4b, c, e, f).

Terminal velocity is therefore apparently unrelated to competitive ability, either in the British herbaceous flora as a whole or within individual communities. Seed mass, on the other hand, is strongly related to competitive ability. This is hardly surprising; plant height is the largest component of competitive ability (Gaudet & Keddy 1988), while Salisbury (1942) demonstrated clearly that tall plants have bigger seeds. Surprisingly, however, while this relationship is clear in the flora as a whole, it is not apparent within individual plant communities. There is, therefore, little evidence that a trade-off between competitive and colonizing ability is responsible for structuring British plant communities.

Seed-limited abundance

Although numerous studies have involved the experimental sowing of seeds into plant communities, many have used only a few species (e.g. Fowler 1986; Kelly 1989, reviewed in Turnbull *et al.* 2000). Here we examine the outcomes of several studies that have employed at least seven species. The evidence for seed limitation of abundance is strong; in most studies published to date some species, and sometimes the majority, are seed limited (Turnbull *et al.* 2000). The interesting question is therefore: which species are seed limited? Thompson and Baster (1992) sowed 10 Umbelliferae into tall grassland and found that four established successfully, including the most (*Heracleum sphondylium*) and the least (*Conopodium majus*) competitive species. Eriksson and Ehrlén (1992) sowed 14 woodland herbs and shrubs and found that nine were limited at least partly by seed availability, while five were not seed limited. There were clonal herbs, clonal dwarf shrubs and tall shrubs or small

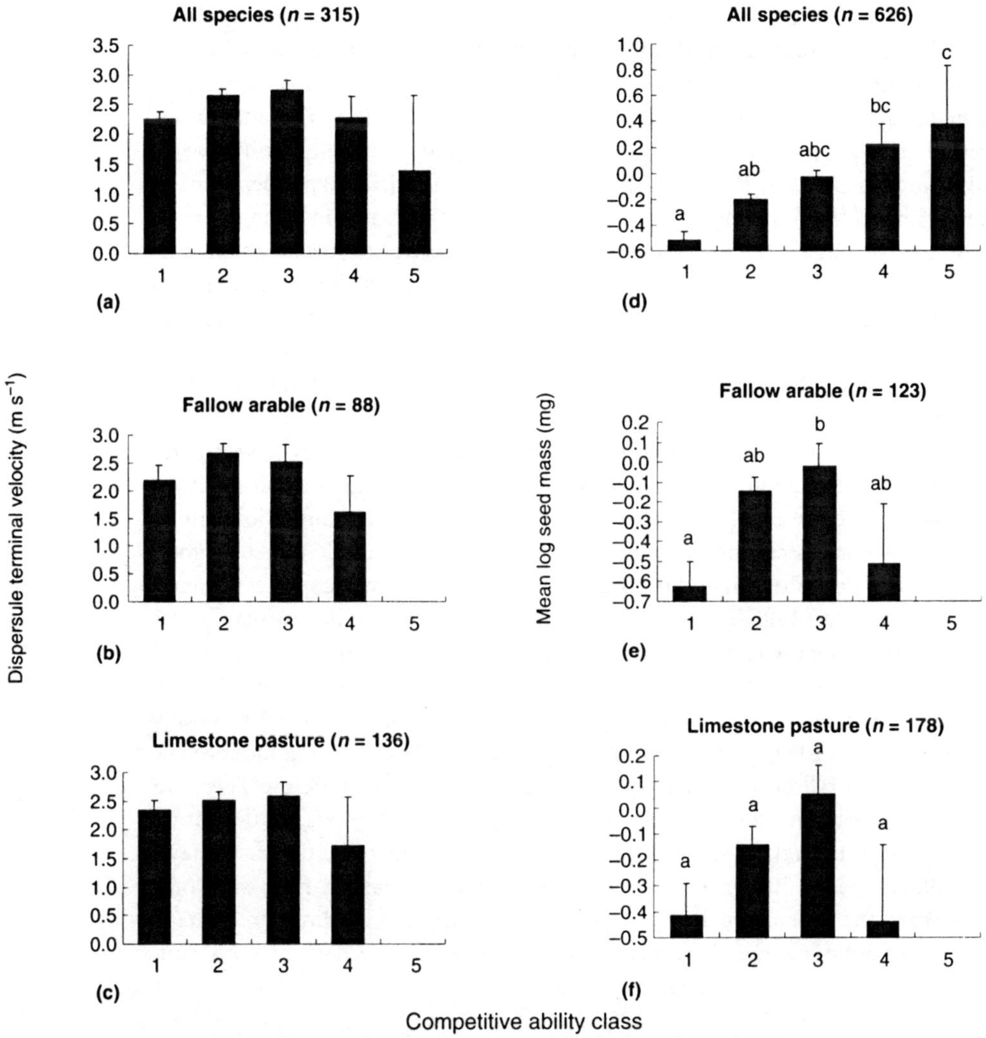

Figure 8.4 Relationship between propagule terminal velocity (a–c), seed size (d–f) and competitive ability in all species for which data are available, and in subsets of species from two contrasted habitats. All species were British native or naturalized herbs. Bars are standard errors. In no case is terminal velocity significantly different between competitive ability classes (one-way ANOVA). All relationships with seed size are significant (one-way ANOVA: d, $P<0.001$; e, $P<0.001$; f, $P=0.006$). Bars with the same letter are not significantly different at $P=0.05$ (Tukey HSD test). *n* is the number of species. Competitive ability classes containing only one species are not shown.

trees in both groups. Six out of seven perennial woodland herbs were found to be seed limited (Ehrlén & Eriksson 2000), while Kiviniemi and Eriksson (1999) found no relationship between competitive ability and adhesive dispersal in 17 grassland plants. Jakobsson and Eriksson (2000) sowed 50 grassland species and obtained evidence for seed-limited recruitment in 45 of them. Using the methods described previously, we were able to calculate the competitive ability of all 50 species (all but one were British natives), and then compare seed-limited recruitment (plants established/seeds sown) for the different classes of competitive ability. This analysis revealed no relationship between competitive ability and seed limitation (one-way ANOVA: $F_{(2,47)} = 1.77$, $P > 0.05$). In perhaps the most satisfactory experiment to date, Tilman (1997) added the seeds of 54 species to native prairie in Minnesota and monitored emerged seedlings for 4 years. The abundance of 19 species was significantly increased by sowing, but there was no evidence that the better competitors were more seed limited. The dominant grasses *Sorghastrum nutans* and *Andropogon gerardii*, by far the most abundant species at the start of the experiment, became less abundant by the end, both in plots where they were sown and in those where they were not.

There is thus no evidence, from perennial-dominated grasslands or woodland ground floras, that good competitors are more seed limited than poor competitors. Similarly, the review of Turnbull *et al.* (2000) concluded that seed limitation is more frequent in early successional habitats and species. Nevertheless, there is some good empirical evidence for this idea from communities of monocarpic species (Gross & Werner 1982; Rees 1995; Turnbull *et al.* 1999). In such communities, we may expect that the most intense competition is between seedlings, and that the role of seed size may be critical in determining the outcome of this competition (Black 1958; Goldberg & Landa 1991). Since large-seeded species suffer reduced fecundity (Turnbull *et al.* 1999), a competition–colonization trade-off among monocarpic species simply reflects the operation of the seed size/number trade-off. There is a positive correlation between seed size and competitive ability (Fig. 8.4d), but we might expect this relationship to have relatively little impact on the structure of perennial communities, because new individuals may establish from seed only rarely, and the main competitive interaction is not between seedlings. In Tilman's (1997) study, for example, there was no relationship between seed size and probability of establishment after 4 years.

Competition and colonization—conclusions

A trade-off between competitive ability and terminal velocity or seed size does not seem to be responsible for determining the composition of British plant communities, and the experimental evidence does not suggest that good competitors are more seed limited than poor competitors. Given the rather plausible theoretical arguments, we might ask why the evidence is so disappointing. Part of the answer may lie in the violation of two key assumptions that underpin the theoretical expectation (Tilman 1994). First, *plants are not immortal.* That is, plants suffer mortality, and this mortality leaves gaps that can be exploited (potentially) by propagules of other species. In practice many good competitors are clonal and therefore effectively immortal. Thus gaps are rapidly monopolized by clonal growth of the existing dominant and are not available to be colonized by seedlings (Hartman 1988; Tappeiner *et al.* 1991; Lieffers *et al.* 1993; Allison 1995; Oleskevich *et al.* 1996).

The second assumption is that *higher allocation to competitive structures (stems, leaves, roots) corresponds to lower reproductive allocation and hence poorer dispersal.* Since plants have limited resources at their disposal, the first part of this assumption is self-evidently true, but the second part is not. Although competitive dominants may have lower relative allocation to reproduction they may, by monopolizing the available resources, achieve very high absolute rates of seed production. For example, a single *Calluna vulgaris* individual may produce over 1.5 million seeds per year (Bullock & Clarke 2000), and *Chamerion angustifolium* and *Phragmites australis* average 76 000 and 1000 seeds per flowering shoot, respectively (Myerscough 1980; Haslam 1972; B. Ekstam, unpubl. data). *Typha* spp. may produce up to 700 000 seeds per inflorescence (Grace & Harrison 1986). It is therefore perhaps not surprising that many large clonal perennials (e.g. *Pteridium aquilinum, Typha latifolia, Phragmites australis, Epilobium hirsutum, Chamerion angustifolium, Petasites hybridus*) are able to combine formidable competitive ability with excellent colonizing ability.

Consequently, we believe that in productive habitats where disturbances are typically small, the competition–colonization mechanism is unlikely to maintain diversity. The competitive dominants eventually occupy most sites and disturbed areas therefore always contain their seeds, making it impossible for the poor competitors to persist via the competition–colonization mechanism; clonal growth, as described above, only makes this situation worse. In such systems coexistence may occur as a result of niche differentiation (Pacala & Rees 1998). One way this can occur is if early successional species have rapid growth under the resource-rich conditions associated with disturbances. These species can dominate recently disturbed sites even though they are eventually excluded by the competitive dominants. This successional niche hypothesis of Pacala and Rees (1998) allows coexistence even when the competitive dominants are not colonization limited.

Further, the model assumes that plants are all the same size. Relaxing this assumption, as in Bolker and Pacala (1999), leads to new mechanisms of coexistence—spatial exploitation strategies where plants rely on small size, rapid growth and local dispersal to exploit gaps between superior competitors. However, there is little evidence for the importance of these strategies at present.

Conclusions

Analyses of trade-offs are beset with problems concerning both methodology and data. If species are used as independent data points in the examination of an evolutionary hypothesis, there is a danger that a modern, ecological explanation is accepted for a pattern that has ancient phylogenetic origins. Failure to find the same relationship in an analysis of PICs does not mean the relationship does not exist, but interpretation needs to be more cautious.

Data issues are more serious. Treatment of dispersal ability as a binary variable is plainly unsatisfactory, but so are the obvious alternatives. The effectiveness of zoochory may depend strongly on the relative abundance of particular dispersers (e.g.

Wenny & Levey 1998), and the generally non-equilibrium relationship between plants and their animal dispersers may preclude any precise ranking of dispersal ability (Herrera 1998). Even wind dispersal, which can be modelled with varying degrees of sophistication, is unlikely ever to be reduced to a single unambiguous number. Variation in climate and topography, and whether one considers short- or long-range dispersal, may all affect relative dispersal abilities of different species (Tackenberg 2001). Colonization ability is even harder to pin down. Dispersal is clearly one component of colonization ability, yet more distant dispersal does not guarantee that more suitable microsites are encountered, nor does it guarantee a higher proportion of successful recruits (e.g. Augspurger & Kitajima 1992). Indeed, there are circumstances in which effective dispersal may be disadvantageous (Cody & Overton 1996). Seed output may be at least as important as dispersal ability, but seed output data are available for relatively few species. Intuitively, it seems reasonable to use seed mass as a surrogate (smaller seeds = more seeds), but this fails to account for the greater absolute quantity of resources available to large plants; we have referred above to the prodigious seed output of many large, highly competitive herbs.

Finally, of course, reaching a high proportion of suitable microsites, by effective dispersal or copious seed output, is itself only a part of colonization ability. The fate of seeds after dispersal may be equally important, and some of the best colonists may be those that combine effective seed dispersal with large seed size, and hence a high probability of seedling establishment. Since large seeds are generally associated with zoochory (Willson *et al.* 1990; Hughes *et al.* 1994); this suggests that there may be a complex relationship between dispersal mode, seed mass and dispersal-related trade-offs.

Therefore it would be wrong to generalize too far from the results presented above. They concern only one mode of dispersal (or colonization), measured in only one (or two) ways, in a largely temperate, herbaceous flora. Nevertheless, the failure to find much evidence of trade-offs between seed dispersal and persistence, or colonization and competition, suggests that they are not ubiquitous, or even widespread. Seed-sowing experiments also fail to support the existence of a competition–colonization trade-off. The very difficulty of defining and measuring dispersal or colonization ability suggests one reason for this slightly surprising conclusion. Dispersal and colonization are useful theoretical concepts, but models rarely suggest how they should (or even could) be measured empirically. Models also routinely fail to consider how dispersal ability might interact with other plant traits, such as life history, plant size and seed mass. Finally, models frequently ignore the rather messy effects of biological reality. The costs of seed persistence (in terms of reduced seed survival) or superior competitive ability (in terms of reduced reproductive allocation) are treated as fixed, when in fact such costs may be largely illusory. The devil, as with much else, is in the detail.

References

Adler, F.R. & Mosquera, J. (2000) Is space necessary? Interference competition and limits to biodiversity. *Ecology* **81**, 3226–3232.

Allison, S.K. (1995) Recovery from small-scale anthropogenic disturbances by northern California salt-marsh plant assemblages. *Ecological Applications* **5**, 693–702.

Andersen, M.C. (1992) An analysis of variability in seed settling velocities of several wind-dispersed Asteraceae. *American Journal of Botany* **79**, 1087–1091.

Askew, A.P., Corker, D., Hodkinson, D.J. & Thompson, K. (1997) A new apparatus to measure the rate of fall of seeds. *Functional Ecology* **11**, 121–125.

Augspurger, C.K. & Kitajima, K. (1992) Experimental studies of seedling recruitment from contrasting seed distributions. *Ecology* **73**, 1270–1284.

Black, J.N. (1958) Competition between plants of different initial seed sizes in swards of subterranean clover (*Trifolium subterraneum* L.) with particular reference to leaf area and the light microclimate. *Australian Journal of Agricultural Research* **9**, 299–318.

Bolker, B.M. & Pacala, S.W. (1999) Spatial moment equations for plant competition: understanding spatial strategies and the advantages of short dispersal. *American Naturalist* **153**, 575–602.

Bullock, J.M. & Clarke, R.T. (2000) Long distance seed dispersal by wind: measuring and modelling the tail of the curve. *Oecologia* **124**, 506–521.

Chase, M.W., Soltis D.E., Olmstead, R.G. *et al.* (1993) Phylogenetics of seed plants: an analysis of nucleotide sequences from the plastid gene rbcL[1]. *Annals of the Missouri Botanical Garden* **80**, 528–580.

Clapham, A.R., Tutin, T.G. & Moore, D.M. (1987) *Flora of the British Isles*, 3rd edn. Cambridge University Press, Cambridge.

Cody, M.L. & Overton, J.M. (1996) Short-term evolution of reduced dispersal in island plant populations. *Journal of Ecology* **84**, 53–61.

Crawford, R.M.M. (1989) *Studies in Plant Survival.* Blackwell Scientific Publications, Oxford.

Crawley, M.J. (1997) Life history and environment. In: *Plant Ecology*, 2nd edn (ed. M.J. Crawley), pp. 73–131. Blackwell Science, Oxford.

Cresswell, E. & Grime, J.P. (1981) Induction of a light requirement during seed development and its ecological consequences. *Nature* **291**, 583–585.

Cronquist, A. (1988) *The Evolution and Classification of Flowering Plants*, 2nd edn. New York Botanical Garden, New York.

Del Moral, R. & Wood, D.M. (1993) Early primary succession on the volcano Mount St-Helens. *Journal of Vegetation Science* **4**, 223–234.

Díaz, S. & Cabido, M. (1997) Plant functional types and ecosystem function in relation to global change. *Journal of Vegetation Science* **8**, 463–474.

Dwzonko, Z. & Loster, S. (1992) Species richness and seed dispersal to secondary woods in southern Poland. *Journal of Biogeography* **19**, 195–204.

Earle, F.R. & Jones, Q. (1962) Analysis of seed samples from 113 plant families. *Economic Botany* **16**, 221–250.

Ehrlén, J. & Eriksson, O. (2000) Dispersal limitation and patch occupancy in forest herbs. *Ecology* **81**, 1667–1674.

Eriksson, O. & Ehrlén, J. (1992) Seed and microsite limitation of recruitment in plant-populations. *Oecologia* **91**, 360–364.

Fernald, M.L. (1987) *Grey's Manual of Botany.* Timber Press, Portland.

Fischer, S.F., Poschlod, P. & Beinlich, B. (1996) Experimental studies of the dispersal of plants and animals on sheep in calcareous grasslands. *Journal of Applied Ecology* **33**, 1206–1222.

Fowler, N.L. (1986) Density-dependent population regulation in a Texas grassland. *Ecology* **67**, 545–554.

Funes, G., Basconcelo, S., Díaz, S. & Cabido, M. (1999) Seed size and shape are good predictors of seed persistence in soil in temperate mountain grasslands of Argentina. *Seed Science Research* **9**, 341–345.

Gaudet, C.L. & Keddy, P.A. (1988) A comparative approach to predicting competitive ability from plant traits. *Nature* **334**, 242–243.

Gimingham, C.H. (1972) *Ecology of Heathlands.* Chapman & Hall, London.

Gleeson, S.K. & Tilman, D. (1990) Allocation and the transient dynamics of succession on poor soils. *Ecology* **71**, 1144–1155.

Goldberg, D.E. & Landa, K. (1991) Competitive effect and response—hierarchies and correlated traits in the early stages of competition. *Journal of Ecology* **79**, 1013–1030.

Gorski, T., Gorska, K. & Nowicki, J. (1977) Germination of seeds of various species under leaf canopy. *Flora* **166**, 249–259.

Grace, J. (1981) Some effects of wind on plants. In: *Plants and their Atmospheric Environment* (eds J. Grace, E.D. Ford & P.G. Jarvis), pp. 31–56. Blackwell Scientific Publications, Oxford.

Grace, J.B. & Harrison, J.S. (1986) The biology of Canadian weeds. 73. *Typha latifolia* L., *Typha angustifolia* L. and *Typha* x *glauca* Godr. *Canadian Journal of Plant Science* **66**, 361–379.

Greene, D.F. & Johnson, E.A. (1989) A model of wind dispersal of winged or plumed seeds. *Ecology* **70**, 339–347.

Greene, D.F. & Johnson, E.A. (1996) Wind dispersal of seeds from a forest into a clearing. *Ecology* **77**, 595–609.

Grime, J.P. (1986) The circumstances and characteristics of spoil colonisation within a local flora. *Philosophical Transactions of the Royal Society of London, Series B* **314**, 637–654.

Grime, J.P., Mason, G., Curtis, A.V. *et al.* (1981) A comparative study of germination characteristics in a local flora. *Journal of Ecology* **69**, 1017–1059.

Grime, J.P., Hunt, R. & Krzanowski, W.J. (1987) Evolutionary physiological ecology of plants. In: *Evolutionary Physiology Ecology* (ed. P. Calow), pp. 105–125. Cambridge University Press, Cambridge.

Grime, J.P., Hodgson, J.G. & Hunt, R. (1988) *Comparative Plant Ecology: a Functional Approach to Common British Plants.* Unwin Hyman, London.

Gross, K.L. & Werner, P.A. (1982) Colonizing abilities of 'biennial' plant species in relation to ground cover: implications for their distributions in a successional sere. *Ecology* **63**, 921–931.

Gutterman, Y. (1993) *Seed Germination in Desert Plants.* Springer Verlag, Berlin.

Hartman, J.M. (1988) Recolonization of small disturbance patches in a New England salt marsh. *American Journal of Botany* **75**, 1625–1631.

Harwood, J.L. (1980) Plant acyl lipids: structure, distribution, and analysis. In: *Lipids: Structure and Function. Vol. 4. The Biochemistry of Plants: a Comprehensive Treatise* (eds P.K. Stumpf & E.E. Conn), pp. 2–55. Academic Press, New York.

Harvey, P.H. & Pagel, M.D. (1991) *The Comparative Method in Evolutionary Biology.* Oxford University Press, Oxford.

Haslam, S.M. (1972) Biological flora of the British Isles. *Phragmites communis* Trin. (*Arundo phragmites* L., ?*Phragmites australis* (Cav.) Trin. ex Steudel). *Journal of Ecology* **60**, 585–610.

Hendry, G.A.F. & Thorpe, P.C. (1993) Seed organic reserves. In: *Methods in Comparative Plant Ecology* (eds G.A.F. Hendry & J.P. Grime), pp. 196–199. Chapman & Hall, London.

Hendry, G.A.F., Thompson, K., Moss, C.J., Edwards, E. & Thorpe, P.C. (1994) Seed persistence: a correlation between seed longevity in the soil and ortho-dihydroxyphenol concentration. *Functional Ecology* **8**, 658–664.

Herrera, C.M. (1998) Long-term dynamics of Mediterranean frugivorous birds and fleshy fruits: a 12-year study. *Ecological Monographs* **68**, 511–538.

Hitchcock, A.S. (1971) *Manual of the Grasses of the United States,* Vols I and II. Dover Publications, Mineola, NY.

Hitchcock, C.L. & Cronquist, A. (1973) *Flora of the Pacific North West.* University of Washington Press, Seattle.

Hodgson, J.G., Grime, J.P., Hunt, R. & Thompson, K. (1995) *The Electronic Comparative Plant Ecology.* Chapman & Hall, London.

Hodgson, J.G., Wilson, P.J., Hunt, R., Grime, J.P. & Thompson, K. (1999) Allocating C-S-R plant functional types: a soft approach to a hard problem. *Oikos* **85**, 282–296.

Hughes, L., Dunlop, M., French, K. *et al.* (1994) Predicting dispersal spectra—a minimal set of hypotheses based on plant attributes. *Journal of Ecology* **82**, 933–950.

Jakobsson, A. & Eriksson, O. (2000) A comparative study of seed number, seed size, seedling size and recruitment in grassland plants. *Oikos* **88**, 494–502.

Jansen, R.K., Holsinger, K.E., Michaels, H.J. & Palmer, J.D. (1990) Phylogenetic analysis of chloroplast DNA restriction site data at higher taxonomic levels: an example from the Asteraceae. *Evolution* **44**, 2089–2105.

Jones, Q. & Earle, F.R. (1966) Chemical analyses of seeds. II. Fat and protein content of 759 species. *Economic Botany* **20**, 127–155.

Jongejans, E. & Schippers, P. (1999) Modelling seed dispersal by wind in herbaceous species. *Oikos* **87**, 362–272.

Keeley, J.E. & Fotheringham, C.J. (1998) Mechanism of smoke-induced seed germination in a postfire chaparral annual. *Journal of Ecology* **86**, 27–36.

Kelly, D. (1989) Demography of short-lived plants in chalk grassland. II. Control of mortality and fecundity. *Journal of Ecology* **77**, 770–784.

Kiviniemi, K. & Eriksson, O. (1999) Dispersal, recruitment and site occupancy of grassland plants in fragmented habitats. *Oikos* **86**, 241–253.

Lamont, B.B., Le Maitre, D.C., Cowling, R.M. & Enright, N.J. (1991) Canopy seed storage in woody plants. *Botanical Review* **57**, 278–311.

Leishman, M.R. & Westoby, M. (1992) Classifying plants into groups on the basis of associations of individual traits—evidence from Australian semi-arid woodlands. *Journal of Ecology* **80**, 417–424.

Leishman, M.R., Westoby, M. & Jurado, E. (1995) Correlates of seed size variation—a comparison among 5 temperate floras. *Journal of Ecology* **83**, 517–529.

Levin, D.A. (1974) The oil content of seeds: an ecological perspective. *American Naturalist* **108**, 193–206.

Lieffers, V.I., Macdonald, S.E. & Hogg, E.H. (1993) Ecology of and control strategies for *Calamagrostis canadensis* in boreal forest sites. *Canadian Journal of Forest Research* **23**, 2070–2077.

Lokesha, R., Hegde, S.G., Uma Shaanker, R. & Ganeshaiah, K.N. (1992) Dispersal mode as a selective force in shaping the chemical composition of seeds. *American Naturalist* **140**, 520–525.

Marks, P.L. (1983) On the origin of the field plants of the northeastern United States. *American Naturalist* **122**, 210–228.

McDougall, W.B. (1973) *Seed Plants of Northern Arizona.* Museum of Northern Arizona, Flagstaff, AZ.

Meyer, S.E. & Carlson, S.L. (2001) Achene mass variation in *Ericameria nauseosus* (Asteraceae) in relation to dispersal ability and seedling fitness. *Functional Ecology* **15**, 274–281.

Milberg, P. & Andersson, L. (1997) Seasonal variation in dormancy and light sensitivity in buried seeds of eight annual weed species. *Canadian Journal of Botany* **75**, 1998–2004.

Murray, K.G. (1988) Avian seed dispersal of 3 neotropical gap-dependent plants. *Ecological Monographs* **58**, 271–298.

Myerscough, P.J. (1980) Biological flora of the British Isles. *Epilobium angustifolium* L. *Journal of Ecology* **68**, 1047–1074.

Nakashizuka, T., Iida, S., Suzuki, W. & Tanimoto, T. (1993) Seed dispersal and vegetation development on a debris avalanche on the Ontake volcano, Central Japan. *Journal of Vegetation Science* **4**, 537–542.

Oleskevich, C., Shamoun, S.F. & Punja, Z.K. (1996) The biology of Canadian weeds. 105. *Rubus strigosus* Michx, *Rubus parviflorus* Nutt, and *Rubus spectabilis* Pursh. *Canadian Journal of Plant Science* **76**, 187–201.

Pacala, S.W. & Rees, M. (1998) Models suggesting field experiments to test two hypotheses explaining successional diversity. *American Naturalist* **152**, 729–737.

Peco, B., Ortega, M. & Levassor, C. (1998) Similarity between seed bank and vegetation in Mediterranean grassland: a predictive model. *Journal of Vegetation Science* **9**, 815–828.

Pons, T.L. (1989) Breaking of seed dormancy by nitrate as a gap detection mechanism. *Annals of Botany* **63**, 139–143.

Purvis, A. & Rambaut, A. (1995) Comparative analysis by independent contrasts (CAIC): an Apple Macintosh application for analysing comparative data. *Computer Applications in Biosciences* **11**, 247–251.

Rees, M. (1993) Trade-offs among dispersal strategies in the British flora. *Nature* **366**, 150–152.

Rees, M. (1995) Community structure in sand dune annuals: is seed weight a key quantity? *Journal of Ecology* **83**, 857–864.

Rees, M. (1996) Evolutionary ecology of seed dormancy and seed size. *Philosophical Transactions of the Royal Society of London, Series B* **351**, 1299–1308.

Russi, L., Cocks, P.S. & Roberts, E.H. (1992) Seed bank dynamics in a Mediterranean grassland. *Journal of Applied Ecology* **29**, 763–771.

Salisbury, E.J. (1942) *The Reproductive Capacity of Plants.* G. Bell & Sons, London.

Sheldon, J.C. & Burrows, F.M. (1973) The dispersal effectiveness of the achene-pappus units of selected Compositae in steady winds with convection. *New Phytologist* **72**, 665–675.

Shipley, B. & Dion, J. (1992) The allometry of seed production in herbaceous angiosperms. *American Naturalist* **139**, 467–483.

Shipley B., Keddy, P.A., Moore, D.R.J. & Lemky, K. (1989) Regeneration and establishment strategies of emergent macrophytes. *Journal of Ecology* **77**, 1093–1110.

Slack, C.R. & Browse, J. (1985) Synthesis of storage lipids in developing seeds. In: *Seed Physiology*, Vol. 1 (ed. D.R. Murray), pp. 209–243. Academic Press, London.

Stace, C.A. (1997) *New Flora of the British Isles*, 2nd edn. Cambridge University Press, Cambridge.

Stöcklin, J. & Bäumler, E. (1996) Seed rain, seedling establishment and clonal growth strategies on a glacier foreland. *Journal of Vegetation Science* **7**, 45–56.

Tackenberg, O. (2001) *Methoden zur Bewertung gradueller Unterschiede des Ausbreitungspotentials von Pflanzarten: Modellierung des Windausbreitungpotentials und regelbasierte Ableitung die Fernausbreitungpotentials.* PhD thesis, University of Marburg, Germany.

Tappeiner, J., Zasada, J., Ryan, P. & Newton, M. (1991) Salmonberry clonal and population-structure—the basis for a persistent cover. *Ecology* **72**, 609–618.

Thanos, C.A. & Rundel, P.W. (1995) Fire-followers in chaparral: nitrogenous compounds trigger seed germination. *Journal of Ecology* **83**, 207–216.

Thompson, K. (1987) Seeds and seed banks. *New Phytologist* **106** (Suppl.), 23–34.

Thompson, K. & Baster, K. (1992) Establishment from seed of selected Umbelliferae in unmanaged grassland. *Functional Ecology* **6**, 346–352.

Thompson, K. & Grime, J.P. (1983) A comparative study of germination responses to diurnally-fluctuating temperatures. *Journal of Applied Ecology* **20**, 141–156.

Thompson, K., Band, S.R. & Hodgson, J.G. (1993) Seed size and shape predict persistence in soil. *Functional Ecology* **7**, 236–241.

Thompson, K., Bakker, J.P. & Bekker, R.M. (1997) *The Seed Banks of North West Europe: Methodology, Density and Longevity.* Cambridge University Press, Cambridge.

Thompson, K., Bakker, J.P., Bekker, R.M. & Hodgson, J.G. (1998) Ecological correlates of seed persistence in soil in the NW European flora. *Journal of Ecology* **86**, 163–169.

Tilman, D. (1994) Competition and biodiversity in spatially structured habitats. *Ecology* **75**, 2–16.

Tilman, D. (1997) Community invasibility, recruitment limitation, and grassland biodiversity. *Ecology* **78**, 81–92.

Tilman, D., Lehman, C.I. & Yin, C. (1997) Habitat destruction, dispersal, and deterministic extinction in competitive communities. *American Naturalist* **149**, 407–435.

Turnbull, L.A., Rees, M. & Crawley, M.J. (1999) Seed mass and the competition/colonization trade-off: a sowing experiment. *Journal of Ecology* **87**, 899–912.

Turnbull, L.A., Crawley, M.J. & Rees, M. (2000) Are plant populations seed-limited? A review of seed sowing experiments. *Oikos* **88**, 225–238.

Van Dorp, D., van den Hoek, W.P.M. & Daleboudt, C. (1996) Seed dispersal capacity of six perennial grassland species measured in a wind tunnel at varying wind speed and height. *Canadian Journal of Botany* **74**, 1956–1963.

Vàzquez-Yanes, C. & Orozco-Segovia, A. (1990) Ecological significance of light controlled seed germination in two contrasting tropical habitats. *Oecologia* **83**, 171–175.

Venable, D.L. & Brown, J.S. (1988) The selective interactions of dispersal, dormancy, and seed size as adaptations for reducing risk in variable environments. *American Naturalist* **131**, 360–384.

Wenny, D.G. & Levey, D.J. (1998) Directed seed dispersal by bellbirds in a tropical cloud forest. *Proceedings of the National Academy of Sciences of the USA* **95**, 6204–6207

Went, F.W. (1949) Ecology of desert plants. II. The effect of rain and temperature on germination and growth. *Ecology* **30**, 1–13.

Willis, J.C. (1985) *A Dictionary of the Flowering Plants and Ferns*, 8th student edn. Cambridge University Press, Cambridge.

Willson, M.F., Rice, B.L. & Westoby, M. (1990) Seed dispersal spectra: a comparison of temperate plant communities. *Journal of Vegetation Science* **1**, 547–562.

Woodland, D.W. (1990) *Contemporary Plant Systematics.* Prentice-Hall, Paramus, NJ.

Chapter 9

Manipulating your host: host–pathogen population dynamics, host dispersal and genetically modified baculoviruses

Greg Dwyer and Rosie S. Hails

Introduction

Mathematical models have played an important role in improving our understanding of the effects of dispersal on population dynamics (Tilman & Kareiva 1997). Perhaps the best known example of a model that combines dispersal and population dynamics is Fisher's model, which can be used to describe the spread of a population that experiences density-dependent growth (Murray 1993):

$$\frac{\partial N}{\partial t} = rN\left(1 - \frac{N}{K}\right) + D\frac{\partial^2 N}{\partial x^2} \qquad (9.1)$$

where N is population density, r is the population growth rate, K is the carrying capacity, D is a dispersal parameter, t is time and x is spatial location on a line. If in this model we introduce a small density of organisms at $x = 0$, then the model will eventually show a wave of population density that moves across the landscape with constant shape and rate of spread (Fig. 9.1). The rate of spread of this wave, c, depends on both the population growth rate r and the dispersal parameter D, according to:

$$c = 2\sqrt{rD} \qquad (9.2)$$

In this chapter, we use models to describe how the dispersal of insects affects the population dynamics of insect–baculovirus interactions, and especially how dispersal affects the likelihood that genetically engineered virus strains may out-compete wild-type strains. A key difference between our work and Fisher's, however, is that the population dynamics of insect–pathogen interactions are quite complicated, so we focus more on population dynamics than on dispersal. Nevertheless, dispersal is a critical part of the story.

Basic biology of baculoviruses

Insects are attacked by a wide range of pathogens, including baculoviruses, a large diverse group of double-stranded DNA viruses. Most baculoviruses have been isolated from lepidopteran larvae, and can be easily formulated as environmentally

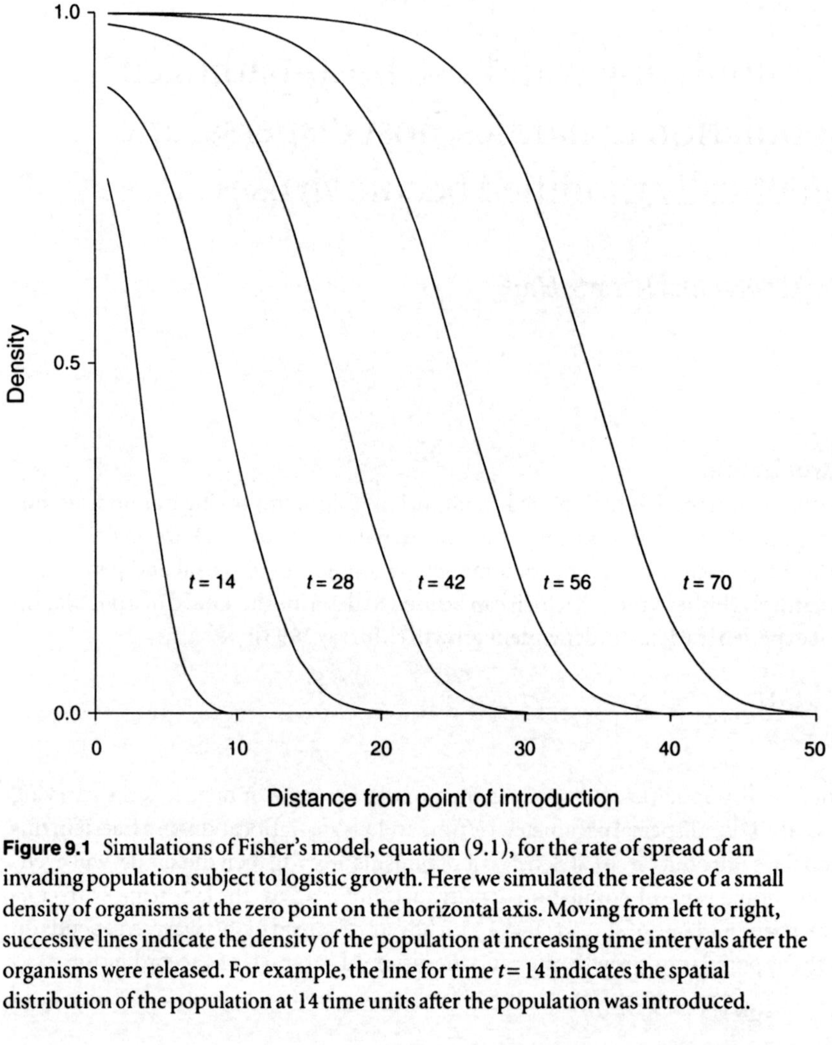

Figure 9.1 Simulations of Fisher's model, equation (9.1), for the rate of spread of an invading population subject to logistic growth. Here we simulated the release of a small density of organisms at the zero point on the horizontal axis. Moving from left to right, successive lines indicate the density of the population at increasing time intervals after the organisms were released. For example, the line for time $t = 14$ indicates the spatial distribution of the population at 14 time units after the population was introduced.

benign biopesticides against lepidopteran pests (Moscardi 1999). The outcome of an established infection is usually death in the larval stage, resulting in the host disintegrating and liberating the virus in the form of occlusion bodies over the foliage of the host insect's food plant. The infection is then transmitted to other hosts when they accidentally consume the occlusion bodies. Baculoviruses generally take days or weeks to kill their hosts, however, and this slow speed of kill has limited their use in pest management. Some baculoviruses, therefore, have been genetically modified to kill more rapidly by the insertion of genes for insect-specific toxins (Stewart *et al.* 1991; Tomalski & Miller 1991; Possee *et al.* 1997; van Beek & Hughes 1998) or by the deletion of existing viral genes that limit the speed of kill (O'Reilly & Miller 1991).

Because of the possibility of unanticipated environmental impacts of releasing genetically modified organisms into the environment, before genetically modified baculoviruses can be used widely in agriculture, we must understand how they might affect the ecology of the interaction between pest insects and wild-type viruses (Dushoff & Dwyer 2001; Hails *et al.* 2001). To reach this understanding, we must in turn understand basic baculovirus ecology.

For example, a key issue is how the occlusion bodies survive in the environment. Occlusion bodies of the best known group of baculoviruses, the nuclear polyhedrosis viruses, are typically comprised of a large number of virus particles surrounded by a proteinaceous coat. This coat reduces but does not eliminate the degrading effect of UV, microbial enzymes and so forth, so that the survival of occlusion bodies varies among microhabitats. For example, on leaf surfaces, occlusion bodies may degrade in a matter of hours, but just below the soil surface, or in the crevices of tree bark, they may persist in active form for years (Jaques 1974; Podgwaite *et al.* 1979). This kind of persistence in the environment is believed to be an important mechanism by which pathogens persist between epidemics, but it has been little studied (but see Kukan & Myers 1999). Mechanisms by which these pathogens disperse and persist between host populations are similarly poorly understood.

In contrast, considerable progress has been made in understanding dispersal within populations. As we will describe, advances in molecular genetics mean that we can now attribute defined virus phenotypes to specific virus genes, including genes involved in the transmission of a pathogen from one infected larva to another, which is essentially local dispersal. Also, simple models have been used to successfully predict within-population dispersal, assuming horizontal transmission and dispersal of infected larvae as the primary mechanisms of dispersal. We are therefore in a position to tentatively predict the effects of releasing genetically engineered baculoviruses into the environment.

Dispersing from the inside out: genes and individuals

Destruction of the host

Because the transmission process in baculoviruses requires that progeny occlusion bodies be consumed by host larvae, so in turn the release of occlusion bodies from infected cadavers and the dispersal of baculoviruses from their hosts usually involves liquefaction of the cadaver (Fig. 9.2). At the molecular level, these processes have been studied intensively using the nuclear polyhedrosis virus of *Autographa californica*. Chitinase and cathepsin, the latter a cysteine protease, are two baculovirus gene products that facilitate the release of virus after death by causing lysis of the cadaver (Slack *et al.* 1995; Hawtin *et al.* 1997). A third gene, p10, codes for a protein involved in cell lysis. On destruction of the cell, chitinase and cathepsin, previously locked up in the endoplasmic reticulum and cytoplasmic vesicles, respectively, are released to act upon the insect cuticle (Thomas *et al.* 1998). Both chitinase and cathepsin are required to ensure that the insect integument is dissolved

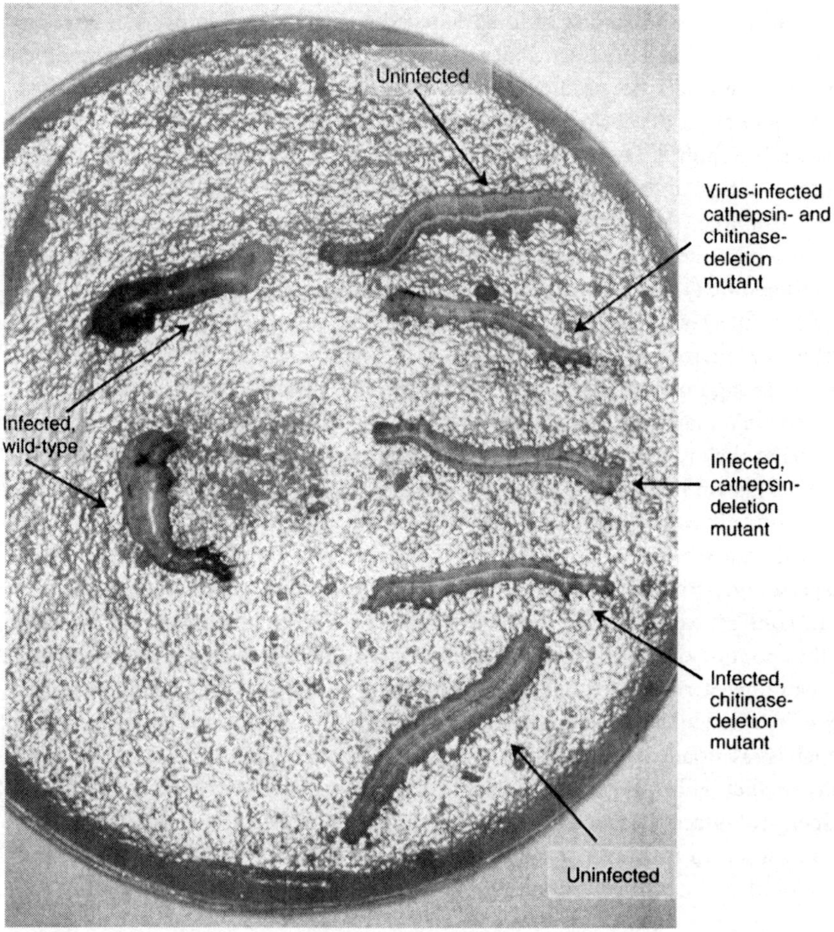

Figure 9.2 Larvae of *Autographa californica*, alfalfa looper, some of which were infected either with wild-type or genetically modified nuclear polyhedrosis virus strains. Larvae on the left-hand side of the picture were infected with the wild-type. The top and bottom larvae on the right-hand side of the picture are uninfected. The remaining larvae on the right-hand side were infected with one of three deletion mutants: in order from top to bottom, the mutants are a double-deletion of both cathepsin and chitinase genes, a cathepsin-deletion mutant and a chitinase-deletion mutant. Notice that larvae that were infected with the wild-type were liquefied. In contrast, larvae that were infected with either of the deletion mutants or with the double-deletion mutant remained intact, and so resemble the uninfected larvae. Both cathepsin and chitinase genes are thus required for liquefaction.

(Fig. 9.2), as they code for enzymes which attack alternating layers of proteinaceous and chitinous components in the insect integument. Specific baculovirus genes can thus be linked directly to phenotypic traits that have a major influence on the dispersal of the virus in the field.

The link between genotype and phenotype, however, is not always so straightforward. For example, the granulosis virus that infects *Plodia interpunctella*, called PiGV, fails to liquefy its host after death. Instead, the host remains largely intact, and transmission occurs when cadavers are cannibalized (Knell *et al.* 1998). Because the ancestral state is to cause liquefaction, the lack of liquefaction in this virus suggests that there must be fitness costs of liquefaction. One possible advantage of *not* liquefying is that remaining in a cadaver gives the occlusion bodies a better chance of surviving.

Dispersal of infected insects and 'Wipfelkrankheit'

A key issue in what follows is that baculoviruses also influence the behaviour of their hosts, although the fitness advantages of this manipulation for the virus are poorly understood. For example, when larvae of *Lymantria monacha* are infected with a nuclear polyhedrosis virus, they migrate to the top branches of pine trees, earning the disease the name 'Wipfelkrankheit' or 'tree top disease' (Wahl 1909, cited by Entwistle & Evans 1985). Similar behaviour has been confirmed and quantified for other host–baculovirus relationships (Vasconcelos *et al.* 1996; Goulson 1997). For example, *Mamestra brassicae* larvae dying from a nuclear polyhedrosis virus infection are far more likely to be found at the apex of cabbage leaves than anywhere else on the plant (Vasconcelos *et al.* 1996). Larvae of *M. brassicae* also exhibit higher rates of dispersal when infected with the virus than do uninfected larvae (Vasconcelos *et al.* 1996; Goulson 1997). Although the mechanism underlying this behaviour is unknown, the enhanced mobility appears by the fourth day postinfection, when the virus has begun to replicate in nerve cells, fat bodies, muscles and ganglia (Keddie *et al.* 1989). It has therefore been hypothesized that the viral infection disrupts the metabolism of the nervous system, leading to alterations in behaviour (Entwistle & Evans 1985).

One possible fitness benefit for the virus of upward host movement is that occlusion bodies that are released from higher on the host plant may be more widely available to infect the healthy insects below. Indeed, experiments have shown that when larvae of the gypsy moth *Lymantria dispar* are infected with a nuclear polyhedrosis virus, they are found slightly higher in the canopy than larvae that are uninfected (Murray & Elkinton 1992). Further experiments have shown that, in the presence of simulated rainfall, infected gypsy moth cadavers on branches that are higher up can infect insects on branches below them (D'Amico & Elkinton 1995), and infected gypsy moths are on average found slightly higher in the canopy than are uninfected gypsy moths (Murray & Elkinton 1992). Changes in the movement rate of infected hosts relative to uninfected hosts may, therefore, change the transmission rate of baculoviruses. This is important for risk assessment because insects infected with genetically modified baculoviruses often behave differently to insects infected with wild-type baculoviruses.

Genetically modified baculoviruses

A major goal of genetic modification of baculoviruses has been to shorten their speed of kill by the insertion of toxin genes that are insect-specific, and thus harmless to vertebrates, and cause death rapidly (e.g. Possee *et al.* 1997). For baculoviruses, shorter time to death means a reduced generation time, so that all else being equal genetically modified baculoviruses would be expected to out-compete wild-type strains, in turn altering the natural interaction between host and pathogen. The genes in question, however, also influence at least two other life-history parameters of the virus. First, the yield of virus from infected cadavers, measured as the number of occlusion bodies per infected larva, is usually reduced because viruses that kill faster have less time in which to replicate (Kunimi *et al.* 1996; Burden *et al.* 2000; Hernandez-Crespo *et al.* 2001). Moreover, in addition to differences in yield, some gene products further reduce transmission because they cause paralysis before death (Hails *et al.* 2002). This paralysis can cause as much as 60% of the recombinant-infected larvae to fall from the foliage (Cory *et al.* 1994; Hoover *et al.* 1995), as opposed to wild-type-infected larvae that die and disintegrate directly on foliage. Wild-type strains thus have a fitness advantage that may compensate for the fitness disadvantage of killing more slowly. Because of these fitness trade-offs, it is uncertain whether genetically modified baculoviruses will out-compete wild-type strains in the field. One approach to predicting the effect of releasing genetically engineered baculoviruses is therefore to use parameterized mathematical models. In the next section, we review some relevant models.

Models of insect–pathogen interactions

As we have described, understanding the effects of dispersal on insect–baculovirus interactions requires a thorough understanding of the ecology of these diseases. In reviewing models of host–pathogen interactions, we therefore begin by focusing on disease ecology with no consideration of dispersal. One of the earliest host–pathogens models is from Kermack and McKendrick's (1927) work, in which the dynamics of an epidemic follow the equations:

$$\frac{dS}{dt} = -\beta SI \tag{9.3}$$

$$\frac{dI}{dt} = \beta SI - (\alpha + \nu)I \tag{9.4}$$

where S is the density of uninfected (S for 'susceptible') hosts, I is the density of infected hosts, β is the rate at which the disease is transmitted horizontally, and $\alpha + \nu$ is the rate at which infected hosts are removed from the population due to death (α) or recovery (ν). A basic assumption of this model is that the epidemic occurs over a short enough timescale that host reproduction is essentially irrelevant to the dynamics of the epidemic. This assumption is reasonable for the human epidemics that Kermack and McKendrick focused on, whereas for many animal diseases, one is interested in whether the disease can regulate the host population and whether or not the disease might drive host population fluctuations. To extend the Kermack and

McKendrick approach to pathogens of insects, Anderson and May (1981) therefore added host reproduction, as well as a new equation that describes the density of the occlusion bodies outside the hosts, to get:

$$\frac{dS}{dt} = a(S+I) - bS - \nu SP \tag{9.5}$$

$$\frac{dI}{dt} = \nu SP - (\alpha + b)I \tag{9.6}$$

$$\frac{dP}{dt} = \lambda I - \nu(S+I)P - \mu P \tag{9.7}$$

Here P is the density of occlusion bodies in the environment, a is the reproductive rate of the hosts, b is the rate at which hosts die of non-disease causes, ν is the rate at which the disease is transmitted horizontally, α is the rate at which infected hosts die of the disease, λ is the rate at which infected hosts produce occlusion bodies and μ is the rate at which occlusion bodies break down in the environment.

Anderson and May (1981) attempted to use this model to explain the ecology of the interactions between forest-defoliating insects and their pathogens. Many of the insects in question show long-period, large-amplitude fluctuations in density, for reasons that historically have been unclear (Varley *et al.* 1973). Anderson and May's model similarly shows long-period, low-amplitude fluctuations for apparently realistic values of some of the parameters (some parameters were unknown). Whether or not this explanation is correct is debatable (Myers 1988, 1993; Elkinton & Leibhold 1990; Bowers *et al.* 1993; Hunter & Dwyer 1998; Hunter & Price 1998). Nevertheless, the occurrence of baculoviruses in many defoliators, and the propensity of the model to show cycles, together suggest that insect–baculovirus population dynamics are likely to be complex.

More realistic models

Most mathematical disease models have been constructed to provide qualitative understanding, rather than predictive ability *per se*. Such models are said to provide 'strategic' guidance, rather than the 'tactical' guidance provided by complex simulation models that contain hundreds or thousands of parameters (Onstad *et al.* 1990; Briggs *et al.* 1995). The improved fit to data given by models with more parameters, however, is often outweighed by the increased uncertainty associated with the inclusion of the additional parameters (Hilborn & Mangel 1997; Burnham & Anderson 1999). In light of such statistical considerations, simple models may provide useful descriptions of field data on insect–pathogen interactions. This is a key issue in evaluating the environmental impacts of releasing modified baculoviruses, because in predicting the impact of releasing a given strain, we need to know whether or not the model that we are using is supported by the data. It turns out, however, that the models that we have discussed can be useful with only minor modifications. The resulting models accurately describe data for a particular insect–virus interaction, that between the gypsy moth and its nuclear polyhedrosis virus. Nevertheless, the biology of different baculoviruses is similar enough that it is likely that these models will provide a useful description for a number of different insect–pathogen systems.

To begin with, Dwyer and colleagues (Dwyer & Elkinton 1993; Dwyer *et al.* 1997) combined features of the Anderson and May and Kermack and McKendrick models to arrive at a description of the dynamics of single baculovirus epidemics. First, the Anderson and May model assumes that transmission occurs continuously, where as most baculoviruses infect only larvae and, like the gypsy moth, many insect hosts have only one generation per year. One may therefore ignore host reproduction, as in the Kermack and McKendrick model, while retaining the assumption of the Anderson and May model that infection occurs when hosts encounter infectious cadavers rather than infected hosts. Also, because epidemics are terminated by host pupation, the time between infection and death is often a substantial fraction of the length of the epidemic, and so cannot be ignored. Finally, variability among gypsy moths in their susceptibility to the virus can be substantial (Dwyer *et al.* 1997), and must also be considered. The resulting model is then:

$$\frac{dS}{dt} = -\bar{\nu}SP\left(\frac{S(t)}{S(0)}\right)^{V} \tag{9.8}$$

$$\frac{dP}{dt} = \bar{\nu}P(t-\tau)S(t-\tau)\left(\frac{S(t-\tau)}{S(0)}\right)^{V} - \mu P \tag{9.9}$$

where τ is the time between infection and death, $\bar{\nu}$ is the mean transmission rate at the beginning of the epidemic and V is the squared coefficient of variation of the distribution of transmission rates ν. Because transmission determines the rate at which healthy hosts become infected, the average transmission rate is a measure of the average susceptibility of the population. This model is derived from a more complex model in which one explicitly keeps track of the entire distribution of transmission rates (Dwyer *et al.* 2000). In the more complex model, both the mean and variance of the distribution of transmission rates decline as the epidemic progresses. To arrive at equations (9.8) and (9.9), we have assumed that, as the mean and variance of the distribution decline, the coefficient of variation (the ratio of the variance to the square of the mean) remains the same. This assumption provides a good approximation to the more complex model if the initial distribution of transmission rates follows a gamma distribution, and a somewhat poorer approximation if the initial distribution follows distributions with fatter tails (J. Dushoff, pers. comm.). Because we do not have enough data to reject the gamma distribution, and because the approximation is convenient, in what follows we use the approximation.

This model reproduces the data at both the large scale of hectares at which epidemics occur in naturally occurring populations, and at the small scale of centimetres at which the host actually encounters the virus. Specifically, the average horizontal transmission parameter $\bar{\nu}$ and the coefficient of variation V can be estimated from small-scale experiments involving no more than $0.5\ m^2$ of foliage, and the resulting estimate is close to estimates calculated by fitting the model directly to large-scale data (Dwyer *et al.* 1997). In fact, using point estimates of $\bar{\nu}$ and V as calculated from small-scale transmission experiments in the model gives excellent predictions of the timing and intensity of natural virus epidemics (Dwyer *et al.* 2000). The model, including the approximation, therefore appears to be accurate enough to be usable in risk assessment.

For risk assessment, however, we must also consider timescales longer than a single epidemic. The two additional important features of the biology of the gypsy moth virus that therefore must be considered are first that there is only one larval generation per year, so the virus must survive for many months between larval periods. The major means by which the virus survives is by contaminating egg masses (Murray & Elkinton 1990), thereby infecting larvae as they emerge from the egg in the spring. Second, because infected hosts do not reproduce (Murray *et al.* 1991), only surviving hosts can contribute to the population in the following year. Our long-term model is thus:

$$N_{t+1} = \lambda N_t(1 - I(N_t, Z_t)) \tag{9.10}$$

$$Z_{t+1} = fN_t I(N_t, Z_t) + \gamma Z_t \tag{9.11}$$

Here N_t and Z_t are the densities of healthy hosts and infectious cadavers at the beginning of generation t, respectively, f is the survival rate of cadavers produced in the previous season's epidemic, and γ is the survival rate of the cadavers produced in earlier generations. Because we suspect that the virus generally does not survive for more than a single generation, in what follows we assume that $\gamma = 0$. The function $I(N_t, Z_t)$ is the fraction of hosts that become infected, as described by equations (9.8) and (9.9). Like the Anderson and May model, the model equations (9.10) and (9.11) show long-period, large-amplitude fluctuations in host population density that roughly reproduce long time series of forest-defoliator abundances for realistic parameter values (Fig. 9.3). The main difference is that, as we have described, equations (9.10) and (9.11) have also been tested with experimental transmission data and with observational epidemic data (Dwyer *et al.* 2000). We therefore have confidence that this model can play a role in risk assessment of genetically engineered baculoviruses.

An important methodological issue in what follows is that if we use equations (9.8) and (9.9) to describe the fraction of hosts surviving the epidemic $I(N_t, Z_t)$, the resulting model can be understood only through computer simulations. By approximating $I(N_t, Z_t)$, however, we arrive at a model that can be understood not just through simulations but also through mathematical analyses that permit a deeper understanding. The approximate version of $I(N_t, Z_t)$ that we use is given by the following implicit expression (Dwyer *et al.* 2000):

$$1 - I = \left(1 + \frac{V\bar{\nu}}{\mu}(N_t I + \eta Z_t)\right)^{(1/V)} \tag{9.12}$$

Note that we have introduced the additional parameter η, the initial average transmission rate (and hence susceptibility) of hatching larvae relative to later stage

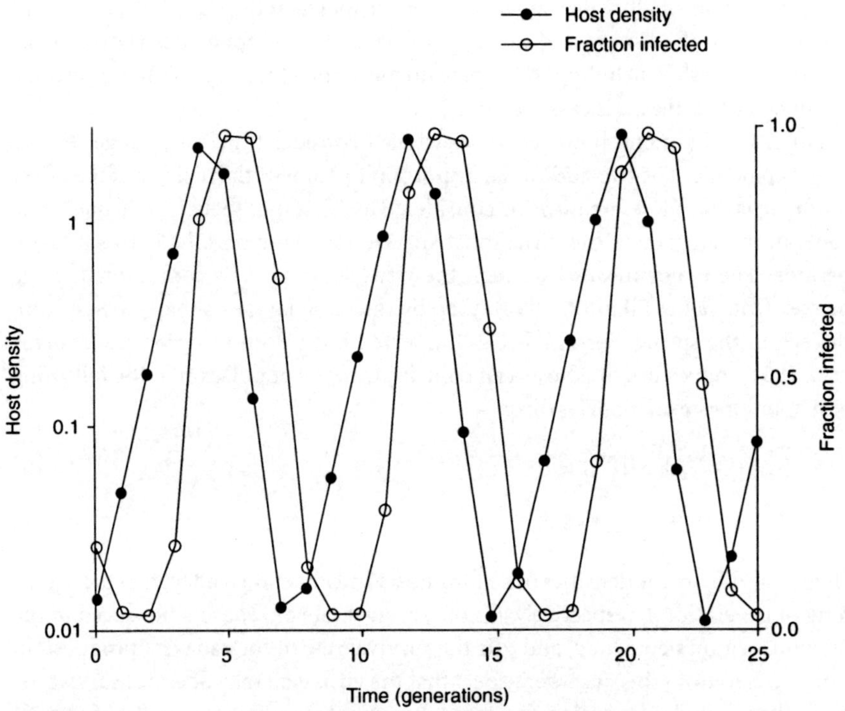

Figure 9.3 A simulation of a rescaled version of the host–pathogen population dynamic model from equations (9.10–9.12). Rescaling according to $\hat{N} = \dfrac{\mu N}{\nu}$ and $\hat{Z} = \dfrac{\mu Z}{\nu}$ allows us to replace equations (9.10–9.12) with the equations:

$$\hat{N}_{t+1} = \lambda \hat{N}_t (1 - I) \tag{9.26}$$

$$\hat{Z}_{t+1} = \phi \hat{N}_t I \tag{9.27}$$

$$1 - I = \left(1 + V\left(\hat{N}_t I + \hat{Z}_t\right)\right)^{1/V} \tag{9.28}$$

where the compound parameter $\phi = f\eta$, or the product of the probability of the pathogen's overwinter survival f times the susceptibility of hatching larvae relative to that of later instar larvae η. Here fecundity $\lambda = 40$, heterogeneity in susceptibility $V = 0.4$ and $\phi = 0.01$.

larvae, to allow for the relatively higher susceptibility of hatching larvae. To arrive at equation (9.12), we have made the seemingly unrealistic assumption that time $t \to \infty$, as if the annual epidemics run forever. Although this assumption is superficially unrealistic, the approximation turns out to be a good one as long as the length of the epidemic is at least five times the speed of kill τ. Consequently, for epidemics that run for at least 5τ time units, the dynamics of equations (9.10–9.12) are quite similar whether we use equations (9.8) and (9.9) or equation (9.12). In what follows, we will use this approximation to help us understand models of insect–baculovirus interactions that incorporate dispersal.

Allowing for spatial structure and dispersal

The models that we have described can be extended easily to allow for spatial structure and dispersal. To begin with, one of us (G.D.) has extended the reaction–diffusion approach exemplified by equation (9.1) to insect–baculovirus interactions (Dwyer 1992, 1994). The resulting model shows waves of infection that travel through the insect population, much as in Fig. 9.1, and again the spread rate increases linearly with the square root of the host dispersal rate. An interesting difference, however, is that small changes in transmissibility have a much larger impact on the transmission rate of an insect virus than increases in reproductive rate have on the rate of spread of a logistically growing population. Details of the ecology of the host–pathogen interaction thus have a big impact on spread.

Travelling wave models, however, do not consider how dispersal may affect population dynamics. In this section, we therefore introduce models that show how dispersal can change the population dynamics of host–pathogen interactions (for an overview, see Wilson 1998). Our ultimate interest is in understanding how the modification of insect dispersal by genetically engineered baculoviruses affects insect–baculovirus population dynamics and competition among strains. First, however, we begin by attempting to understand how dispersal can affect population dynamics in the case of a single pathogen strain. A convenient place to start is with the predator–prey models presented by Kot (1989), which can be written as:

$$N_{t+1}(x) = \int k_N(x-y) f_N(N_t(y), Z_t(y))dy \tag{9.13}$$

$$Z_{t+1}(x) = \int k_Z(x-y) f_Z(N_t(y), Z_t(y))dy \tag{9.14}$$

Here f_N and f_Z are functions that describe the growth of the prey (N) and predator (Z) populations, respectively. Like the Fisher model, this model assumes that space is one-dimensional. The functions $k_N(x-y)$ and $k_Z(x-y)$, the 'dispersal kernels', therefore describe the distance that the prey and predator organisms disperse. For example, $k_N(x-y)$ is the probability that a prey organism will disperse from spatial location x to spatial location y, which is a function only of the distance between x and y.

We can use equation (9.13) and (9.14) to describe insect–baculovirus interactions by using the right-hand sides of equations (9.10) and (9.11) to represent the growth functions $f_N(N_t(y), Z_t(y))$ and $f_Z(N_t(y), Z_t(y))$, with equation (9.12) describing the fraction of hosts that become infected each year. This is equivalent to assuming that dispersal and virus infection occur at different points in the life cycle. This assumption is a good approximation to the biology of the gypsy moth–virus interaction in North America and Europe, where females are flightless so that the most important round of dispersal occurs when hatching larvae balloon off the egg mass. We further assume that the dispersal kernels are described by so-called 'double exponential' distributions, according to:

$$k_N = \frac{1}{2}\alpha \exp(-\alpha|x-y|) \tag{9.15}$$

$$k_Z = \frac{1}{2}\beta \exp(-\beta|x-y|) \tag{9.16}$$

where α and β are the average dispersal distances of hosts N and pathogens Z, respectively.

This model is interesting because it allows us to clearly see the effects of allowing for dispersal in insect–pathogen models such as equations (9.10)–(9.12). As described in greater detail by Kot (1989), under some circumstances dispersal can directly cause population fluctuations, a phenomenon known as 'diffusive instability' (Murray 1993). For our purposes, however, the interesting feature of this model is that dispersal can turn the cycles seen in Fig. 9.3 into stable, equilibrium population dynamics. Specifically, for equations (9.13) and (9.14), long-period cycles can only occur if:

$$\hat{k}_N \hat{k}_Z D > 1 \tag{9.17}$$

where $\hat{k}_N$ and $\hat{k}_Z$ are the Fourier transforms of the dispersal kernels k_N and k_Z, respectively, and D is the determinant of the Jacobian or 'community matrix' for the non-spatial model (Kot 1989). For the specific case of negative exponential kernels, stability is guaranteed if the ratio between the mean dispersal rates of host and pathogen α/β is:

$$\frac{\alpha}{\beta} > 2D - 1 + \frac{1}{2}\sqrt{D(D-1)} \tag{9.18}$$

$$\frac{\alpha}{\beta} < 2D - 1 - \frac{1}{2}\sqrt{D(D-1)} \tag{9.19}$$

Figure 9.4 shows simulations of the models with and without dispersal, to show a case in which dispersal damps out population fluctuations. Note that this effect is independent of the boundary conditions or the spatial extent of the population (Kot 1989). To see the effects of dispersal over a wider range of parameter values, Fig. 9.5 shows ranges of parameter values in which limit cycles or point equilibria occur. As one of us has shown elsewhere (Dwyer *et al.* 2000), for the model without dispersal, stability is more likely to occur if fecundity is low or if heterogeneity in susceptibility is high. For the model with dispersal, however, differences in the average dispersal distance of host and pathogen are also stabilizing. An important caveat is that these results hold for the model in which equation (9.12) describes the annual epidemic. Because this model is more stable than the model in which the annual epidemic is shorter (Dwyer *et al.* 2000), for more realistic models the effect of dispersal may be to turn unstable cycles into stable cycles. Irrespective of these details, however, the strong effects of dispersal on models with a single pathogen strain suggests that it is important to consider dispersal when assessing the risk of releasing engineered baculoviruses.

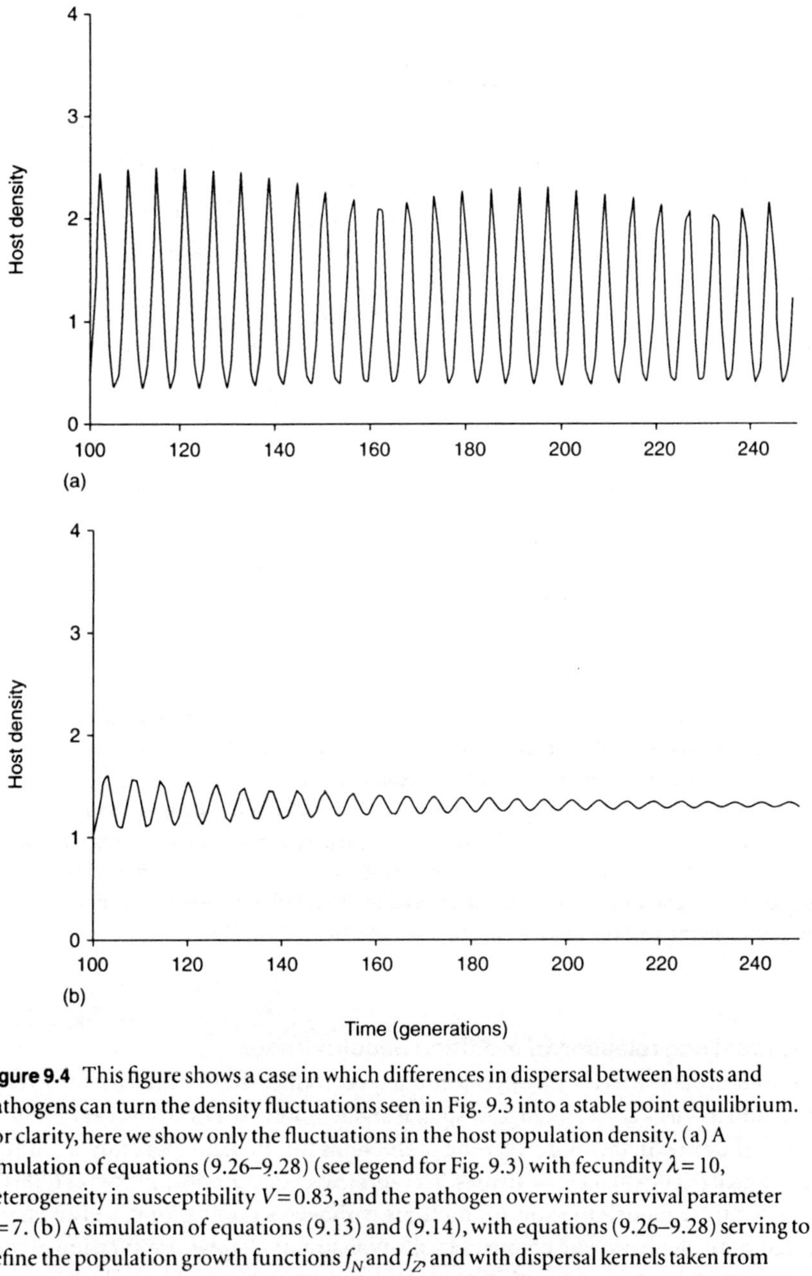

Figure 9.4 This figure shows a case in which differences in dispersal between hosts and pathogens can turn the density fluctuations seen in Fig. 9.3 into a stable point equilibrium. For clarity, here we show only the fluctuations in the host population density. (a) A simulation of equations (9.26–9.28) (see legend for Fig. 9.3) with fecundity $\lambda = 10$, heterogeneity in susceptibility $V = 0.83$, and the pathogen overwinter survival parameter $\phi = 7$. (b) A simulation of equations (9.13) and (9.14), with equations (9.26–9.28) serving to define the population growth functions f_N and f_Z, and with dispersal kernels taken from equations (9.15) and (9.16). Parameter values here are the same as in (a), with the addition that the average dispersal distance of the host $\alpha = 50$, and the average dispersal distance of the pathogen $\beta = 1$. Notice that the addition of dispersal causes the unstable population dynamics seen in (a) to eventually reach a stable equilibrium.

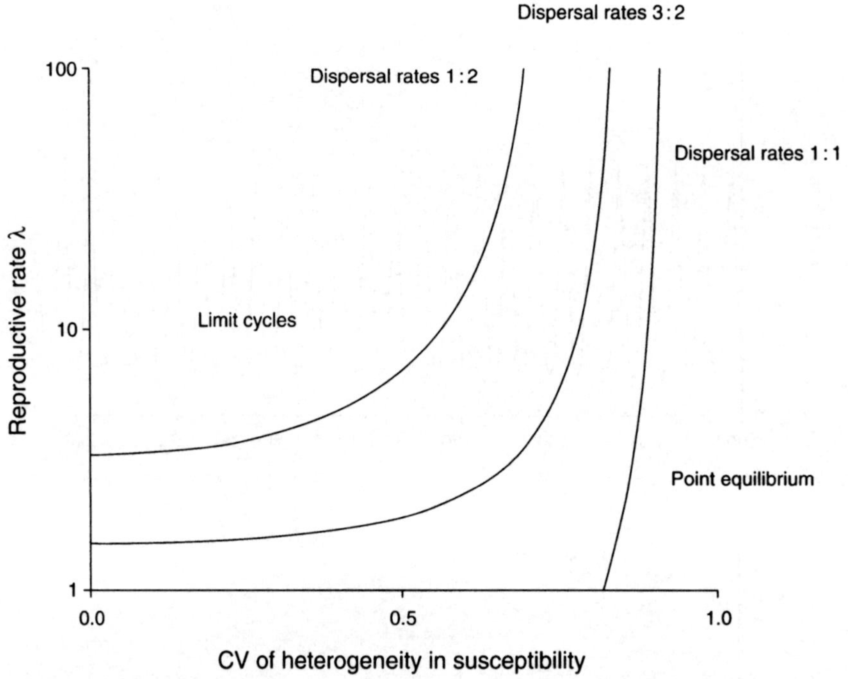

Figure 9.5 Effects of fecundity λ, heterogeneity in susceptibility C and average dispersal distance ratio on the stability of the long-term host–pathogen model. Here we have set the pathogen overwintering parameter $\phi = 5$. The region in which limit cycles occurs gives the kind of behaviour seen in Fig. 9.4a, while the region in which point equilibria occur gives the kind of behaviour seen in Fig. 9.4b. Notice that it is not necessary to specify whether the ratio of average dispersal distances is the dispersal distance of the host relative to that of the pathogen, or vice versa. Any difference in the average dispersal distances of host and pathogen thus increases the likelihood of stable population dynamics.

Dispersal and releases of modified baculoviruses

The models that we have described accurately reproduce the dynamics of the gypsy moth and its virus across a range of spatial and temporal scales (Dwyer *et al.* 1997, 2000). They might, therefore, provide useful tools for predicting the impact of releasing genetically engineered viruses. Towards this end, Dushoff and Dwyer (2001) extended these models to allow for multiple pathogen strains, asking, will an engineered strain be able to out-compete a wild-type strain? Here we use a version of the same model to see how competition among pathogen strains is affected by host dispersal and spatial structure. Before proceeding, however, we note that the model equations (9.10) and (9.11) accurately reproduce gypsy moth–virus dynamics *without* the inclusion of dispersal or spatial structure. It is unclear whether the model is accurate because dispersal is not that important or because the data are not spatially referenced and therefore cannot reject a model that ignores dispersal. In fact, dispersal may not be that important for gypsy moth population dynamics because first-stage gypsy moth larvae have very high dispersal relative to the scale of typical gypsy moth habitats (Dwyer & Elkinton 1995).

Assessing the risks of releasing engineered baculoviruses nevertheless requires an understanding of the importance of spatial structure and dispersal, for several reasons. First, the introduction of a genetically engineered baculovirus is likely to occur over a small area, so that dispersal would be important for several generations after release. Second, as we have described, it is often the case that hosts infected with engineered strains move less than hosts infected with wild-type strains (Cory *et al.* 1994; Hails *et al.* 2001). This kind of differential movement is likely to have important impacts on the spatial patchiness of the two strains, and may in fact determine whether an engineered strain is likely to out-compete a wild-type strain. Finally, and most importantly, we are interested in the effects of releasing engineered strains *in general*, rather than for the particular case of the gypsy moth virus. In fact, for many of the insects of concern, larvae disperse by crawling rather than by ballooning, and in such cases dispersal may be more important for population dynamics than is the case with the gypsy moth. Indeed, baculoviruses that are targeted for genetic engineering usually infect pests of row crops. It is therefore worth remembering that our model assumes that dispersal and virus transmission occur at different points in the life cycle, which is probably incorrect for most insects that attack row crops. Consequently, the models that we present here are preliminary efforts. Also, our models are not yet detailed enough to allow for the vertical movement changes of 'Wipfelkrankheit', so we focus on differences in movement rates across the landscape rather than within the canopy of vegetation.

Given these caveats, we can extend our single-epidemic model to allow for two pathogens, as follows:

$$\frac{dS}{dt} = -S\left(\frac{S(t)}{S(0)}\right)^{V} (\bar{\nu}_W P_W + \bar{\nu}_G P_G) \tag{9.20}$$

$$\frac{dP_W}{dt} = \bar{\nu}_W P_W(t-\tau_W) S(t-\tau_W)\left(\frac{S(t-\tau_W)}{S(0)}\right)^{V} - \mu P_W \tag{9.21}$$

$$\frac{dP_G}{dt} = \bar{\nu}_G P_G(t-\tau_G) S(t-\tau_G)\left(\frac{S(t-\tau_G)}{S(0)}\right)^{V} - \mu P_G \tag{9.22}$$

Note first that these equations describe the dynamics of the host and pathogen at one point in space. Otherwise, this model is essentially the same as equations (9.8) and (9.9), except that we now have two pathogen strains, W for 'wild type' and G for 'genetically modified'. In this model, the two strains thus differ in their speed of kill τ and mean transmission rate ν, but are otherwise identical. In reality, it is likely that modified and wild-type strains will differ in other ways as well, but to date measurements have only been made of these two fitness components, and so again this model acts as a useful starting point.

To complete the model, we incorporate dispersal, host reproduction and pathogen survival by extending equations (9.13) and (9.14) to allow for two pathogen strains, each with its own dispersal kernel, according to:

$$N_{t+1}(x) = \int k_N(x-y) f_N(N_t(y), Z_{W,t}(y), Z_{G,t}(y))dy \tag{9.23}$$

$$Z_{W,t+1}(x) = \int k_W(x-y) f_W(N_t(y), Z_{W,t}(y), Z_{G,t}(y))dy \tag{9.24}$$

$$Z_{G,t+1}(x) = \int k_G(x-y) f_G(N_t(y), Z_{W,t}(y), Z_{G,t}(y))dy \tag{9.25}$$

Again we assume that the dispersal kernels k_N, k_W and k_G are double exponentials. Also, f_W and f_G are the growth functions for the wild-type and genetically modified pathogens, respectively.

We simulated this model by first running it for 150 generations with only the wild-type pathogen present, and then introducing infectious cadavers of the genetically modified strain at a density equal to 1% of the density of wild-type cadavers. As one would expect, and as described by Dushoff and Dwyer (2001) for the model without dispersal, the engineered strain can invade if its transmission rate is not too much lower than the wild-type, because of the fitness advantage of killing faster (Figs 9.6 and 9.7). As the transmission rate of the engineered strain increases relative to that of the wild-type, we encounter first a region of coexistence, and then a region in which the engineered strain is able to competitively exclude the wild-type.

The ways in which dispersal affects this basic picture are qualitatively similar to the ways in which dispersal affects the population dynamics of the model with only one pathogen strain. That is, what matters in the one-strain model is that there are differences in dispersal between the host and pathogen. Likewise, in the model with two strains, what matters for the ability of the engineered strain to invade is that there are differences in dispersal between the host and the two pathogen strains, or between the two pathogen strains themselves. For example, Fig. 9.6 shows that if the pathogens have the same dispersal rate but the host's dispersal rate is substantially lower, the region of coexistence is only slightly larger. Similar results hold if the host's dispersal rate is higher than that of the pathogens, as long as the hosts infected with the two strains have the same dispersal rate. Differences in dispersal between the pathogen strains, however, can have a very strong effect on the ability of the engineered strain to invade. As Fig. 9.7 shows, if the host and the wild-type pathogen have the same dispersal rate, but the dispersal rate of the engineered strain is substantially lower, then the engineered strain can invade even if its transmission rate is quite a bit lower than that of the wild type.

The relevance of these effects can be seen when we consider estimates of relative transmission rates and speeds of kill for particular engineered strains. First, in Figs 9.6 and 9.7, the data for *egt*– is from transmission experiments with an engineered strain of the gypsy moth nuclear polyhedrosis virus for which the *egt* gene has been replaced with a biologically inert β-galactosidase reporter gene. The *egt* gene glycosylates the moulting hormone ecdysteroid, thereby preventing moulting if infection occurs early in the instar (Slavicek *et al.* 1999). Deletion of the *egt* gene produces a

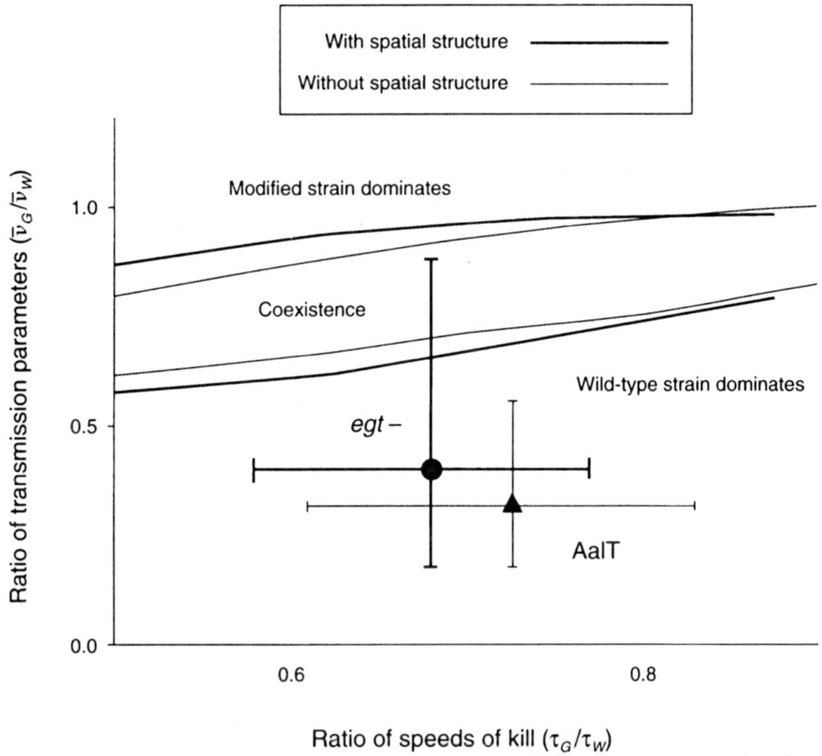

Figure 9.6 Regions of coexistence and competitive exclusion for the host–pathogen model with dispersal and two pathogen strains. The vertical axis shows the ratio of the initial average transmission rate of the genetically modified strain $\bar{\nu}_G$ to that of the wild type $\bar{\nu}_W$. The horizontal axis shows the ratio of the speed of kill of the genetically modified strain τ_G to that of the wild type τ_W. Values of other parameters are as follows: transmission rate of wild type $\bar{\nu}_W = 1$, the decay rates of both pathogens $\mu = 1$, and speed of kill of wild type $\tau_W = 2$ weeks. For the spatial model, the average dispersal distance of the host $\alpha = 1$, while the average dispersal distances of the two pathogens $\beta_W = \beta_G = 50$, so that the host has a dispersal rate equal to one-fiftieth of that of the two pathogen strains. The data point labelled *egt*– was calculated from a strain of the gypsy moth virus that was engineered to eliminate expression of the *egt* gene (data from Dwyer *et al.* 2001). The data point labelled AaIT was calculated from data for a strain of *Autographa californica* virus engineered through the insertion of a gene that expresses an insect-specific toxin, infecting larvae of *Trichoplusia ni* (data from Hails *et al.* 2001). Error bars indicate bootstrapped 95% confidence intervals. See Dwyer *et al.* (1997) for more details of how the transmission rates and error bars were calculated.

virus strain that kills more rapidly, but, as is typical of engineered strains, the resulting *egt*– strain produces fewer occlusion bodies and so is less infectious (Slavicek *et al.* 1999; Dwyer *et al.* 2001). Figure 9.6 suggests that such a strain would be unlikely to out-compete a wild-type strain, which is evolutionarily reasonable given that so

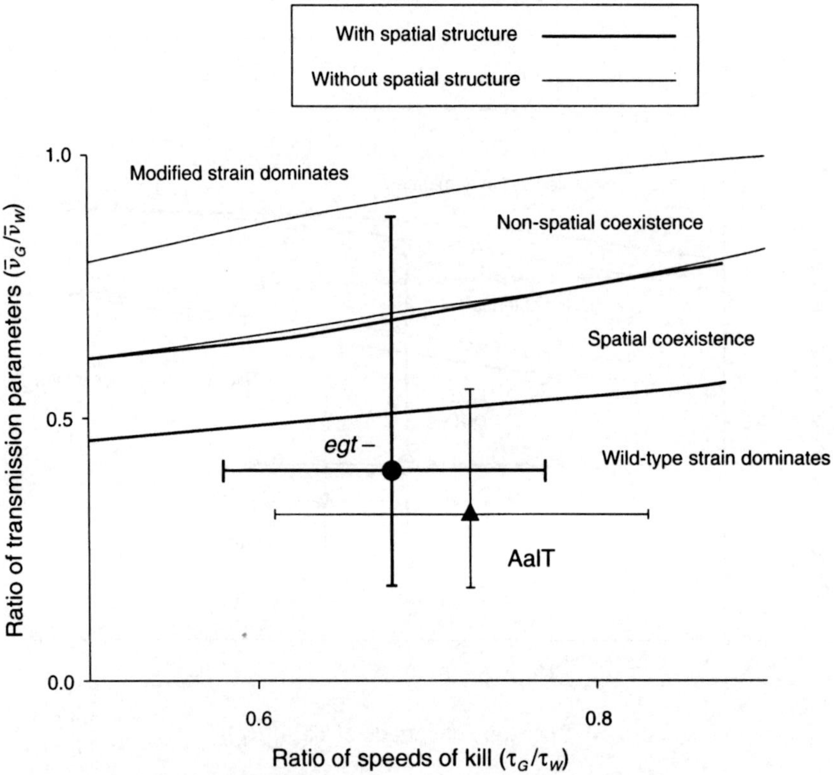

Figure 9.7 As in Fig. 9.6, except that the average dispersal distance of the host $\alpha = 50$, the average dispersal distance of the wild-type pathogen $\beta_W = 50$ and the average dispersal distance of the genetically modified pathogen $\beta_G = 1$. The engineered strain thus has an average dispersal distance equal to one-fiftieth the average dispersal distances of the host and the wild-type pathogen. Notice that this difference in dispersal distances greatly increases the likelihood that the engineered strain will persist.

far as is known most wild-type strains possess the *egt* gene. In contrast, Fig. 9.7 suggests that this strain would be able to invade, but this conclusion assumes a difference in dispersal between the two pathogens that is unlikely to hold for *egt*– deleted strains.

The biology of the scorpion-toxin construct, known as AaIT, however, is rather different. The data for AaIT come from experiments in which larvae of the cabbage pest *Trichoplusia ni* were infected with a virus of *Autographa californica* modified by the insertion of an insect-specific toxin gene from the scorpion *Androctonus australis* (Cory *et al.* 1994; Hails *et al.* 2002). As with *egt*–, the AaIT construct kills more rapidly but has a lower transmission rate relative to its wild-type equivalent, and Fig. 9.6 similarly suggests that the construct has little chance of invading. Unlike *egt*–, however, the AaIT construct changes its host's dispersal rate, because it paralyses its host well before death occurs. The average dispersal distance of insects infected with this strain is therefore likely to be substantially less than the average dispersal distance of insects infected with the wild type. Figure 9.7 shows that the 95% confidence interval on the transmission rate of this strain relative to that of the wild type slightly overlaps the region within which coexistence occurs. This suggests that there is at least an outside possibility that this strain may be able to invade natural populations. This conclusion is tentative, given that the model in question is more appropriate to forest defoliators than to crop insects, and given that little is known of the long-term survival of the AaIT strain in nature. Nevertheless, our results suggest that the reduced dispersal rates typical of insects infected with engineered virus strains may allow such strains to persist in the environment for a long time.

References

Anderson, R.M. & May, R.M. (1981) The population dynamics of microparasites and their invertebrate hosts. *Philosophical Transactions of the Royal Society, Series B* **291**, 451–524.

Bowers, R.G., Begon, M. & Hodgkinson, D.E. (1993) Host–pathogen population cycles in forest insects—lessons from simple-models reconsidered. *Oikos* **67**, 529–538.

Briggs, C.J., Hails, R.S., Barlow, N.D. & Godfray, H.C.J. (1995) The dynamics of insect–pathogen interactions. In: *Ecology of Infectious Diseases in Natural Populations* (eds B.T. Grenfell & A.P. Dobson), pp. 295–326. Cambridge University Press, Cambridge.

Burden, J.P., Hails, R.S., Windass, J.D., Suner, M.M. & Cory, J.S. (2000) Pathogenicity, virulence and productivity of a baculovirus expressing the Itch Mite toxin TxP-1 in second and fourth instar larvae of *Trichoplusia ni*. *Journal of Invertebrate Pathology* **75**, 226–236.

Burnham, K.P. & Anderson, D.R. (1998) *Model Selection and Inference: an Information–Theoretic Approach*. Springer Verlag, New York.

Cory, J.S., Hirst, M.L., Williams, T. *et al.* (1994) Field trial of a genetically improved baculovirus insecticide. *Nature* **370**, 138–140.

D'Amico, V. & Elkinton, J.S. (1995) Rainfall effects on transmission of gypsy moth (Lepidoptera: Lymantriidae) nuclear polyhedrosis virus. *Environmental Entomology* **24**: 1144–1149.

Dushoff, J. & Dwyer, G. (2001) Evaluating the risks of engineered viruses: modeling pathogen competition. *Ecological Applications* **11**, 1602–1609.

Dwyer, G. (1992) On the spatial spread of insect viruses: theory and experiment. *Ecology* **73**, 479–494.

Dwyer, G. (1994) Density-dependence and spatial structure in the dynamics of insect pathogens. *American Naturalist* **143**, 533–562.

Dwyer, G. & Elkinton, J.S. (1993) Using simple models to predict virus epizootics in gypsy moth populations. *Journal of Animal Ecology* **62**, 1–11.

Dwyer, G. & Elkinton, J.S. (1995) Host dispersal and the spatial spread of insect pathogens. *Ecology* **76**, 1262–1275.

Dwyer, G., Elkinton, J.S. & Buonaccorsi, J.P. (1997) Host heterogeneity in susceptibility and disease dynamics: tests of a mathematical model. *American Naturalist* **150**, 685–707.

Dwyer, G., Dushoff, J., Elkinton, J.S. & Levin, S.A. (2000) Pathogen-driven outbreaks in forest defoliators revisited: building models from experimental data. *American Naturalist* **156**, 105–120.

Dwyer, G., Dushoff, J., Elkinton, J.S., Burand, J.P. & Levin, S.A. (2001) Host heterogeneity in susceptibility: lessons from an insect virus. In: *Adaptive Dynamics of Infectious Diseases: in Pursuit of Virulence Management* (eds U. Dieckmann, A.J. Johan, M. Metz, W. Sabelis & K. Sigmund). Cambridge University Press, Cambridge.

Elkinton, J.S. & Leibhold, A.M. (1990) Population dynamics of gypsy moth in North America. *Annual Review of Entomology* **35**, 571–596.

Entwhistle, P.F. & Evans, H.F. (1985) Viral control. *Comprehensive Insect Physiology, Biochemistry and Pharmacology*, **12**, 347–412.

Goulson, D. (1997) Wipfelkrankheit: modification of host behaviour during baculoviral infection. *Oecologia* **109**, 219–228.

Hails, R.S., Hernandez-Crespo, P., Sait, S.M., Donnelly, C.A., Green, B.M. & Cory, J.S. (2002) Transmission patterns of natural and recombinant baculoviruses. *Ecology* **83**, 906–916.

Hawtin, R.E., Zarkowska, T., Arnold, K. *et al.* (1997) Liquefaction of *Autographa californica* nucleopolyhedrovirus-infected insects is dependent on the integrity of virus-encoded chitinase and cathepsin genes. *Virology* **238**, 243–253.

Hernandez-Crespo, P., Sait, S.M., Hails, R.S. & Cory, J.S. (2001) Behaviour of a recombinent baculovirus in lepidopteran hosts with different susceptibilities. *Applied and Environmental Microbiology* **67**, 1140–1146.

Hilborn, R. & Mangel, M. (1997) *The Ecological Detective: Confronting Models with Data.* Princeton University Press, Princeton, NJ.

Hoover, K., Schultz, C.M., Lane, S.S. *et al.* (1995) Reduction in damage to cotton plants by a recombinant baculovirus that knocks moribund larvae of *Heliothis virescens* off the plant. *Biological Control* **5**, 419–426.

Hunter, A.F. & Dwyer, G. (1998) Insect population explosions synthesized and dissected. *Integrative Biology* **1**, 166–177.

Hunter, M.D. & Price, P.W. (1998) Cycles in insect populations: delayed density dependence or exogenous driving variables? *Ecological Entomology* **23**, 216–222.

Jacques, R.P. (1974) Occurrence and accumulation of granulosis virus of *Pieris rapae* in treated field plots. *Journal of Invertebrate Pathology* **23**, 351–359.

Keddie, B.A., Aponte, G.W. & Volkman, L.E. (1989) The pathway of infection of *Autograph-californica* nuclear polyhedrosis-virus in an insect host. *Science* **243**, 1728–1730.

Kermack, W.O. & McKendrick, A.G. (1927) A contribution to the mathematical theory of epidemics. *Proceedings of the Royal Society of London, Series A* **115**, 700–721.

Knell, R.J., Begon, M. & Thompson, D.J. (1998) Transmission of *Plodia interpunctella* granulosis virus does not conform to the mass action model. *Journal of Animal Ecology* **67**, 592–599.

Kot, M. (1989) Diffusion-driven period-doubling bifurcations. *Biosystems* **22**, 279–287.

Kukan, B. & Myers, J.H. (1999) Dynamics of viral disease and population fluctuations in western tent caterpillars (Lepidoptera: Lasiocampidae) in southwestern British Columbia, Canada. *Environmental Entomology* **28**, 44–52.

Kunimi, Y., Fuxa, J.R. & Hammock, B.D. (1996) Comparison of wild type and genetically engineered nuclear polyhedrosis viruses of *Autographa californica* for mortality, virus replication and polyhedra production in *Trichoplusia ni* larvae. *Entomologia Experimentalis et Applicata* **81**, 251–257.

Moscardi, F. (1999) Assessment of the application of baculoviruses for the control of Lepidoptera. *Annual Review of Entomology* **44**, 257–289.

Murray, K.D. & Elkinton, J.S. (1990) Transmission of nuclear polyhedrosis virus to gypsy moth (Lepidoptera: Lymantriidae) eggs via contaminated substrates. *Environmental Entomology* **19**, 662–665.

Murray, K.D. & Elkinton, J.S. (1992) Vertical distribution of nuclear polyhedrosis virus-infected gypsy-moth (Lepidoptera: Lymantriidae) larvae and effects on sampling for estimation of disease prevalence. *Journal of Economic Entomology* **85**, 1865–1872.

Murray, K.D., Shields, K.S., Burand, J.P. & Elkinton, J.S. (1991) The effect of gypsy moth metamorphosis on the development of nuclear polyhedrosis virus infection. *Journal of Invertebrate Pathology* **57**, 352–361.

Murray, J.D. (1993) *Mathematical Biology*, 2nd edn. Springer Verlag, Berlin.

Myers, J.H. (1988) Can a general hypothesis explain population cycles of forest Lepidoptera? *Advances in Ecological Research* **18**, 179–284.

Myers, J.H. (1993) Population outbreaks in forest Lepidoptera. *American Scientist* **81**, 240–251.

Onstad, D.W., Maddox, J.V., Cox, D.J. & Kornkven, E.A. (1990) Spatial and temporal dynamics of animals and the host-density threshold in epizootiology. *Journal of Invertebrate Pathology* **55**, 76–84.

O'Reilly, D.R. & Miller, L.K. (1991) Improvement of a baculovirus pesticide by deletion of the *egt* gene. *Bio/Technology* **9**, 1086–1089.

Podgwaite, J.D., Shields, K.S., Zerillo, R.T. & Bruen, R.B. (1979) Environmental persistence of the nucleopolyhedrosis virus of the gypsy moth, *Lymantria dispar*. *Environmental Entomology* **8**, 528–536.

Possee, R.D., Barnett, A.L., Hawtin, R.E. & King, L.A. (1997) Engineered baculoviruses for pest control. *Pesticide Science* **51**, 462–470.

Slack, J.M., Kuzio, J. & Faulkner, P. (1995) Characterization of V-CATH, a cathepsin l-like proteinase expressed by the baculovirus *Autographa-californica* multiple nuclear polyhedrosis virus. *Journal of General Virology* **76**, 1091–1098.

Slavicek, J.M., Popham, H.J.R. & Riegel, C.I. (1999) Deletion of the *Lymantria* dispar multicapsid nucleopolyhedrovirus ecdysteroid UDP-glucosyl transferase gene enhances viral killing speed in the last instar of the gypsy moth. *Biological Control* **16**, 91–103.

Stewart, L.M.D., Hirst, M., Ferber, M.L., Merryweather, A.T., Cayley, P.J. & Possee, R.D. (1991) Construction of an improved baculovirus insecticide containing an insect-specific toxin gene. *Nature* **352**, 85–88.

Tilman, D. & Kareiva, P. (1997) *Spatial Ecology: the Role of Space in Population Dynamics and Interspecific Interactions*. Princeton University Press, Princeton, NJ.

Thomas, C.J., Brown, H.L., Hawes, C.R. *et al.* (1998) Localization of a baculovirus-induced chitinase in the insect cell endoplasmic reticulum. *Journal of Virology* **72**, 10207–10212.

Tomalski, M.D. & Miller, L.K. (1991) Insect paralysis by baculovirus mediated expression of a mite neurotoxin gene. *Nature* **352**, 82–85.

Van Beek, N.A.M. & Hughes, P.R. (1998) The response time of insect larvae infected with recombinant baculoviruses. *Journal of Invertebrate Pathology* **72**, 338–347.

Varley, G.C., Gradwell, G.R. & Hassell, M.P. (1973) *Insect Population Ecology; an Analytical Approach*. University of California Press, Berkeley.

Vasconcelos, S.D., Cory, J.S., Wilson, K.R., Sait, S.M. & Hails, R.S. (1996) Modified behaviour in baculovirus-infected lepidopteran larvae and its impact on the spatial spread of inoculum. *Biological Control* **7**, 299–306.

Wilson, W.G. (1998) Resolving discrepancies between deterministic population models and individual-based simulations. *American Naturalist* **153**, 116–134.

Chapter 10

Gene flow and the evolutionary ecology of passively dispersing aquatic invertebrates

Beth Okamura and Joanna R. Freeland

Introduction

Apart from a relatively small number of ancient inland bodies of water, most freshwater lakes originated during the last glaciation and will eventually disappear due to processes such as infilling with sediments and encroachment of marginal vegetation (Wetzel 1975). Therefore, to ensure long-term persistence, most freshwater organisms must at least occasionally disperse to and colonize new sites, and the widespread distribution of many species is indicative of such events (Darwin 1859; Pennak 1989). However, as dispersal is notoriously difficult to assess, its extent and frequency are poorly understood. This ignorance significantly compromises our understanding of both population structure and population dynamics.

This chapter will review the evidence for, and assess the biological significance of, intersite dispersal and gene flow by invertebrates that inhabit lakes and ponds. We begin by broadly considering how variation in the frequency and scale of dispersal affects population and community structures. We briefly review dispersal mechanisms of freshwater invertebrates, and then focus on our own research programme on dispersal, gene flow and the metapopulation ecology of freshwater bryozoans. This is followed by a review of dispersal in freshwater zooplankton, an approach that allows us to compare and contrast the significance and extent of dispersal and gene flow for groups that differ in ecology and life history.

Frequency and scale of dispersal

Research over the last decade on the spatial ecology of populations and communities has revealed that a profound influence may be exerted by the frequency and scale of dispersal. Much of the research at the population level has focused on metapopulation dynamics, whereas community ecologists have focused on the relative importance of regional verus local influences on patterns of species richness.

Dispersal and metapopulation structure

Since many organisms occur in a series of discrete and isolated sites, the metapopu-

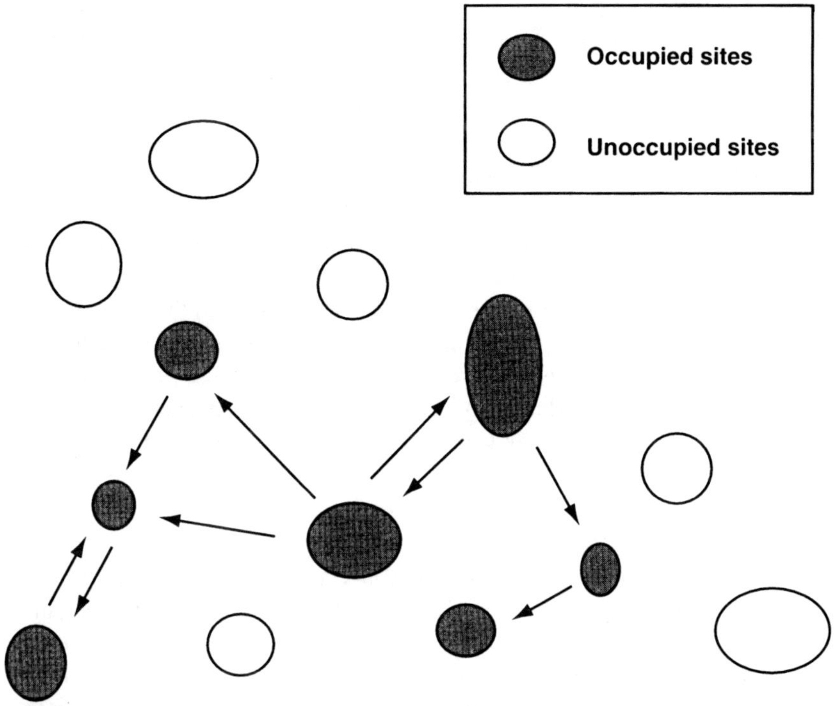

Figure 10.1 Schematic representation of a classic metapopulation. Populations within local sites go extinct and are recolonized through dispersal (shown by arrows) from other currently occupied sites within the metapopulation. The pattern of site occupation varies over time with dispersal playing a crucial role in linking subpopulations within the metapopulation.

lation concept (Levins 1969, 1970) has been increasingly applied by ecologists to describe the dynamics of subdivided populations. Levins' classic metapopulation model demonstrated that, in a collection of subdivided populations, the fraction of habitat patches occupied at any one time results from a balance of the rate at which local populations go extinct and the rate of colonization of empty patches (Fig. 10.1) (Hanski 1989). Since all local populations are subject to extinction, persistence applies only at regional scales (Hanski & Simberloff 1997). Dispersal lies at the heart of metapopulation dynamics since it determines which habitat patches collectively comprise a metapopulation at any given time.

Although metapopulation dynamics are believed to apply to many populations, conformation to the classic metapopulation model appears to be relatively rare (Harrison & Taylor 1997; Bullock *et al.*, this volume). Further refinements of the classic model include effects such as variation in patch size or quality on extinction probabilities (e.g. island–mainland and source–sink metapopulations), the influence of interspecific interactions on rates of local extinction and colonization, and spatially dependent variation in the probability of dispersal between sites (e.g. stepping stone and other spatially explicit models) (for reviews see Hanski & Simberloff 1997; Hanski 1999). The latter developments can account for effects of the vagility of species and landscape structure. Thus, whether a series of patches represent separate, independent populations, true metapopulations or 'patchy populations' (unitary populations within which there are discrete patches of resources; Menéndez & Thomas 2000) will be influenced by the frequency and scale of dispersal across the landscape.

The frequency and scale of dispersal and accompanying gene flow will also influence patterns of genetic differentiation (Slatkin 1987). Thus, patchy populations should converge towards panmixia given frequent gene flow, and there will be little genetic differentiation among local populations. At intermediate levels, gene flow will link subpopulations within a metapopulation, and moderate levels of genetic differentiation are expected due to, for example, founder effects, genetic drift and local selection. Finally, in independent populations genetic differentiation should ultimately lead to speciation.

Dispersal and patterns of species richness

Consideration of dispersal at the community level provides the basis for regional versus local patterns of species richness. Patterns of species composition among sites may reflect the extent to which dispersal and local interactions limit local diversity (for reviews see Cornell & Karlson 1997; Srivastava 1999). If dispersal rates greatly exceed local extinction probabilities for many species, then nearly all species in the region that are capable of invading the site should be present. Local processes will then determine species diversity and composition within patches. In this case, local species richness will approach an upper asymptote with increasing regional richness (Fig. 10.2) (Cornell & Lawton 1992). However, if dispersal is rare, then many species may be absent. Local species composition in this case will depend on a site's history of colonization, and local interactions may be relatively unimportant (Ricklefs 1987; Cornell & Lawton 1992; Cornell & Karlson 1997). Species composition will then be undersaturated, and local species richness will increase continuously with increasing richness of the regional pool of species (Fig. 10.2). Such communities are considered to be dispersal limited and under strong regional control. Linear, positive relationships in support of dispersal limitation have been discerned in a diversity of systems in marine, freshwater and terrestrial environments (Cornell & Karlson 1997; Srivastava 1999).

How do freshwater invertebrates disperse?

Intersite dispersal of freshwater invertebrates can be achieved by active or passive means (for review see Bilton *et al.* 2001). Active dispersal occurs through self-generated movements of organisms, mainly exemplified by aerial flight of adult insects. Passive dispersal is achieved by means of external agents, including floods, wind and animal vectors. Most non-insect invertebrates of lakes and ponds colonize new sites through the passive dispersal of small, dormant propagules (Williams

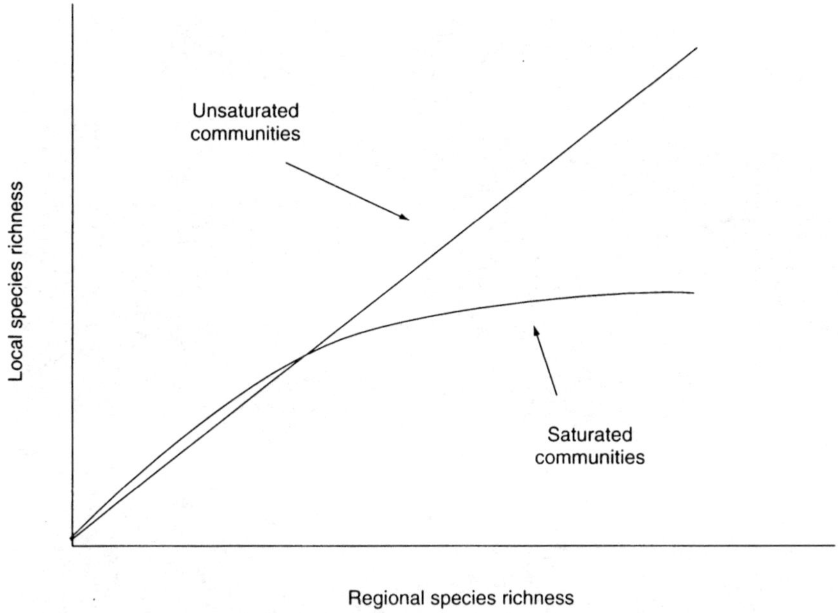

Figure 10.2 In unsaturated communities local species richness is predicted to be a fixed proportion of regional species richness. In saturated communities, local richness may increase with regional richness but will reach an upper limit at higher regional richness. (Adapted from Srivastava 1999.)

1987). These propagules are typically highly resistant to desiccation and extremes of temperature and often show apparent adaptations to increase the likelihood of attachment to animal vectors. Thus, many propagules display features such as hooks, spines and sticky surfaces (Fig. 10.3); avoidance of deposition in bottom sediments through buoyancy; and accumulations of large numbers at times that coincide with peaks of waterfowl annual migrations (Bilton *et al.* 2001).

Dispersal, life history and population genetic structure in freshwater bryozoans

The life history of freshwater bryozoans

Freshwater bryozoans are benthic, hermaphroditic, colonial invertebrates and are common, but overlooked, residents of freshwater habitats (Wood 1991). Their life cycle involves a high degree of clonal reproduction through colony growth by budding of new zooids, colony fission or fragmentation, and the asexual production of large numbers of resistant propagules, called statoblasts. Statoblasts represent both an overwintering and dispersal stage (Fig. 10.3). Small colonies emerge from statoblasts when favourable conditions return in the following growing season. Sexual reproduction is brief in duration and results in short-lived, swimming larvae that can

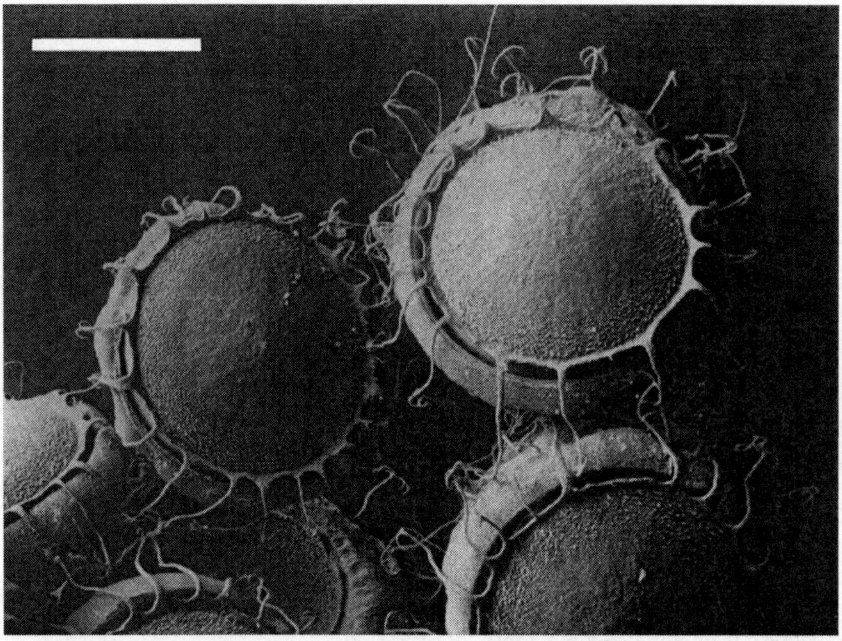

Figure 10.3 Scanning electron micrograph of statoblasts of *Cristatella mucedo*. Scale bar = 400 μm.

disperse within a site before metamorphosing into a small colony. The early seasonal occurrence of sex in bryozoans is unusual. The majority of freshwater invertebrate groups undergo a late season sexual phase which is viewed to be adaptive in that it results in the production of maximal genetic variation at a time that coincides with unpredictable conditions (Lynch & Spitze 1994). Thus, when conditions deteriorate, zooplankton taxa undergo a sexual phase to produce dormant, resistant embryos that will hatch out the following season.

Population genetic studies of *Cristatella mucedo*

The bryozoan *Cristatella mucedo* occurs as elongate, gelatinous colonies in lakes and ponds throughout the Holarctic. This widespread distribution has allowed our population genetics studies to be conducted over large spatial scales; thus approximately 30 colonies of *C. mucedo* were collected from each of 12 water bodies across a transect of approximately 2500 km in northwest Europe and from each of eight water bodies across a transect of approximately 1500 km in central North America (Fig. 10.4). Each colony was collected from separate substrata to avoid collecting replicate genotypes from aggregations of genetically identical colonies that can develop through colonization by a single statoblast or larva and subsequent colony growth and fission. Note, however, that dissemination of floating statoblasts can potentially spread the same clonal genotype throughout a site. Microsatellite profiles

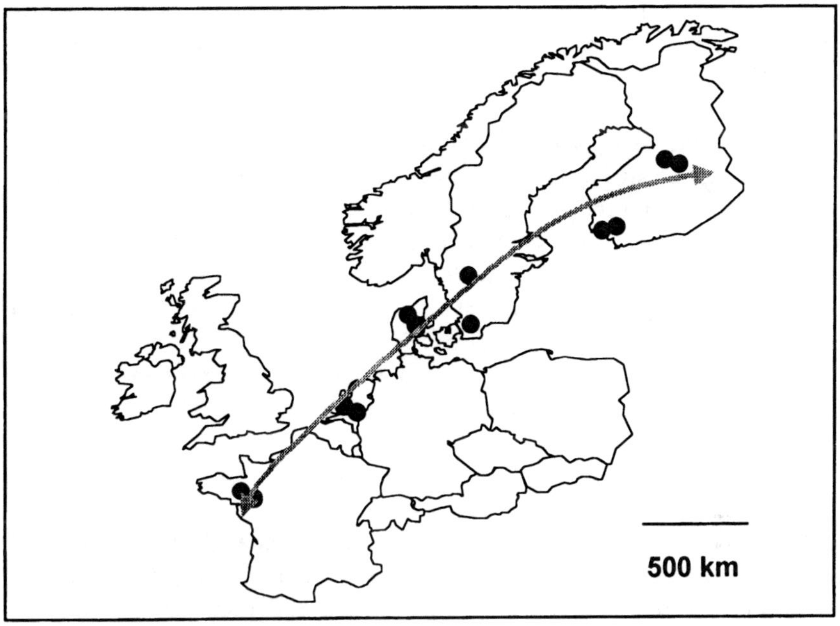

(a)

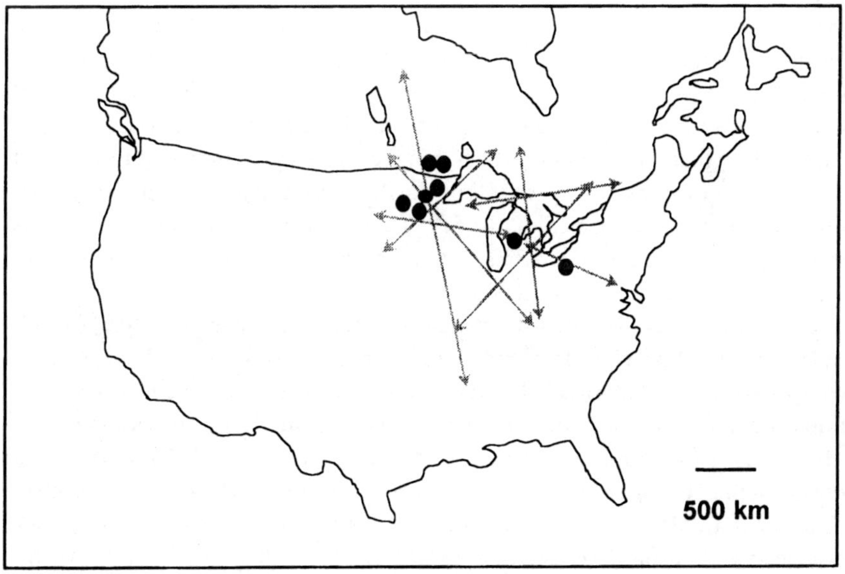

(b)

Figure 10.4 (a) Map showing the location of 12 sites sampled for *Cristatella mucedo* in northwest Europe with track of main waterfowl annual migratory route traversing the region sampled shown by arrow. (b) Map showing the location of eight sites sampled in central North America with pathways of numerous, divergent annual migratory waterfowl routes traversing the region shown by arrows.

were generated for all colonies, and a subset of these was further genetically characterized using mitochondrial DNA sequence data (for further details, see Freeland *et al.* 2000c).

Analysis of gene flow and population structure using microsatellites

Within each continental region, we characterized populations based on five microsatellite loci (Freeland *et al.* 2000a, 2000b). However, only three of these loci were readily amplified from both North American and European *C. mucedo* (Freeland *et al.* 2000c), and we base much of our discussion on patterns revealed by these three loci. This selective presentation of data is justified because direct comparisons of microsatellite data from different loci may be hampered by inherently variable mutation rates. Furthermore, the results of separate population genetic analyses based on the full set of five microsatellite loci (Freeland *et al.* 2000a, 2000b) provided similar conclusions to analyses based on the three common loci (Freeland *et al.* 2000c). To support this latter statement, we refer to conclusions based on the full data sets (Freeland *et al.* 2000a, 2000b) when appropriate.

Traditional means of gaining evidence for dispersal have relied on estimates of gene flow (*Nm*) based on F_{ST} approaches. However, it is now widely appreciated that such estimates may be flawed owing to violation of assumptions such as population equilibrium, no spatial variation in migration rates and population sizes, and no mutation (Whitlock & McCauley 1999; Raybould *et al.*, this volume). In view of these problems we conducted non-parametric discriminant function analyses as an independent means of obtaining evidence for gene flow amongst populations within each continent (Freeland *et al.* 2000c). These analyses classified 32.5% of European colonies (representing 51.3% of multilocus genotypes, MLGs) to populations other than the one from which they were collected, compared to 18.3% of colonies (representing 24% of MLGs) in North America. When discriminant function analyses were performed on the full data sets of five microsatellite loci, 14% of European colonies were misclassified and only 8% were misclassified in North America (Freeland *et al.* 2000a, 2000b).

The occurrence of identical MLGs (identified by the three microsatellites) in multiple populations provided further support for higher levels of gene flow in Europe compared to North America. In Europe, nine MLGs were each found in two populations and one MLG was found in four populations. In North America, only two MLGs were found in multiple (in both cases two) populations. We should point out the possibility that these MLG distributions would be uninformative about levels of gene flow if the MLGs arose by processes other than recent common ancestry. However, in an earlier study, we specifically tested this possibility by conducting a computer simulation to assess the likelihood that two MLGs (based on five microsatellite loci), found in European lakes separated by 700 km, resulted from a chance recombination of alleles (Freeland *et al.* 2000a). The likelihood of such a chance recombination of alleles at the two sites was exceedingly small ($P < 0.001$). While we did not conduct such simulations to determine the likelihoods that all the MLGs arose by chance in the various sites discussed here, our previous simulation study and concordance with other inferences of gene flow (see below) together provide evidence that MLG distributions give a useful indication of relative levels of gene flow amongst the populations sampled.

Although the potential inaccuracies of F_{ST}-based estimates of gene flow must be born in mind, *Nm* values conformed with the above conclusions of greater gene flow amongst European than North American populations. Overall levels of *Nm* based on ρ (an analogue of F_{ST} developed for use with microsatellite data; Goodman 1997) were higher among European (range = 0.11 – 14.18, mean ± SD = 1.21 ± 0.31) than North American populations (range = 0.03 – 0.87, mean ± SD = 0.20 ± 0.04). When *Nm* was derived from F_{ST}, the difference was reduced but values were still considerably higher for European populations (Europe: range = 0.01 – 4.21, mean ± SD = 0.70 ± 0.007; North America: range = 0.19 – 0.73, mean ± SD = 0.44 ± 0.03). The relatively higher *Nm* estimates for European populations are mirrored by analyses based on the five microsatellite loci for each continent (Freeland *et al.* 2000a, 2000b).

In summary, consistent results provided by our three separate approaches (discriminant function assignments, distributions of MLGs and *Nm* estimates) together provide strong evidence for ongoing gene flow amongst populations in northwest Europe and considerably reduced to non-existent levels of gene flow amongst populations in central North America.

As estimates of *Nm* are inversely correlated with values of F_{ST} or ρ, the higher estimates of gene flow in Europe are naturally equated with lower, but none the less substantial, levels of population subdivision in Europe compared with North America. High levels of genetic differentiation for populations in both continents were similarly supported by analysis of the full data set of five microsatellite loci, with overall values being lower in Europe (Freeland 2000a, 2000b). Mantel tests showed no pattern of isolation by distance in North America or Europe when either ρ or F_{ST} values were used (Fig. 10.5). The lack of isolation by distance was also supported by analyses of full data sets based on five microsatellite loci (Freeland *et al.* 2000a, 2000b).

Data sets based on both three and five microsatellite loci revealed that none of the 39 populations on either continent were in Hardy–Weinberg equilibrium (Freeland *et al.* 2000a, 2000b, 2000c). If random sampling from clonally reproducing populations was responsible for an absence of Hardy–Weinberg equilibrium then we would expect to see instances of both observed heterozygosity (H_o) deficits and excesses. However, in all 39 *C. mucedo* populations, deviations from Hardy–Weinberg proportions were consistently due to H_o deficits, which were reflected in high estimates of F_{IS} (Freeland *et al.* 2000a, 2000b, 2000c). Both inbreeding and the Wahlund effect (see later) are likely to contribute to low H_o values. Self-fertilization and inbreeding may be promoted by the short distances traversed by sperm spawned in the water column, the development of clonal aggregations through colony growth and fission, and the sessile nature of colonies. However, microsatellite analysis has indicated that over half of the larval progeny produced by representative colonies from one site are the products of cross-fertilization since at least one allele was not present in parental colonies (J. Freeland, unpubl. data).

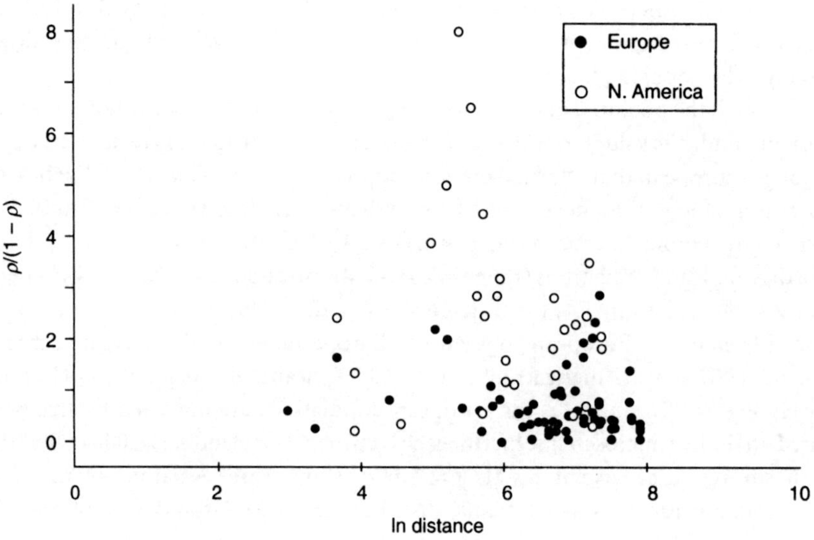

Figure 10.5 Genetic relatedness (as $\rho/(1-\rho)$ values) versus distances (as ln distance values) between sites for European and North American populations of *Cristatella mucedo.*

Microsatellite analysis revealed similar levels of within-population diversity in Europe and North America as measured by values of H_e (expected heterozygosity) and I (Shannon–Weaver diversity index) (Freeland *et al.* 2000a, 2000b, 2000c). There was a relatively higher clonal diversity (number of unique MLGs/number of colonies) in North America as a whole compared to Europe ($\chi^2 = 13.281$, $df = 1$, $P < 0.001$) as well as a higher allelic diversity (the number of alleles/number of colonies sampled) ($z = 4.239$; $P \leq 0.001$) (Fig. 10.6).

Analysis of historical dispersal events through mtDNA sequence data

We complemented our microsatellite analyses by sequencing a region of 16S rDNA (479–480 base pairs) from three randomly chosen MLGs per site, identified from the full set of five microsatellites. We also sequenced single colonies from 10 other North American sites and from five other European sites from which only a small number of colonies were collected.

The 34 North American colonies yielded 16 haplotypes with a maximum sequence divergence of 0.036 (Freeland *et al.* 2000c). The 41 European colonies yielded only three haplotypes with a maximum sequence divergence of 0.006 (Freeland *et al.* 2000c) (Fig. 10.7). The maximum divergence between North American and European sequences was 0.026. The relative haplotype diversity per site was significantly greater in North America than in Europe ($t = 5.01$, $df = 18$, $P < 0.001$) (Fig. 10.8a), but the mean number of sites per haplotype was significantly greater in Europe (analysis of log-transformed data: $t = 2.93$, $df = 17$, $P = 0.009$) (Fig. 10.8b).

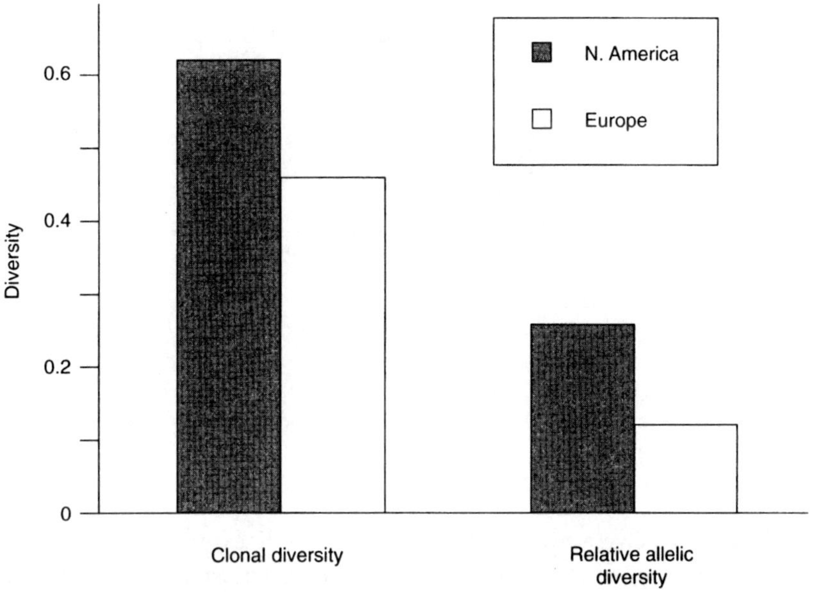

Figure 10.6 Clonal diversity (unique multilocus genotypes/number of colonies) and relative allelic diversity (number of alleles/number of colonies sampled) in North American and European populations of *Cristatella mucedo.*

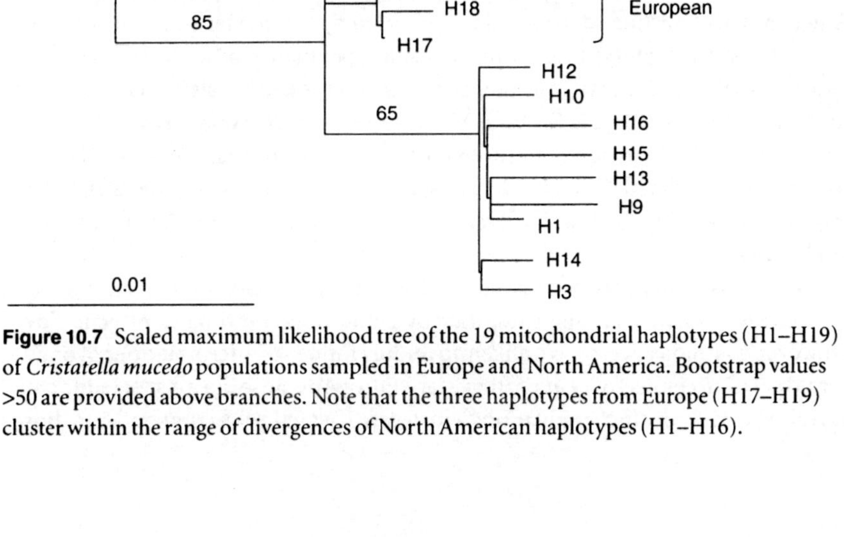

Figure 10.7 Scaled maximum likelihood tree of the 19 mitochondrial haplotypes (H1–H19) of *Cristatella mucedo* populations sampled in Europe and North America. Bootstrap values >50 are provided above branches. Note that the three haplotypes from Europe (H17–H19) cluster within the range of divergences of North American haplotypes (H1–H16).

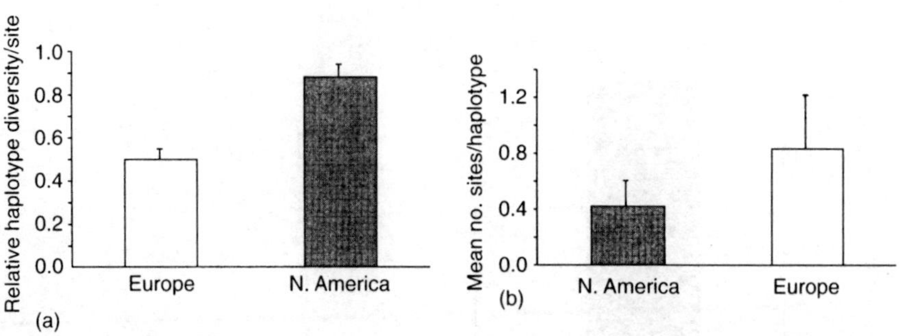

Figure 10.8 (a) Relative haplotype diversity for North American and European populations. Data are plotted as the relative diversity of haplotypes in sites for which haplotypes were obtained for three multilocus genotypes (see text for further discussion). (b) Mean number of sites per haplotype for North American and European populations of *Cristatella mucedo*. Data plotted as log $(x+1)$.

The degree of divergence amongst the haplotype lineages in North America indicates that many of these haplotypes must have been maintained throughout glacial–interglacial cycles, as much of the region sampled was covered by ice 10 000–20 000 years ago. The co-occurrence of divergent mitochondrial lineages within the same population today must reflect repeated postglacial colonization events from glacial refugia, occasional dispersal during postglacial range extension or ongoing dispersal that was not detected by our genetic analyses (see discussion below).

The widespread distribution of haplotypes in Europe may be interpreted as affirmation that levels of gene flow are higher on that continent. However, as the 16S rDNA sequence is expected to evolve at a substantially slower rate than microsatellite loci, biogeographical patterns inferred from mitochondrial sequences may reflect historical events. The paucity of mitochondrial haplotypes in Europe compared with North America and their relatively low degree of divergence strongly suggest a historical bottleneck, possibly following intercontinental dispersal from North America. This direction of dispersal is supported by the greater similarity of some North American haplotypes to European haplotypes than to other North American sequences (Fig. 10.7). Intercontinental dispersal events have also been identified in zooplankton taxa (Berg & Garton 1994; Taylor *et al.* 1996; Weider *et al.* 1999a). An alternative scenario, that relatively few haplotypes were maintained in Europe throughout the glacial–interglacial cycles, is unlikely as this would be reflected by a deep, rather than shallow, coalescence of the European and North American haplotypes.

A recent common ancestor shared amongst European haplotypes could explain the genetic similarity among sites, thereby confounding estimates of gene flow. However, common ancestry is unlikely to unduly influence interpretations of ongoing population connectivity in northwestern Europe because the microsatellite data reveal genetically distinct populations with levels of genetic diversity similar to those in North America, as well as a number of unique alleles. While the microsatellite loci may be expected to retain higher levels of diversity following a population bottleneck as compared with mitochondrial DNA, as nuclear DNA is effectively four times the size of mitochondrial DNA, this would translate into population-specific genetic signatures only after an adequate amount of time has elapsed. In Europe there has evidently been sufficient time since divergence from a common ancestor for population-specific genetic signatures to have evolved, a process that will have been accelerated by the extinction–recolonization dynamics inherent to a metapopulation (see below).

Synthesis of genetic data

Although many populations of *C. mucedo* in northwest Europe are linked by gene flow, they are none the less genetically distinct. These results conform to theoretical evidence that founder effects and genetic bottlenecks will result in population differentiation within a metapopulation (Wade & McCauley 1988). A variety of evidence supports metapopulation dynamics in *C. mucedo* populations in Europe, including: apparent extinction and colonization events in UK populations (Okamura 1997b); historical records of multiple disappearances (up to some 30 years) and reappearances of statoblasts in sediment cores from two UK sites (cores dated from 1740 and from 1963 to present) (R. Shaw *et al.*, unpubl. data); large fluctuations in abundance within UK and Austrian populations (Wöss 1994; Vernon *et al.* 1996); infection by myxozoan and microsporidian parasites that may cause populations to fluctuate and occasionally go extinct (Okamura 1997a); and genetic evidence of a bottleneck in a Scottish population based on microsatellite data (Freeland *et al.* 2000a) and of genetic drift and founder effects in populations analysed by RAPD (random amplification of polymorphic DNA) in southern England (Hatton-Ellis *et al.* 1998). A metapopulation in Europe is further suggested by microsatellite data which revealed similar within-population levels of genetic diversity on both continents, whereas overall levels of genetic diversity were considerably lower in Europe. This conforms to the prediction that the effective size of a metapopulation should be smaller than the sum of its component subpopulations (Hedrick & Gilpin 1997).

Our results could be explained by sampling regimes. The European populations were located along a transect that corresponds to an annual migratory route converged on by many waterfowl species (Fig. 10.4a) (Scott & Rose 1996). Waterfowl are obvious agents for long-distance dispersal of propagules of freshwater invertebrates (for review see Bilton *et al.* 2001), and sites located along a common annual migratory route may be expected to maintain a level of connectivity. Furthermore, traits such as small size, marginal hooks and spines (Fig. 10.3), and protective, chitinous valves are likely to promote transport of *C. mucedo* statoblasts by waterfowl through entanglement in feathers, or internally, if consumed. Indeed, moulted feathers with adherent statoblasts are often encountered around lake margins (B. Okamura, pers. comm.), while experiments demonstrating the viability of *C. mucedo* and other statoblasts ingested by ducks (Brown 1933; Charalambides *et al.*, unpubl. data) and the presence of *C. mucedo* statoblasts in waterfowl digestive tracts and faeces (A. Green, unpubl. data) provide evidence for internal transport. In contrast, the sites sampled in North America were located along multiple annual migratory routes (Fig. 10.4b) and relatively low levels of connectivity among populations may be explained by the divergent nature of routes and the locations of waterfowl stop-over sites across the region sampled. None the less, multiple haplotypes within North American populations provide evidence for widespread postglacial dispersal, and it is possible that ongoing dispersal continues among populations that were not sampled.

A final consideration is the possibility of at least two cryptic *Cristatella* species in North America, as is suggested by the segregation of combined mtDNA and microsatellite data into two distinct classes of genotypes (Freeland *et al.* 2000c) and the apparent existence of two genome sizes in North American *C. mucedo* (Potter 1979). Combining data from two species would artificially reduce our estimates of gene flow.

Evidence for dispersal in zooplankton taxa

We now review evidence for dispersal in several zooplankton groups by considering the results of population genetic and ecological studies.

Population genetic studies of zooplankton

Copepods, and especially cladocerans, are the most extensively studied freshwater zooplankton taxa with respect to patterns of population genetics. Many species are characterized by endemic distributions and highly genetically differentiated populations (Boileau & Hebert 1988, 1991; Crease *et al.* 1990; Boileau *et al.* 1992; Hebert & Wilson 1994; Hebert & Finston 1996; Taylor *et al.* 1998). In addition, the genetic dissimilarity of neighbouring populations reveals no pattern of isolation by distance (Hebert & Moran 1980; Hebert *et al.* 1989; Weider 1989; Hebert & Finston 1996; Vanoverbeke & De Meester 1997), and estimates of gene flow among populations are generally very low (Hebert & Moran 1980; Crease *et al.* 1990; Boileau *et al.* 1992). Recent studies of rotifers and ostracods demonstrate similar patterns of regional endemism, significant geographical structuring and reduced gene flow (Gómez *et al.* 2000; Schön *et al.* 2000). However, some cladoceran populations are separated by only small genetic differences even over very large spatial scales (Innes 1990; Cerny & Hebert 1993; Hann 1995). These exceptions suggest that dispersal as a life-history trait has attained varying levels of importance in different zooplankton species and populations. None the less, it is widely recognized that former views of broad species distributions achieved through exceptional dispersal capabilities of zooplankton must now be tempered (see Schwenk *et al.* 2000 and references therein).

Cladocerans show phenotypic stasis associated with genetically divergent lineages, and many traditionally recognized cladoceran taxa have been revealed as an array of cryptic species (Taylor *et al.* 1998; Schwenk *et al.* 2000). Such cryptic species complexes presumably result from low levels of gene flow leading to genetic divergence and speciation. Speciation mechanisms include microevolution following range fragmentation of an ancestral species by physical barriers, disruptive selection, and interspecific hybridization and introgression that results from reproductive compatibility despite marked genetic divergence (for review see Schwenk *et al.* 2000). The rapid evolution of mean resistance to nutritionally poor or toxic cyanobacteria by *Daphnia galeata* following eutrophication-driven changes in phytoplankton assemblages provides further graphic evidence of significant and precipitous genetic change within zooplankton populations (Hairston *et al.* 1999b), as does the evolution of body size in *Daphnia* in response to fluctuating selection within a season (Tessier *et al.* 1992), and genetic changes in a *Daphnia* population in response to changes in predation pressure over a 30-year period (Cousyn *et al.* 2001).

Although regular, ongoing gene flow is apparently rare amongst most zooplankton populations within regions, genetic evidence indicates that dispersal can be important in promoting range expansions from postglacial refugia, in limiting zooplankton at very broad spatial scales (Hebert & Hann 1986; Boileau & Hebert 1991; Stemberger 1995), and in introducing exotic species (Hairston *et al.* 1999a; Duffy *et al.* 2000) as further discussed below. As for bryozoans, genetic evidence implicates waterfowl as dispersal agents, with distributions of genetic lineages showing concordance with annual migratory routes (Weider *et al.* 1996; Weider & Hobæk 1997; Taylor *et al.* 1998).

Ecological studies of zooplankton

Recent ecological studies provide evidence that patterns of zooplankton dispersal can play an important role in regulating community structure and function. Jenkins and Buikema (1998) found highly variable patterns of colonization by multiple zooplankton species (rotifers, cladocerans, copepods and phantom midges) and subsequent variation in community development in a series of environmentally similar, artificial ponds. Only 23% of zooplankton taxa from the regional species pool were present in all ponds after 1 year. Variation in zooplankton community composition among ponds was due to the behaviour of 75% of taxa in the regional species pool, with individual zooplankton species varying widely in their colonization success and timing. Overall, ponds differed in zooplankton community structure as measured by diversity-based (colonization and accrual curves, presence/absence, species richness) and population-based (density and biomass) data (Fig. 10.9). This study obtained evidence that, although dispersal is a rate-limiting process in determining zooplankton community structure, community development reflects contingencies of colonization histories in local sites. Thus, historical events produce priority effects which can have a lasting influence on subsequent community structure (see also Holland & Jenkins 1998).

Shurin (2000) conducted experimental studies of dispersal limitation in zooplankton by introducing rotifer and crustacean species from the region into established zooplankton communities in ponds and assessing the invasion resistance of these intact communities. Overall, success of invasion was low, as >91% of the introduced species immediately went extinct. However, some introduced species managed to colonize sites successfully, although they remained rare. Greater rates of

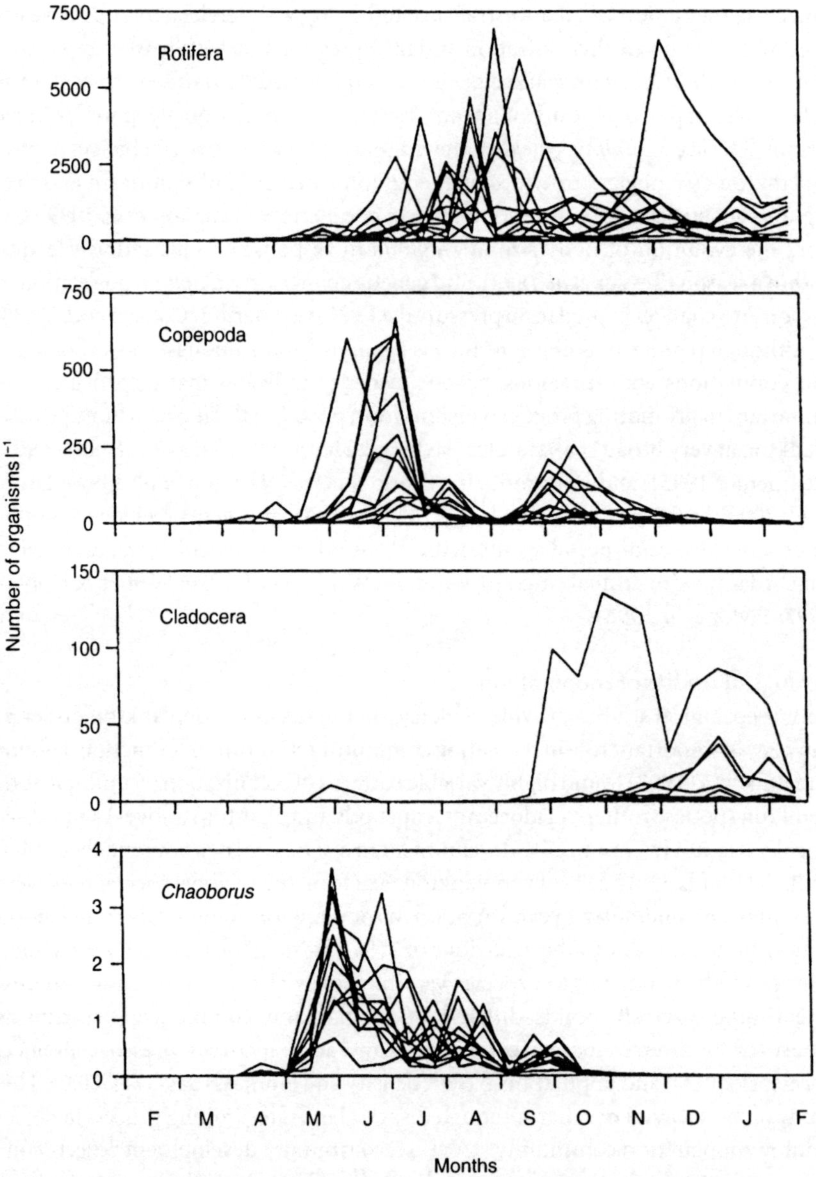

Figure 10.9 Zooplankton taxa (phantom midges (*Chaoborus*), cladoceran, copepod, rotifer) density over time in a series of 12 experimental ponds whose colonization patterns were monitored over a 1-year period. Each line represents a pond. (From Jenkins & Buikema 1998, with permission.)

invasion were observed when the abundance of native species was experimentally reduced, providing evidence that interactions with resident species were responsible for excluding potential invaders. These results indicate a minor role for dispersal limitation in structuring pond zooplankton communities, and complement the priority effects identified by Jenkins and Buikema (1998) in showing that established residents can inhibit colonization.

In an examination of patterns revealed by 25 studies of regional and local species richness in freshwater crustacean zooplankton, Shurin *et al.* (2000) came to the conclusion that the relative strengths of local and regional processes in determining species richness depend on the regional scale studied. Dispersal limitation was concluded to be important at very large spatial scales, while local processes determine zooplankton diversity within a biogeographical region.

Synthesis of genetic and ecological investigations of zooplankton

The above studies suggest that gene flow amongst zooplankton populations may be hampered by the difficulty of invading established communities. Microevolution within sites may further exacerbate problems of successful colonization following dispersal if established residents are better adapted to exploit local conditions than potential colonists (for further discussion, see De Meester 1996). On the face of it, this is at odds with instances of invasions by exotic species (e.g. Berg & Garton 1994; Havel *et al.* 1995) and the colonization of new biogeographical regions by occasional long-distance dispersal events (Hebert & Hann 1986; Boileau & Hebert 1991; Stemberger 1995). This discrepancy may be explained, in some cases, if dispersal to new biogeographical regions reflects range expansion from glacial refugia into unsaturated communities (e.g. Boileau & Hebert 1991; Weider & Hobæk 1997; Weider *et al.* 1999a, 1998b). In addition, invasion by exotics may often be explained by changes in habitat conditions. For instance, invasion by pollution-tolerant, exotic *Daphnia* species followed habitat changes due to pollution (Hairston *et al.* 1999a; Duffy *et al.* 2000). In other cases, invasions may be temporary, as suggested by apparent postinvasion extinction in some sites (Havel *et al.* 1995). Finally, attributes of some invasive species may specifically promote invasion and exploitation of established communities. Such attributes include short generation times, being habitat and trophic generalists (Ehrlich 1986) and filling an empty niche (Johnson & Carlton 1996; see also Cohen, this volume).

While local adaptation may reconcile seemingly low levels of gene flow with repeated observations of rapid colonization of new sites, a consideration that should be addressed is the possibility that gene flow estimates may be unreliable (Whitlock & McCauley 1999). In contrast to many assumptions about gene flow calculations, population genetic structure seldom reflects a precise relationship between gene flow and genetic drift (Bohonak 1999). For example, estimates of gene flow may be particularly biased if populations have arisen from recent (re)colonization, a process that requires dispersal, but which may not reveal evidence of gene flow because of stochastic processes associated with founder effects (e.g. Nurnberg & Harrison 1995). However, it is worth noting that despite the inherent difficulties in inferring gene flow from genetic data, gene flow estimates are seldom so overwhelmed by factors such as population history or disequilibrium as to be uninformative (Bohonak 1999).

Of course practical issues may also influence detection of gene flow, for instance if populations are characterized using relatively invariant markers such as allozymes or some mitochondrial sequence. In addition, many zooplankton populations are extremely large and inhabit numerous water bodies. As a result, for logistic reasons, sampled individuals may not provide a comprehensive assessment of population structure. We can expect that as new methods of genetic analyses evolve and further studies are completed, our perception of gene flow among populations may continue to change.

Temporal gene flow via propagule banks

There is accumulating evidence that dormant propagules produced by many freshwater invertebrates are retained within sediments where they remain viable over prolonged periods of time and can thus act in the same way as plant 'seed banks', promoting temporal gene flow within local populations. Such resting egg banks have been found in cladocerans (Caceres 1998), copepods (Hairston *et al.* 1995) and rotifers (Gómez & Carvalho 2000). In addition, statoblast banks of *C. mucedo* have been inferred from population genetic studies which suggest repopulation from statoblasts released from sediments within a site, rather than from colonizers originating elsewhere (Freeland *et al.* 2001). The extent to which statoblast banks in *C. mucedo* may counteract extinction events, thereby confounding interpretations of metapopulation dynamics, remains to be determined (see also Bullock *et al.*, this volume). However, few statoblasts have remained viable beyond 6 years in laboratory studies, and most appear to be inviable after 2 years (Bushnell & Rao 1974; Mukai 1982; Wood 1991). The general pattern is for statoblast viability to decline with age. Criteria based on size and taxonomic similarity suggest a viability of 4–8 years for statoblasts of *C. mucedo.* In contrast, the duration of viability of zooplankton eggs is in the order of decades to several hundreds of years (e.g. Hairston *et al.* 1995). It is to be expected that the dynamics of local populations of *C. mucedo* will be influenced by a mixture of both spatial and short-term (<10 years) temporal gene flow.

Emergence from propagule banks may be particularly significant in years when reproduction is poor. Thus, gains achieved during good years can compensate for losses sustained in poor years, thereby providing a storage effect in temporally varying environments (De Stasio 1990; Hairston *et al.* 1996). These repeated releases of lineages can also maintain local genetic variation, thereby influencing the rate of evolution in response to switches in selection pressures (Hairston & De Stasio 1988; Ellner *et al.* 1999) which may be reflected in genetic shifts in propagule banks (Weider *et al.* 1997; Cousyn *et al.* 2001). The presence of heterozygote deficiencies may reflect a temporal Wahlund effect if the resting propagule bank contains an admixture of propagules produced by temporally isolated populations differing in their allele frequencies (Freeland *et al.* 2000a; Gómez & Carvalho 2000), and indeed such a temporal Wahlund effect may contribute to the high levels of F_{IS} in *C. mucedo* populations referred to earlier.

For both zooplankton and bryozoans, temporal gene flow provided by hatching of propagules that have been retained in sediments may often be of greater importance to population genetic structure than spatial gene flow. These taxa rely on passive dispersal to achieve colonization of new sites, and such dispersal is likely to be a relatively rare event, contributing in part to the low to non-existent levels of gene flow that are typically observed (see earlier discussions). Periodic reintroduction of dormant propagules from the sediment may be a much more reliable means of achieving dispersal. Temporal gene flow may, therefore, play a pivotal role in both the maintenance of genetic diversity and the avoidance of extinction in local populations. In addition, in taxa such as *C. mucedo* that produce asexual dormant propagules, temporal gene flow may promote the long-term persistence of clones within sites. Finally, although the storage effect provided by propagule banks may slow the rate of microevolution in response to fluctuating selection (Hairston & De Stasio 1988; Ellner *et al.* 1999), propagules banks may nevertheless promote local adaptation by enhancing the long-term persistence of populations within sites.

Dispersal, microevolution, life history and ecological patterns: lessons from bryozoans and zooplankton

Population genetic studies of freshwater invertebrates of lakes and ponds are beginning to provide a link between microevolution within populations and the resultant dynamics of those populations. Investigations of zooplankton taxa commonly reveal extensive local adaptation reflected in high levels of genetic differentiation amongst populations and cryptic speciation. Ecological studies of established and experimental zooplankton communities show that colonization by new species is often precluded, despite dispersal, because communities are often saturated (Jenkins & Bukema 1998; Shurin 2000; Shurin *et al.* 2000). This combined information from ecological and genetic studies suggests that microevolutionary processes within sites determine the relative importance of regional versus local influences on zooplankton population and community structure. It also suggests that the apparent absence of metapopulation structures associated with zooplankton taxa can be explained if local adaptation precludes ongoing gene flow between sites.

In contrast, genetic investigations of populations of *C. mucedo* reveal ongoing gene flow that maintains a metapopulation across northwest Europe. In this metapopulation, the genetic differentiation which characterizes local populations may be greatly influenced by founder effects during (re)colonization, followed by inbreeding, genetic drift and temporal gene flow via statoblast banks. None the less, the identification of multilocus genotypes at multiple sites suggests the persistence of general purpose genotypes that are adapted to the broad range of environmental conditions present within this metapopulation, although the gradual erosion of clonal diversity within a season is suggestive of local clonal selection (Hatton-Ellis 1997; Freeland *et al.* 2001).

North American populations of *C. mucedo* show intriguing differences from populations in northwest Europe. In particular, they reveal lower levels of gene flow, lack of evidence for a metapopulation structure, and the possibility of cryptic speciation. There are a number of potential explanations for the high levels of genetic differentiation among these North American populations. These explanations include sampling subpopulations from different metapopulations and processes that characterize zooplankton taxa such as infrequent spatial gene flow and local adaptation. Identifying the major determinants of genetic structure in North American populations will require further investigation.

It is notable that genetic and ecological data indicate a variable role of sexual reproduction in bryozoan and zooplankton populations. *C. mucedo* undergoes a brief, early period of sexual activity which entails at least some outcrossing (J. Freeland, unpubl. data.), although a sexual phase appears to be foregone in some populations at least in some years (Okamura 1997b). None of the *C. mucedo* populations that have been characterized were in Hardy–Weinberg equilibrium, and marked observed heterozygosity deficiencies provide evidence for both inbreeding and the Wahlund effect. In contrast, populations of many zooplankton taxa undergo sexual reproduction at the end of the season and generally conform to Hardy–Weinberg proportions (De Meester 1996; Gómez & Carvalho 2000). Since sex may play an important role in contributing to local adaptation through the generation of genetic diversity, there is an intriguing possibility that variation in the timing and expression of sexual reproduction may explain variation in levels of population divergence and local adaptation. In zooplankton taxa, regular sexual production of resting stages will generate significant genetic diversity in the population and hence will promote adaptive change. In bryozoans, the asexual production of resting stages will at best only maintain extant levels of genetic diversity into the next season. Further factors that may depress adaptive change in bryozoans are inbreeding, irregular sex and the erosion of genetic diversity through the growing season by local selection prior to the asexual production of overwintering statoblasts.

Conclusions

Population genetic and ecological studies of freshwater invertebrates are beginning to reveal how the frequency and scale of dispersal and gene flow relate to ecological and evolutionary patterns. Occasional, ongoing dispersal appears to maintain a widespread metapopulation structure that unites some bryozoan populations. However, in many zooplankton taxa colonization of new sites appears to be relatively infrequent and evolutionary divergence following postglacial expansion has resulted in local adaptation and cryptic speciation. Further study of freshwater invertebrate taxa will help to unravel the specific influences of gene flow in space and time, and may also clarify the conditions that promote dispersal and metapopulation structure versus population subdivision and accelerated adaptive change.

References

Berg, D.J. & Garton, D.W. (1994) Genetic differentiation in North American and European populations of the cladoceran *Bythotrephes*. *Limnology and Oceanography* **39**, 1503–1516.

Bilton, D., Freeland, J.R. & Okamura, B. (2001) Dispersal in freshwater invertebrates. *Annual Review of Ecology and Systematics* **32**, 159–181.

Bohonak, A.J. (1999) Dispersal, gene flow, and population structure. *Quarterly Review of Biology* **74**, 21–45.

Boileau, M.G. & Hebert, P.D.N. (1988) Genetic differentiation of fresh-water pond copepods at Arctic sites. *Hydrobiologia* **167**, 393–400.

Boileau, M.G. & Hebert, P.D.N. (1991) Genetic consequences of passive dispersal in pond-dwelling copepods. *Evolution* **45**, 721–733.

Boileau, M.G., Hebert, P.D.N. & Schwartz, S.S. (1992) Nonequilibrium gene frequency divergence: persistent founder effects in natural populations. *Journal of Evolutionary Biology* **5**, 25–39.

Brown, C.J.D. (1933) A limnological study of certain fresh-water Polyzoa with special reference to their statoblasts. *Transactions of the American Microscopical Society* **52**, 271–314.

Bushnell, J.H. & Rao, K.S. (1974) Dormant or quiescent stages and structures among the Ectoprocta: physical and chemical factors affecting viability and germination of statoblasts. *Transactions of the American Microscopical Society* **93**, 524–543.

Caceres, C.E. (1998) Interspecific variation in the abundance, production, and emergence of *Daphnia* diapausing eggs. *Ecology* **79**, 1699–1710.

Cerny, M. & Hebert, P.D.N. (1993) Genetic diversity and breeding system variation in *Daphnia pulicaria* from North American lakes. *Heredity* **71**, 497–507.

Cornell, H.V. & Karlson, R.H. (1997) Local and regional processes as controls of species richness. In: *Spatial Ecology: The Role of Space in Population Dynamics and Interspecific Interactions* (eds D. Tilman & P. Kareiva), pp. 250–268. Princeton University Press, Princton, NJ.

Cornell, H.V. & Lawton, J.H. (1992) Species interactions, local and regional processes, and limits to the richness of ecological communities: a theoretical perspective. *Journal of Animal Ecology* **61**, 1–12.

Cousyn, C., De Meester, L., Colbourne, J.K., Bendonck, L., Verschuren, D. & Volckaert, F. (2001) Rapid local adaptation of zooplankton behavior to changes in predation pressure in absence of neutral genetic changes. *Proceedings of the National Academy of Sciences of the USA* **98**, 625–626.

Crease, T.J., Lynch, M.M. & Spitze, K. (1990) Heirarchical analysis of population genetic variation in mitochondrial and nuclear genes of *Daphnia pulex*. *Molecular Biology and Evolution* **7**, 444–458.

Darwin, C. (1859) *On the Origin of Species by Means of Natural Selection*. John Murray, London.

De Meester, L. (1996) Local genetic differentiation and adaptation in freshwater zooplankton populations: patterns and processes. *Ecoscience* **3**, 385–399.

De Stasio Jr, B.T. (1990) The role of dormancy and emergence patterns in the dynamics of a freshwater zooplankton community. *Limnology and Oceanography* **35**, 1070–1090.

Duffy, M.A., Perry, L.J., Kearns, C.M., Weider, L.J. & Hairston Jr, N.G. (2000) Paleogenetic evidence for a past invasion of Onondaga Lake, New York, by exotic *Daphnia curvirostris* using mtDNA from dormant eggs. *Limnology and Oceanography* **45**, 1409–1414.

Ehrlich, P.R. (1986) Which animal will invade? In: *Ecology of Biological Invasions of North America and Hawaii* (eds H.A. Mooney & J.A. Drake), pp. 79–95. Springer Verlag, New York.

Ellner, S.P., Hairston Jr, N.G., Kearns, C.M. & Babaï, D. (1999) The roles of fluctuating selection and long-term diapause in microevolution of diapause timing in a freshwater copepod. *Evolution* **53**, 111–122.

Freeland, J.R., Noble, L.R. & Okamura, B. (2000a) Genetic consequences of the metapopulation biology of a facultatively sexual freshwater invertebrate. *Journal of Evolutionary Biology* **13**, 383–395.

Freeland, J.R., Noble, L.R. & Okamura, B. (2000b) Genetic diversity of North American populations of *Cristatella mucedo*, inferred from microsatellite and mitochondrial DNA. *Molecular Ecology* **9**, 1375–1389.

Freeland, J.R., Romualdi, C. & Okamura, B. (2000c) Gene flow and genetic diversity: a comparison of freshwater bryozoan populations in Europe and North America. *Heredity* **85**, 498–508.

Freeland, J.R., Rimmer, V.K. & Okamura, B. (2001) Genetic changes within freshwater bryozoan populations suggest temporal gene flow from statoblast banks. *Limnology and Oceanography* **46**, 1121–1129.

Gómez, A. & Carvalho, G.R. (2000) Sex, parthenogenesis and genetic structure of rotifers: microsatellite analysis of contemporary and resting egg bank populations. *Molecular Ecology* **9**, 203–214.

Gómez, A., Carvalho, G.R. & Lunt, D.H. (2000) Phylogeography and regional endemism of a passively dispersing zooplankter: mitochondrial DNA variation in rotifer resting egg banks. *Proceedings of the Royal Society of London, Series B* **267**, 2189–2197.

Goodman, S.J. (1997) R-st calc: a collection of computer programs for calculating estimates of genetic differentiation from microsatellite data and determining their significance. *Molecular Ecology* **6**, 881–885.

Hairston Jr, N.G. & De Stasio Jr, B.T. (1988) Rate of evolution slowed by a dormant propagule pool. *Nature* **336**, 239–242.

Hairston Jr, N.G., Van Brunt, R.A. & Kearns, C.M. (1995) Age and survivorship of diapausing eggs in a sediment egg bank. *Ecology* **76**, 1706–1711.

Hairston Jr, N.G., Ellner, S. & Kearns, C.M. (1996) Overlapping generations: the storage effect and the maintenance of biotic diversity. In: *Populations Dynamics in Ecological Space and Time* (eds O.E. Rhodes, R.K. Chesser & M.H. Smith), pp. 109–145. Chicago University Press, Chicago.

Hairston Jr, N.G., Perry, L.J., Bohonak, A.J., Fellows, M.Q. & Kearns, C.M. (1999a) Population biology of a failed invasion: paleolimnology of *Daphnia exilis* in upstate New York. *Limnology and Oceanography* **44**, 477–486.

Hairston Jr, N.G., Lampert, W., Cáceres, C.E. *et al.* (1999b) Rapid evolution revealed by dormant eggs. *Nature* **401**, 446.

Hann, B.J. (1995) Genetic variation in *Simocephalus* (Anomopoda, Daphniidae) in North America: patterns and consequences. *Hydrobiologia* **307**, 9–14.

Hanski, I. (1989) Metapopulation dynamics: does it help to have more of the same? *Trends in Ecology and Evolution* **66**, 335–343.

Hanski, I. (1999) *Metapopulation Ecology*. Oxford University Press, Oxford.

Hanski, I. & Simberloff, D. (1997) The metapopulation approach, its history, conceptual domain, and application to conservation. In: *Metapopulation Biology. Ecology, Genetics, and Evolution* (eds I. Hanski & M.E. Gilpin), pp. 5–26. Academic Press, San Diego.

Harrison, S. & Taylor, A.D. (1997) Empirical evidence for metapopulation dynamics. In: *Metapopulation Biology. Ecology, Genetics, and Evolution* (eds I. Hanski & M.E. Gilpin), pp. 27–42. Academic Press, San Diego.

Hatton-Ellis, T.W. (1997) *The molecular ecology of Cristatella mucedo (Bryozoa: Phylactolaemata) in space and time.* PhD thesis, University of Bristol, Bristol.

Hatton-Ellis, T., Noble, L.R. & Okamura, B. (1998) Genetic variation in a freshwater bryozoan. I. Populations in the Thames Basin. *Molecular Ecology* **7**, 1575–1585.

Havel, J.E., Mabee, M.R. & Jones, J.R. (1995) Invasion of the exotic cladoceran *Daphnia lumholtzi* into North American reservoirs. *Canadian Journal of Fisheries and Aquatic Sciences* **52**, 151–160.

Hebert, P.D.N. & Finston, T.L. (1996) Genetic differentiation in *Daphnia obtusa*: a continental perspective. *Freshwater Biology* **35**, 311–321.

Hebert, P.D.N. & Hann, B.J. (1986) Patterns in the composition of Arctic tundra pond microcrustacean communities. *Canadian Journal of Fisheries and Aquatic Sciences* **43**, 1416–1425.

Hebert, P.D.N. & Moran, C. (1980) Enzyme variability in natural populations of *Daphnia carinata* King. *Heredity* **45**, 313–321.

Hebert, P.D.N. & Wilson, C.C. (1994) Provincialism in plankton: endemism and allopatric speciation in Australian *Daphnia*. *Evolution* **48**, 1333–1349.

Hebert, P.D.N., Beaton, M.J., Schwartz, S.S. & Stanton, D.J. (1989) Polyphyletic origins of asexuality in *Daphnia pulex*. 1. Breeding system variation and levels of clonal diversity. *Evolution* **43**, 1004–1015.

Hedrick, P.W. & Gilpin, M.E. (1997) Genetic effective size of a metapopulation. In: *Metapopulation Biology* (eds I. Hanski & M.E. Gilpin), pp. 166–182. Academic Press, San Diego.

Holland, T.A. & Jenkins, D.G. (1998) Comparison of processes regulating zooplankton assemblages in new freshwater pools. *Hydrobiologia* **387**, 207–214.

Innes, D.J. (1990) Geographic patterns of genetic differentiation among sexual populations of *Daphnia pulex*. *Canadian Journal of Zoology* **69**, 995–1003.

Jenkins, D.G. & Buikema Jr, A.L. (1998) Do similar communities develop in similar sites? A test with zooplankton structure and function. *Ecological Monographs* **68**, 421–443.

Johnson, L.E. & Carlton, J.T. (1996) Post-establishment spread in large-scale invasions: dispersal mechanisms of the zebra mussel *Dreissena polymorpha*. *Ecology* **77**, 1686–1690.

Levins, R. (1969) Some demographic and genetic consequences of environmental heterogeneity for biological control. *Bulletin of the Entomological Society of America* **15**, 237–240.

Levins, R. (1970) Extinction. *Lecture Notes in Mathematical Life Sciences* **2**, 75–107.

Lynch, M. & Spitze, K. (1994) Evolutionary genetics of *Daphnia*. In: *Ecological Genetics* (ed. L.A. Real), pp. 109–128. Princeton University Press, Princeton, NJ.

Menéndez, R. & Thomas, C.D. (2000) Metapopulation structure depends on spatial scale in the host-specific moth *Wheeleria spilodactylus* (Lepidoptera: Pterophoridae). *Journal of Animal Ecology* **69**, 935–951.

Mukai, H. (1982) Development of freshwater bryozoans (Phylactolaemata). In: *Developmental Biology of Freshwater Invertebrates* (eds F.W. Harrison & R.R. Cowden), pp. 535–576. Alan R. Liss, New York.

Nurnberg, B. & Harrison, R.G. (1995) Spatial population structure in the whirligig beetle *Dineutus assimilis*: evolutionary inferences based on mitochondrial DNA and field data. *Evolution* **49**, 266–275.

Okamura, B. (1997a) Genetic similarity, parasitism, and metapopulation structure in a cyclically clonal bryozoan. In: *Evolutionary Ecology of Freshwater Animals* (eds B. Streit, T. Stäedler & C. Lively), pp. 294–320. Birkhäuser Verlag, Basel, Switzerland.

Okamura, B. (1997b) The ecology of subdivided populations of a clonal freshwater bryozoan in southern England. *Archiv für Hydrobiologie* **141**, 13–34.

Pennak, R.W. (1989) *Freshwater Invertebrates of the United States. Protozoa to Mollusca*, 3rd edn. John Wiley & Sons, New York.

Potter, R. (1979) Bryozoan karyotypes and genome sizes. In: *Advances in Bryozoology* (eds G.P. Larwood & M.B. Abbott), pp. 11–32. Academic Press, London.

Ricklefs, R.E. (1987) Community diversity: relative roles of local and regional processes. *Science* **235**, 167–171.

Schön, I., Gandolfi, A., Di Masso, E. *et al.* (2000) Persistence of asexuality through mixed reproduction in *Eucypris virens* (Crustacea, Ostracoda). *Heredity* **84**, 161–169.

Schwenk, K., Posada, D. & Hebert, P.D.N. (2000) Molecular systematics of European Hyalodaphnia: the role of contemporary hybridization in ancient species. *Proceedings of the Royal Society of London, Series B* **267**, 1833–1842.

Scott, D.A. & Rose, P.M. (1996) *Atlas of Anatidae Populations in Africa and Western Eurasia*. Wetlands International Publication No. 41. Wetlands International, Wageningen, the Netherlands.

Shurin, J.B. (2000) Dispersal limitation, invasion resistance, and the structure of pond zooplankton communities. *Ecology* **81**, 3074–3086.

Shurin, J.B., Havel, J.E., Leibold, M.A. & Pinel-Alloul, B. (2000) Local and regional zooplankton species richness: a scale-independent test for saturation. *Ecology* **81**, 3062–3073.

Slatkin, M. (1987) Gene flow and the geographic structure of natural populations. *Science* **236**, 787–792.

Srivastava, D.S. (1999) Using local–regional richness plots to test for species saturation: pitfalls and potentials. *Journal of Animal Ecology* **68**, 1–16.

Stemberger, R.S. (1995) Pleistocene refuge areas and post-glacial dispersal of copepods of the northeastern United States. *Canadian Journal of Fisheries and Aquatic Sciences* **52**, 2197–2210.

Taylor, D.J., Hebert, P.D.N. & Colbourne, J.K. (1996) Phylogenetics and evolution of the *Daphnia longispina* group (Crustacea) based on 12S rDNA sequence and allozyme variation. *Molecular Phylogeny and Evolution* **5**, 495–510.

Taylor, D.J., Finston, T.L. & Hebert, P.D.N. (1998) Biogeography of a widespread freshwater crustacean: pseudocongruence and cryptic endemism in the North American *Daphnia laevis* complex. *Evolution* **52**, 1648–1670.

Tessier, S.J., Young, A. & Leibold, M.A. (1992) Population dynamics and body size selection in *Daphnia. Limnology and Oceanography* **37**, 1–13.

Vanoverbeke, J. & De Meester, L. (1997) Among-populational genetic differentiation in the cycle parthenogen *Daphnia magna* (Crustacea, Anomopoda) and its relation to geographic distance and clonal diversity. *Hydrobiologia* **360**, 135–142.

Vernon, J.G., Okamura, B., Jones, C.S. & Noble, L.R. (1996) Temporal patterns of clonality and parasitism in a population of freshwater bryozoans. *Proceedings of the Royal Society of London, Series B* **263**, 1313–1318.

Wade, M.J. & McCauley, D.E. (1988) Extinction and recolonization: their effect on the genetic differentiation of local populations. *Evolution* **42**, 995–1005.

Weider, L.J. (1989) Population genetics of *Polyphemus pediculus* (Cladocera: Polyphemidae). *Heredity* **62**, 1–10.

Weider, L.J. & Hobæk, A. (1997) Postglacial dispersal, glacial refugia, and clonal structure in Russian/Siberian populations of the arctic *Daphnia pulex* complex. *Heredity* **78**, 363–372.

Weider, L.J., Hobæk, A., Crease, T.J. & Stibor, H. (1996) Molecular characterization of clonal population structure and biogeography of arctic apomictic *Daphnia* from Greenland and Iceland. *Molecular Ecology* **5**, 107–118.

Weider, L.J., Lampert, W., Wessels, M., Colbourne, J.K. & Limburg, P. (1997) Long-term genetic shifts in microcrustacean egg bank associated with anthropogenic changes in Lake Constance ecosystem. *Proceedings of the Royal Society of London, Series B* **264**, 1613–1618.

Weider, L.J., Hobæk, A., Colbourne, J.K., Crease, T.J., Defresne, F. & Hebert, P.D.N. (1999a) Holarctic phylogeography of an asexual species complex: I. mtDNA variation in arctic *Daphnia. Evolution* **53**, 777–792.

Weider, L.J., Hobæk, A., Hebert, P.D.N. & Crease, T.J. (1999b) Holarctic phylogeography of an asexual species complex: II. Allozymic variation in arctic *Daphnia. Molecular Ecology* **8**, 1–13.

Wetzel, R.G. (1975) *Limnology.* W.B. Saunders, Philadelphia.

Whitlock, M.C. & McCauley, D.E. (1999) Indirect measures of gene flow and migration: F-st ≠ 1/(4Nm + 1). *Heredity* **82**, 117–125.

Williams, D.D. (1987) *The Ecology of Temporary Waters.* Croom Helm, London.

Wood, T.S. (1991) Bryozoans. In: *Ecology and Classification of North American Freshwater Invertebrates* (eds J.H. Thorp & A.P. Covich), pp. 481–499. Academic Press, New York.

Wöss, E.R. (1994) Seasonal fluctuations of bryozoan populations in five water bodies with special emphasis on the life cycle of *Plumatella fungosa* (Pallas). In: *Biology and Palaeobiology of Bryozoans* (eds P.J. Hayward, J.S. Ryland & P.D. Taylor), pp. 211–214. Olsen & Olsen, Fredensborg, Denmark.

Part 3

Dispersal and spatial processes

Chapter 11

Niche colonization and the dispersal of bacteria and their genes in the natural environment

Mark J. Bailey and Andy K. Lilley

Introduction

In this chapter we have limited our deliberations in order to present a logical consideration of the factors that constrain or drive the dispersal of bacteria. In this respect we have focused on the relevance of dispersal to bacterial populations in relation to biological scale. For example, the distribution and survival of bacteria and their genes in the terrestrial environment: natural selection determines the ability to persist and colonize and in simple terms is the objective end result of successful dispersal. But are these local, habitat driven events, or global effectors? Bacteria are ubiquitous in nature and can be found at the very extremes of the biosphere. Recent estimates indicate that there may be 4–6 × 10^{30} prokaryotic cells representing some 3.5–5.5 × 10^{17} g of cellular carbon on earth, effectively half of the biomass of plants (Whitman *et al.* 1998). Bacteria are central to the functioning of terrestrial habitats, they play a key role in mineralization, decomposition and biogeochemical cycling by their own inherent activities or by interaction with higher trophic predators within the microbial loop. Natural bacterial communities consist of diverse assemblages of species in a wide range of states ranging from rapidly to slowly growing, senescent and dead. Importantly, many micro-organisms form quiescent resting structures or spores that directly aid survival and assist dispersal. Despite this apparent physiological heterogeneity, bacterial populations have traditionally been defined as bulk entities in relation to physiological process measurements and some hold the view that they are globally distributed and that everything is everywhere. The concept that everything is everywhere and the environment selects was first examined at the turn of the 20th century (Beijerinck 1901) and to some extent still holds true depending on the scale of measurement. The extent of relative distribution and diversity is a fundamental issue, not only in terms of evolution and the biological influence of dispersal, but also when defining biodiversity and ecosystem function. Simply put, the extent of global distribution can be supported or refuted in accordance with the quality of the data used to characterize and compare geographically disparate isolates. A complicating factor is the difficulty of speciation in bacteria. Estimates of the numbers of species vary from a few hundreds of thousands to tens of millions, but current classification systems can be misleading. The common use of phylogenies based on 16S SSU ribosomal RNA sequence analysis is of benefit in simplifying description and identification, but it is inadequate for describing the extent of prokaryotic genetic diversity. Indeed the application of phenotypic and genetic methods confirms the extent of microbial diversity within close phylogenetic groups and the evolutionary pressures that lead to the selection of ecotypes and specialists.

Specialization and selection of individuals may involve defined functions, but the overall diversity of the microflora allows for heterogeneity in ecosystem function in response to imposed changes in the local environment. This ability of a community to exploit resources may be defined in terms of dispersal within the system, where individuals disperse by proliferation and clonal expansion of the cell itself and the genes carried. Further, dispersal of the population of cells in the environment can be mediated by physical conditions. For example, soil and plant bacteria can readily be dispersed by flooding, rain splash, in aerosols or as dust particles, through the rubbing of leaves, pollinating or phytophagous animals and insects, and as a consequence of disease. The distances that they move may be huge, and probably pan-global. But whether they survive and proliferate is also a key factor to successful dispersal. For pathogens successful dispersal, leading to proliferation, depends on contact and infection with a susceptible host (Hirano & Upper 2000). Barriers to dispersal include host susceptibility (resistance, physical damage), the presence of suitable vectors (i.e. biting insects) or the appropriate means of physical transfer (wind dispersal or aerosols). Non-pathogens still require dispersal to an exploitable niche where the immigrant can out-compete residents. Ultimately, successful invasion following dispersal is a function of the phenotypic plasticity of the cell (Rainey *et al.* 1994). These are all obvious considerations, but bacteria have the additional capacity to extend their genotype by the acquisition of genes, not by sexual recombination, but by access to the horizontal gene pool (Thomas 2000). Gene transfer is mediated by conjugative (self-transmissible plasmids), bacteriophage and direct (the uptake of and recombination with DNA) transformation. Horizontal gene transfer may be universal within bacteria, but because the transfer of mobile genetic elements is an active process there are limitations to the extent of dispersal dependent on the physiological state of both the donor and recipient bacteria. The extent of transfer is often constrained by the range of host bacteria that are permissive to the replication and maintenance of the acquired mobile genetic element. None the less, gene dispersal can and does occur to new populations including those beyond the boundaries of the genus. Therefore, in ecological terms the dispersal of the functional attributes of bacteria carried within its gene pool may not necessarily be limited by the physical constraints of the cell.

It is important, therefore, to consider spread and dispersal in terms of whether bacterial populations can be defined as either endemic or cosmopolitan. This definition was considered in some detail by Cho and Teidje (2000) who investigated the biogeography of *Pseudomonas sensu stricto* and identified 85 unique genotypes in a sample of 248 isolates from 10 independent transect sites on four continents. Interestingly, they found that genotypes did not overlap among sites but observed 'strict site endemism'. Essentially the genetic distances between isolates were correlated with the distances between their sites of origin (see also Raybould *et al.*, this volume). In this respect it is reasonable to conclude that these bacteria are not globally mixed, a point supported by Fulthorpe *et al.* (1998) who recorded high levels of endemicity in 3-chlorobenzoate-degrading soil bacteria. In the habitat we targeted to study the population genetics of pseudomonads, the sugar beet phytosphere, detailed investigations have revealed considerable diversity in this group of bacteria and in the horizontal gene pool they carry. For example, in a sample of 473 fluorescent pseudomonad isolates, sampled approximately every 30–40 days over three consecutive growing seasons, a total of 115 distinct ribotypes (genotypes) were recorded (Ellis *et al.* 1999). Although the abundance of isolates was log-normally distributed, individual ribotypes were detected more often than would be predicted by such a distribution. The number of ribotypes identified in each sample was directly related to the size of that sample. These observations demonstrate that the genetic distances between individual ribotypes are probably fixed and selected for by the habitat, and exhibit a dynamic, non-random, continuous turnover within the population. The seasonal pattern of isolate reoccurrence was consistent with the dispersal and selection of individual groups, proliferating and displacing other groups as they are better adapted to the prevailing conditions in competition for resources. Although it is easy to postulate that successional sweeps of populations occur (Bergstrom *et al.* 2000), it is less easy to propose a suitable model that accounts for their continued persistence season to season. We need to know whether all, or most, of the components of a community are pervasive and persist in low numbers. If they are then redundancy must be extensive and individuals must have adopted survival mechanisms to facilitate their long-term survival. We need also to consider how, in the immediate habitat and at greater scales, inocula arrive and physically disperse during colonization. Indeed, how significant a role does immigration play in the establishment and maintenance of microbial communities?

In Fig. 11.1 we present a simplified model identifying three distinct levels of dispersal relevant to bacteria and their genes. To illustrate this we have included the example of mercuric resistance (*mer*) genes which confer resistance to highly toxic mercuric ions (Hg^{2+}), are widely dispersed, if not ubiquitous, in nature (Yurieva *et al.* 1997) and because they provide a practical model for our studies. The *mer* genes are found organized in operon structures containing seven to nine genes. Mercuric resistance (Hg^r) is conferred by the production of proteins for isolating, transporting and reducing Hg^{2+} to metal mercury Hg^0 which is relatively non-toxic. However, the precise role that this resistance plays in the ecology of bacterial populations is unknown.

The first level of dispersal described in Fig. 11.1 is intragenomic where genes, e.g. the *mer* operon, may be moved by transposition within a genome. This dispersal can result in new copies of the transposon and the genes it carries being inserted in the chromosome and other extrachromosomal DNA such as plasmids and bacterial viruses (phages). The second level of dispersal is described by plasmids and bacteriophage. These mobile genetic elements function as vectors that mobilize DNA to new bacterial hosts. In this overview we concentrate on the impact of conjugative,

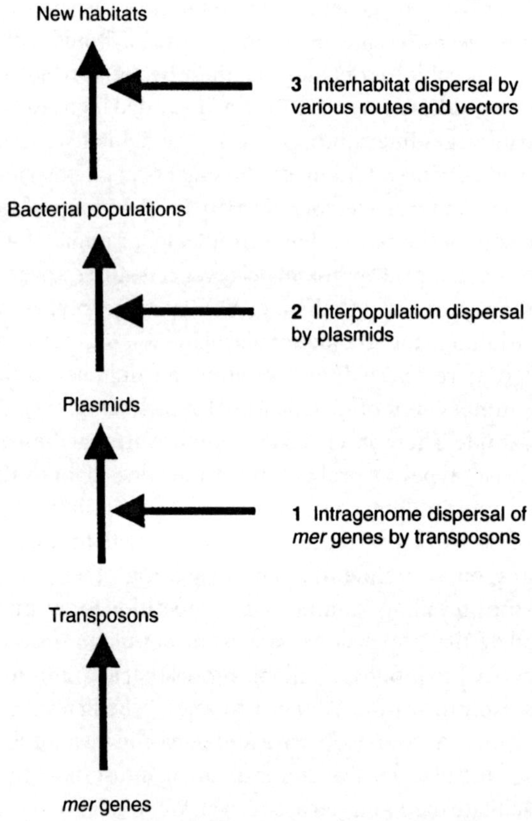

Figure 11.1 A hierarchial model of *mer* gene dispersal. This scheme shows *mer* genes carried on transposons, which are mobile elements, and in this case are carried on plasmids, which in turn disperse to new bacterial populations, which in turn are vectored or move to new habitats.

self-transmissible plasmids as mediators of genetic variation and gene dispersal. However, for further detail of the role of bacteriophage in host population dynamics in the sugar beet phytosphere we recommend Ashleford *et al.* (1999a, 1999b, 2000). Conjugative plasmids carry operator genes necessary for their own replication, and auxiliary genes that provide selective advantage under specific conditions, such as those of the *mer* operon. These plasmids represent a vital part of the horizontal gene pool, which is a key component of bacterial communities providing genotypic plasticity to new populations. Their host range may be narrow or broad allowing transfer to occur between bacteria separated by considerable taxonomic distances. Essentially, mobile genetic elements such as plasmids may be the vectors of traits that drive evolution by the subsequent transfer of novel genes, or variation to existing genes, by recombination into the chromosome. Plasmids are essentially ubiquitous. Estimates of their distribution vary, but typically 10–40% of all bacterial isolates examined carry self-replicating extrachromosomal genetic elements. They persist and many freely disperse within and between bacterial populations. Later, we demonstrate that plasmids of indigenous bacterial communities carry genes that not only respond to local environmental signals, but that they carry specific traits that confer adaptive advantage to the host bacteria. Such selective advantage effectively aids the dispersal of the plasmid, as the host–plasmid combination proliferates under selection. As a consequence of increased density the probability of the plasmid transferring to a new host during successional sweeps is enhanced, thus ensuring survival by optimizing conditions for dispersal.

The third level of dispersal defines how bacteria themselves may move or be vectored to new habitats. At each level of dispersal it is likely that different regulatory mechanisms operate. Each level of dispersal impacts on the higher level, introduces variation and is influenced by the levels of selection imposed. Generally, we would argue that *effective* dispersal (at each level) in the biological context is a function of selection, where local adaptation and effective competition for persisting strains favours niche selection and the development of ecotypes.

To understand the dispersal of bacteria and their genes we have targeted the non-pathogenic, ubiquitous, numerically dominant resident fluorescent pseudomonads and their associated mobile genetic elements that are typical residents of the leaf and root surfaces of plants. This is primarily because this is a well studied group that demonstrates considerable genetic diversity despite their apparent niche adaptation and specialization as plant-associated bacteria (Speirs *et al.* 2000). Studies have concentrated on a particular strain, *Pseudomonas fluorescens* SBW25, which was isolated from the leaves of sugar beet grown at the University Farm, Wytham, Oxfordshire, UK (Bailey *et al.* 1995; Rainey & Bailey 1996). In field and laboratory experiments *P. fluorescens* SBW25 has been shown to be an effective colonizer of both the phyllosphere and rhizosphere following seed inoculation. Figure 11.2 illustrates the concept of the ecological scale at which investigations are conducted, and demonstrates the distribution of this bacteria on seedlings.

Our studies have demonstrated that SBW25 is highly competitive and disperses over the entire surface of growing sugar beet plants. It reached population densities equivalent to the indigenous fluorescent pseudomonads and, optimally, represented approximately 80% of all pseudomonads present on immature leaves (Thompson *et al.* 1995b; Lilley & Bailey 1997a). In addition we have demonstrated that large, conjugative plasmids, that confer resistance to both inorganic and organic mercurial compounds (Hg^r), are indigenous to plant-associated pseudomonads (Lilley *et al.* 1996), transfer at high frequency between populations of bacteria colonizing the plant surface (Lilley *et al.* 1994; Lilley & Bailey 1997a) and also provide periodic selective advantage to the bacteria that carry them (Bailey *et al.* 1996; 2001; Lilley & Bailey 1997b).

In the following text, the three levels of dispersal considered in Fig. 11.1 are illustrated with data from studies of the dispersal of *mer* genes, plasmids and pseudomonads in root, leaf and soil habitats. Three specific and linked investigations are presented to demonstrate the importance of dispersal in maintaining and

Figure 11.2 Illustration of the biological scale over which the dispersal of bacteria and their genes needs to be considered. (a) The field plot of sugar beet grown at the University Farm, Wytham, Oxfordshire, UK. Inocula were introduced as a seed dressing to three plots of 100 plants. The treated plots were surrounded by bare soil and two blocks of guard rows. No dispersal of the inocula, *Pseudomonas fluorescens* SBW25, to the guard rows was detected (Thompson *et al.* 1995b). (b) Dispersal of SBW25 modified to express the green fluorescent protein following seed inoculation. This confocal image shows *P. fluorescens* SBW25 (*rrn*BP1::*gfp*) colonizing a 2-day-old seedling. (Courtesy of C. Ramos and S. Molin, DTU, Copenhagen.)

establishing bacterial communities in the phyllosphere of plants. Firstly we consider data arising from the sequence analysis of *mer* genes and the role of mobile genetic elements in intracellular dispersal of these to new DNA loci and molecules. The second example demonstrates the central role of conjugative plasmids on the intercellular and interpopulation dispersal of genes and the effect this has on bacterial survival by extending the phenotype to enhance local adaptation (Lilley & Bailey 1997a, 1997b). Thirdly, the role of passive contact in the spatial and interplant dispersal of leaf-colonizing bacteria, mediated by phytophagous insects, is demonstrated (Lilley *et al.* 1997).

Intragenomic dispersal

Mechanisms exist which can mobilize DNA sequences between DNA molecules within a cell. These mechanisms involve mobile genetic elements which mediate the relocation of DNA sequences or the insertion of new copies of a sequence to new loci. A common class of these mobile genetic elements are the transposons which include the DNA sequences for self-recognition (the insertion sequences IS that flank the functional transposon) and the genes for self-replication and/or insertion (transposases) into new sites in the genome (for a review, see Sherratt 1995). Some transposons insert at specific sites, but many have minimal recognition sites and effectively integrate at random. Transposons are dependent on host replication for their maintenance, but regulate their own spread and copy number within a genome. In this respect they may be viewed as molecular parasites or selfish elements (Brookfield 1995; Plasterk 1995). Another important feature of transposable elements is that they acquire and incorporate other genes by recombination and disperse them to new loci. This dispersal can redistribute genes within and between chromosomes, plasmids and phages within a cell. To return to our example of *mer*, it is well known that genes encoding resistance to mercuric compounds are typically associated with transposons, and that these transposons are frequently located in plasmids. We have established that three of the most common plasmids groups (I, III and IV) resident in the pseudomonad populations at our single field site, which are otherwise genetically distinct, carry an identical mercury resistance transposon (K. Bruce, unpubl. data). Clearly, recent transposition events have dispersed these genes. The *mer* gene–plasmid associations recorded represent only a subset of those generated that are able to persist in the community. By intragenomic transposition to the self-transferring plasmids, genes have been dispersed to new populations, and to new plasmids, enhancing the potential of transfer to a greater diversity of host bacteria.

Dispersal of bacteria and their genes through microbial communities

A major problem in understanding population processes occurring in bacterial ecology and evolution is knowing whether the bacterial populations are clonal or reticulate evolutionary units. For some time it was believed that gene flow and recombination among bacteria occurred at such low frequencies, relative to mutation, that it was considered evolutionarily insignificant. However, gene flux and recombination between different strains or species can result in genetic diversification, promoting the expansion of the network structure of the species. In this respect, if bacterial populations behave as independent evolutionary units, adaptive genomic changes facilitate clonal expansion which results in the emergence of a distinct variety. Conversely, if they behave as reticulate evolutionary units, genomic adaptive changes occurring within a single bacterium can be transferred horizontally to many or all members of the species and any related bacteria that can maintain or act as recipients of the mobile genetic element. In principle free gene transfer and unrestricted recombination between different genomic types would result in an unbounded continuum of divergence and speciation. This does not appear to be the case. But, depending on local circumstances, bacteria may alternate from one form to the other; that is, either propagate clonally for long periods of time, or shift into a networked structure as a result of the acquisition of adaptive traits carried on plasmids (Bergstrom *et al.* 2000; Lilley *et al.* 2000; Ochman *et al.* 2000; Bailey *et al.* 2001).

It is worth noting that the fitness costs associated with competition, bacterial densities, relative growth rates and plasmid transfer that prevail in nature are generally considered to be too high for plasmids to persist by parasitic spread alone (Simonsen 1991; Smets *et al.* 1993, 1994). It is logical to assume that plasmids have evolved effective operator mechanisms for plasmid maintenance and dispersal which enhance vertical persistence (passage from parental to daughter cells) coincident with the ability to transfer horizontally. Therefore, under conditions of genetic fluidity, it is also probable that plasmids acquire traits which raise or contribute to host fitness, which link directly to their improved vertical persistence. Simply, carriage of phenotypes beneficial to host persistence and adaptation may account for coevolution of host bacteria and mobile genetic elements. Such coevolution may well be mutualistic, and can be illustrated by the diversity of plasmids and the traits described (Thomas 2000). A number of studies in the phytosphere have considered this and a few have identified important phenotypes carried by the plasmids of plant-associated bacteria; the ability to fix nitrogen, synthesize plant growth regulators, confer resistance to antibiotics, catabolize plant sugars, organic acids and hydrocarbons, and direct cellular chemotaxis are of clear adaptive advantage in the phytosphere (see Bailey *et al.* 1996).

Furthermore, in order to survive the physical extremes of temperature, ultraviolet irradiation, nutrient limitations and water availability typical of the leaf surface, bacteria need to exhibit considerable genetic and phenotypic plasticity (Lindow 1994; Rainey *et al.* 1994; Haubold & Rainey 1996; Rainey & Travisano 1998). To develop these considerations of host chromosomal plasticity it is logical that plasmids function by directly and sporadically extending the available gene pool (Lilley *et al.* 2000). Plasmids therefore provide bacteria with an evolutionary advantage, namely the ability to adapt rapidly to different environments. Through this horizontal gene flux there is greater dispersal of non-essential genes carried on plasmids than of those fixed in chromosomes. This allows recipients to deal with variations in local conditions in an ecological timeframe. Simply, plasmids function as an extended gene pool providing a source of variability and play a central role in the ecology and evolution of bacteria.

Dispersal of plasmids by self-transfer in the natural environment

The adaptation of bacterial populations to change is well known. That the speed of adaptation can be impressive is shown by the response of bacterial populations to antibiotics, novel pollutants and heavy metals. This capacity is largely accounted for by the horizontal transfer of genes between populations. In some cases, transfer may be in direct response to selection. In most cases, however, the horizontal transfer of genes is an example of dispersal, generating diversity and variation upon which selection may act. A major mechanism for the cell-to-cell, that is within- and between-population, dispersal of genes is their incorporation in plasmids. Plasmids, as described above, are DNA molecules that regulate their own replication and transfer to other cells.

We have studied the transfer of plasmids between pseudomonad populations colonizing crops and wild plants, and return here to our example of the *mer* genes. Studies of *mer* plasmids have identified three genetically distinct groups (I, III and IV) that persist in, and periodically transfer between, pseudomonad populations of leaves and roots at the Oxfordshire field site. The plasmids studied are large, ranging in size between 150 and 400 kilo base pairs (kbp), and carry additional, but as yet unidentified genes for a variety of traits beneficial to both the plasmid and the host.

The majority of microcosm experiments (Trevors *et al.* 1987; Lilley *et al.* 2000) demonstrate that plant surfaces are sites of enhanced transfer activity. *In situ* field experiments with sugar beet confirmed that contact with the root surface promoted gene transfer at greater frequencies than could be recorded under laboratory conditions using the plasmids isolated from the same site (Lilley *et al.* 1994). However, these experiments were performed between donor and recipient bacteria introduced for short periods. To make more realistic assessments long term, *in situ* field experiments were performed to study gene dispersal between established, natural populations. Again to simulate natural conditions native strains and plasmids indigenous to the site of study were used to ensure the colonization of plant surfaces under natural competitive conditions. Field releases were performed with *P. fluorescens* SBW25 (described above), genetically modified to carry antibiotic, colourimetric and lactose utilization genes which facilitate isolation and enumeration (Bailey *et al.* 1995). Inocula were grown under laboratory conditions and washed to remove endogenous nutrients before introduction as a seed dressing. The population dynamics of the inocula were monitored over complete growing seasons of sugar beet, and the acquisition of native plasmids, conferring mercuric resistance, assessed at every sampling point (Fig. 11.3). The uptake of plasmids from the

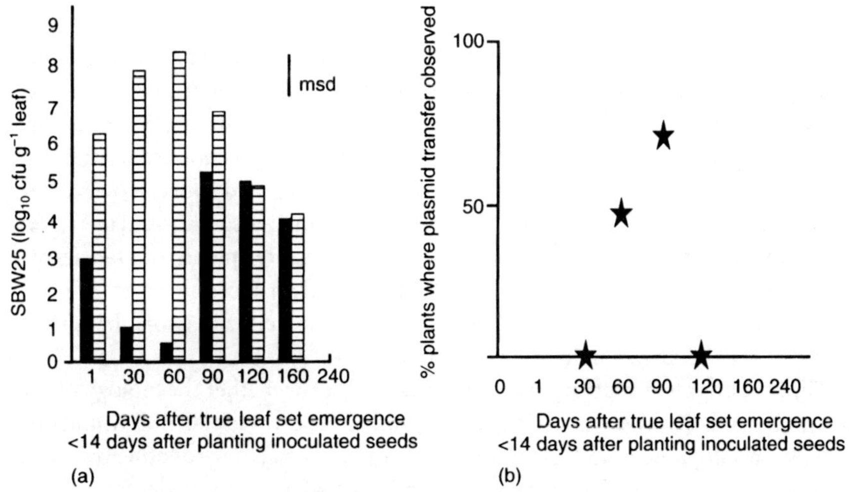

Figure 11.3 Influence of plasmid dispersal and maintenance on the population dynamics of *Pseudomonas fluorescens* SBW25 introduced as a seed dressing. Bacterial density was determined throughout the growing season of field-grown sugar beet (Fig. 11.2) on the emerging leaves ($n = 9$). (a) Population dynamics of SBW25 inocula (striped bars), SBW25 (pQBR103) inocula (solid bars), and (b) the dispersal of conjugative plasmids from the indigenous microflora to SBW25 inocula (percentage of plants sampled, $n = 9$, where the SBW25 population had acquired plasmids). These data demonstrate the efficient dispersal of SBW25 to leaves, the periodic cost and benefit of carrying a plasmid (pQBR103) to the bacterial inocula, and that plasmid transfer, in ecological time, occurs between natural bacterial communities. msd, minimum significant difference.

natural community was detected coincidentally on roots and leaves only from the mid-season to the maturation of the crop (Fig. 11.3). The proportion of *P. fluorescens* SBW25 that acquired plasmids varied between 0.005% and almost 100% according to date and plant tissue. These findings were replicated in a second experiment the following year (Lilley & Bailey 1997b).

In both experiments, the released strain continued to be isolated through the autumn, however the subpopulation which had acquired indigenous plasmids declined below the limits of detection. The decline was probably due to decreased fitness associated with the burden of plasmid carriage. The apparent periodic abundance of bacteria that had acquired plasmids suggests dispersal either under conditions that favour horizontal transfer of the plasmid, or as a result of vertical spread as the host strain and plasmid proliferate. Of course, both horizontal and vertical factors may coincidentally contribute to plasmid dispersal. This may be exaggerated where horizontal transfer to a population is rare, but the combined products of transfer, for example the plasmid entering a new host background, could increase relative fitness and proliferation. To test these considerations another release was conducted using marked *P. fluorescens* SBW25 carrying plasmid pQBR103. In Fig. 11.3 the population dynamics of plasmid-free and plasmid-containing inocula are compared. As the seedlings developed and the plants grew, plasmid carriage substantially reduced colonizing fitness to the extent that 60 days after planting the population density of the plasmid-free variant was seven orders of magnitude greater. However, over the following 100 days, as the plants matured from mid-summer into autumn, the plasmid-carrying variant completely overcame any fitness costs and proliferated to densities equivalent to those of the plasmid-free strain (Fig. 11.3). This period of fitness recovery was also observed in greenhouse studies and coincided with the observed periods of plasmid transfer noted above. However, by comparison this emerging plasmid-carrying population persisted and did not suffer the immediate decline observed in bacteria that were recipients of transferred plasmids. In conclusion, plasmids confer traits selected for at certain stages of plant development, demonstrating that both horizontal and vertical factors drive their dispersal and the spread of the attendant of *mer* genes.

These results and observations raised questions as to whether the biology of the plasmids studied were specific to sugar beet, and whether these plasmids would disperse from a single host, and if so to what extent and to what range of other bacteria? To study the nature of this dispersal an experimental system was developed that allowed the transfer of a plasmid to be tracked from one host to another so that its realized host range could be determined. The plasmid was marked by the insertion of antibiotic resistance and chromogenic genes and was released in the host background, *P. fluorescens* SBW25. Transfer was assessed using a host-specific phage to kill SBW25 so that recipient bacteria, from plant and soil samples, that had acquired the plasmid could be isolated and identified. As it was not practical to conduct this experiment under field conditions the release was conducted using the Ecotron facility (Lawton *et al.* 1993) at the Centre for Population Biology, Imperial College, Silwood Park. Model field-margin weed communities, herbivores, their predators and a soil fauna were created in 16 replicated chambers (mesocosms). These were maintained under environmentally realistic conditions of temperature, rainfall and illumination. While a facsimile of field conditions, these communities were considered sufficiently complex to accurately mimic bacterial behaviour in the open environment.

The released bacteria carrying the marked plasmid were applied to *Stellaria media* plants as a seed dressing and grown in the mesocosms. Transfer of the marked plasmid from the introduced strain (donor) to indigenous bacteria was detected after 70 days, coincidentally on roots and leaves as described for sugar beet above. Up to 75% of plants sampled carried indigenous bacteria that had acquired the marked plasmid at transfer frequencies equivalent to 4% of the donor population. The realized host range of this plasmid was confined to the fluorescent pseudomonads, but the genetic diversity of the recipients was large and representative of the normal high diversity of pseudomonad populations present on the plants. The population dynamics of the inocula carrying the plasmids was similar to those observed for the sugar beet field experiments where the plasmid-bearing variants suffered initial reduced fitness which recovered as the plants matured. Again this increase in fitness coincided with the period of horizontal plasmid transfer.

The plasmids studied were observed to be dispersing the *mer* genes in an environment in which there is believed to be no selection for mercuric resistance. These *mer* genes were: (i) dispersed by a mechanism which is probably independent of the *mer* genes; and (ii) probably linked to selection for other genes and traits of which the *mer* operon takes advantage. It is also possible that the *mer* operon codes for functions other than the obvious ability to survive in polluted environments. None the less the persistence of the genes and the elements that function as their vectors are dependent on the survival and dispersal of the host.

Dispersal of pseudomonad populations in the phytosphere by phytophagous caterpillars

A large number of investigations have reported difficulty in designing protocols for successfully inoculating plants with plant-growth-promoting bacteria. In many instances only transient populations have been observed after application, and this poor performance has been attributed to many factors. However, *P. fluorescens* SBW25 is an effective colonizer, which not only colonizes plants following seed inoculation but also disperses over the entire plant surface, establishing populations of high density. Of note, however, were the observations that little or no dispersal of these colonizing bacteria, from source sugar beet plants to surrounding inocula-free plants (sugar beet and a variety of weeds), occurred during glass-house or field experiments (see Fig. 11.2a; the blocks of guard rows surrounding the nine treatment blocks were sampled) (Thompson *et al.* 1995a, 1995b). In associated studies, where *P. fluorescens* SBW25 was applied to wheat, some transient dispersal was observed in a sandy loam soil; this was attributed to the high concentration of inocula applied, *c.* 1×10^{13} colony-forming units (cfu) ha^{-1}, and physical dispersal due to a heavy rainstorm soon after drilling (De Leij *et al.* 1995). In both these studies the inocula failed to survive in the absence of viable plant material and dispersal was further limited as the bacterium is susceptible to low temperatures, in particular ground frosts. Under artificial conditions there have been reports of direct transfer, for example by leaf to leaf contact, simulated by rubbing leaves together (particularly under conditions of high humidity or rainfall) (McCartney & Butterworth 1992; Hirsch & Spokes 1994; Thompson *et al.* 1995a) where bacteria survive for limited periods. These factors underline a lack of understanding of the natural routes of dispersal and that considerable obstacles exist to the establishment of viable populations (Wilson & Lindow 1993).

To address these issues a field study was conducted (Lilley *et al.* 1996) to establish whether phytophagous insects could disperse plant-associated bacteria, and whether such dispersal could result in viable populations colonizing new plants. The rationale was to exploit the large populations of SBW25 that establish following seed inoculation throughout the growing season in the immature growing leaves at the heart of the sugar beet canopy (Fig. 11.3) and determine, in mature plants, whether dispersal of a natural community of the bacteria was possible. The vector assessed, to study dispersal to untreated plants, was the larvae of the cabbage moth, *Mamestra brassicae* (L). *M. brassicae* (L.) is a polyphagous defoliator of many crops, including sugar beet, and a number of wild plants. On sugar beet crops, the larvae were most frequently found feeding on or resting among the younger leaves at the heart of the rosette (Fig. 11.4). High densities of larvae were applied so that the food resources available on a single plant would be rapidly exhausted, forcing the feeding caterpillars to migrate to the adjacent untreated plants.

The bacterial inocula, *P. fluorescens* SBW25 (described above), was washed to remove endogenous nutrients and applied, *c.* 1 × 106 cfu $seed^{-1}$, before planting to establish a natural community on the plant surface. Control seeds without bacteria were similarly prepared and planted. Sugar beet plants were planted under standard field conditions (Thompson *et al.* 1995b). Seventy-three days after planting, plants were transferred to netted (2 mm gauge nylon) enclosures (1 m^3) (Fig. 11.4a), and arranged with a single inner plant and three outer plants with their leaves just touching. Plants were left for a further week to acclimatize. Leaf samples were then collected and assayed for the presence of inocula. *P. fluorescens* SBW25 was only found on those plants that had been seed inoculated; no dispersal was recorded to the sentinel plants. Twenty third-instar *M. brassicae* were placed on the emerging leaves of the inner sugar beet plants in five treatments, each with nine replicates using 45 enclosures and a randomized block design (see Table 11.1 for the combinations of *P. fluorescens* SBW25 and *M. brassicae* used in the five treatments). These treatments test the dispersal of bacteria: (i) plant to plant, and (ii) gut to plant by *M. brassicae* larvae.

The foraging activities of the phytophagous insect larvae dispersed the marked pseudomonads to new leaf sets developing on field-grown sugar beet plants (Fig. 11.5). *P. fluorescens* SBW25 was detected on the leaves of uninoculated plants by the third day of caterpillar activity. Considerable damage to the plants was observed by the feeding insects (Fig. 11.4b). The caterpillars migrated from the plants and

Figure 11.4 Details from a field study into the role of *Mamestra brassicae* larvae in the dispersal of pseudomonads in the sugar beet phyllosphere (Lilley *et al.* 1997). Sugar beet plants were grown in netted cages (a). A central plant (with or without bacterial inocula) was grown for 73 days, then three untreated plants of the same age were positioned around it before treatment with (b) or without (c) 20 third-instar *M. brassicae* larvae (see Table 11.1).

Table 11.1 Treatments used in the examination of the role of phytophagous insects in the dispersal of leaf-colonizing populations of a fluorescent pseudomonad.

Treatment	Inner plant grown from	Larvae introduced
1	*P. fluorescens* SBW25-inoculated seed pellet	No larvae
2	*P. fluorescens* SBW25-inoculated seed pellet	Untreated larvae
3	Untreated seed pellet	Untreated larvae
4	*P. fluorescens* SBW25-inoculated seed pellet	*M. brassicae* fed *P. fluorescens* SBW25
5	Uninoculated seed pellet	*M. brassicae* fed *P. fluorescens* SBW25

pupated within 18 days, at which point the central plant was removed from each treatment. To determine whether the dispersed bacteria were able to establish and colonize the leaves of the recipient plants, a final sample was collected 42 days after the release of the caterpillars and 24 days after pupation (Fig. 11.5). The population densities of the transferred, marked bacterium were not significantly different from those recorded for plants that had been seed inoculated with marked bacteria. This unequivocally demonstrated that, under optimal field conditions, not only that insect-mediated dispersal of phyllosphere bacterial populations occurred, but also

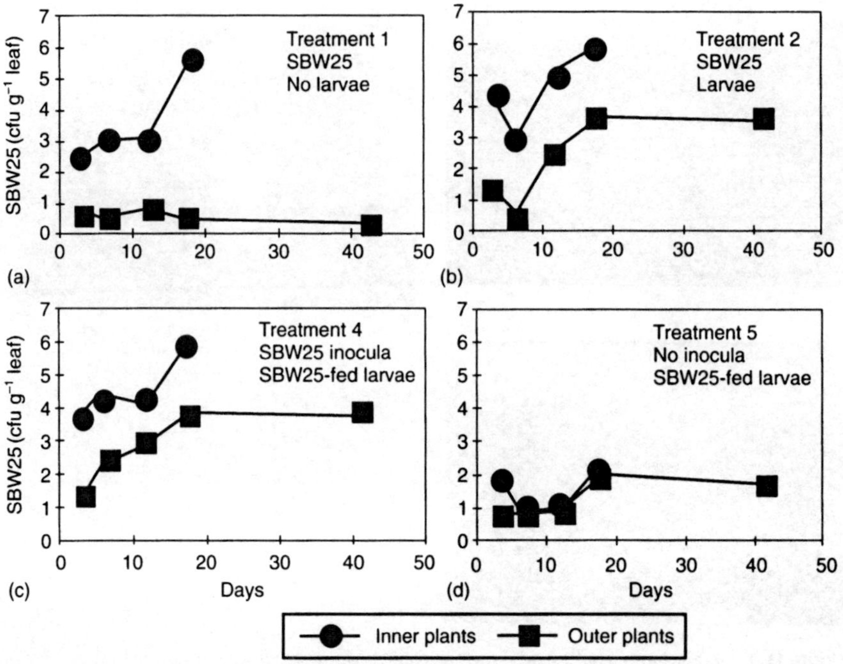

Figure 11.5 Dispersal of *Pseudomonas fluorescens* SBW25 from the leaves of inner sugar beet plants to the leaves of outer sugar beet plants by *Mamestra brassicae* larvae and dispersal from the guts of larvae to inner and outer plants (treatment 5). Plants were field grown from seeds treated with *P. fluorescens* SBW25 and from uninoculated seeds. After 73 days the plants were transferred to netted field enclosures (nine replicated per treatment) in which lone inner (treatment) plants were surrounded by three untreated plants and left to acclimatize for a week (see Fig. 11.4). At day 0, third-instar larvae, untreated or prefed with *P. fluorescens* SBW25, were added to the inner plants (20 per plant). The treatment combinations are given in Table 11.1 and were chosen to evaluate larval dispersal of pseudomonads by physical contact and gut passage. Developing leaves at the heart of each plant were sampled and assayed for *P. fluorescens* SBW25. Damage to plants by feeding was observed throughout the study with the central infested plant defoliated by the 12th day. Following pupation, 18 days after insect introduction, the central plant was removed and no further leaf grazing was observed with all the new leaves on the outer edge remaining undamaged. (a–c) Where the inner plant had been grown from *P. fluorescens* SBW25-treated seeds (treatments 1, 2 and 4), the bacterium colonized the inner plant. Transfer to the outer plants was negligible in the absence of larvae (treatment 2) but common where larvae were included (treatments 1 and 4). (d) Dispersal and establishment via the gut of *P. fluorescens* SBW25-fed larvae was poor (treatment 5).

that the transferred bacteria were able to compete with the resident microflora, proliferate and establish a new persisting community following transfer.

Of interest are the observations that plants treated with larvae fed on bacteria were not colonized to a significant extent (Fig. 11.5c, d: treatments 4 and 5, respectively).

In laboratory tests approximately 20% of the marked strain survived passage through the gut or third-instar *M. brassicae* larvae, i.e. meals of *c.* 3.5×10^6 marked pseudomonads per larvae resulted in *c.* 7.1×10^5 pseudomonads surviving gut passage and being detected in frass. On day 18, other invertebrates that could be captured in the enclosures were assessed for carriage of *P. fluorescens* SBW25. This provided a necessary comparison of the role of the natural insect fauna as potential vectors of passive dispersal. Small numbers of the marked bacteria were detected on the cuticle of late-instar larvae of three moth species: *M. brassicae, Phylogophora meticulosa* and *Discestra trifolii.* These captured insects were reared to their adult moth stage in the laboratory but no *Pseudomonas fluorescens* SBW25 was detected in emergent moths. The bacterium was similarly detected on samples of ladybirds (Coccinellidae, Coleoptera) but not on assorted spiders or chrysomelids (Coleoptera). At 42 days invertebrates collected from the outer plants were also assayed and the marked bacterium was also detected on aphids, ladybirds and crane flies (Tipulidae, Diptera). The marked bacterium was common on crane flies and aphids, which were typically sampled from the centre of the leaf rosette among the emergent leaves of the sugar beet. Crane flies and aphids from all plants carried the marked strain on their legs. Despite the apparent passive contamination of these insects they were not effective in the dispersal and establishment of inocula in control treatment 1 (Fig. 11.5).

These studies demonstrated that *M. brassicae* larvae dispersed and established populations of a marked pseudomonad on the leaves of neighbouring plants. A number of investigators have attempted to determine which factors influence the ability of dispersed (or inoculated) bacteria to establish viable populations on leaves (Pedersen & Leser 1992; Lindow 1994). Notably, survival of dispersed bacteria was significantly less when grown in laboratory cultures than when the inocula was washed from colonized leaves (Wilson & Lindow 1993). This may help explain the high levels of establishment attained following insect dispersal. That is by occupying similar zones of subsequent plants, they disperse bacteria to new sites to which the bacteria are both generally adapted and at that time physiologically attuned to.

Conclusions

Three levels of dispersal have been considered: genes, mobile genetic elements and bacterial cells. Experimental investigations confirm that linkage between these factors are not static, either in respect of the diversity of their vectors or in the host associations which arise following their dispersal. The mobile genetic elements apparently ebb and flow within and between bacterial populations, where horizontal and vertical factors drive dispersal. The dispersal of genes by plasmids, within and between populations, may in fact be common, but problems arise when trying to develop a model of such dispersal. The dispersal of mobile elements and genes can provide opportunities for bacterial populations to sample and utilize novel genes. Similarly, habitats may be considered to sample those bacteria dispersed into them. However, at all three levels of dispersal, diversity is generated in the context of fluid and dynamic interactions in which mobile genetic elements often have transient associations with populations. These populations in turn may disperse to habitats where community diversity is frequently marked by substantial functional redundancy. To predict the consequences of dispersal, we need to understand the genetic components of fitness. This will lead to a better appreciation of how processes for generating diversity contribute to the behaviour of bacterial populations and community establishment. Resolving this interaction between population ecology and population genetics promises to be the most exciting and rewarding current endeavour for microbial ecologists and geneticists alike.

Typically, the study of bacterial genetics has focused on the analysis of the genetic divergence of chromosomal loci. These assessments have found that horizontal transfer often does not destroy linkage equilibrium, and so the accumulation of transferred genes has been estimated to occur at a relatively low rate. For example, one detailed study of enteric bacteria estimated that only 16 kbp of new DNA sequence are 'fixed' in the chromosome (essentially the vertically transmitted genome) per million years of evolution (Lawence & Ochman 1998). This evolutionary estimate implies relative genetic isolation, but does not account for the adaptive, ecological role that mobile genetic elements, particularly plasmids, play in the transient dispersal and fixation of novel traits. In the example of the large pseudomonad-associated plasmids described above they regularly disperse c. 300 kbp of 'information' at each transfer event. This represents 20 times the genetic information calculated to integrate into the chromosome in a million years. None the less, it must be emphasized that plasmid transfer is often transient, providing periodic selective advantage. The horizontal gene pool also provides a fundamental role in bacterial evolution, in presenting bacteria with the opportunity to 'sample' dispersed genes. What is less clear is whether genes on mobile genetic elements can be considered 'fixed' as defined for the chromosome, or whether they represent a separate gene pool that interacts with the chromosomal lineage and the mobile genetic elements and function as dynamic, yet distinct, units of evolution.

Plasmids can and do disperse large sets of genes within and between populations. Through dispersal, the gene pool is extended without fixed host–plasmid relationships necessarily being established. At each level of dispersal (the gene, the mobile element and the recombinant recipient cell) variety is generated on which selection acts. This can be hard to model, particularly when trying to determine how much of the variation observed is attributable to the dispersal and exploitation of adaptive traits carried on the horizontal gene pool. Clearly a greater understanding of microbial ecology and the population genetics of bacteria and their associated horizontal gene pool is needed before these issues can be addressed in more detail. Greater effort in the sequence analyses of complete bacterial genomes (plasmids and chromosomes) is required. This effort must include functional and comparative genomic analyses of closely related isolates and isolates from the natural environment that share a common gene pool. These data sets will help us understand the role of the horizontal gene pool and may resolve, in both ecological and evolutionary terms, why mobile genetic elements are retained in bacteria.

References

Ashleford, K.E., Day, M.J., Bailey, M.J., Lilley, A.K. & Fry, J.C. (1999a) *In situ* population dynamics of bacterial viruses in a terrestrial environment. *Applied and Environmental Microbiology* **65**, 169–174.

Ashleford, K.E., Fry, J.C., Bailey, M.J., Jeffries, A. & Day, M.J. (1999b) Characterisation of six bacteriophages of *Serratia liquefaciens* CP6 isolated from the sugar beet phytosphere. *Applied and Environmental Microbiology* **65**, 1959–1965.

Ashleford, K.E., Norris, S.J., Fry, J.C., Bailey, M.J. & Day, M.J. (2000) Seasonal population dynamics and interactions of competing bacteriophages and their host in the rhizosphere. *Applied Environmental Microbiology* **66**, 4193–4199.

Bailey, M.J., Lilley, A.K., Thompson, I.P., Rainey, P.B. & Ellis, R.J. (1995) Site directed chromosomal marking of a fluorescent pseudomonad isolated from the phytosphere of sugar beet; stability and potential for marker gene transfer. *Molecular Ecology* **4**, 755–764.

Bailey, M.J., Lilley, A.K. & Diaper, J.D. (1996) Gene transfer in the phyllosphere. In: *Aerial Plant Surface Microbiology* (eds C.E. Morris, P. Nicot & C. Nguyen-the), pp. 103–123. Plenum Press, New York.

Bailey, M.J., Rainey, P.B., Zhang, X.-X. & Lilley, A.K. (2001) Population dynamics, gene transfer and gene expression in plasmids, the role of the horizontal gene pool in local adaptation at the plant surface. In: *Microbiology of Aerial Plant Surfaces* (eds S. Lindow & V. Elliot), pp. 171–189. American Phytopathological Society Press, St Paul, MN.

Beijerinck, M.W. (1901) Everything is everywhere, the environment selects. Cited in: *Microbial Ecology: Fundamentals and Applications* (1998) (eds R.M. Atlas & R.B. Bartha), p. 11. Benjamin/Cummings Science Publishing, Manlo Park, CA.

Bergstrom, C.T., Lipsitch, M. & Levin, B.R. (2000) Natural selection, infectious transfer and the existence conditions for bacterial plasmids. *Genetics* **155**, 1505–1519.

Brookfield, J.F.Y. (1995) Transposable elements as selfish DNA. In: *Mobile Genetic Elements* (ed. D.J. Sherratt), pp. 130–153. IRL Press/Oxford University Press, Oxford.

Cho, J.C. & Tiedje, J.M. (2000) Biogeography and degree of endemicity of fluorescent *Pseudomonas* strains in soil. *Applied and Environmental Microbiology* **66**, 5448–5456.

De Leij, F.A.A.M., Sutton, E.J., Whipps, J.M., Fenlon, J.S. & Lynch, J.M. (1995) Field release of a genetically modified *Pseudomonas fluorescens* on wheat: establishment, survival and dissemination. *Nature/Biotechnology* **13**, 1488–1992.

Ellis, R.J., Thompson, I.P. & Bailey, M.J. (1999) Temporal fluctuations in the pseudomonad population associated with sugar beet leaves. *FEMS Microbiology Ecology* **28**, 345–356.

Fulthorpe, R.R., Rhodes, A.N. & Tiedje, J.M. (1998) High levels of endemicity of 3-chlorobenzoate-degrading soil bacteria. *Applied and Environmental Microbiology* **64**, 1620–1627.

Haubold, B. & Rainey, P.B. (1996) Genetic and ecotypic structure of a fluorescent *Pseudomonas* population. *Molecular Ecology* 5, 747–761.

Hirano, S.S & Upper, C.D. (2000) Bacteria in the leaf ecosystem with emphasis on *Pseudomonas syringae*—a pathogen, ice nucleus, and epiphyte. *Microbiology and Molecular Biology Reviews* **64**, 624–653.

Hirsch, P.R. & Spokes, J.D. (1994) Survival and dispersion of genetically modified rhizobia in the field and genetic interactions with native strains. *FEMS Microbiology Ecology* **15**, 147–160.

Lawrence, J. & Ochman, H. (1998) Molecular archaeology of the *Escherichia coli* genome. *Proceedings of the National Academy of Sciences of the USA* **95**, 9413–9417.

Lawton, J.H., Naeem, S., Woodfin, R.M. *et al.* (1993) The Ecotron: a controlled environmental facility for the investigation of population and ecosystems processes. *Philosophical Transactions of the Royal Society, Series B* **341**, 181–194.

Lilley, A.K. & Bailey, M.J. (1997a) Impact of pQBR103 acquisition and carriage on the phytosphere fitness of *Pseudomonas fluorescens* SBW25: burden and benefit. *Applied and Environmental Microbiology* **63**, 1584–1587.

Lilley, A.K. & Bailey, M.J. (1997b) The acquisition of indigenous plasmids by a genetically marked pseudomonad population colonising the phytosphere of sugar beet is related to local environmental conditions. *Applied and Environmental Microbiology* **63**, 1577–1583.

Lilley, A.K., Fry, J.C., Day, M.J. & Bailey, M.J. (1994) *In situ* transfer of an exogenously isolated plasmid between indigenous donor and recipient pseudomonad spp. in sugar beet rhizosphere. *Microbiology* **140**, 27–33.

Lilley, A.K., Bailey, M.J., Day, M.J. & Fry, J.C. (1996) Diversity of mercury resistance plasmids obtained by exogenous isolation from the bacteria of sugar beet in three successive seasons. *FEMS Microbiology Ecology* **20**, 211–227.

Lilley, A.K., Hails, R.S., Cory, J.S. & Bailey M.J. (1997) The dispersal and establishment of pseudomonad populations in the phyllosphere of sugar beet by phytophagous caterpillars. *FEMS Microbiology Ecology* **24**, 151–157.

Lilley, A.K., Young, J.P. & Bailey, M.J. (2000) Bacterial population genetics: do plasmids maintain diversity and adaptation? In: *The Horizontal Gene Pool: Bacterial Plasmids and Gene Spread* (ed C.M. Thomas), pp. 287–300. Harwood Academic Publishers, New York.

Lindow, S.E. (1994) Novel methods for identifying bacterial mutants with reduced epiphytic fitness. *Applied and Environmental Microbiology* **59**, 1586–1592.

McCartney, H.A. & Butterworth, J. (1992) Effects of humidity on the dispersal of *Pseudomonas syringae* from leaves by water splash. *Microbial Releases* **1**, 187–190.

Ochman, H., Lawrence, J.G. & Groisman, E.A. (2000) Lateral gene transfer and the nature of bacterial innovation. *Nature* **405**, 299–304.

Pedersen, J.C. & Leser, T.D. (1992) Survival of *Enterobacter cloacae* on leaves and in soil detected by immunofluorescence microscopy in comparison with selective plating. *Microbial Releases* **1**, 95–102.

Plasterk, R.H.A. (1995) Mechanisms of DNA transposition. In: *Mobile Genetic Elements* (ed. D. J. Sherratt), pp. 18–37. IRL Press/Oxford University Press, Oxford.

Rainey, P.B. & Bailey, M.J. (1996) Physical and genetic map of the *Pseudomonas fluorescens* SBW25 chromosome. *Molecular Microbiology* **19**, 521–533.

Rainey, P.B. & Travisano, M. (1998) Adaptive radiation in a heterogeneous environment. *Nature* **394**, 69–72.

Rainey, P.B., Bailey, M.J. & Thompson, I.P. (1994) Phenotypic and genotypic diversity of fluorescent pseudomonads isolated from field-grown sugar beet. *Microbiology* **140**, 2315–2331.

Sherratt, D.J. (ed.) (1995) *Mobile Genetic Elements.* IRL Press/Oxford University Press, Oxford.

Simonsen, L. (1991) The existence conditions for bacterial plasmids—theory and reality. *Microbial Ecology* **22**, 187–205.

Smets, B.F., Rittmann, B.E. & Stahl, D.A. (1993) The specific growth-rate of *Pseudomonas putida* PAW1 influences the conjugal transfer rate of the Tol plasmid. *Applied and Environmental Microbiology* **59**, 3430–3437.

Smets, B.F., Rittmann, B.E. & Stahl, D.A. (1994) Stability and conjugal transfer kinetics of a Tol plasmid in *Pseudomonas aeruginosa* PAO1162. *FEMS Microbiology Ecology* **15**, 337–349.

Speirs, A.J., Buckling, A. & Rainey, P.B. (2000) The causes of *Pseudomonas* diversity. *Microbiology* **146**, 2345–2350.

Thomas, C.M. (ed.) (2000) *The Horizontal Gene Pool: Bacterial Plasmids and Gene Spread.* Harwood Academic Publishers, New York.

Thompson, I.P., Ellis, R.J. & Bailey, M.J. (1995a) Autecology of a genetically modified fluorescent pseudomonad on sugar beet. *FEMS Microbiology Ecology* **17**, 1–14.

Thompson, I.P., Lilley, A.K., Ellis, R.J., Bramwell, P.A. & Bailey, M.J. (1995b) Survival, colonisation and dispersal of genetically modified *Pseudomonas fluorescens* SBW25 in the phytosphere of field grown sugar beet. *Nature/Biotechnology* **13**, 1493–1497.

Trevors, J.T., Barkay, T. & Bourquin, A.W. (1987) Gene-transfer among bacteria in soil and aquatic environments—a review. *Canadian Journal of Microbiology* **33**, 191–198.

Whitman, W.B., Coleman, D.C. & Wiebe, W.J. (1998) Prokaryotes: the unseen majority. *Proceedings of the National Academy of Sciences of the USA* **95**, 6578–6583.

Wilson, M. & Lindow, S.E. (1993) Release of recombinant micro-organisms. *Annual Review of Microbiology* **47**, 913–944.

Yurieva, O., Kholodii, G., Minakhin, L. *et al.* (1997) Intercontinental spread of promiscuous mercury-resistance transposons in environmental bacteria. *Molecular Microbiology* **24**, 321–329.

Chapter 12

Dispersal behaviour and population dynamics of vertebrates

Harry P. Andreassen, Nils C. Stenseth and Rolf A. Ims

Introduction

Within the field of population biology, dispersal behaviour may be seen as the glue linking a variety of ecological, evolutionary and behavioural issues (Stenseth & Lidicker 1992). This is also the view taken in this chapter, where we specifically aim at linking dispersal behaviour with population ecology—and where dispersal links together individual behaviours and the emerging population dynamics. Nevertheless, behavioural and population biologists have typically taken two almost entirely different approaches to the study of dispersal, both with regard to the questions asked and the study procedure.

Behavioural ecologists have typically been concerned with understanding the proximate and ultimate causes of dispersal (e.g. Bengtsson 1978; Greenwood 1980; Liberg & von Schantz 1985; Pusey 1987). Population biologists on the other hand have typically considered dispersal as a key process for understanding population dynamics, spatial synchrony and population genetics (e.g. Stenseth 1983; both approaches have been extensively summarized in Stenseth & Lidicker 1992; Clobert *et al.* 2001; and this volume). Furthermore, while behavioural ecologists typically base their conclusions on empirical studies (but see, for example, Holt & Barfield 2001; Perrin & Goudet 2001), population biologists tend to base their conclusions on theoretical modelling studies (including metapopulation dynamics; Hanski & Gilpin 1997). As part of their efforts, population biologists aim at better understanding the effects of dispersal on the spatiotemporal dynamics of populations (including the genetic structure of subdivided populations). These two approaches have been living quasi-isolated, side by side, without really being thoroughly integrated. Indeed, many of the theoretical population biological models make assumptions which are not supported by empirical studies of dispersal behaviour. In this chapter we attempt, limiting ourselves to the literature on vertebrate dispersal, to summarize the key questions asked within these two approaches; and, as part of this aim, to outline the gaps in the knowledge needed to integrate these different traditions. As an integral part of our account, we will summarize the dispersal studies from an experimental model system on which we have worked (the Evenstad project, which involves, as documented below, both behavioural and population biology approaches to the study of dispersal). We will do this by developing a spatiotemporal population model for the root vole *Microtus oeconomus*. The functions of this model will then be defined using insights gained from the studies on the model system.

Dispersal

Dispersal may be seen as composed of three sequential, but behaviourally different phases (see also South *et al.*, this volume): *emigration*, *transfer* and *immigration* (or, emigration, migration and immigration, *sensu* Ims & Yoccoz 1997). These phases are not only different behaviourally, but also opposite processes as seen from a population biology point of view; i.e. emigration and immigration are the loss and gain of animals, respectively, and transfer (or migration) is the glue linking these two processes together. Below we briefly review the questions asked and sometimes answered by behavioural and population biologists with regard to these three phases of dispersal.

Emigration

The phase of *emigration* initiates the dispersal process as the individual leaves what used to be its area of residence. This is the phase considered principally in the behavioural literature (i.e. exploring which proximate cues individuals use when deciding whether or not to disperse, and thereby aiming at improving our scientific understanding of the ultimate factors causing dispersal).

A large array of proximate factors of natal dispersal have been proposed (see Lidicker & Stenseth 1992), ranging from characteristics of individuals and their siblings (e.g. Bekoff 1977; Ims 1987, 1989, 1990; Bondrup-Nielsen 1993; Jacquot & Vessey 1995; Gundersen & Andreassen 1998; Dufty & Belthoff 2001), local social factors (e.g. kin relations and intraspecific density: Lidicker 1975; McShea 1990; Wolff 1992, 1993; Bollinger *et al.* 1993; McGuire *et al.* 1993; Gundersen & Andreassen 1998; Andreassen & Ims 2001; Sutherland *et al.*, this volume), habitat quality including interspecific interactions such as predation and parasitism (e.g. Lidicker 1975; Wolff & Lidicker 1980; Andreassen & Ims 1990), habitat patch shape (e.g. Stamps *et al.* 1987; Harper *et al.* 1993; Ims 1995; Fauske *et al.* 1997; Gundersen & Andreassen 1998), landscape structure (e.g. habitat corridors: Wegner & Merriam 1979; Hansson 1987; Szacki 1987; Zhang & Usher 1991; Fitzgibbon 1993; Andreassen *et al.* 1996a, 1996b; Bennett 1999; Haddad 1999; Wiens 2001), landscape-level spatial variation in density (Gundersen *et al.* 2001; Andreassen & Ims 2001) and temporal variation in habitat quality (Lurz *et al.* 1997).

Whereas a multitude of proximate factors have been proposed as initiating dispersal, there seems to be less creativity in the literature regarding ultimate factors. The ultimate causes of dispersal may be limited to inbreeding avoidance (e.g. Pusey 1987) and various kinds of resource competition (e.g. Liberg & von Schantz 1985; but see Boulinier *et al.* 2001; Gandon & Michalakis 2001; Lambin *et al.* 2001; Perrin & Goudet 2001 for a more extensive discussion on the ultimate causes of dispersal including intraspecific competition, kin competition and interspecific relations).

In the literature considering population biology, the emigration phase is generally assumed not to have any other effect than the loss of an individual in a given area whatsoever the proximate or ultimate factors might be. In addition, all patches are often assumed to be equally likely to generate emigrants (Hanski & Gilpin 1997).

Transfer

The phase of *transfer* follows directly from emigration and comprises movement until the individual settles in a new area. Behaviourally, the main question of interest is how individuals orientate themselves within a landscape during the transfer phase (i.e. searching behaviour, *sensu* Bell 1991). However, empirical data on behaviour during transfer is almost absent in the vertebrate literature (but see Kenward *et al.*, this volume). This paucity of studies in the vertebrate literature seems to be due to scientists ignoring the transfer process, as well as the methodical and logistic problems associated with following dispersing vertebrates.

Studies of vertebrate transfer behaviour have typically been carried out within a psychological or neurological context (see, for example, Alyan & Jander 1994 and references therein), with no or marginal relevance to the ecological questions about dispersal. However, it is quite reasonable to assume that transfer movement is physiologically influenced by olfactory (Stoddart 1980), visual (Zollner & Lima 1997, 1999; Andreassen *et al.* 1998) or auditory stimuli. Such cues may indeed give information about landscape structure (Stamps *et al.* 1987; Hardt & Forman 1989; Hobbs 1992; Forman 1995; Andreassen *et al.* 1998; Zollner & Lima 1999; for reviews, see also Ims 1995; Wiens 2001), the microhabitat (e.g. movement facilitation by tracks and runways: Harper & Batzli 1996) or the presence of animals of the same or different species (Bennett 1990; Andreassen *et al.* 1996a). It is most likely that a multitude of factors affects the decisions made by an individual during transfer.

Population biologists have mainly discussed the risk of transfer, and hence the loss of animals from the system studied. Few studies have, however, demonstrated that the transfer movement really is a risky behaviour (but see Steen 1994; Andreassen & Ims 1998, 2001). Short occasional sallies (*sensu* Bondrup-Nielsen 1985) outside an established home range might resemble transfer movements, but may not properly be considered as dispersal. If transfer movements are risky (as is often assumed), occasional sallies are probably also risky. Hence, we will never know whether an individual is actually dispersing before it has immigrated into a new area. This problem obviously troubles empirical studies of transfer movements during dispersal.

Orientation during transfer towards a specific new residence site has been completely ignored in the population literature. Most often the models created by population biologists have the assumption that individuals disperse in a diffusion, or random walk, process (e.g. Turchin 1998; Wiens 2001).

Immigration

The phase of *immigration* is typically considered as habitat selection following searching, entering, exploring and, finally, the decision to settle in a new area (Stamps 2001). How far apart the previous and new home ranges must be for the movement to be considered dispersal, is a matter of controversy (Shields 1987; Stenseth & Lidicker 1992; Kenward *et al.*, this volume). A general rule allowing several spatial scales (i.e. body size, home range size and dispersal ability) is that the postdispersal home range does not overlap the predispersal home range (Stenseth & Lidicker 1992).

The immigration process is hardly ever studied empirically as a behavioural response, except in the context of habitat selection (Kramer *et al.* 1997; Stamps 2001). Stamps (2001) outlines conspecific attraction (Stamps 1988, 1991; Gundersen *et al.* 1999) and habitat imprinting (i.e. individuals selecting a habitat similar to that which they left; Tordoff *et al.* 1998) as two proximate mechanisms for immigration, whereas Hestbeck (1982) suggests that a social fence (i.e. exclusion by an existing social group of intraspecific individuals) might limit immigration (see also Stenseth 1986; Andreassen & Ims 2001; Gundersen *et al.* 2001). However, a proximate mechanism should preferably be linked with fitness consequences to understand ultimate factors, and hence the evolution of dispersal (e.g. the resident or immigrant fitness hypothesis; Anderson 1989; Gliwicz 1993). Different studies comparing survival and reproduction of residents and immigrants have found both higher fitness of residents (Tinckle *et al.* 1993; Wiggett & Boag 1993) and no difference (Wauters *et al.* 1994; Johannesen & Andreassen 1998).

From the point of view of population biologists, immigration is simply an increase of numbers. The process by which immigration areas are selected is seldom taken into account. For instance, in metapopulation models all patches are usually assumed to have the same likelihood of receiving immigrants. Only very few studies have searched for other demographic consequences of immigration such as increased reproduction (e.g. due to increased genetic variability) or lower recruitment or survival due to social disturbance (Charnov & Finerty 1980; Lambin & Krebs 1991; Wolff 1995). Empirically, Swenson *et al.* (1997) have shown that the turnover of males reduces recruitment and survival of the resident population.

The Evenstad project

The Evenstad Research Station

The Evenstad Research Station, southeastern Norway, was established in 1989 as an experimental area dedicated to the study of landscape ecological questions within the framework of experimental model systems (EMSs) (Ims & Stenseth 1989; Ims *et al.* 1993; Wiens *et al.* 1993). EMSs are used to explore empirically hypotheses at a relatively small spatial and temporal scale. The experiments have all been conducted in enclosed plots (Fig. 12.1). As usual in scientific experiments, we have attempted to control for factors not under study. Hence, experiments have all been performed by releasing laboratory-raised root voles into enclosed areas empty of mammalian predators and intra- and interspecific competitors other than those added experimentally. The habitat in the various plots comprises dense meadow vegetation which has been maintained as optimal root vole habitat since the initiation of the

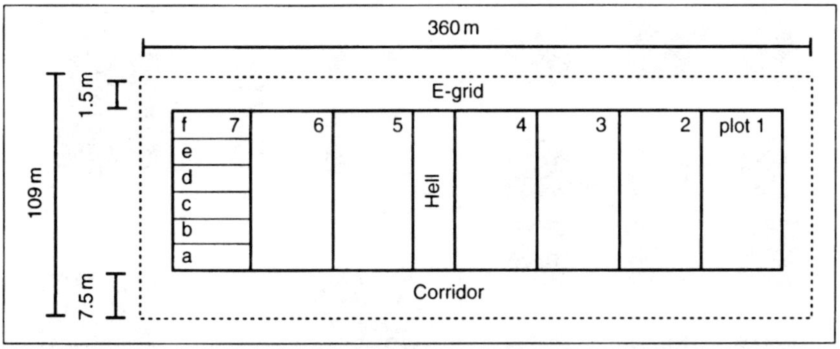

Figure 12.1 A diagram of the experimental areas at Evenstad Research Station. The outer fence (dotted line) is wire mesh with an electric wire to exclude mammalian predators. The inner fences (complete lines) are made of metal sheets. The main population experiments each year have taken place in plots 1 to 7. However, in 1994, plot 7 was subdivided into six fenced plots (a–f) to allow experiments to be done at a lower level of organization than the population (e.g. deme patch, litter, individual). The 'E-grid' and 'corridor' were established in 1991 for the study of behaviour in linear systems. 'Hell' has been an unsuccessful experimental area.

experiment. The flat farmland and meadow vegetation at Evenstad have made the manipulation of habitat a simple task, involving frequent mowing and herbiciding. The matrix (<5 cm tall vegetation; i.e. a lawn) surrounding the plots or habitat patches may be traversed by root voles, but not inhabited by them as it contains no cover (Fig. 12.2). The aim has been to compare behaviour and demography in various kinds of habitat configurations. For each configuration we have used at least four population replicates.

The experimental study organism

The root vole *Microtus oeconomus* has proved to be a very useful organism with which to investigate landscape ecological questions. It is a habitat specialist (Tast 1966). Entire populations may be confined within relatively small areas, and the ease with which it can be trapped alive makes it an easy organism to monitor with respect to the development of the populations and behaviour of individuals. Root voles can be fitted with radiocollars with no known side effects (Johannesen *et al.* 1997), so that individual space use can be studied precisely (e.g. Hansteen *et al.* 1997). Laboratory colonies are easy to create and maintain, allowing more controlled studies of life-history traits (Ims 1994, 1997; dos Santos *et al.* 1995), as well as the production of standardized experimental populations for field experiments. Finally, root voles have a well-developed, but quite flexible, social system characterized by selective intrasexual territoriality (Andreassen *et al.* 1998): a feature that is important for experimental landscape ecological studies addressing the mechanisms underlying population development.

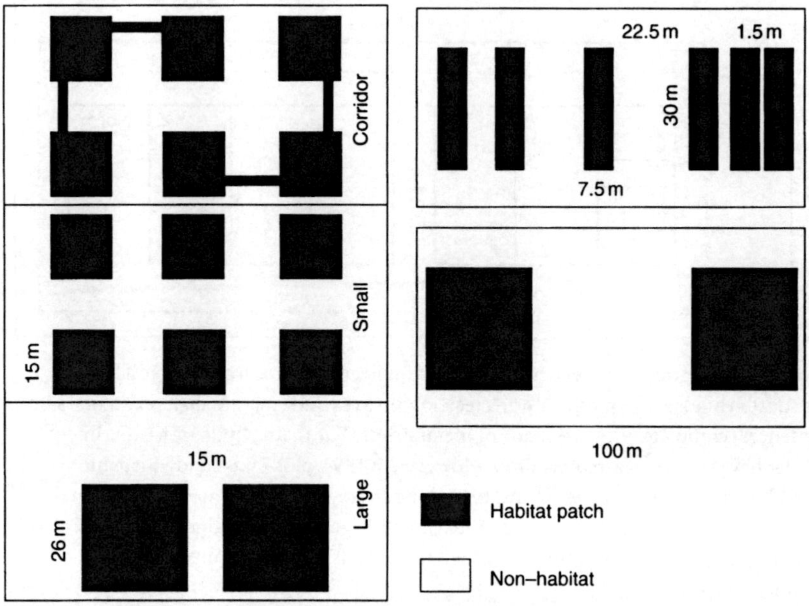

Figure 12.2 Types of habitat configuration used as experimental manipulations at Evenstad (one habitat configuration per plot (50 × 100 m)). Habitat patches (black) consist of dense and tall meadow vegetation surrounded by a non-habitable matrix of very short grass (<5 cm). In 1990/91 we tested the effects of habitat patch size and connectivity on root vole population development (left panel), using four population replicates (plots) of each habitat configuration. These small patches with or without corridors were used to obtain data for model parameterization (see text). The right panel illustrates other habitat configurations used at Evenstad in later years. For demographic responses of the populations see Ims and Andreassen (1999b) and Aars and Ims (1999).

Population and behavioural aspects of dispersal at Evenstad

Studies in the Evenstad project have shown that dispersal is highest when habitat patches are small and narrow. Furthermore, corridors may also enhance dispersal (Fauske *et al.* 1997; Aars & Ims 1999; Andreassen & Ims 2001). However, these processes are beyond the scope of this chapter, which focuses on linking dispersal behaviour and population dynamics processes.

As stated above, dispersal is generally thought to be caused ultimately by inbreeding avoidance and/or resource competition (e.g. Liberg & von Schantz 1985; Pusey 1987). Our studies at Evenstad have shown that the timing of dispersal is often connected to ontogeny, as young animals disperse shortly after sexual maturation, but prior to breeding (Gundersen & Andreassen 1998; Gundersen *et al.* 1999; Andreassen & Ims 2001). Apart from the fact that males tend to disperse to a higher degree than females, as predicted by the inbreeding avoidance hypothesis (Gundersen & Andreassen 1998; Aars & Ims 1999; Andreassen & Ims 2001), we have found few characteristics which relate to the probability that an individual will disperse.

However, in a small-scale experiment (in terms of space and complexity of the study unit; the latter being a family group with an unrelated male) we showed that young animals tended to avoid their mother and were most often reproductively inactive when residing in their mother's habitat patch. In this experiment, pregnant females dispersed short distances, in accordance with the avoidance of resource competition hypothesis. However, females dispersed long distances prior to breeding, suggesting inbreeding avoidance (Gundersen & Andreassen 1998).

Animals emigrated mostly from patches with low densities and tended to immigrate into patches of even lower density (Andreassen & Ims 2001). In addition, there were fewer individuals of the same sex and age as the immigrant in the selected immigration patch than would be predicted from the average of the available patches. Hence, dispersing individuals seemed to be actively searching and selecting new patches, depending on the density and social structure of voles in each patch. Elsewhere we have shown that subadult individuals may be motivated to select patches with sexually mature individuals of the opposite sex (Gundersen *et al.* 1999).

Emigration and immigration is thus higher when vole densities in a patch are low. However, density at the population level is a function of the densities in the separate patches within the population (Andreassen & Ims 2001). Thus, emigration from low-density patches will decrease the overall population density and low-density immigration patches will become available within the same population. It is therefore difficult to conclude whether it is the availability of low-density patches, or the low density in the home patch, that drives individuals to emigrate/immigrate. In the population-level studies at Evenstad, immigration takes place predominantly in the spring and this involves the early cohorts. However, cohort and season are usually confounded with density as the latter increases over the breeding season (see also Aars *et al.* 1999). In a study in which voles were removed constantly from one large patch within the experimental populations, this sink patch (*sensu* Pulliam 1988) induced immigration throughout the season (Gundersen *et al.* 2001). This suggests that immigration is reduced by a social fence mechanism in the patch encountered (Hestbeck 1982). Emigration, however, seems to be based on proximate cues detected at the local emigration patch, as the same negative density-dependent emigration exists even for those animals that emigrate but do not then immigrate into other patches (e.g. animals trapped along the plot fencelines). Furthermore, the probability of emigrating from high-density patches is low even in populations with high variation in patch densities (Andreassen & Ims 2001).

Experiments at Evenstad have demonstrated that the availability of empty habitat patches causes individuals to continue to disperse throughout the season (Gundersen & Andreassen 1998; Andreassen & Ims 2001; Gundersen *et al.* 2001). Actually, this 'drive to disperse' has been largely neglected in the dispersal literature. As individuals disperse and populations expand in space, most individuals will actually be descendants from colonists which may explain this 'drive to disperse'. Although we have not yet found any fitness difference between immigrants and residents (Johannesen & Andreassen 1998), the occurrence of dispersal in spite of risks during transfer indicates that under many circumstances it must be adaptive (the immigrant fitness hypothesis; see Anderson 1989; Gliwicz 1993). Such risks are suggested by the fact that final population sizes in the experimental patch configurations shown in Fig. 12.2 were negatively correlated to the frequency of dispersal movements (i.e. interpatch movements; Ims & Andreassen 2000). This result might be explained by poor survival of dispersing individuals due to avian predation (Aars *et al.* 1999; Ims & Andreassen 2000; Andreassen & Ims 2001).

Root voles tend to disperse to the closest patches available (Ims & Andreassen 1999a; Andreassen & Ims 2001); that is, in a stepping stone fashion. However, these movements do not seem to synchronize population densities between patches at a local scale (i.e. plot; Ims & Andreassen 1999a; Andreassen & Ims 2001). The year-to-year variation in population growth within the enclosures at Evenstad, however, correlates very well to changes in vole abundance in the region around the Evenstad study site (Ims & Andreassen 2000). This link between the local processes and regional dynamics may be mediated by avian predators which synchronize vole population dynamics in a manner suggested by theoretical models (Ims & Steen 1990).

Avian predation may, however, not only cause a simple linear decrease in population size. The most mobile voles, in terms of movements about the matrix, are adult males and, to a lesser degree, subadults (Andreassen & Ims 2001). Hence, adult males and subadults will be the first to be lost from the population due to avian predation. We have also shown that the turnover of males, resulting from immigration of males that replace lost males, lowers recruitment, probably due to infanticidal behaviour (H. Andreassen, unpubl. data). Lactating females that lose their offspring tend to shift habitat patch (Andreassen & Ims 2001), and will in turn become vulnerable to predation. Moreover, female mortality will for unknown reasons increase as a response to high turnover of adult males, if females are aggregated in space (H. Andreassen, unpubl. data). These different causes of mortality tend to generate empty patches in the landscape. As shown above, dispersal movements are most prevalent at low population densities, so empty patches will in turn increase matrix movements followed by increased risk of avian predation. Hence, there will be a feedback process between predation and matrix movements that will increase predation rates. However, the population will not go extinct due to stochasticity in net recruitment.

A tentative model of root vole spatiotemporal dynamics

Besides demonstrating that root voles move principally to the closest patches available, we have also performed experiments in order to learn more about the transfer phase. These demonstrate that root voles use habitat corridors and avoid open areas, but if forced into open areas will move towards nearby visual cues (patches less than 5 m away), and, in particular, small-scale patches located in the direction of distant, large landscape features (Andreassen *et al.* 1996a, 1996b, 1998). These results, together with those described above, have given rise to a tentative model explaining the spatiotemporal dynamics of a root vole (meta-)population (Figs 12.3 and 12.4). Hence, having gone from individuals to populations in the above discussion, we will

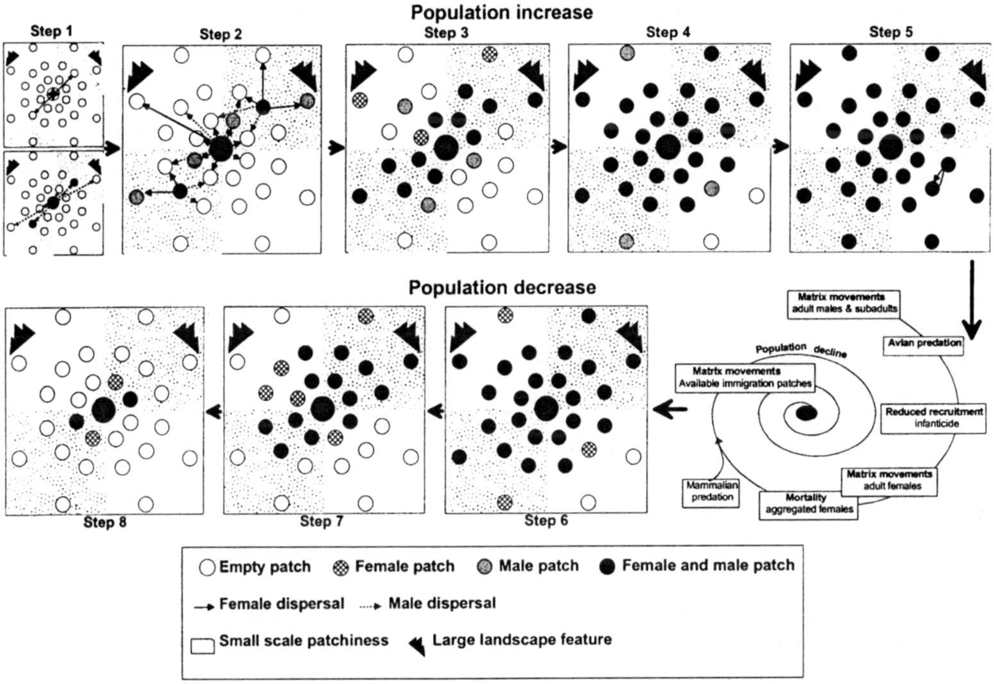

Figure 12.3 The spatiotemporal development of a root vole population in a patchy landscape. The model depicted does not take seasonality into consideration, thus it could take place during one or two reproductive seasons, as long as population structure does not change due to seasonal changes. See text for a further description.

now consider the population responses, and link these to observed individual behavioural responses.

Premises of the model

Figure 12.3 illustrates the development of a root vole population in a patchy landscape during both the increase and decrease phase of a population cycle. Below we describe the different steps presented in Fig. 12.3.

Increase phase of the population

Step 1: Individuals start to disperse from a source patch. Some young females disperse, either short distances due to resource competition or long distances due to inbreeding avoidance. Most young males disperse long distances some days later (Andreassen & Ims 2001) due to inbreeding avoidance (Gundersen & Andreassen 1998; Aars & Ims 1999; Aars *et al.* 1999). Due to some long-distance dispersal, there will be a disjunct distribution of the individuals within the population in space during the increase phase of the population.

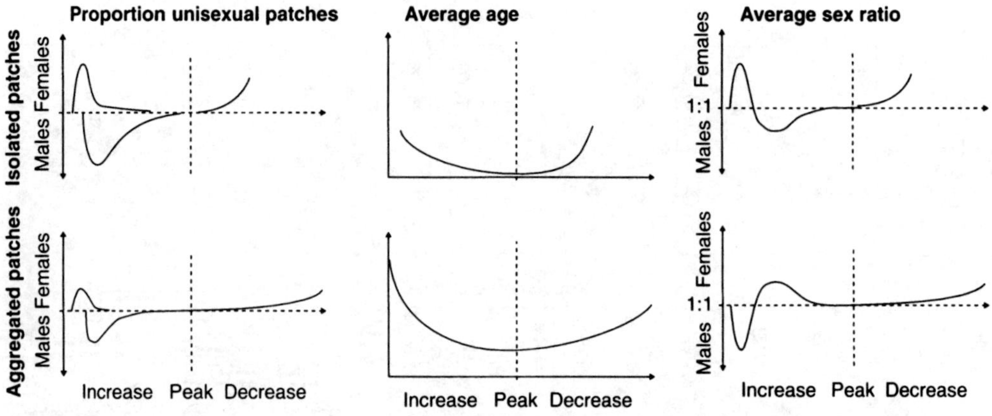

Figure 12.4 The predictions according to the model depicted in Fig. 12.3. The axes represent relative values to show the development of the proportion of colonized patches which are unisexual female or male and the average age and sex ratio of animals inhabiting a patch.

Steps 2, 3 and 4: The subsequent cohorts will continue to disperse as long as patches are available for immigration. The sequence of patch colonization will be determined by both patchiness and patch isolation (i.e. interpatch distance): (i) patches embedded in small-scale patchiness will be colonized before patches embedded in open matrix, but, of the latter, those with large-scale visual cues will be colonized before those without any visual cues; and (ii) aggregated patches will be colonized before isolated patches. There will be unisexual female or male patches in the short term until they are detected by individuals of the opposite sex searching for mates (Ims 1990; Gundersen *et al.* 1999).

Peak phase of the population

Step 5: All the patches are colonized. There will still be matrix movements, but immigration is inhibited due to a social fence (Aars & Ims 1999; Aars *et al.* 1999; Andreassen & Ims 2001; Gundersen *et al.* 2001). Avian predator densities increase with increasing vole densities. A cascade of effects will follow from avian predation of adult males, i.e. reduced recruitment due to infanticide by turnover of males, increased female movement in the matrix when they lose their litter and increased mortality of females which are aggregated in space (Andreassen & Ims 2001). These effects will increase through time as matrix movements increase due to a decrease in densities (Andreassen & Ims 2001). At some stage the increase in density of mammal predators will increase the mortality of voles and cause extinction of local patches. This may further increase matrix movements.

Decrease phase of the population

Steps 6, 7 and 8: Matrix movements are most risky in isolated patches because exposure to predation is increased (Andreassen & Ims 1998). The temporal pattern in the spatial distribution of animals during the decrease phase will thus mirror the increase phase. However, unisexual male patches will rarely occur as males are most prone to avian predation.

Predictions emerging from the model

From this model we may deduce several predictions regarding the sex and age structure in isolated and aggregated patches (Fig. 12.4).

Increase phase of the population

The proportion of unisexual patches will be higher and persist longer in isolated than in aggregated patches. As females start dispersal, unisexual female patches will appear earlier in the increase phase than will unisexual male patches, but the former will persist for a shorter time because males disperse, move and search for mates more than do females. The average age of individuals in the patches will decrease during the increase phase as new cohorts are recruited, but will be higher in aggregated patches as these will be the first colonized (and have the oldest individuals in the population). The average sex ratio in the patches will be female-biased for a short time just after the commencement of dispersal as females disperse earlier than the males. Soon the sex ratio will become male-biased as males disperse more than females. As animals disperse principally from aggregated towards isolated patches, aggregated patches will have an inverse sex ratio development due to the loss of males through dispersal.

Peak phase of the population

There will be no unisexual patches, the average age of individuals in isolated patches will be somewhat lower than in the aggregated patches because these are the first colonized, and the average sex ratio of patches will be roughly equal.

Decrease phase of the population

Because adult males and subadults perform most matrix movements, they are most likely to be lost from the population. In addition, the turnover of males decreases recruitment. Therefore, in general patches will become female only as males are lost from the population, average age will increase as recruitment fails and subadults are predated, and the average sex ratio in the patches will become female-biased. As long-distance matrix movements increase predation risk, predation effects will be most pronounced and start earlier in the isolated than in the aggregated patches.

A model framework linking dispersal and population dynamics

On the basis of this conceptual model we can express the insights achieved in the Evenstad project in a mathematical population model. Only a fragmentary sketch of this model is presented; we will develop this model at a later date.

Consider a habitat complex as shown in Fig. 12.2. Let there be n patches which a group of individuals may inhabit (a subpopulation). For simplicity we will not distinguish males and females. Let the population size within patch i at time t (on a weekly timescale) be given as N_t^i. A population model for this habitat complex may then be given as:

$$N_t^1 = R_{t-1}^1(\cdot)N_{t-1}^1 - \varphi_{\text{emi}}(N_{t-1}^1, \underline{N}_{\underline{t-1}})N_{t-1}^1 + \varphi_{\text{immi}}^1(N_{t-1}^1, \underline{N}_{\underline{t-1}}) \tag{12.1}$$

$$N_t^2 = R_{t-1}^2(\cdot)N_{t-1}^2 - \varphi_{\text{emi}}(N_{t-1}^2, \underline{N}_{\underline{t-1}})N_{t-1}^2 + \varphi_{\text{immi}}^2(N_{t-1}^2, \underline{N}_{\underline{t-1}}) \tag{12.2}$$

. . .

$$N_t^i = R_{t-1}^1(\cdot)N_{t-1}^i - \varphi_{\text{emi}}(N_{t-1}^i, \underline{N}_{\underline{t-1}})N_{t-1}^i + \varphi_{\text{immi}}^i(N_{t-1}^i, \underline{N}_{\underline{t-1}}) \tag{12.3}$$

. . .

$$N_t^n = R_{t-1}^n(\cdot)N_{t-1}^n - \varphi_{\text{emi}}(N_{t-1}^n, \underline{N}_{\underline{t-1}})N_{t-1}^n + \varphi_{\text{immi}}^n(N_{t-1}^n, \underline{N}_{\underline{t-1}}) \tag{12.4}$$

where φ_{emi} is the specific emigration rate from a patch, and φ_{immi}^i is total immigration rate to a patch given as:

$$\begin{aligned} &\varphi_{\text{immi}}^i(N_{t-1}^i, \underline{N}_{\underline{t-1}}) \\ &= k_{\text{trans}}(N_{t-1})\left\{k_{\text{estab}}^i(N_{t-1}^i)\Big/\left[\sum_j k_{\text{estab}}^j(\underline{N_{t-1}^j})\right]\right\}\sum_{j\neq i}\varphi_{\text{emi}}(N_{t-1}^j, \underline{N}_{\underline{t-1}}) \end{aligned} \tag{12.5}$$

with the sum taken over all patches in the habitat complex. The relationship given by equation (12.5) assumes that there is no loss over and above that accounted for by mortality during the transfer process (k_{trans}) and during establishment (k_{estab}^i); but note that the migrating individuals are distributing themselves selectively among the available patches in the habitat complex.

Furthermore, $R_{t-1}^i(\cdot)$ is the specific net growth rate within patch i, incorporating birth and mortality within the patch; the $(\cdot)$ within the argument of this function indicates that these vital rates may involve both density-dependent and density-independent processes. Within the literature on time series population dynamics modelling (see, for example, Stenseth *et al.* 1996), this function is typically given as:

$$R_{t-1}^i(\cdot) = \exp[a + b\ln(N_{t-1}^i) + \varepsilon_t] \tag{12.6}$$

where a and b are constants, and ε_t is some time-independent, but possibly spatially dependent, noise process (environmental and/or demographic noise).

Population ecologists are typically concerned with the R and the φ_{emi} functions (where the latter is usually lumped with mortality so as to obtain an overall ecological mortality rate) and, when writing models, ignore immigration (φ_{immi}^i). Behavioural ecologists focusing on dispersal, on the other hand, primarily focus on φ_{emi} and φ_{immi}^i, ignoring the R component of the population dynamics.

An order one population model (e.g. Grenfell *et al.* 1998) would look like:

$$N_t = \lambda(N_{t-1}, \varepsilon_t)N_{t-1} \tag{12.7}$$

However, assuming a Gompertz type of model (Stenseth *et al.* 1996; Stenseth 1999) (using a log scale so $x_t = \log(N_t)$) we obtain:

$$x_t = \alpha + (1+\beta)x_{t-1} + \varepsilon_t \tag{12.8}$$

where α is related to the maximum net growth rate within a patch (defined by R and φ_{emi}), β is the density dependency within the system as defined by the density dependencies of R and φ_{emi}, and ε_t is as defined above.

Letting the data tell us what the functions should look like

One important objective of this chapter—over and above being a synthesis and review of the results from the Evenstad project—is to define the functions entering into the population dynamics model defined by equation (12.1) on the basis of behavioural studies carried out within an experimental study on small rodents, where the root vole is used as the experimental organism. In this section we will synthesize the experimental data from the Evenstad project within the framework of the model defined in the previous section—this being essentially the approach taken by Stenseth and colleagues in several recent population studies (see, for example, Stenseth 1999). For this purpose we have reanalysed the data according to the above model. A weekly timescale is adopted (see, for example, Ims & Andreassen 1999b; Andreassen & Ims 2001).

The specific net growth rate within patches, $R^i_{t-1}(\cdot)$, is negatively density dependent (Fig. 12.5). The specific emigration function, $\varphi_{emi}(N^i_{t-1}, \underline{N_{t-1}})$, as is indicated by the function arguments, is dependent both on the density within the patch which animals are leaving (N^i_{t-1}) as well as on the overall density structure within the habitat complex (indicated by the vector $\underline{N_{t-1}}$) (Fig. 12.6). As can be seen, the emigration rate is the highest from low-density patches, and particularly when there is a great deal of variation in density among the patches.

The immigration function as defined in equation (12.5), is related to the overall emigration from all patches to other patches, as well as survival during the movement from one patch to another (k_{trans}) and the ease by which an animal can enter a particular patch (k^i_{estab}). Here we will consider the latter two components. We do not, however, know the exact values of k_{trans} and k_{estab} because we only know that a complete dispersal (ending with settlement in a new patch) has taken place when immigration has been successful. We know, however, that matrix movements are risky and result in a lower survival rate than that of resident animals (Andreassen & Ims 2001). The probability of establishment, we have shown experimentally, is dependent on the density of the immigration patch (G. Gundersen, unpubl. data). In Fig. 12.7, however, we show the negative effect of density on the immigration function.

As far as we are aware, this is the first time detailed individual-based population studies have been interpreted within the functions of a full population dynamics model. The aim of this exercise is to demonstrate that it is possible. More specifically, it is worth noticing that the emigration rate has been demonstrated to be a function of the density in the current patch as well as in the overall habitat complex—just as suggested in the verbal model presented above.

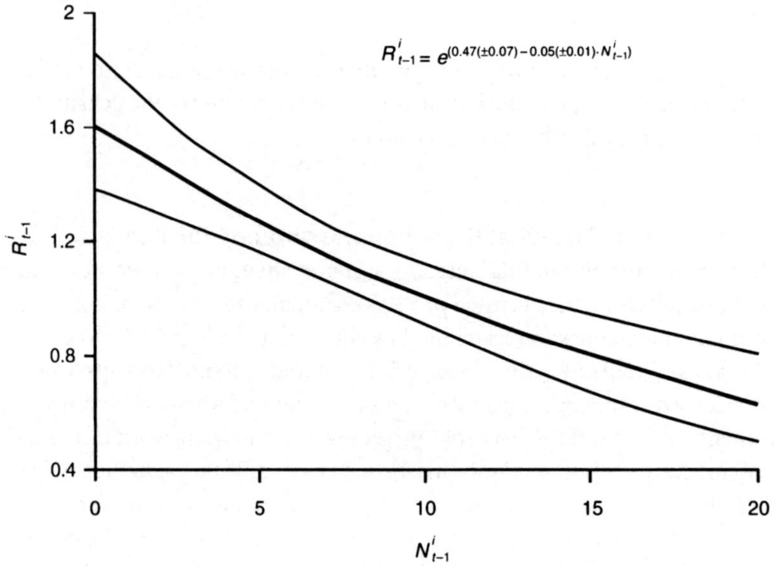

Figure 12.5 The predicted (±95% confidence limit) density-dependent net growth rate at Evenstad. The results are obtained from a total of eight plots studied in 1990/91 with six small habitat patches each (four plots also contained corridors; see Fig. 12.2) (Ims & Andreassen 1999a, 1999b; Andreassen & Ims 2001). R^i_{t-1} is the specific net growth rate within patch i; N^i_{t-1} is the density of animals at time $t-1$ within patch i.

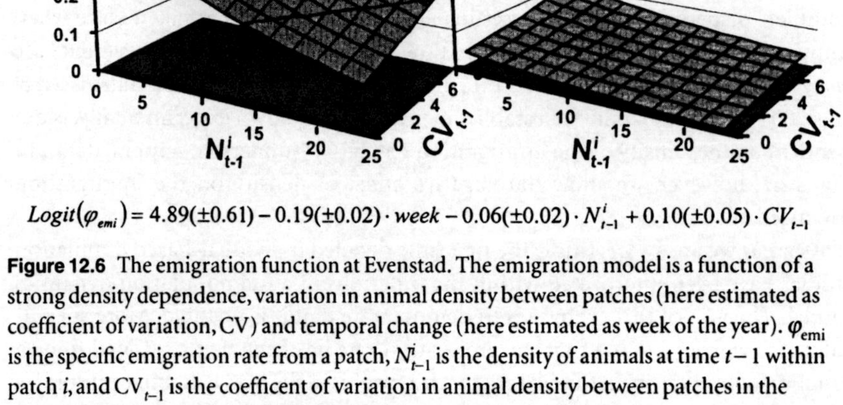

$$Logit(\varphi_{emi}) = 4.89(\pm 0.61) - 0.19(\pm 0.02) \cdot week - 0.06(\pm 0.02) \cdot N^i_{t-1} + 0.10(\pm 0.05) \cdot CV_{t-1}$$

Figure 12.6 The emigration function at Evenstad. The emigration model is a function of a strong density dependence, variation in animal density between patches (here estimated as coefficient of variation, CV) and temporal change (here estimated as week of the year). φ_{emi} is the specific emigration rate from a patch, N^i_{t-1} is the density of animals at time $t-1$ within patch i, and CV_{t-1} is the coefficent of variation in animal density between patches in the system. (Same data source as in Fig. 12.5.)

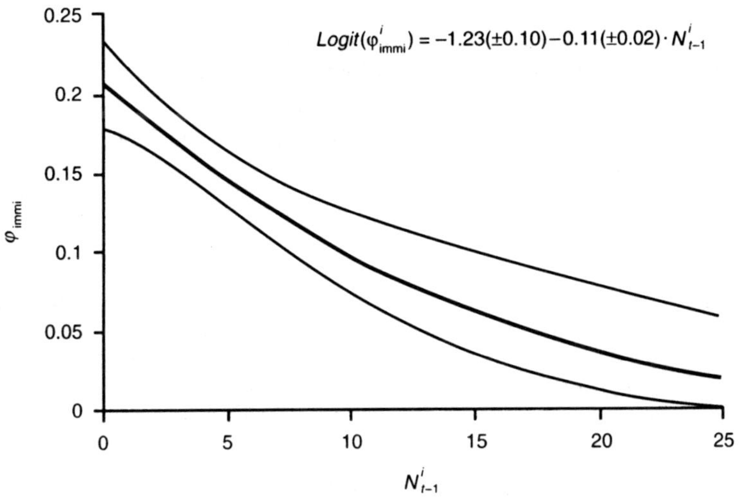

Figure 12.7 The predicted (±95% confidence limit) density-dependent immigration function at Evenstad. φ_{immi} is the total immigration rate into a patch and N^i_{t-1} is the density of animals at time $t-1$ within patch 1. (Same data source as in Fig. 12.5.)

Altogether these functions suggest that there is positive feedback causing low-density patches to drop to even lower densities without contributing to increased density in any of the higher-density patches. How can such a system be maintained? The answer is, that in itself, it will not. Dispersal in this system does not contribute to the maintenance of population size within a habitat complex; it only leads to the redistribution of animals within the habitat complex. What seems to contribute to the maintenance of the overall population size within the habitat complex is the within-patch net growth rate ($R^i_{t-1}(\bullet)$) and its high degree of stochasticity (ε_t).

Future developments and conclusions

The modelling framework suggested here may be a worthwhile one for linking behavioural and population biology more closely. Indeed, with such a model we might be able to explore the population dynamics consequences of dispersal. In a narrow sense this would be worthwhile to understand the Evenstad study system more fully, and we have demonstrated that this is indeed possible. Building upon this, we might extend the small-scale experiment of Evenstad to a larger geographical scale for the root vole and other small rodents. Such an extension will certainly need much careful consideration and further empirical data analyses—all of which are beyond the scope of this chapter.

In this chapter we have reviewed the general literature on the various phases of dispersal in vertebrates. Against this background we have summarized the insights derived from the Evenstad experimental study on the root vole, and reinterpreted these data within the context of a mathematically formulated population dynamics model. Although there might not be as much data for other biological systems, such an approach may, we think, serve as a basis for gluing together much of the available information on behavioural and population dynamics for individual species. Such an exercise will surely provide a better basis for drawing conclusions regarding both basic and applied issues relating to dispersal.

References

Aars, J. & Ims, R.A. (1999) The effect of habitat corridors on rates of transfer and inbreeding between vole demes. *Ecology* **80**, 1648–1655.

Aars, J., Johannesen, E. & Ims, R.A. (1999) Demographic consequences of movements in subdivided root vole populations. *Oikos* **85**, 204–216.

Alyan, S. & Jander, R. (1994) Short-range homing in the house mouse, *Mus musculus*: stages in the learning of directions. *Animal Behaviour* **48**, 285–298.

Anderson, P.K. (1989) *Dispersal in Rodents: a Resident Fitness Hypothesis.* Special Publication No. 9. The American Society of Mammalogists, USA.

Andreassen, H.P. & Ims, R.A. (1990) Responses of female grey-sided voles *Clethrionomys rufocanus* to malnutrition: a combined laboratory and field experiment. *Oikos* **59**, 107–114.

Andreassen, H.P. & Ims, R.A. (1998) The effects of experimental habitat destruction and patch isolation on space use and fitness parameters in female root vole *Microtus oeconomus*. *Journal of Animal Ecology* **67**, 941–952.

Andreassen, H.P. & Ims, R.A. (2001) Dispersal movements in patchy vole populations: the role of habitat patch structure and demography. *Ecology* **82**, 2911–2926.

Andreassen, H.P., Halle, S. & Ims, R.A. (1996a) Optimal width of movement corridors for root voles: not too narrow and not too wide. *Journal of Applied Ecology* **33**, 63–70.

Andreassen, H.P., Ims, R.A. & Steinset, O.K. (1996b) Discontinuous habitat corridors: effects on male root vole movements. *Journal of Applied Ecology* **33**, 555–560.

Andreassen, H.P., Hertzberg, K. & Ims, R.A. (1998) Space use responses to habitat fragmentation and connectivity in the root vole *Microtus oeconomus*. *Ecology* **79**, 1223–1235.

Bekoff, M. (1977) Mammalian dispersal and the ontogeny of individual phenotypes. *American Naturalist* **111**, 715–732.

Bell, W.J. (1991) *Searching Behaviour.* Chapman & Hall, London.

Bengtsson, B.O. (1978) Avoiding inbreeding: at what cost? *Journal of Theoretical Biology* **73**, 439–444.

Bennett, A.F. (1990) *Habitat Corridors. Their Role in Wildlife Management and Conservation.* Department of Conservation and Environment, Victoria Arthur Ryah Institute for Environmental Research, Melbourne.

Bennett, A.F. (1999) *Linkages in the Landscapes: the Role of Corridors and Connectivity in Wildlife Conservation.* IUCN Publications, Cambridge.

Bollinger, E.K., Harper, S.J. & Barrett, G.W. (1993) Inbreeding avoidance increases dispersal movements of the meadow vole. *Ecology* **74**, 1153–1156.

Bondrup-Nielsen, S. (1985) An evaluation of the effects of space use and habitat patterns on dispersal in small mammals. *Annales Zoologici Fennici* **22**, 373–383.

Bondrup-Nielsen, S. (1993) Early malnutrition increases emigration of adult female meadow voles, *Microtus pennsylvanicus*. *Oikos* **67**, 317–320.

Boulinier, T., McCoy, K.D. & Sorci, G. (2001) Dispersal and parasitism. In: *Dispersal* (eds J. Clobert, E. Danchin, A.A. Dhondt & J.D. Nichols), pp. 169–179. Oxford University Press, Oxford.

Charnov, E.L. & Finerty, J.P. (1980) Vole population cycles: a case for kin-selection? *Oecologia* **45**, 1–2.

Clobert, J., Danchin, E., Dhondt, A.A. & Nichols, J.D. (eds) (2001) *Dispersal.* Oxford University Press, Oxford.

Dos Santos, E.M., Andreassen, H.P. & Ims, R.A. (1995) Differences in tolerance to inbreeding between two geographically distinct strains of root voles, *Microtus oeconomus*. *Ecography* **18**, 238–247.

Dufty Jr, A.M. & Belthoff, J.R. (2001) Proximate mechanisms of natal dispersal: the role of body conditions and hormones. In: *Dispersal* (eds J. Clobert, E. Danchin, A.A. Dhondt & J.D. Nichols), pp. 217–229. Oxford University Press, Oxford.

Fauske, J., Andreassen, H.P. & Ims, R.A. (1997) Spatial organisation in a small population of the root vole *Microtus oeconomus* in a linear habitat. *Acta Theriologica* **42**, 79–90.

Fitzgibbon, C.D. (1993) The distribution of grey squirrel dreys in farm woodland: the influence of wood area, isolation and management. *Journal of Applied Ecology* **30**, 736–742.

Forman, R.T.T. (1995) *Land Mosaics: the Ecology of Landscape and Regions*. Cambridge University Press, Cambridge.

Gandon, S. & Michalakis, Y. (2001) Multiple causes of the evolution of dispersal. In: *Dispersal* (eds J. Clobert, E. Danchin, A.A. Dhondt & J.D. Nichols), pp. 155–167. Oxford University Press, Oxford.

Gliwicz, J. (1993) Dispersal in bank voles: benefits to emigrants or to residents? *Acta Theriologica* **38**, 31–38.

Greenwood, P.J. (1980) Mating systems, philopatry and dispersal in birds and mammals. *Animal Behaviour* **28**, 1140–1162.

Grenfell, B.T., Wilson, K., Finkenstadt, B.F. *et al.* (1998) Noise and determinism in synchronized sheep dynamics. *Nature* **394**, 674–677.

Gundersen, G. & Andreassen, H.P. (1998) Causes and consequences of natal dispersal in root voles, *Microtus oeconomus*. *Animal Behaviour* **56**, 1355–1366.

Gundersen, G., Moe, J.A., Andreassen, H.P., Carlsen, R.G. & Gundersen, H. (1999) Intersexual attraction in natal dispersing root voles *Microtus oeconomus*. *Acta Theriologica* **44**, 283–290.

Gundersen, G., Johannesen, E., Andreassen, H.P. & Ims, R.A. (2001) Source–sink dynamics: how sinks affect demography of sources. *Ecology Letters* **4**, 14–21.

Haddad, N.M. (1999) Corridor use predicted from behaviors at habitat boundaries. *American Naturalist* **153**, 215–227.

Hanski, I. & Gilpin, M.E. (1997) *Metapopulation Biology: Ecology, Genetics and Evolution*. Academic Press, London.

Hansson, L. (1987) Dispersal routes of small mammals at an abandoned field in central Sweden. *Holarctic Ecology* **10**, 154–159.

Hansteen, T.L., Andreassen, H.P. & Ims, R.A. (1997) Effects of spatiotemporal scale on autocorrelation and home range estimators. *Journal of Mammalogy* **61**, 280–290.

Hardt, R.A. & Forman, R.T.T. (1989) Boundary form effects on woody colonization of reclaimed surface mines. *Ecology* **70**, 1252–1260.

Harper, S.J. & Batzli, G.O. (1996) Monitoring use of runways by voles with passive integrated transponders. *Journal of Mammalogy* **77**, 364–369.

Harper, S.J., Bollinger, E.K. & Barrett, G.W. (1993) Effects of habitat patch shape on population dynamics of meadow voles (*Microtus pennsylvanicus*). *Journal of Mammalogy* **74**, 1045–1055.

Hestbeck, J.B. (1982) Population regulation of cyclic mammals: the social fence hypothesis. *Oikos* **39**, 157–163.

Hobbs, R.J. (1992) The role of corridors in conservation: solution or bandwagon. *Trends in Ecology and Evolution* **7**, 389–392.

Holt, R.D. & Barfield, M. (2001) On the relations between the ideal free distribution and the evolution of dispersal. In: *Dispersal* (eds J. Clobert, E. Danchin, A.A. Dhondt & J.D. Nichols), pp. 83–95. Oxford University Press, Oxford.

Ims, R.A. (1987) Responses in the spatial organization and behavior to manipulations of the food resource in the vole *Clehtrionomys rufocanus*. *Journal of Animal Ecology* **56**, 485–596.

Ims, R.A. (1989) Kinship and origin effects on dispersal and space sharing in *Clethrionomys rufocanus*. *Ecology* **70**, 607–616.

Ims, R.A. (1990) Determinants of natal dispersal and space use in the grey-sided vole, *Chletrionomys rufocanus*: a combined laboratory and field experiment. *Oikos* **57**, 106–113.

Ims, R.A. (1994) Litter sex ratio variation in colonies of two geographically distinct strains of the root vole *Microtus oeconomus*. *Ecography* **17**, 141–146.

Ims, R.A. (1995) Movement patterns related to spatial structures. In: *Mosaic Landscapes and Ecological Processes* (eds L. Hansson, L. Fahrig & G.

Merriam), pp. 85–109. Chapman & Hall, London.

Ims, R.A. (1997) Determinants of geographic variation in growth and reproductive traits in the root vole. *Ecology* **78**, 461–470.

Ims, R.A. & Andreassen, H.P. (1999a) Demographic synchrony in fragmented populations. In: *The Ecology of Small Mammals at the Landscape Level: Experimental Approaches* (eds G.W. Barrett & J.D. Peles), pp. 129–146. Springer Verlag, Berlin.

Ims, R.A. & Andreassen, H.P. (1999b) Effects of experimental habitat fragmentation and connecivity on root vole demography. *Journal of Animal Ecology* **68**, 839–852.

Ims, R.A. & Andreassen, H.P. (2000) Spatial synchronization of vole population dynamics by predatory birds. *Nature* **408**, 194–197.

Ims, R.A. & Steen, H. (1990) Geographical synchrony in microtine population cycles: a theoretical evaluation of the role of nomadic avian predators. *Oikos* **57**, 381–387.

Ims, R.A. & Stenseth, N.C. (1989) Divided the fruit-flies fall. *Nature* **342**, 21–22.

Ims, R.A. & Yoccoz, N.G. (1997) Studying transfer in metapopulations: emigration, migration and immigration. In: *Metapopulation Biology: Ecology, Genetics, and Evolution* (eds I. Hanski & M. Gilpin), pp. 247–265. Academic Press, London.

Ims, R.A., Rolstad, J. & Wegge, P. (1993) Predicting space use responses to habitat fragmentation: can voles *Microtus oeconomus* serve as an experimental model system EMS for capercaille grouse *Tetrao urogallus* in boreal forest? *Biological Conservation* **63**, 261–268.

Jacquot, J.J. & Vessey, S.H. (1995) Influence of the natal environment on dispersal of white-footed mice. *Behavioral Ecology and Sociobiology* **37**, 407–412.

Johannesen, E. & Andreassen, H.P. (1998) Survival and reproduction of resident and immigrant female root voles (*Microtus oeconomus*). *Canadian Journal of Zoology* **76**, 763–766.

Johannesen, E., Andreassen, H.P. & Steen, H. (1997) Effect of radiocollars on survival of root voles. *Journal of Mammalogy* **78**, 638–642.

Kramer, D.L., Rangeley, R.W. & Chapman, L.J. (1997) Habitat selection: patterns of spatial distribution from behavioural decisions. In: *Behavioural Ecology of Teleost Fishes* (ed. J.J. Godin), pp. 37–79. Oxford University Press, Oxford.

Lambin, X. & Krebs, C.J. (1991) Can changes in female relatedness influence microtine population dynamics. *Oikos* **61**, 126–132.

Lambin, X., Aars, J. & Pierney, S.B. (2001) Dispersal, intraspecific competition, kin competition and kin facilitation: a review of empirical evidence. In: *Dispersal* (eds J. Clobert, E. Danchin, A.A. Dhondt & J.D. Nichols), pp. 110–122. Oxford University Press, Oxford.

Liberg, O. & von Schantz, T. (1985) Sex-biased philopatry and dispersal in birds and mammals: the Oedipus hypothesis. *American Naturalist* **126**, 129–135.

Lidicker Jr, W.Z. (1975) The role of dispersal in the demography of small mammals. In: *Small Mammals: their Productivity and Population Dynamics* (eds F.B. Golley, K. Pertrusewicz & L. Ryszkowski), pp. 103–128. Cambridge University Press, London.

Lidicker Jr, W.Z. & Stenseth, N.C. (1992) To disperse or not to disperse: who does it and why? In: *Animal Dispersal: Small Mammals as a Model* (eds N.C. Stenseth & W.Z. Lidicker Jr), pp. 21–36. Chapman & Hall, London.

Lurz, P.W.W., Garson, P.J. & Wauters, L.A. (1997) Effects of temporal and spatial variation in habitat quality on red squirrel dispersal behaviour. *Animal Behaviour* **54**, 427–435.

McGuire, B., Getz, L.L., Hofmann, J.E., Pizzuto, T. & Frase, B. (1993) Natal dispersal and philopatry in prairie voles (*Microtus ochrogaster*) in relation to population density, season, and natal social environment. *Behavior Ecology and Sociobiology* **32**, 293–302.

McShea, W.J. (1990) Social tolerance and proximate mechanisms of dispersal among winter groups of meadow voles, *Microtus pennsylvanicus. Animal Behaviour* **39**, 346–351.

Perrin, N. & Goudet, J. (2001) Inbreeding, kinship, and the evolution of natal dispersal. In: *Dispersal* (eds J. Clobert, E. Danchin, A.A. Dhondt & J.D. Nichols), pp. 123–142. Oxford University Press, Oxford.

Pulliam, H.R. (1988) Sources, sinks, and population regulation. *American Naturalist* **132**, 652–661.

Pusey, A.E. (1987) Sex-biased dispersal and inbreeding avoidance in birds and mammals. *Trends in Ecology and Evolution* **2**, 295–299.

Shields, W.M. (1983) Optimal inbreeding and the evolution of philopatry. In: *The Ecology of Animal*

Movement (eds I.R. Swingland & P.J. Greenwood), pp. 132–159. Clarendon Press, Oxford.

Stamps, J.A. (1988) Conspecific attraction and aggregation in territorial species. *American Naturalist* **131**, 329–347.

Stamps, J.A. (1991) The effect of conspecifics on habitat selection in territorial species. *Behavioral Ecology and Sociobiology* **28**, 29–36.

Stamps, J.A. (2001) Habitat selection by dispersers: integrating proximate and ultimate approaches. In: *Dispersal* (eds J. Clobert, E. Danchin, A.A. Dhondt & J.D. Nichols), pp. 230–242. Oxford University Press, Oxford.

Stamps, J.A., Buechner, M. & Krishnan, V.V. (1987) The effects of edge permeability and habitat geometry on emigration from patches of habitat. *American Naturalist* **129**, 533–552.

Steen, H. (1994) Low survival of long distance dispersers of the root vole (*Microtus oeconomus*). *Annales Zoologici Fennici* **31**, 271–274.

Stenseth, N.C. (1983) Causes and consequences of dispersal in small mammals. In: *The Ecology of Animal Movement* (eds I.R. Swingland & P.J. Greenwood), pp. 63–101. Clarendon Press, Oxford.

Stenseth, N.C. (1986) On the interaction between stabilizing social factors and destabilizing trophic factors in small rodent populations. *Theoretical Population Biology* **29**, 365–384.

Stenseth, N.C. (1999) Population cycles in voles and lemmings: density dependence and phase dependence in a stochastic world. *Oikos* **87**, 427–461.

Stenseth, N.C. & Lidicker Jr, W.Z. (1992) *Animal Dispersal: Small Mammals as a Model.* Chapman & Hall, London.

Stenseth, N.C., Bjørnstad, O.N. & Falck, W. (1996) Is spacing behaviour coupled with predation causing the microtine density cycle? A synthesis of current process-oriented and pattern-oriented studies. *Proceedings of the Royal Society of London, Series B* **263**, 1423–1435.

Stoddart, D.M. (1980) *The Ecology of Vertebrate Olfaction.* Chapman & Hall, London.

Swenson, J.E., Sandegren, F., Soderberg, A., Bjarvall, A. & Franzen, R. (1997) Infanticide caused by hunting of male bears. *Nature* **386**, 450–451.

Szacki, J. (1987) Ecological corridors as a factor determining the structure and organisation of a bank vole population. *Acta Theriologica* **32**, 31–44.

Tast, J. (1966) The root vole, *Microtus oeconomus* (Pallas), as an inhabitant of seasonally flooded land. *Annales Zoologici Fennici* **3**, 127–171.

Tinckle, D.W., Dunham, A.E. & Congdon, J.D. (1993) Life history and demographic variation in the lizard *Sceloporus graciosus*: a long term study. *Ecology* **74**, 2413–2429.

Tordoff, H.B., Martell, M.S. & Redig, P.T. (1998) Effect of fledge site on choice of nest site by midwestern peregrine falcons. *Loon* **70**, 127–129.

Turchin, P. (1998) *Quantitative Analysis of Movement: Measuring and Modeling Population Redistribution in Animals and Plants.* Sinneaur Associates, Sunderland, MA.

Wauters, L., Mathysen, E. & Dhondt, A.A. (1994) Survival and lifetime reproductive success in dispersing and resident red squirrels. *Behavioral Ecology and Sociobiology* **34**, 197–201.

Wegner, J.F. & Merriam, G. (1979) Movement of birds and small mammals between wood and adjoining farmland habitats. *Journal of Applied Ecology* **16**, 349–358.

Wiens, J.A. (2001) The landscape context of dispersal. In: *Dispersal* (eds J. Clobert, E. Danchin, A.A. Dhondt & J.D. Nichols), pp. 96–109. Oxford University Press, Oxford.

Wiens, J.A., Stenseth, N.C., Van Horne, B. & Ims, R.A. (1993) Ecological mechanisms and landscape ecology. *Oikos* **66**, 369–380.

Wiggett, D.R. & Boag, D.A. (1993) Annual reproductive success in three cohorts of Columbian ground squirrels: founding immigrants, subsequent immigrants, and natal residents. *Canadian Journal of Zoology* **71**, 1577–1584.

Wolff, J.O. (1992) Parents suppress reproduction and stimulate dispersal in opposite-sex juvenile white-footed mice. *Nature* **359**, 409–410.

Wolff, J.O. (1993) What is the role of adults in mammalian juvenile dispersal? *Oikos* **68**, 173–176.

Wolff, J.O. (1995) Friends and strangers in vole population cycles. *Oikos* **73**, 411–414.

Wolff, J.O. & Lidicker Jr, W.Z. (1980) Population ecology of the taiga vole, *Microtus xanthognathus*, in interior Alaska. *Canadian Journal of Zoology* **58**, 1800–1812.

Zhang, Z. & Usher, M.B. (1991) Dispersal of wood mice and bank voles in an agricultural landscape. *Acta Theriologica* **36**, 239–245.

Zollner, P.A. & Lima, S.L. (1997) Landscape-level perceptual abilities in white-footed mice: perceptual range and the detection of forested habitat. *Oikos* **80**, 51–60.

Zollner, P.A. & Lima, S.L. (1999) Search strategies for landscape-level interpatch movements. *Ecology* **80**, 1019–1030.

Chapter 13

Dispersal and the spatial dynamics of butterfly populations

Rob J. Wilson and Chris D. Thomas

Introduction

Dispersal determines, for example, rates of colonization, immigration, gene flow and habitat choice. Thus, considerable effort has been spent estimating rates of dispersal for taxa of conservation concern like butterflies, and for other taxa which are pests or disease vectors. This chapter addresses three issues as they pertain specifically to butterfly dispersal. First, what are the approaches and difficulties in measuring dispersal? Second, in the face of the difficulties of obtaining a full empirical understanding of dispersal in any species, what sort of dispersal information is relevant to particular questions? Third, when can we take a pattern-based approach to dispersal, and when must we attempt to understand the processes giving rise to those patterns? The first two issues apply to virtually all studies of animal and plant dispersal; the third applies to most animal studies, and also to animal-mediated pollination and seed dispersal of plants. We argue that, rather than seek a 'perfect' understanding of dispersal, the type of data collected should be appropriate for the question being asked. Direct measurements of where most individuals move, such as mark–release–recapture and the tracking of individuals, are informative for questions of local population structure, dynamics and conservation management. But direct measures do not provide a satisfactory understanding of large-scale patterns, since they routinely underestimate rates of long-distance movement ('the tail of the distribution'), and extrapolation from short-distance measures is unlikely to predict very long-distance movements accurately. For an understanding of large-scale patterns and processes, indirect measures such as patterns of colonization provide more realistic estimates of dispersal rates. These measures of dispersal are relevant for understanding landscape-scale dynamics and distribution changes.

In insects, flight represents the main means of dispersal. In concentrating on butterflies, we should first ask why flightlessness is so rare in the Lepidoptera, when flight polymorphisms or complete loss of flight have secondarily evolved in many other insect groups (Southwood 1962; Harrison 1980; Roff 1986, 1990; Wagner & Liebherr 1992). The answer seems obvious, though difficult to prove. Most lepidopteran larvae are phytophagous, and complete their development on or near the food plant where they hatched. But adult Lepidoptera must fly to exploit the nutritious liquid foods for which their mouthparts are specialized, because these liquid foods (nectar, fruit, sap, carrion, dung) tend to be both temporary and scattered around habitats. The only non-flying insects to exploit these resources to any great extent are much more generalized in their mouthparts and feeding range (ants). Adult dragonflies and damselflies also lack flightless forms, needing to fly actively to hunt for their prey (Roff 1990). Thus, both Lepidoptera and Odonata possess quite different feeding niches in their immature and adult stages, and flight is practically essential to their adult modes of life. In contrast, most of the insect groups that exhibit distinct wing polymorphisms (e.g. Coleoptera and Orthoptera) have much more similar immature and adult feeding niches, with many individuals capable of carrying out their entire life cycle within easy walking distance of the point where they hatched.

This background is important in the context of this chapter because it defines the range of evolutionary options available to a butterfly, and hints at certain features of lepidopteran population structure and dynamics. The specialized larval host plants of many species of Lepidoptera are highly aggregated in *patches* of suitable habitat, where more or less discrete breeding populations occur, but adult butterflies may migrate relatively long distances between habitat patches. As a result, lepidopteran population structures often present some of the necessary attributes of *metapopulations*, in which local populations may become extinct from time to time and empty patches may be colonized (Hanski 1991, 1999). Consequently, butterflies have been important taxa in the development of metapopulation theory (Thomas & Hanski 1997), and we address butterfly dispersal largely in the context of metapopulation biology. In particular, we show that although most adult butterflies fly relatively short distances (within habitat patches), their capacity for active flight means that a few individuals may move much further. These relatively rare events are difficult to quantify but may have important consequences for lepidopteran population structure and dynamics.

Direct and indirect measurements of dispersal

Dispersal may be measured directly; or indirectly by measuring its consequences. Direct measurements include following individuals (e.g. Root & Kareiva 1984; Schultz 1998; Haddad 1999a, 1999b; Conradt *et al.* 2000), marking them and recapturing them later (e.g. Ford 1945; Ehrlich 1961; Gilbert & Singer 1973, 1975; Watt *et al.* 1977; Southwood 1978; Brakefield 1982; Hanski *et al.* 1994; Hill *et al.* 1996; Bennetts *et al.* 2001) or tagging them electronically and detecting them automatically (e.g. Roland *et al.* 1996; Osborne *et al.*, this volume). 'Consequences' measured may be patterns of genetic variation, population variability, colonization and distribution changes (e.g. Brakefield 1991; Avise 1994; Dennis & Shreeve 1996, 1997; Shigesada & Kawasaki 1997; Hill *et al.* 1999; Mallet 2001; Rousset 2001).

Where most individuals move

Direct measurement of dispersal has great benefits because, in principle, it does not make any of the assumptions implicit in any of the indirect approaches. In practice, direct approaches usually provide reasonable or good measures of short-distance movement, but fail to detect the tail of the dispersal distribution (the longest movements) with any accuracy.

The only truly unbiased direct measures of dispersal come from studies of radio-tagged animals, monitored by satellites. Satellite-based systems are not yet feasible for butterflies because the batteries required are much too heavy. All other direct measurements of dispersal are likely to incorporate some degree of distance-weighted sampling bias.

Individuals that can be followed without disturbance can provide unbiased measures of movement over relatively short distances, but it will rarely be possible to follow enough individuals to reveal information on the small fraction of individuals which move very long distances (but see correlated random walk approaches, below). Electronic transponders can be attached to large butterflies, but these can only be detected from relatively short distances (up to hundreds of metres in uncluttered environments; Osborne *et al.*, this volume), so they are mainly used to aid detection of individuals within their habitats, rather than to detect long-distance migrants (Roland *et al.* 1996). With current technologies, the transponder approach is likely to be too expensive and slow to detect the longest distance movements, since it would require potentially thousands of individuals to be tagged.

In mark–release–recapture (MRR) studies, the longest distance dispersers may leave the study area entirely, thus becoming undetectable (Barrowclough 1978; Mason 1990). Similarly, many studies may be systematically biased because habitat is aggregated, and short-distance movements are therefore more detectable than long-distance movements (Porter & Dooley 1993; Koenig *et al.* 1996). It is possible to search for long-distance dispersers beyond the marking zone, but it can be difficult to find them amongst large populations of unmarked individuals. One solution is to release individuals into landscapes where a species does not exist, and record the dispersal rate (Fahrig & Paloheimo 1987; Kuussaari *et al.* 1996; Lewis *et al.* 1997; Menéndez & Thomas 2000). Rare individuals or their offspring may then be spotted relatively easily. However, it is only possible to study some species in this way (for example, because of restrictions on movement of rare species or pests), and, by definition, it is not possible to study species in their natal habitats. Moreover, the release approach reduces the problem, but does not eliminate it, since habitat and sampling may still be aggregated, and long-distance movements may still go undetected.

The problem of distance-weighted sampling bias is particularly acute in the many butterfly species that live in patchy habitats, where the area of habitat is a small fraction of the total area of the landscape. In these fragmented landscapes, we need to ask ourselves how instructive the empirical measurements of dispersal distances actually are. To what extent are the observed distributions of movements determined by the landscape or study area dimensions, by the dispersal traits of the study organism, or by the interaction of the two?

These issues can be illustrated using MRR data on the brown argus butterfly *Aricia agestis*, which occupies patches of unimproved calcareous grassland where its

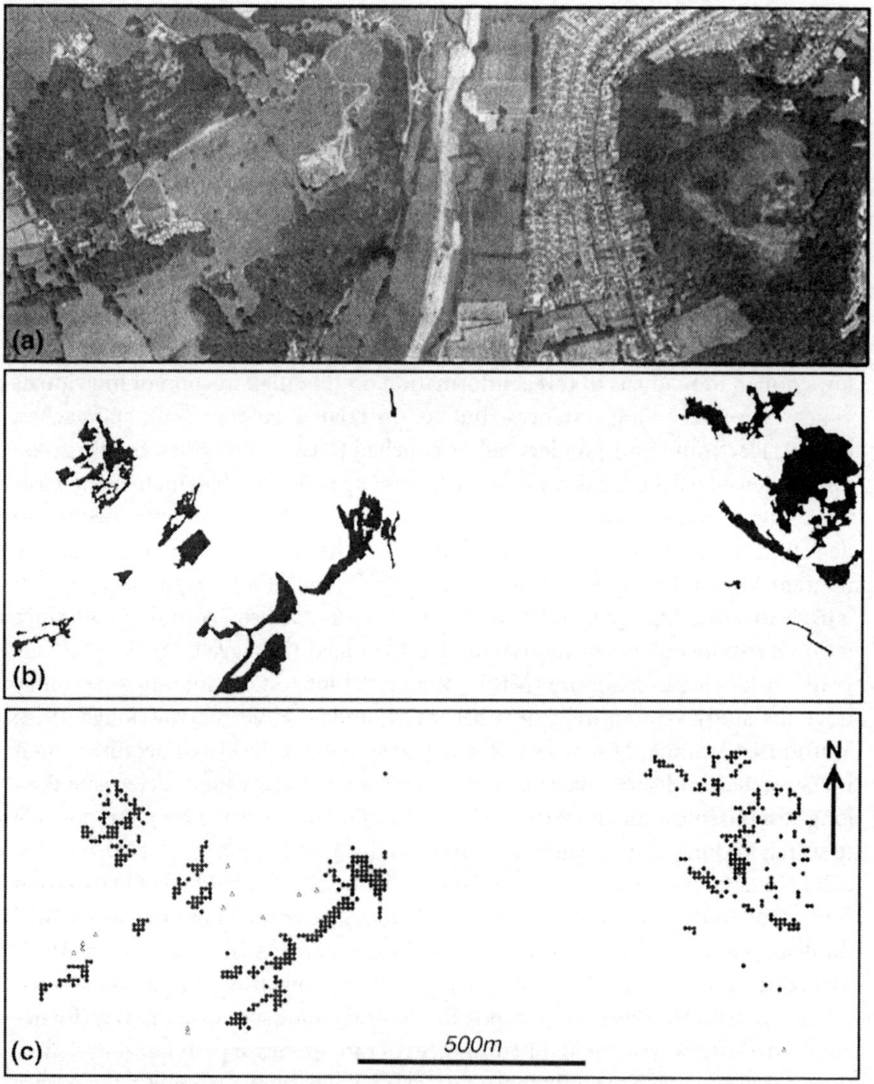

Figure 13.1 The *Aricia agestis* mark–release–recapture network (1.8 × 0.7 km) on the Creuddyn Peninsula in northwest Wales (53° 18′ N, 3° 46′ W). (a) Aerial photograph (NERC Airborne Remote Sensing Programme 1998). (b) Habitat patch network. (c) Locations of 549 separate 10 × 10 m grid cells where butterflies were captured: captures within habitat patches (filled circles) and captures outside habitat (open triangles).

larval host plant *Helianthemum nummularium* grows. We marked butterflies in a 1.25 km² area, in which the habitat network of 15 patches contained a total area of approximately 9 ha (Fig. 13.1). MRR studies are very labour intensive: concentrating our study within the habitat patches, and working with a species with fairly low population density, this study area was the largest that it was practical for a team of up to five field workers to cover. Searching the >90% of the area outside habitat patches was not feasible, and only a handful of *A. agestis* individuals were detected outside the habitat patches in a broader study that spanned eight generations of the species. We marked 1156 butterflies on 19 days over a 32-day period, and recaptured 391 individuals a total of 596 times. Most recaptured butterflies had flown very short distances (50% less than 60 m, 90% less than 250 m), but six (1.5%) flew more than 500 m, and one was recaptured 950 m from where it was first marked (Fig. 13.2a). A negative exponential model fitted to this distribution of dispersal distances ($I = e^{-kD}$, the cumulative proportion of dispersers (I) moving distance D km or further), had a slope (k) of −9.75 (± SE 0.26, $r^2 = 0.98$, $P < 0.0001$) (Fig. 13.2b, solid line). This equation estimates that 0.8% of butterflies would fly 500 m or further (empirically, 1.5% flew ≥500 m).

The problem with fitting this kind of equation to the empirical distribution of distances moved is that the movements recorded are a biased sample of total movements. Butterflies that leave one habitat patch and do not reach another, or which leave the study area entirely, are undetectable. We employed a simple method to estimate the likely proportions of dispersing individuals that had been missed by our sampling regime, by assuming that the probability of recording movements of a given distance depended on the proportion of the landscape at that distance which was inside, as opposed to outside, the habitat patches where capture was concentrated (see also Barrowclough 1978; Mason 1990; Mason *et al.* 1995). We calculated the ratio of habitat to non-habitat (*detectability*) in 25 m distance intervals away from each of the 549 separate 10 m grid references where butterflies were captured (Fig. 13.1c), and divided the observed (raw) number of movements in each distance interval by the detectability of movements in that interval: this gave the *adjusted* 'number of moves' per interval. We then calculated the cumulative proportion of adjusted moves as far or further than each distance interval, and repeated the negative exponential model (*finite area adjusted* (FAA) model—Fig. 13.2b, dashed line). This, like the raw model, was weighted by the actual cumulative number of dispersers at each distance. The adjusted model had a slope (k) of −3.43 (± SE 0.04, $r^2 = 0.99$, $P < 0.0001$), which estimated that 18% of butterflies would fly 500 m or further, a 22.5-fold increase relative to the estimate for the raw data. Note that this discrepancy occurred well within the limits imposed by the maximum dimensions of the study area (1800 m). We also fitted inverse power functions to raw and adjusted data (see Hill *et al.* 1996), but, in this instance, these produced poorer fits to the data both before and after adjustment.

The FAA approach attempts to adjust for biased sampling within the MRR network, but it cannot estimate the rate of movement beyond the maximum dimensions of the study area. More individuals are likely to leave smaller than larger study areas (Barrowclough 1978; Koenig *et al.* 1996), and these individuals may further bias estimates of dispersal. To illustrate the possible effects of maximum study area dimensions on dispersal estimates for *A. agestis*, we calculated negative exponential regressions separately using movements shorter than 100, 200, 300, 400, 500, 600

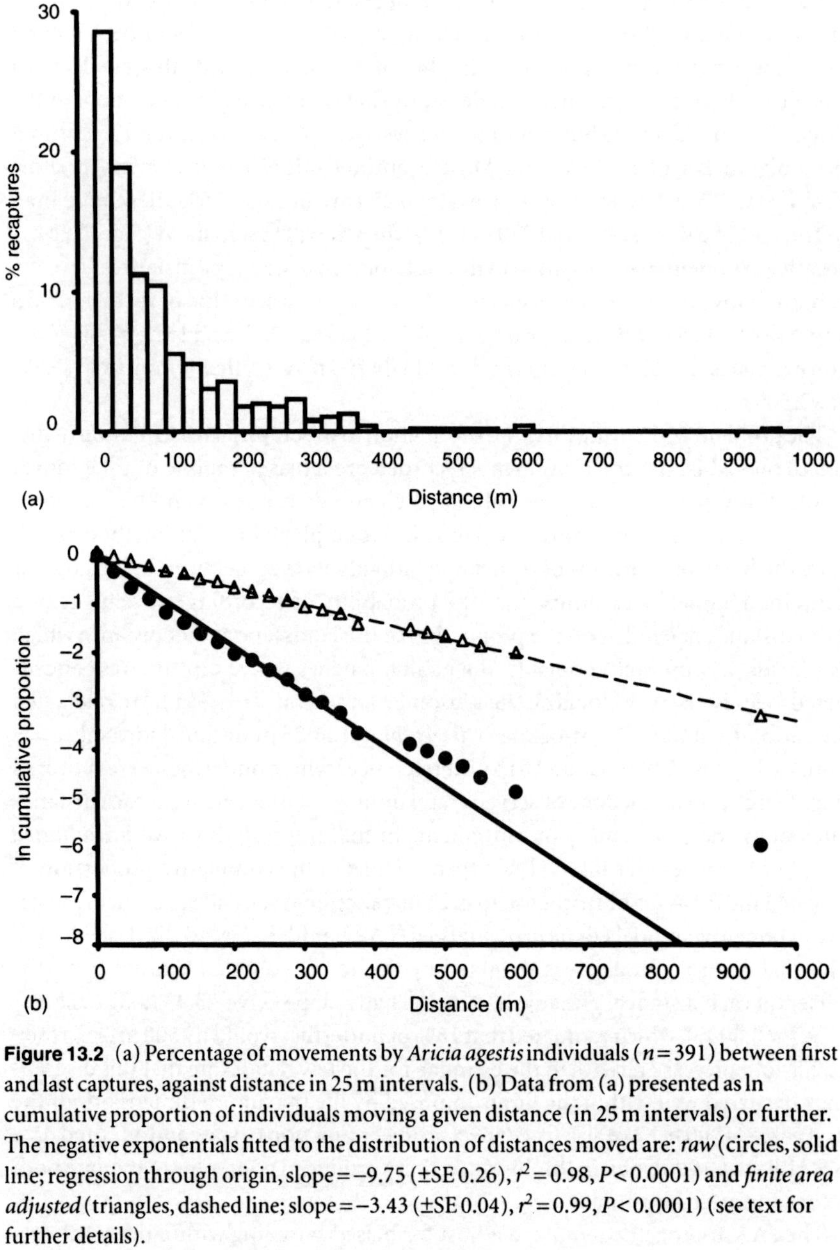

Figure 13.2 (a) Percentage of movements by *Aricia agestis* individuals ($n = 391$) between first and last captures, against distance in 25 m intervals. (b) Data from (a) presented as ln cumulative proportion of individuals moving a given distance (in 25 m intervals) or further. The negative exponentials fitted to the distribution of distances moved are: *raw* (circles, solid line; regression through origin, slope $= -9.75$ (±SE 0.26), $r^2 = 0.98$, $P < 0.0001$) and *finite area adjusted* (triangles, dashed line; slope $= -3.43$ (±SE 0.04), $r^2 = 0.99$, $P < 0.0001$) (see text for further details).

and 1000 m, respectively, as if these were the maximum movements that could possibly have been recorded. Using both raw and adjusted data, this approach showed that larger study areas would give longer estimates of movement (Fig. 13.3); ignoring long-distance dispersal events, for example because a study area is too small to detect them, can seriously underestimate rates of long-distance dispersal.

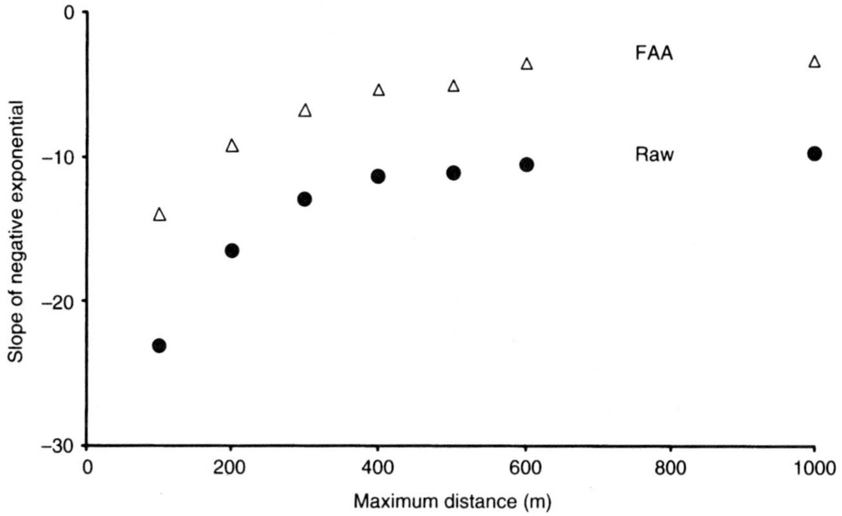

Figure 13.3 Raw (filled circles) and finite area adjusted (FAA) (open triangles) negative exponential dispersal estimates calculated for *Aricia agestis* from study areas with maximum dimensions of 100, 200, 300, 400, 500, 600 and 1000 m.

Distributions of raw and adjusted movement distances provide a crude estimate of the maximum proportion of dispersing individuals that are lost in the non-habitat matrix between habitat patches. Thus, if rates of dispersal by *A. agestis* were identical within and between habitat patches, it is likely that only about 4% of individuals which dispersed 500 m or more would successfully locate other habitat patches (0.8% from the raw estimate divided by 18% from the adjusted estimate). In fact, this percentage may be much higher, because long-distance, between-habitat movements may involve different behaviours to short-distance, within-habitat movements:

1 Movement within habitat patches ('foraging' for food, mates or breeding sites) may have shorter step lengths and tighter turning angles than movement outside habitat patches (Odendaal *et al.* 1989; Schultz 1998).

2 Butterflies may tend to turn back when they reach habitat boundaries or leave habitat patches (Cappuccino & Kareiva 1985; Sutcliffe & Thomas 1996; Schultz 1998; Haddad 1999a, 1999b).

3 Butterflies may 'home in' on habitat patches from relatively long distances (Harrison 1989; Conradt *et al.* 2000; see also Compton, this volume).

Thus, 18% can be regarded as a probable upper bound on the proportion of *A. agestis* individuals actually moving 500 m or more.

Rates of exchange among patches within metapopulations have been measured directly using MRR studies, leading to fairly consistent conclusions that patches which are further from each other exchange fewer individuals than patches which are closer together, and that smaller patches have higher rates of per capita emigration and immigration than large patches (Hill *et al.* 1996; Kuussaari *et al.* 1996; Sutcliffe *et al.* 1997; Thomas & Hanski 1997; Baguette *et al.* 2000). Both patterns are a clear consequence of skewed distributions of dispersal distances (most individuals fly short distances), while the latter may also be attributed to greater encounter rates of butterflies with patch boundaries in small patches, which have higher perimeter: area ratios than larger patches (Thomas *et al.* 1998, 2000). High rates of per capita emigration from small habitat patches or patches with low population density may be an important source of mortality in small populations (Kuussaari *et al.* 1998, Thomas *et al.* 1998).

In the *A. agestis* MRR, exchange rates between patches declined as patches became further apart, but there was no significant relationship between patch area and rates of emigration or immigration, probably because of habitat differences between patches (more exposed patches tended to have higher emigration rates, perhaps because individuals were generally concentrated in sheltered areas near to habitat boundaries) (Wilson 1999). However, when the MRR study area was divided into grid cells of different sizes (arbitrarily defined by the Ordnance Survey national grid), we found the expected pattern: high immigration and emigration fractions for small habitat areas, and high fractions of individuals recaptured in the cells where they were marked ('residents') for large areas (Fig. 13.4, triangles). The actual immigration and emigration fractions for natural habitat patches were clustered below the data calculated for arbitrary habitat areas, whereas the resident fractions of sites were clustered above the values for the arbitrary habitat areas (Fig. 13.4, crosses). Thus, *A. agestis* stayed in recognizable habitat patches to a greater extent than in arbitrarily defined cells of equivalent area. This pattern, and others which have shown that butterflies are more likely to fly through habitat than non-habitat (e.g. Sutcliffe & Thomas 1996), may result from individuals turning back at habitat boundaries (Haddad 1999a, 1999b).

Thus, the movements observed in MRR studies result from complex interactions between the area, quality and configuration of habitat patches, and from the behaviour of study species. Consequently, distributions of dispersal distances from one MRR study do not necessarily provide clear information on likely rates of colonization of remote habitat patches, or rates of exchange between separate populations in other habitat networks with different habitat quality and configuration. Modelling approaches are currently being developed to lead to general and comparable conclusions about dispersal from MRR studies (e.g. Lebreton *et al.* 1992; Hanski *et al.* 2000). These modelling approaches may yield most insight when combined with detailed knowledge of the behaviour and biology of individual species (e.g. Petit *et al.* 2001). Based on various assumptions, each of these approaches estimates how many long-distance movements might have been missed, and they have distinct advantages over the simple presentation and analysis of raw data. They, and future developments of the same types of approach, are likely to be particularly useful in providing increasingly realistic estimates of the distances moved by the majority of individuals. However, direct measurement and analysis is unlikely to provide reliable estimates of the distances moved by the few individuals that move the furthest.

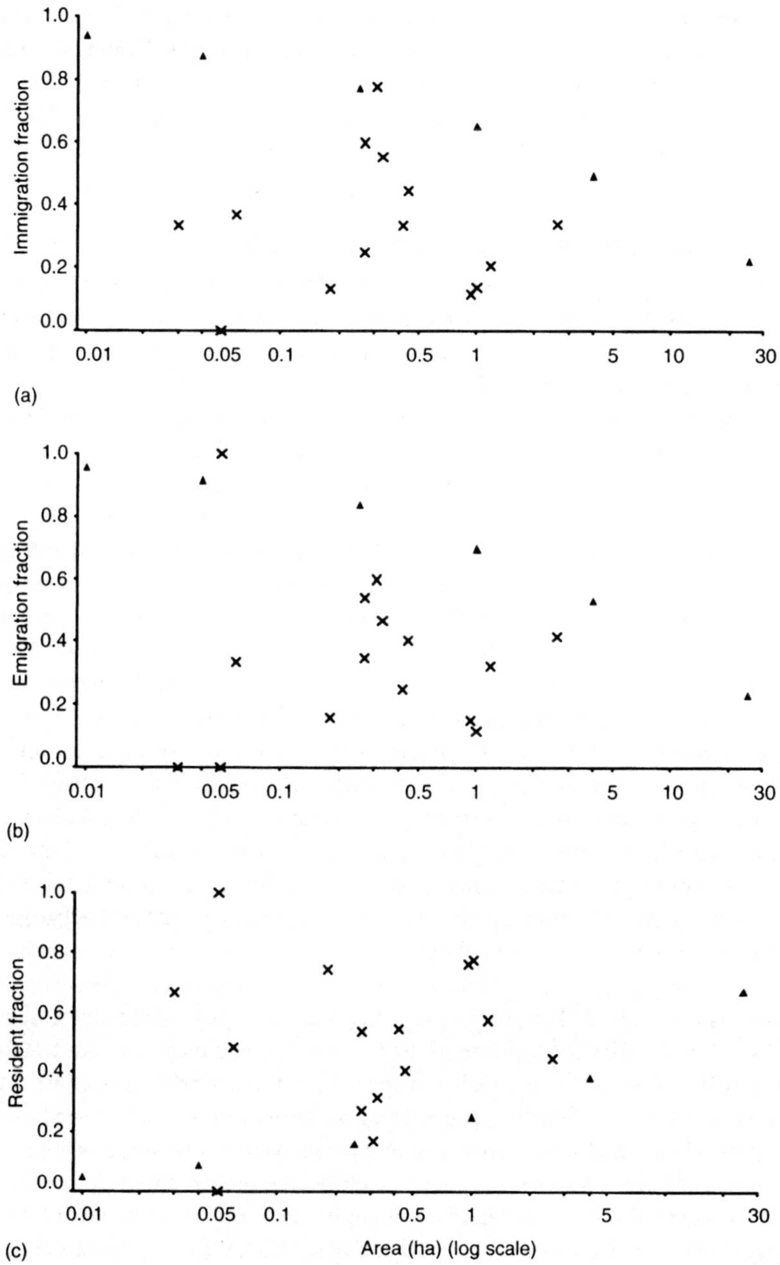

Figure 13.4 The effects of habitat and area on migration of *Aricia agestis.* (a) Immigration, (b) emigration and (c) resident fractions plotted against area of habitat patches (crosses) and arbitrarily defined grid cells of different sizes (filled triangles). The migrant fractions are calculated as in Sutcliffe and Thomas (1996) and Wilson (1999).

Direct empirical estimates of movement over the longest distances will continue to suffer from sampling error (small numbers of individuals recorded), and estimates beyond the maximum dimensions of the study area can only be made by projecting beyond any of the data collected. Indirect estimates of the longest distances achieved are discussed later.

Consequences of where most individuals move

The previous section describes how it is possible to obtain reasonable estimates of the movements of most individuals in patchy environments. These have important implications for population dynamics and practical conservation management, and this section describes a few examples.

Synchrony in the fluctuations of animal populations is an important consideration when designing population sampling strategies and deciding on the sizes of habitat management units. Synchrony in extinction risk is a major determinant of metapopulation persistence (Gilpin 1990; Harrison & Quinn 1990; Gilpin & Hanski 1991; Hanski 1998; Kendall *et al.* 2000). Widespread climatic variation can cause synchrony in population fluctuations (potentially over very long distances), as can spatially correlated land management (at various scales), and the dispersal of individuals (at scales relevant to dispersal) (Pollard 1991; Hanski & Woiwod 1993; Ranta *et al.* 1995; Sutcliffe *et al.* 1996; Koenig 1999). Within species, butterfly populations often fluctuate in partial synchrony over hundreds to thousands of kilometres, presumably because climatic variation is correlated over these distances. However, dispersal plays a role over shorter distances, and also over relatively long distances in the most dispersive Lepidoptera (Hanski & Woiwod 1993; Williams & Liebhold 2000). Sutcliffe *et al.* (1996) found that relatively immobile butterfly species (sedentary plus intermediate, using the categories of Pollard and Yates 1993) like *A. agestis*, have synchrony enhanced (over the climatically induced background synchrony) in populations up to about 2 km apart, whereas this effect extends to 4 km for more mobile species. Thus, dispersal-induced synchrony affects areas of approximately 12 and 50 km^2, respectively. In a separate study, one of the least mobile British butterflies, *Plebejus argus*, had synchrony enhanced to about 600 m (Thomas 1991a) even though median dispersal distances for this species are only 28 m (males) and 8 m (females), and <5% of recaptured individuals moved more than 100 m (Lewis *et al.* 1997). These distances suggest either that very few migrants are capable of inducing synchrony or, more likely, that movements of larger numbers over shorter distances couple adjacent populations, that cumulatively give rise to larger scale spatial autocorrelation in dynamics. This linkage in population fluctuations suggests that there could be advantages in developing regional-scale habitat management programmes.

Not all habitat areas within a landscape are likely to be equally productive in terms of the breeding success of a particular species. On average, there is likely to be a flow of individuals from productive locations (population sources) to other habitats that are net consumers of individuals, increasing population densities in the vicinity of population sources (Pulliam 1988; Thomas *et al.* 1996; Thomas & Kunin 1999). In some cases the less productive habitats may lack host plants (Schultz 1998) or prove fatal for larvae (Rodríguez *et al.* 1994), but in other cases the less productive habitat may simply support lower population densities in the absence of dispersal (Thomas *et al.* 1996). These dynamics may be important determinants of overall patterns of birth and death across heterogeneous landscapes, including nature reserves. Dispersal into relatively poor habitats and non-breeding areas gives the appearance that the landscape contains a larger quantity of habitat than is really the case, and that the species can survive in a wider range of habitat types than is possible. Both types of misinterpretation have serious implications for conservation, underestimating the rarity of an endangered species and potentially leading to inappropriate conservation management programmes.

The distances over which one finds dispersal-enhanced population densities in unproductive habitats will depend on the distances that individuals disperse away from population sources, and the rate at which populations decline in poor habitats. In the case of the blue butterfly *Cyaniris semiargus*, the sink habitat was fatal to larvae, so the butterfly only occurred in sink habitats up to one generation's worth of dispersal away from population sources (Rodríguez *et al.* 1994). This density-enhancement effect may have stretched to many generations' worth of dispersal distances in a metapopulation of the checkerspot *Euphydryas editha*, in which the poorer habitat simply supported a lower carrying capacity than the population sources. Few data are available to assess the spatial scales over which these source–sink dynamics are important (Boughton 2000), but we would expect them to be comparable to (when poor habitat is not much worse than the source-type habitat) or less than (when the poor habitat is very poor) the scales at which dispersal enhances population synchrony.

Many species of insect occupy transient as well as patchy habitats. Here, dispersal is essential if metapopulations are to persist, tracking changes in the spatial distribution of suitable breeding habitats (Warren 1987; Thomas 1991b; Thomas 1995). In such species, persistence is likely to depend on the regular creation (naturally or through active management) of new habitats that are within the 'normal' dispersal range of the species. The faster the habitat turnover, and the lower the population densities achieved within populated habitat patches, the greater the importance of creating new patches close to existing ones. Imperfect as direct, empirical measures of dispersal may be, they are still likely to provide a useful guide in the context of managing landscapes for species that inhabit transient habitats.

Footprints of long-distance dispersal

Indirect methods of assessing dispersal cannot accurately reveal the proportions of individuals moving unusually long distances. However, their consequences may be assessed and can be used to project future scenarios and to provide insight into large-scale population processes.

The principal indirect methods are deductions based on: (i) static distributions of which patches of habitat (Thomas *et al.* 1992) and offshore islands (Dennis & Shreeve 1996, 1997) are, or are not, occupied; (ii) which empty patches of habitat are colonized over a given time period (Thomas & Jones 1993; Hanski *et al.* 1994, 1995, 1996); (iii) where stray individuals are seen (Shreeve 1992; Dennis & Shreeve 1996, 1997); (iv) expansion rates of species into empty habitats (Hill *et al.* 1999); and (v) spatial patterns of genetic variation (Mallet 2001).

Most of these topics have been reviewed elsewhere (Thomas & Hanski 1997; Mallet 2001). We will merely make a few points here. The distances over which colonizations of empty habitat are observed over periods of several years, up to 10 years, are often extremely similar to the maximum isolation distances of occupied habitat patches recorded in single surveys. As a rule of thumb, maximum colonization and isolation distances are very roughly 10 times further than the distances that the furthest individuals are found to move in extensive MRR studies. Maximum recapture distances are often hundreds of metres, up to about 1 km, whereas maximum colonization distances are usually several kilometres, up to 10 km. For example, the maximum distance moved by an individual in a MRR study of the skipper butterfly *Hesperia comma* was 1090 m (Hill *et al.* 1996), whereas the maximum distance colonized over 9 years was 8.65 km (Thomas & Jones 1993). Occupancy of offshore islands by various British butterflies implies some colonization over even greater distances (Dennis & Shreeve 1996, 1997), but this is to be expected because of the longer timescale of extinction from entire islands than from single habitat patches.

Shreeve (1992, 1995) has argued that dispersal rates are actually much higher than measured in MRR studies, based partly on the evidence of the occurrence of various species on islands. In part this is correct, because of the distance-weighted sampling bias of many published studies. However, the remaining discrepancy represents the important consequences of the rare long-distance dispersal events that result in colonization. Because successful colonization may require more than simply the arrival of one individual (e.g. arrival in suitable habitat; surviving Allee effects), colonization rates may still underestimate the occurrence of migrants of unusually long distances.

Many butterfly species in Britain and Europe are now expanding northwards at their northern range margins, most probably in response to regional climate warming (Parmesan *et al.* 1999; Asher *et al.* 2001). This provides an additional approach based on the assumption that, at a coarse scale, the expansion can be modelled as if it was a diffusion process (Van den Bosch *et al.* 1990; Lensink 1997). Applying this method to the speckled wood butterfly *Pararge aegeria* in Britain provided an estimate of 2.7 km year^{-1} in the 20th century (Hill *et al.* 1999). This expansion rate reflects population growth rates and landscape patterns, as well as the range of actual dispersal distances. The difficulty in applying the approach more widely is the incompleteness of historical records (part of the expansion rate reflects increasing recorder effort), but the approach may become increasingly useful in the future.

Consequences of long-distance dispersal

In metapopulations, landscape-level persistence of a species depends on there being sufficient new colonization of empty habitat patches to offset extinctions caused by, for example, chance population demography, environmental variation and changes to the habitat (Hanski 1991, 1999). In most metapopulation models, persistence of the population system increases with an increasing migration rate (Gilpin & Hanski 1991; Hanski & Gilpin 1997; Hanski 1999), unless mortality during migration is so high that migrants act as a drain on adult populations in individual patches and throughout entire networks (Hanski & Zhang 1993; Thomas & Hanski 1997; Thomas *et al.* 1998, 2000). In this regard, a long (or fat) tail to the distribution of dispersal distances is likely to favour metapopulation persistence. These individuals may bring about colonizations of empty habitat patches (normally only the more isolated patches are empty), without providing a major mortality factor across the whole metapopulation.

Given enough data (requiring the ability to recognize and monitor habitat before it is colonized), the chance events that surround individual colonization events even out, providing clear spatial patterns of colonization. Colonization events are most frequent for large patches of high-quality habitat that are relatively close to potential population sources (Thomas *et al.* 1992; Thomas *et al.* 2001b). These patterns may be used to project future colonization (e.g. Thomas & Jones 1993; Thomas & Hanski 1997). This approach provides considerable insight into landscape-level population processes and determinants of persistence that would be very unlikely to be deduced from any amount of MRR.

The advantages of adopting a phenomenological approach to long-distance migration and colonization can be illustrated by a study of expansion of *Hesperia comma* in the South Downs hills of southern England. *H. comma*'s habitat network in the South Downs has increased in size over the past 25 years, associated with changes in grazing management and increased range of aspects that the species is now able to inhabit (Thomas & Jones 1993; Thomas *et al.* 2001a). The species' distribution has expanded because there are now more patches to colonize, shorter distances between them and more populations generating emigrants (Thomas *et al.* 2001a). We modelled the expansion using Hanski's incidence function (metapopulation) model (IFM), which includes a negative exponential to describe colonization distances (Hanski 1994, 1999; Moilanen 1999). Three approaches to parameterizing mobility were taken: (i) using MRR ($\alpha = 9.84$; Hill *et al.* 1996); (b) using an IFM migration parameter value that has been used previously for this species ($\alpha = 2$; Hanski 1994); and (iii) allowing the IFM to estimate its own migration parameter (see Moilanen 1999), based on the existing distribution of the butterfly in another part of southern England ($\alpha = 0.445$, 95% CI 0.331–0.697; Thomas *et al.* 2001a). None of these three α values had been derived from the South Downs, so the modelled projections provide genuinely independent tests of the performance of the different parameter estimates.

Empirical and modelled expansions are shown in Fig. 13.5. The MRR estimates of dispersal predicted virtually no expansion (100 simulations, mean expansion

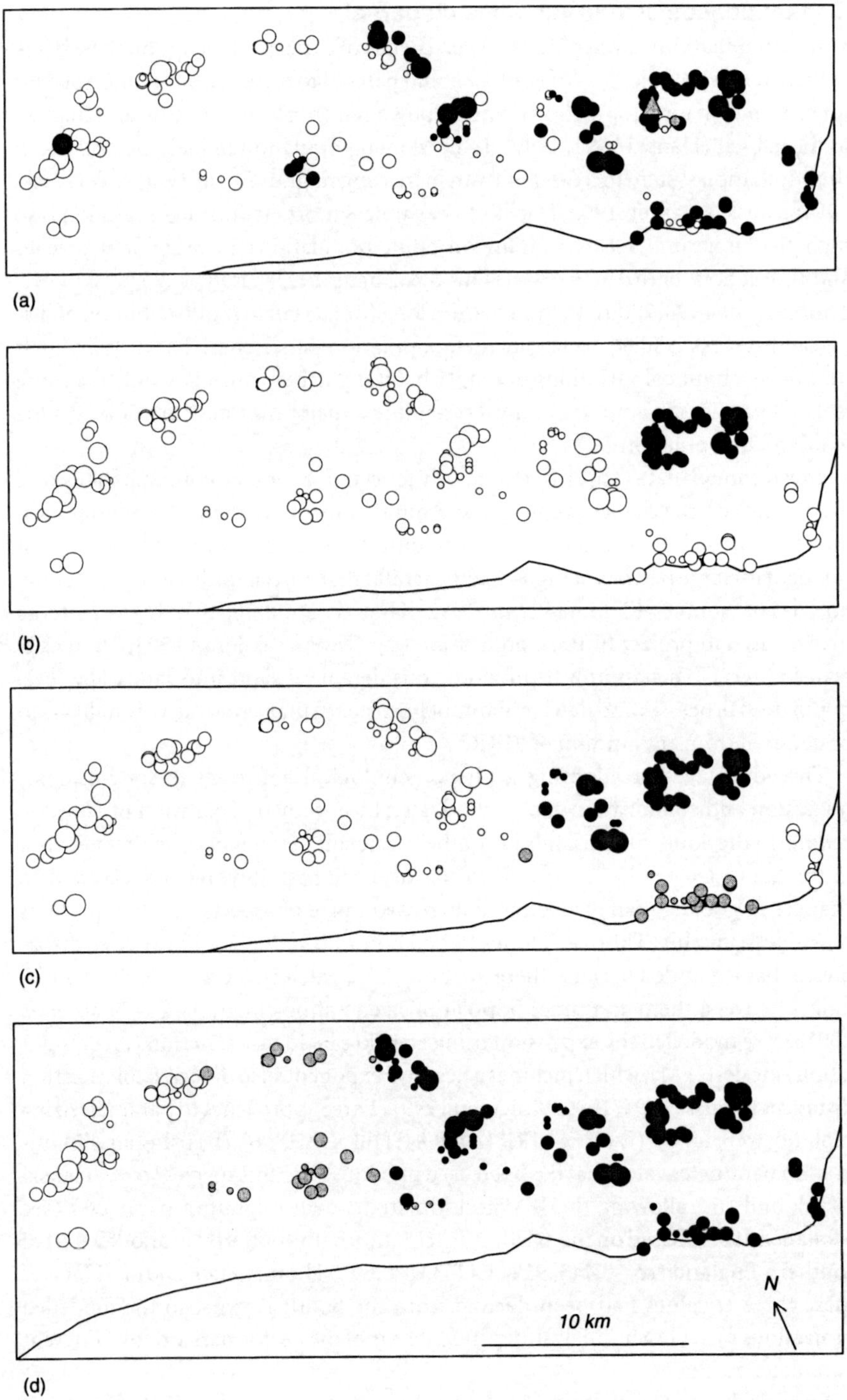
(a)
(b)
(c)
(d)
N
10 km

(furthest 10 patches colonized) 2.80 km ± SD 0.94, maximum expansion 3.48 km), and the previously used parameter value ($\alpha=2$) suggested limited expansion (mean expansion 4.35 km ± 0.44, maximum expansion 5.92 km). In contrast, the pattern-based estimate of α predicted the extent of range expansion closely (mean modelled expansion 14.35 km ± 1.79, maximum expansion 18.72 km; empirical expansion 16.37 km). Thus, the MRR estimate of movement was essentially useless at predicting range expansion over a few tens of kilometres, whereas an indirect estimate of colonization capacity based on spatial patterns of occupied and empty habitats was excellent. In contrast, the MRR estimate gives a realistic measure of numbers of individuals moving tens to hundreds of metres (relevant to management within nature reserves), whereas the pattern-based approach overestimates this by at least an order of magnitude.

To derive realistic patterns of movement that describe the fat-tailed distribution of dispersal more realistically may require the simultaneous use of either two separate dispersal kernels to give appropriate numbers moving both short and long distances (see also Greene & Calogeropoulos, this volume), or of complex mathematical distributions (e.g. Shaw 1995). But because each distribution will vary in response to landscape patterns, and it will inevitably be difficult to estimate the true tail of the dispersal distribution accurately, it is probably an illusion to imagine that there will ever be one distribution that fits all situations (even for a single species). The practical solution is to gather the data that allow movement and colonization potential to be assessed on the spatial and temporal scales that are appropriate to the scientific or management issues under consideration.

Pattern or process?

All of the approaches described so far have been based on analysing and projecting patterns of movement, rather than basing dispersal modelling on the behavioural processes that underlie movement. We can go a long way with a pattern-based ap-

---

Figure 13.5 (*opposite*) Empirical and modelled expansions of *Hesperia comma* in the South Downs (line shows south coast of England; Beachy Head is southeastern point). (a) The range of *H. comma* in 1982 (grey) and 2000 (black), and unoccupied habitat patches in 2000 (white). Empirical expansion is 16.37 km (mean distance from the largest population in 1982 (triangle) to the furthest 10 patches colonized). Using the 2000 habitat network, expansions were modelled using the incidence function model (IFM) parameterized using *H. comma*'s distribution in another network (Surrey). (b) IFM parameters $\alpha=9.84$ (Hill *et al.* 1996), $x=0.15$, $y=0.05$, $e=0.57$; mean expansion 2.80 km ± SD 0.94. (c) IFM parameters $\alpha=2$ (Hanski 1994), $x=0.13$, $y=0.28$, $e=0.61$; mean expansion 4.35 km ± 0.44. (d) IFM parameters $\alpha=0.45$ (Thomas *et al.* 2001a), $x=0.28$, $y=7.26$, $e=0.34$; mean expansion 14.35 km ± 1.79. Black patches were colonized after 18 generations in >50% of 100 simulation runs; grey patches were colonized after 18 generations in <50% of 100 simulation runs; and white patches were never colonized. Symbol sizes exaggerate patch areas: small <0.5 ha, medium 0.5 to <5 ha, large >5 ha.

proach. But this is not entirely satisfactory both from an intellectual perspective (in reality, we know that the patterns are derived from many complex behavioural responses to circumstances) and because a practical understanding of behaviour may allow us to manipulate movement more effectively in achieving management goals.

Diffusion models of dispersal are widely applied, for example by geneticists estimating gene flow and neighbourhood size (e.g. Avise 1994; Turchin 1998; Mallet 2001). Given the rate of movement over a given length of time and average longevity (or mean time until offspring are produced), per-generation dispersal rate can be estimated as the standard deviation of movement distances, in one dimension. However, it needs to be demonstrated, not assumed, that this model is appropriate. A high proportion of animal species show some form of home range behaviour, and this may vary within species depending, for example, on population density, age and gender (e.g. Mason *et al.* 1995). *Aricia agestis* males, but not females, show clear home range behaviour (territorial males wait to intercept passing females in sheltered locations). Accordingly, females show the expected (diffusion-type) pattern of increasing distances moved as the time between marking and recapture increases, whereas males show no such effect (Fig. 13.6). Whether we are worried about this depends on what we are interested in. If we are interested in colonization potential, it is not a problem because most colonization will be carried out by mated females. If we are interested in gene flow among populations, movements by males may also need to be considered. This does not imply that movement must be modelled as an individual behaviour-based process, but that an understanding of behaviour may be required to select an appropriate pattern-based approach.

A useful half-way house between individual behaviour and pattern-based approaches is to deal with movement patterns as correlated random walks, following individuals and recording their movement step lengths and turning angles. In simulation runs, each 'next move' is drawn from the empirically derived set of step lengths and angles. Researchers can build increasing detail into the models, until it can be shown by independent tests of model predictions that they describe movement satisfactorily. With sufficient data, models of step lengths and angles of turning can be used to build population-level models of individual movements in different environments (Kareiva & Shigesada 1983; Root & Kareiva 1984; Schultz 1998; Turchin 1998; Kindvall 1999).

Haddad (1999a, 1999b; Haddad & Baum 1999) provides an excellent example of this approach. He recorded turning angles and step lengths of butterfly movements in forest clearings, and the tendency for butterflies to move away from forest boundaries. Haddad was then able to construct a model which realistically described how butterflies moved through experimental corridors of open habitat faster than expected from purely random movement. This approach often works, but will break down if, for example: (i) field workers are more likely to lose individuals that take long-distance movement steps; or (ii) a very small proportion of movements are responsible for major components of an animal's displacement (e.g. if very rare encounters with home range boundaries result in statistically undetectable direction reversals but are important determinants of total displacement; or if a few move-

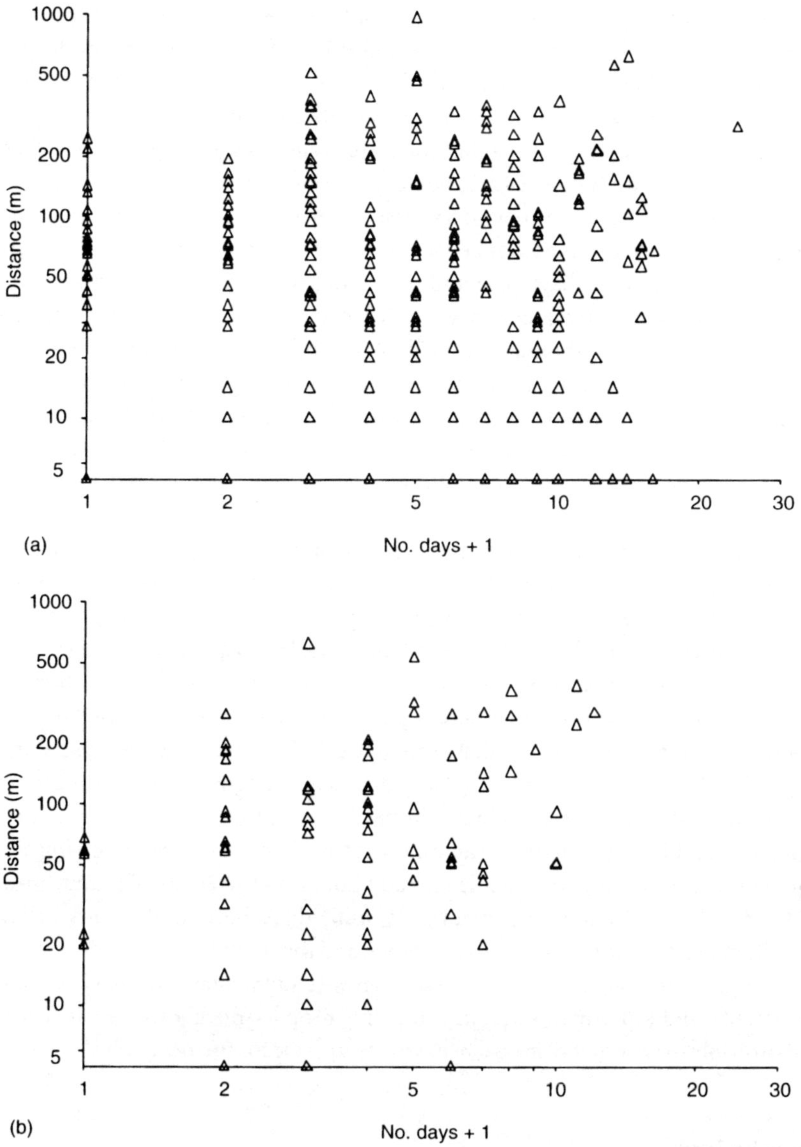

Figure 13.6 Log/log plots of distance moved against days (+1) between first and last captures of *Aricia agestis* for: (a) males ($n = 329$), and (b) females ($n = 88$). (a) Males did not show a significant increase in distance moved over time (intercept = $3.75 \pm$ SE 0.16, $P < 0.001$; slope = 0.07 ± 0.09, $P = 0.436$; $r^2 = 0.002$, $F = 0.61$, $P = 0.436$). (b) Females moved further as the number of days between captures increased (intercept = $3.65 \pm$ SE 0.27, $P < 0.001$; slope = 0.42 ± 0.19, $P = 0.027$; $r^2 = 0.06$, $F = 5.09$, $P = 0.027$).

ments very early in an insect's life are responsible for a high proportion of the total displacement from the point of emergence) (Mallet 1986). Because the model predictions are only likely to be genuinely testable for relatively short-distance movements it is still uncertain how often this approach will have advantages over MRR-type data. It seems unlikely that correlated random walk approaches will accurately predict the tails to species movement distributions.

Despite the practical utility of pattern-based approaches, a deeper understanding of dispersal is likely to come from behavioural studies. Rather than ask 'what is the pattern of dispersal?', behavioural and evolutionary ecologists wish to ask 'how should a dispersing individual behave in order to maximize its fitness?', which is often equivalent to 'how should an individual search for habitat?' Given the pattern-based conclusion that the mortality of migrants may be high, selection for habitat-finding capacity should be strong, at least from time to time. Such ability may greatly reduce overall levels of adult mortality and increase colonization, allowing species to persist in landscapes where they would otherwise become extinct (see Compton, this volume).

We give just one example, but it hints at the complicated issues that are likely to arise in future studies of dispersal. Removing meadow brown *Maniola jurtina* butterflies from their natural habitat, and releasing them either in a heavily grazed and fertilized field or in stubble (both non-habitat), the butterflies orientated successfully back to their habitat from up to about 150 m (Conradt *et al.* 2000). When two groups of butterflies, from two different patches of habitat, were released in non-habitat, the individuals preferentially orientated back to their own home habitat, suggesting that the 'homing' involved a spatial map (or equivalent). Furthermore, individuals released far from habitat did not behave as if undertaking a random walk, but exhibited an apparent searching strategy, flying loops of increasing size, until eventually heading for suitable habitat (Conradt *et al.* 2000). This may mean that at low levels of habitat fragmentation, individuals have no difficulty behaviourally navigating from one patch to another, and that populations within a region are strongly connected. But beyond a certain level of fragmentation, this will cease to be possible, and movement rates may suddenly drop to much lower levels. These, and probably many other dispersal behaviours, are ripe for further study.

Conclusions

It is possible to measure short- to medium-distance movements (up to 500–1000 m) reasonably well, and to manipulate these data to improve estimates of movement. Such approaches will probably never allow us to estimate the proportion of individuals in the tip of the dispersal tail accurately, but this may not be too important in many cases. For example, information on shorter-distance movements has important implications for local population processes and conservation management.

We can estimate long-distance movements by their consequences, in colonizing empty habitat patches and extending geographical ranges. The only major constraint is the time it may take to measure many colonization events, but even analyses of population presence and absence in single surveys can give a strong indication of colonization potential. These deductions provide insight into landscape-scale population processes and distribution changes.

For many purposes, pattern-based approaches provide an adequate answer or prediction. In some cases, however, projections may not be possible until the behavioural underpinning of dispersal is understood, and understanding the processes will sometimes be needed before dispersal rates can be manipulated for practical ends. Studies of when this deeper understanding is required to achieve practical management goals are still in their infancy. Behavioural studies are likely to be rewarding, enhancing our fundamental understanding of the processes of dispersal.

References

Asher, J., Warren, M., Fox, R., Harding, P., Jeffcoate, G. & Jeffcoate, S. (2001) *The Millennium Atlas of Butterflies in Britain and Ireland.* Oxford University Press, Oxford.

Avise, J.C. (1994) *Molecular Markers, Natural History and Evolution.* Chapman & Hall, New York.

Baguette, M., Petit, S. & Quéva, F. (2000) Population spatial structure and migration of three butterfly species within the same habitat network: consequences for conservation. *Journal of Applied Ecology* **37**, 1–10.

Barrowclough, G.F. (1978) Sampling bias in dispersal studies based on finite area. *Bird Banding* **49**, 333–341.

Bennetts, R.E., Nichols, J.D., Lebreton, J.-D. *et al.* (2001) Methods for estimating dispersal probabilities and related parameters using marked animals. In: *Dispersal* (eds J. Clobert, E. Danchin, A.A. Dhondt & J.D. Nichols), pp. 3–17. Oxford University Press, Oxford.

Boughton, D.A. (2000) The dispersal system of a butterfly: a test of source–sink theory suggests the intermediate-scale hypothesis. *American Naturalist* **156**, 131–144.

Brakefield, P.M. (1982) Ecological studies on the butterfly *Maniola jurtina* in Britain. I. Adult behaviour, microdistribution and dispersal. *Journal of Animal Ecology* **51**, 713–726.

Brakefield, P.M. (1991) Genetics and the conservation of invertebrates. In: *The Scientific Management of Temperate Communities for Conservation* (eds I.F. Spellerburg, F.B. Goldsmith & M.G. Morris), pp. 45–80. Blackwell Scientific Publications, Oxford.

Cappuccino, N. & Kareiva, P. (1985) Coping with a capricious environment: a population study of a rare pierid butterfly. *Ecology* **66**, 152–161.

Conradt, L., Bodsworth., E.J. & Thomas, C.D. (2000) Non-random dispersal in the butterfly *Maniola jurtina*: implications for metapopulation models. *Proceedings of the Royal Society of London, Series B* **267**, 1505–1510.

Dennis, R.L.H. & Shreeve T.G. (1996) *Butterflies on British and Irish Offshore Islands.* Gem Publishing, Wallingford, UK.

Dennis, R.L.H. & Shreeve T.G. (1997) Diversity of butterflies on British islands: ecological influences underlying the roles of area, isolation and the size of the faunal source. *Biological Journal of the Linnaean Society* **60**, 257–275.

Ehrlich, P.R. (1961) Intrinsic barriers to dispersal in checkerspot butterfly. *Science* **134**, 108–109.

Fahrig, L. & Paloheimo, J.E. (1987) Interpatch dispersal in the cabbage butterfly. *Canadian Journal of Zoology* **65**, 616–622.

Ford, E.B. (1945) *Butterflies.* Collins, London.

Gilbert, L.E. & Singer, M.C. (1973) Dispersal and gene flow in a butterfly species. *American Naturalist* **107**, 58–72.

Gilbert, L.E. & Singer, M.C. (1975) Butterfly ecology. *Annual Review of Ecology and Systematics* **6**, 365–397.

Gilpin, M.E. (1990) Extinction of finite metapopulations in correlated environments. In: *Living in a Patchy Environment* (eds B. Shorrocks & I.R. Swingland), pp. 177–186. Oxford University Press, Oxford.

Gilpin, M.E. & Hanski, I. (eds) (1991) *Metapopulation Dynamics: Empirical and Theoretical Investigations.* Academic Press, London.

Haddad, N.M. (1999a) Corridor use predicted from behaviors at habitat boundaries. *American Naturalist* **153**, 215–227.

Haddad, N.M. (1999b) Corridor and distance effects on interpatch movements: a landscape experiment with butterflies. *Ecological Applications* **9**, 612–622.

Haddad, N.M. & Baum, K.A. (1999) An experimental test of corridor effects on butterfly densities. *Ecological Applications* **9**, 623–633.

Hanski, I. (1991) Single-species metapopulation dynamics. In: *Metapopulation Dynamics: Empirical and Theoretical Investigations* (eds M. Gilpin & I. Hanski), pp. 17–38. Academic Press, London.

Hanski, I. (1994) A practical model of metapopulation dynamics. *Journal of Animal Ecology* **63**, 151–162.

Hanski, I. (1998) Metapopulation dynamics. *Nature* **396**, 41–49.

Hanski, I. (1999) *Metapopulation Ecology.* Oxford University Press, Oxford.

Hanski, I. & Gilpin, M.E. (eds) (1997) *Metapopulation Biology: Ecology, Genetics, and Evolution.* Academic Press, San Diego.

Hanski, I. & Woiwod, I.P. (1993) Spatial synchrony in the dynamics of moth and aphid populations. *Journal of Animal Ecology* **62**, 656–668.

Hanski, I. & Zhang, D.-Y. (1993) Migration, metapopulation dynamics and fugitive coexistence. *Journal of Theoretical Biology* **163**, 491–504.

Hanski, I., Kuussaari, M. & Nieminen, M. (1994) Metapopulation structure and migration in the butterfly *Melitaea cinxia. Ecology* **75**, 747–762.

Hanski, I., Pakkala, T., Kuussaari, M. & Lei, G. (1995) Metapopulation persistence of an endangered butterfly in a fragmented landscape. *Oikos* **72**, 21–28.

Hanski, I., Moilanen, A., Pakkala, T. & Kuussaari, M. (1996) The quantitative incidence function model and persistence of an endangered butterfly metapopulation. *Conservation Biology* **10**, 578–590.

Hanski, I., Alho, J. & Moilanen, A. (2000) Estimating the parameters of survival and migration of individuals in metapopulations. *Ecology* **8**, 239–251.

Harrison, R.G. (1980) Dispersal polymorphism in insects. *Annual Review of Ecological Systematics* **11**, 95–118.

Harrison, S. (1989) Long-distance dispersal and colonization in the bay checkerspot butterfly. *Ecology* **70**, 1236–1243.

Harrison, S. & Quinn, J.F. (1990) Correlated environments and the persistence of metapopulations. *Oikos* **56**, 293–298.

Hill, J.K., Thomas, C.D. & Lewis, O.T. (1996) Effects of habitat patch size and isolation on dispersal by *Hesperia comma* butterflies: implications for metapopulation structure. *Journal of Animal Ecology* **65**, 725–735.

Hill, J.K., Thomas, C.D. & Huntley, B. (1999) Climate and habitat availability determine 20th century changes in a butterfly's range margin. *Proceedings of the Royal Society of London, Series B* **266**, 1197–1206.

Kareiva, P.M. & Shigesada, N. (1983) Analyzing insect movement as a correlated random walk. *Oecologia* **56**, 234–238.

Kendall, B.E., Bjørnstad, O.N., Bascompte, J., Keitt, T.H. & Fagan, W.F. (2000) Dispersal, environmental correlation, and spatial synchrony in population dynamics. *American Naturalist* **155**, 628–636.

Kindvall, O. (1999) Dispersal in a metapopulation of the bush cricket *Metrioptera bicolor* (Orthoptera: Tettigoniidae). *Journal of Animal Ecology* **68**, 172–185.

Koenig, W.D. (1999) Spatial autocorrelation of ecological phenomena. *Trends in Ecology and Evolution* **14**, 22–26.

Koenig, W.D., Van Vuren, D. & Hooge, P.N. (1996) Detectability, philopatry, and the distribution of dispersal distances in vertebrates. *Trends in Ecology and Evolution* **11**, 514–517.

Kuussaari, M., Nieminen, M. & Hanski, I. (1996) An experimental study of migration in the butterfly *Melitaea cinxia. Journal of Animal Ecology* **65**, 791–801.

Kuussaari, M., Saccheri, I., Camara, M. & Hanski, I. (1998) Allee effect and population dynamics in the Glanville fritillary butterfly. *Oikos* **82**, 384–392.

Lebreton, J.-D., Burnham, K.P., Clobert, J. & Anderson, D.R. (1992) Modeling survival and testing biological hypotheses using marked animals: a unified approach with case studies. *Ecological Monographs* **62**, 67–118.

Lensink, R. (1997) Range expansion of raptors in Britain and the Netherlands since the 1960s: testing an individual-based diffusion model. *Journal of Animal Ecology* **66**, 811–826.

Lewis, O.T., Thomas, C.D., Hill, J.K. *et al.* (1997) Three ways of assessing metapopulation structure in the butterfly *Plebejus argus*. *Ecological Entomology* **22**, 283–293.

Mallet, J. (1986) Dispersal and gene flow in a butterfly with home range behaviours—*Heliconius erato* (Lepidoptera, Nymphalidae). *Oecologia* **68**, 210–217.

Mallet, J. (2001) Gene flow. In: *Insect Movement: Mechanisms and Consequences* (eds I.P. Woiwod, D.R. Reynolds & C.D. Thomas). CAB International, Wallingford, UK.

Mason, P.L. (1990) Dispersal in the alpine grasshopper, *Podisma pedestris* (L.): old and new wine in bottles. *Boletin de Sanidad Vegetal* **20**, 321–326.

Mason, P.L., Nichols, R.A. & Hewitt, G.M. (1995) Philopatry in the alpine grasshopper, *Podisma pedestris*: a novel experimental and analytical method. *Ecological Entomology* **20**, 137–145.

Menéndez, R. & Thomas, C.D. (2000) Metapopulation structure depends on spatial scale in the host-specific moth *Wheeleria spilodactylus* (Lepidoptera: Pterophoridae). *Journal of Animal Ecology* **69**, 935–951.

Moilanen, A. (1999) Patch occupancy models of metapopulation dynamics: efficient parameter estimation using implicit statistical inference. *Ecology* **80**, 1031–1043.

Odendaal, F.J., Turchin, P. & Stermitz, F.R. (1989) Influence of host-plant density and male harassment on the distribution of female *Euphydryas anicia* (Nymphalidae). *Oecologia* **78**, 283–288.

Parmesan, C., Ryrholm, N., Stefanescu, C. *et al.* (1999) Poleward shifts in geographical ranges of butterfly species associated with regional warming. *Nature* **399**, 579–583.

Petit, S., Moilanen, A., Hanski, I. & Baguette, M. (2001) Metapopulation dynamics of the bog fritillary butterfly: movements between habitat patches. *Oikos* **92**, 491–500.

Pollard, E. (1991) Synchrony of population fluctuations: the dominant influence of widespread factors on local butterfly populations. *Oikos* **60**, 7–10.

Pollard, E. & Yates, T.J. (1993) *Monitoring Butterflies for Conservation*. Chapman & Hall, London.

Porter, J.H. & Dooley Jr, J.L. (1993) Animal dispersal patterns: a reassessment of simple mathematical models. *Ecology* **74**, 2436–2443.

Pulliam, H.R. (1988) Sources, sinks, and population regulation. *American Naturalist* **132**, 652–661.

Ranta, E., Kaitala, V., Lindstrom, J. & Linden, H. (1995) Synchrony in population dynamics. *Proceedings of the Royal Society of London, Series B* **262**, 113–118.

Rodríguez, J., Jordano, D. & Fernández-Haeger, J. (1994) Spatial heterogeneity in a butterfly–host plant interaction. *Journal of Animal Ecology* **63**, 31–38.

Roff, D.A. (1986) The evolution of wing dimorphism in insects. *Evolution* **40**, 1009–1020.

Roff, D.A. (1990) The evolution of flightlessness in insects. *Ecological Monographs* **60**, 389–421.

Roland, J., McKinnon, G., Backhouse, C. & Taylor, P.D. (1996) Even smaller radar tags on insects. *Nature* **381**, 120.

Root, R.B. & Kareiva, P.M. (1984) The search for resources by cabbage butterflies (*Pieris rapae*): ecological consequences and adaptive significance of Markovian movements in a patchy environment. *Ecology* **65**, 147–165.

Rousset, F. (2001) Genetic approaches to the estimation of dispersal rates. In: *Dispersal* (eds J. Clobert, E. Danchin, A.A. Dhondt & J.D. Nichols), pp. 18–28. Oxford University Press, Oxford.

Schultz, C.B. (1998) Dispersal behaviour and its implications for reserve design in a rare Oregon butterfly. *Conservation Biology* **12**, 284–292.

Shaw, M.W. (1995) Simulation of population expansion and spatial pattern when individual dispersal distributions do not decline exponentially with distance. *Proceedings of the Royal Society of London, Series B* **259**, 243–248.

Shigesada, N. & Kawasaki, K. (1997) *Biological Invasions: Theory and Practice*. Oxford University Press, Oxford.

Shreeve, T.G. (1992) Monitoring butterfly movements. In: *The Ecology of Butterflies in Britain* (ed. R.L.H. Dennis), pp. 120–138. Oxford University Press, Oxford.

Shreeve, T.G. (1995) Butterfly mobility. In: *Ecology and Conservation of Butterflies* (ed. A.S. Pullin), pp. 37–45. Chapman & Hall, London.

Southwood, T.R.E. (1962) Migration of terrestrial arthropods in relation to habitat. *Biological Reviews* **37**, 171–214.

Southwood, T.R.E. (1978) *Ecological Methods: with Particular Reference to the Study of Insect Populations*, 2nd edn. Methuen, London.

Sutcliffe, O.L. & Thomas, C.D. (1996) Open corridors appear to facilitate dispersal by ringlet butterflies (*Aphantopus hyperantus*) between woodland clearings. *Conservation Biology* **10**, 1359–1365.

Sutcliffe, O.L., Thomas, C.D. & Moss, D. (1996) Spatial synchrony and asynchrony in butterfly population dynamics. *Journal of Animal Ecology* **65**, 85–95.

Sutcliffe, O.L., Thomas, C.D. & Peggie, D. (1997) Area-dependent migration by ringlet butterflies generates a mixture of patchy and metapopulation attributes. *Oecologia* **109**, 229–234.

Thomas, C.D. (1991a) Spatial and temporal variability in a butterfly population. *Oecologia* **87**, 577–580.

Thomas, C.D. (1995) Ecology and conservation of butterfly metapopulations in the fragmented British landscape. In: *Ecology and Conservation of Butterflies* (ed. A.S. Pullin), pp. 46–63. Chapman & Hall, London.

Thomas, C.D. & Hanski, I. (1997) Butterfly metapopulations. In: *Metapopulation Biology: Ecology, Genetics, and Evolution* (eds I. Hanski & M.E. Gilpin), pp. 359–386. Academic Press, San Diego.

Thomas, C.D. & Jones, T.M. (1993) Partial recovery of a skipper butterfly (*Hesperia comma*) from population refuges: lessons for conservation in a fragmented landscape. *Journal of Animal Ecology* **62**, 472–481.

Thomas, C.D. & Kunin W.E. (1999) The spatial structure of populations. *Journal of Animal Ecology* **68**, 647–657.

Thomas, C.D., Thomas, J.A. & Warren, M.S. (1992) Distributions of occupied and vacant butterfly habitats in fragmented landscapes. *Oecologia* **92**, 563–567.

Thomas, C.D., Singer, M.C. & Boughton, D.A. (1996) Catastrophic extinction of population sources in a butterfly metapopulation. *American Naturalist* **148**, 957–975.

Thomas, C.D., Jordano, D., Lewis, O.T., Hill, J.K., Sutcliffe, O.L. & Thomas, J.A. (1998) Butterfly distributional patterns, processes and conservation. In: *Conservation in a Changing World: Integrating Processes into Priorities for Action. Symposium of the Zoological Society of London* (eds G.M. Mace, A. Balmford & J.R. Ginsberg), pp. 107–138. Cambridge University Press, Cambridge.

Thomas, C.D., Baguette, M. & Lewis, O.T. (2000) Butterfly movement and conservation in patchy landscapes. In: *Behaviour and Conservation* (eds L.M. Gosling & W.J. Sutherland), pp. 85–104. Cambridge University Press, Cambridge.

Thomas, C.D., Bodsworth, E.J., Wilson R.J. *et al.* (2001a) Ecological and evolutionary processes at expanding range margins. *Nature* **411**, 577–581.

Thomas, J.A. (1991b) Rare species conservation: case studies of European butterflies. In: *The Scientific Management of Temperate Communities for Conservation* (eds I.F. Spellerburg, F.B. Goldsmith & M.G. Morris), pp. 149–198. Blackwell Scientific Publications, Oxford.

Thomas, J.A., Bourn, N.A.D., Clarke, K.E. *et al.* (2001b) The quality and isolation of habitat patches both determine where butterflies persist in fragmented landscapes. *Proceedings of the Royal Society of London, Series B* **268**, 1791–1796.

Turchin, P. (1998) *Quantitative Analysis of Movement—Measuring and Modeling Population Redistribution in Animals and Plants.* Sinauer, Sunderland, MA.

Van den Bosch, F., Metz, J.A.J. & Dickman, O. (1990) The velocity of spatial population expansion. *Journal of Mathematical Biology* **28**, 529–565.

Wagner, D.L. & Liebherr, J.K. (1992) Flightlessness in insects. *Trends in Ecology and Evolution* **7**, 216–220.

Warren, M.S. (1987) The ecology and conservation of the heath fritillary butterfly, *Mellicta athalia*. II. Adult population structure and mobility. *Journal of Applied Ecology* **24**, 483–498.

Watt, W.B., Chew, F.S., Snyder, L.R.G., Watt, A.G. & Rothschild, D.E. (1977) Population structure of Pierid butterflies I. Numbers and movements of some montane *Colias* species. *Oecologia* **27**, 1–22.

Williams, D.W. & Liebhold, A.M. (2000) Spatial synchrony in spruce budworm outbreaks in eastern north America. *Ecology* **81**, 2753–2766.

Wilson, R.J. (1999) *The spatiotemporal dynamics of three lepidopteran herbivores of Helianthemum chamaecistus.* PhD thesis, University of Leeds, Leeds.

Chapter 14

Plant dispersal and colonization processes at local and landscape scales

James M. Bullock, Ibby L. Moy, Richard F. Pywell, Sarah J. Coulson, Abigail M. Nolan and Hal Caswell

Introduction

In general, adult higher plants do not move from one place to another. Mobility occurs only at the recruitment stage, either through dispersal of seeds or other propagules or (generally much less importantly in terms of spatial displacement) clonal growth. Therefore, recruitment is a colonization process, where colonization is the establishment of plants in areas currently unoccupied by the species. This is such a fundamental point that, in some senses, it is trivial (see also Silvertown *et al.* 2001). However, we shall show in this chapter that considering colonization can be very important in changing the conclusions we draw or the predictions we make about the dynamics of plant ecological systems. We shall emphasize the dispersal aspect of colonization—the mechanism by which plants arrive at new sites—and show that this can be a, if not *the*, dominant factor controlling plant dynamics. This chapter will discuss dispersal and colonization processes at the local and landscape scales, i.e. within-population and community dynamics, and the linked dynamics of spatially distinct populations and communities. Larger-scale processes, such as invasions across regions or countries, are covered in other chapters (Hengeveld & Hemerik; Shigesada & Kawasaki; Cohen; Watkinson & Gill, all this volume). These two scales are illustrated in Fig. 14.1, which shows a landscape of vegetation patches, or communities. However, this map also illustrates the usual anthropocentric interpretation of landscape structure. While it is difficult to study spatial processes without such interpretation, in this chapter we shall attempt to retain the 'plant's-eye-view' (Harper 1977) by considering spatial structure in terms of dispersal patterns. This will highlight some current errors in attempts to test theories on spatial dynamics in the field. We shall also demonstrate that studies of dispersal should be targeted to address particular ecological questions, which may need quite different types of information about plant movement.

We shall discuss the two processes controlling change at both local and landscape scales: (i) the dynamics of established species, whereby the colonization process involves recolonization following 'extinction' (of individuals or populations); and (ii) the spread (invasion) of new species into areas not previously occupied. We will ad-

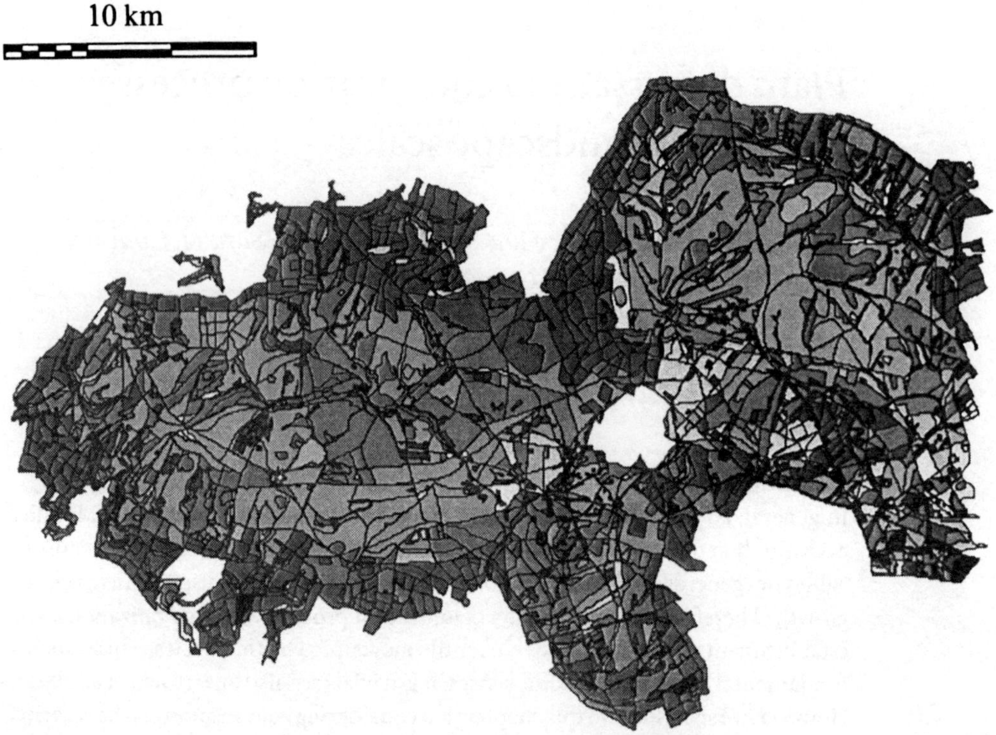

Figure 14.1 An example of a landscape: part of the Salisbury Plain, England, mapped by ground survey. Different tones represent patches of different vegetation types: including types of chalk grassland, mesotrophic grassland, scrub and woodland, and arable fields. This structure is caused by both the history of land use and the underlying soil, geology and topography.

dress subject areas which traditionally have been treated quite separately, but when considered in terms of colonization and dispersal are shown to be linked mechanistically. In particular, dispersal limitation will be shown to be a vitally important process at all scales.

Local dynamics

Studies of the dynamics of single communities or populations have traditionally taken little account of spatial dynamics. While much work on the spatial dynamics of plants concentrates on the landscape (see below) or larger scales (Shigesada & Kawasaki, this volume; Watkinson & Gill, this volume), and current seed-dispersal studies are emphasizing long-distance dispersal (Cain *et al.* 2000; Nathan & Muller-Landau 2000; Greene & Calogeropoulos, this volume), the sedentary nature of growing plants means that dispersal and dispersal limitation must be considered at even the smallest scales.

Dynamics of established species (gap dynamics)

There is an increasing appreciation of the importance of colonization and thus dispersal in local plant population and community dynamics. This can be seen in recent studies of population dynamics (Alvarez-Buylla & García-Barrios 1991; Rees & Paynter 1997) and modelling studies of species coexistence and community dynamics (Pacala & Rees 1998; Bolker & Pacala 1999; He & Mladenhof 1999; Kinzig *et al.* 1999; Gustafsson *et al.* 2000). These studies highlight gap creation and subsequent colonization as the dominant process driving population and/or community dynamics and this is the most important and obvious example of the role of dispersal at this scale. However, other processes must also invoke dispersal, such as the dynamics of source and sink patches within a population (e.g. Keddy 1981; Watkinson *et al.* 1989; Kadmon & Shmida 1990).

It has long been known that the process of gap formation and colonization is the major structuring force in forest dynamics (Brokaw & Busing 2000; Bullock 2000), but there is increasing evidence of the importance of gaps in other community types, such as grasslands (Hobbs & Hobbs 1987; Bullock *et al.* 1995; Moloney & Levin 1996; Eriksson 1997; Bullock 2000). Eriksson (1997), for example, showed that while local competitive interactions were important at the very small scale in a dry grassland, gap colonization explained species composition at the community scale. Bullock *et al.* (2001) tested whether grazing animals affected a grassland community by selective grazing, changing the intensity of competition or creating gaps. The hypotheses were tested by linking them to plant traits and then examining correlations between species traits and relative responses to grazing. Only the gap creation hypothesis was supported. The best correlate with grazing effects was gap colonization ability (Fig. 14.2), measured as the ratio of the abundance of a species in a recently colonized gap compared with its abundance in the closed vegetation.

Gaps are important in community dynamics, but what is the role of dispersal in determining species' gap colonization ability? Traditionally, gap research has concentrated on characterizing species differences in establishment responses to gaps. Bullock (2000) reviewed this research and showed that while there are species differences in establishment responses within communities, these tend to be quantitative rather than qualitative, and they are not large. Brokaw and Busing (2000) concluded that the classic model of forest dynamics—that the driving force is species differences in regeneration responses to gaps—is less important than dispersal limitation. Gap colonization involves getting propagules into a gap (what Bullock (2000) called the 'gap attainment ability') followed by plant establishment from these propagules. Gap attainment involves the seed bank, the sapling bank and/or clonal growth from the gap edge as well as seed dispersal, but the latter has a major role in many communities (Bullock *et al.* 1995; Dalling *et al.* 1998; Kalamees 1999). Species differences in gap attainment ability, when considered for a particular gap, will relate to two variables: the proximity of source plants to the gap and species' life history. Life-history differences, especially the species' dispersal ability (see Thompson *et al.*, this volume) and fecundity, will determine the ability of a species to get seeds into a gap, given a particular source size and proximity. Such differences have been concen-

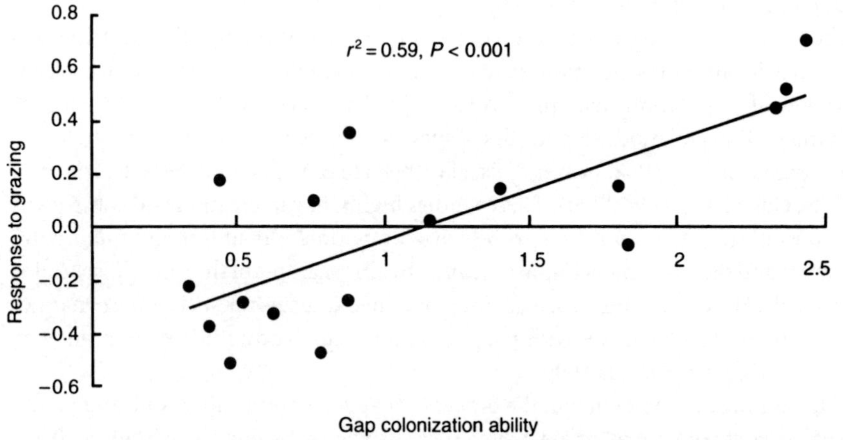

Figure 14.2 The gap colonization ability of 17 species explains their changes in abundance in response to grazing. Gap colonization ability is the ratio of the abundance of the species in a recently colonized gap compared with its abundance in the closed vegetation. The 'response to grazing' is the residual of the ratio of the species' relative abundance under heavy grazing to its abundance under light grazing after analysis of variance, which corrected for phylogeny (Bullock *et al.* 2001). A larger (more positive) value indicates an increasingly positive response to heavier grazing compared with light grazing.

trated upon in theoretical modelling studies (Pacala & Rees 1998; Bolker & Pacala 1999; Kinzig *et al.* 1999) and this results in a form of niche partitioning which leads to equilibrium dynamics (Pacala & Levin 1997). However, empirical studies have had the stumbling block that each gap has a different set of neighbouring adults and it is this spatial context which seems to be the primary determinant of the identity of colonizers. This leads to large within-community variation in the composition of colonized gaps, as shown in Fig. 14.3a where individual species show great differences in the number of seedlings colonizing 64 identically created gaps in a grassland (see also Rusch 1992; Dalling *et al.* 1998). In this example, the variation in composition is largely related to the gap context, so that, for the seven species studied, the number of seedlings colonizing a gap is correlated with the abundance of the species in the immediate neighbourhood of the gap (Fig. 14.3b). Dalling *et al.* (1998) found a similar relationship between gap composition and the proximity of parents in forest gaps. In a different manner, by sowing seeds into grassland gaps, Eriksson (1997) showed that dispersal limitation constrained species' potential to colonize gaps.

This context dependence of gap colonization patterns will push a system more towards the non-equilibrium dynamics modelled by Hurtt and Pacala (1995). Some authors have started to emphasize the role of strong dispersal limitation in determining local gap composition and suggest it masks any clear relationship between gap and community dynamics (Hubbell *et al.* 1999; Brokaw & Busing 2000). However, while gap neighbourhood is important in determining gap composition, there

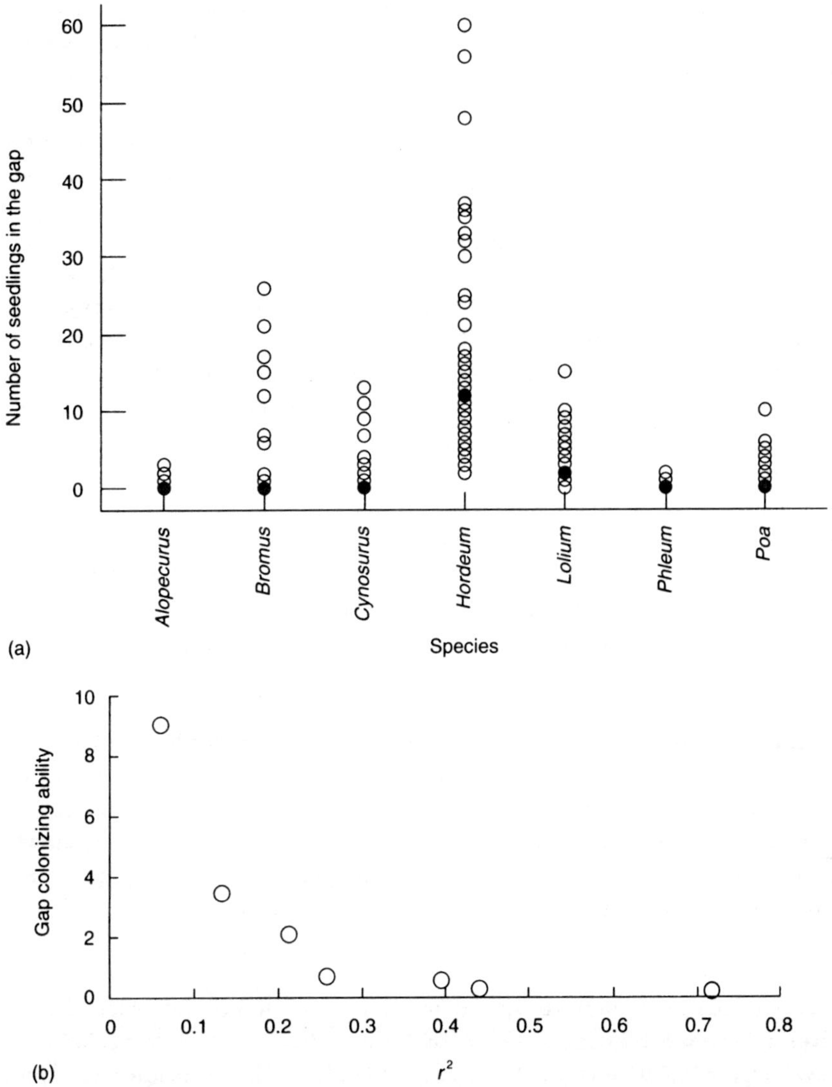

Figure 14.3 Gap colonization ability of grassland species. (a) The variation in composition of colonized gaps within a single grassland. The number of seedlings found for seven grasses varies widely among 64 gaps, which were created identically. The filled circle shows the median. (b) Species less constrained by distance from a gap are better gap colonizers. The r^2 value was calculated for each species by a regression of the number of seedlings in each of 64 gaps on the number of ramets within a 2 cm border zone around each gap. This was a significant relationship for all species ($P < 0.05$). Gap colonization ability is explained in Fig. 14.2. (J. M. Bullock, unpubl. data.)

is evidence for the joint importance of species' life-history differences, particularly in dispersal ability. Dalling *et al.* (1998) showed a relationship between the proximity of adult trees with the probability of gap colonization for some species, especially those with large seeds. However, some small-seeded species did not show this relationship, possibly because they had a more extended seed rain and were less dispersal limited. While our seven grassland species showed significant relationships between local ramet density and the number of seedlings colonizing a gap (Fig. 14.3b), the variation explained ranged between 6% and 72%. This variation may be caused by differences in dispersal ability, so that the ability of good dispersers to colonize gaps is less constrained by the proximity of adults to the gap and thus r^2 is smaller. Interestingly, the gap colonization ability of these species was inversely related to the r^2 values of the proximity-seedling number relationship (Spearman rank correlation $R_s = 1$, $P < 0.001$), suggesting better dispersers are better gap colonizers. Therefore, gap and community dynamics are clearly linked in this grassland system.

While the role of gaps in seedling establishment has been well studied, much less is known about the process by which seeds get to gaps, or species differences in gap attainment ability. A few studies have started to characterize the gap attainment process (Schupp *et al.* 1989; Malo *et al.* 1995; Pakeman *et al.* 1998; Kalamees 1999). There is a need for these studies to be extended and related to gap and community dynamics. Without such knowledge, it is premature to say that the effect of vegetation composition surrounding a gap is the dominant process in determining the identity of colonists. As demonstrated above for our grassland study site, it is likely that species differences in gap attainment ability will also be important and will be superimposed on effects relating to source proximity. While we have concentrated on seed dispersal here, it will be necessary to consider the full range of gap attainment modes that species may exhibit (seed/propagule dispersal, seed bank, clonal growth, sapling banks) in order to understand this process fully.

Spread of new species

Species invade communities by spreading out from one or more initial foci of a few plants (Cousens & Mortimer 1995; Verdú & Garcia Fayos 1996; Parker & Reichard 1997). Figure 14.4 shows the patterns of spread of five forb species introduced into experimental grassland plots (R. F. Pywell, unpubl. data). These were introduced along linear foci (i.e. slot-seeding lines) and their rate of spread away from these foci was monitored over 5 years. Rates of spread differed strongly among species, and the environment (in this case the management system) had species-specific effects on spread. Demographic processes such as establishment and survival are important in this process (Bullock *et al.* 1994; Parker 2000), but there is a need to consider the role of dispersal. Coulson *et al.* (2001) used the grassland linear foci experiment to determine the rate-limiting steps in the invasion of two of the introduced forbs. By 1998 *Rhinanthus minor* had spread further in the hay-cut treatment than in the grazed treatment, but *Leucanthemum vulgare* had spread poorly, with no treatment effects (Fig. 14.5). Standard demographic measures of seedling establishment and plant

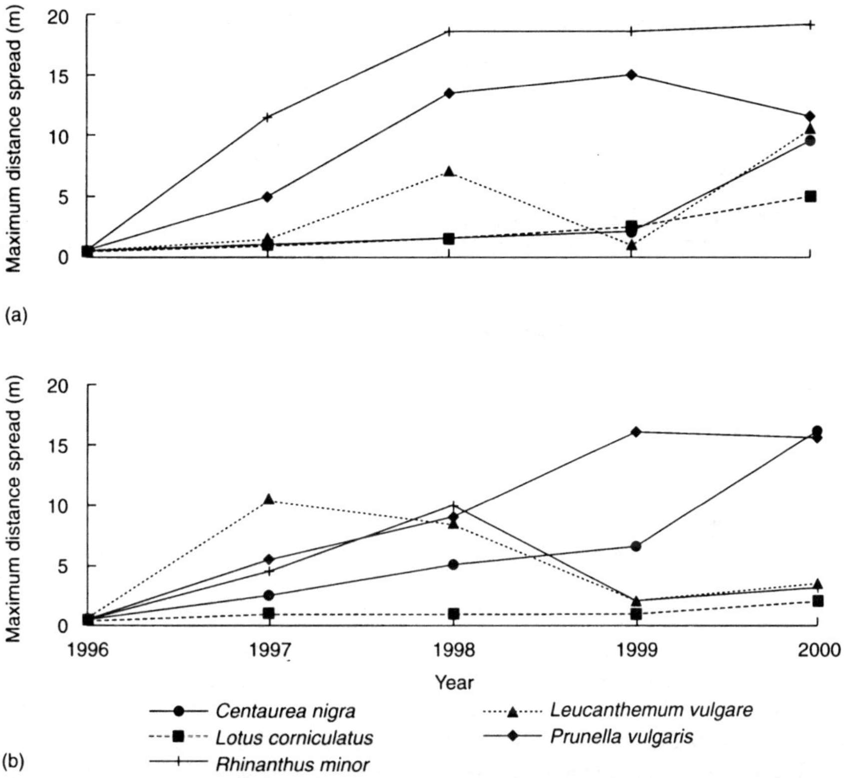

Figure 14.4 Environmental effects on rates of spread of five forb species introduced into a grassland in 1996 (R. F. Pywell, unpubl. data): (a) hay cut and grazed, and (b) grazed only. Data are the maximum distance spread within experimental plots from linear introduction foci (slot seeding lines).

survival showed no treatment effects for either species. However, seed dispersal curves for *Rhinanthus* were much longer in hay-cut treatment than the grazed treatment, because it set seed at the time of cutting and seed was thrown longer distances by the hay-cutting machinery. *Leucanthemum* showed poor dispersal, which was unaffected by hay-cutting because it set seed after the hay was cut. Therefore, management effects on the spread of *Rhinanthus* reflected effects on dispersal, rather than establishment. *Leucanthemum* showed poor dispersal, but good establishment, in all treatments, suggesting its spread may also have been dispersal limited.

It is rare to assess the rate-limiting steps in the spread of species invading communities, but this example suggests it is necessary in order to be able to manage such spread. In this case, the aim was to introduce these species to recreate a species-rich grassland. Coulson *et al.*'s (2001) study suggests that this would best be achieved by measures facilitating dispersal. A more rigorous approach would be to create spatially explicit models of population spread incorporating empirically estimated dis-

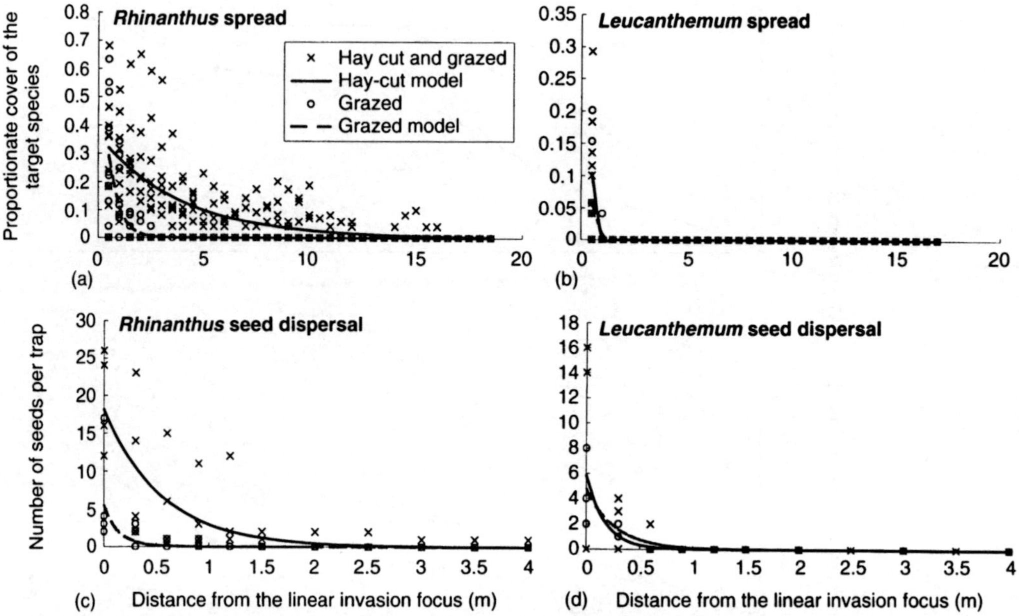

Figure 14.5 Management effects on the spread of *Rhinanthus minor* and *Leucanthemum vulgare* mirror effects on seed dispersal rather than effects on demographic parameters (Coulson *et al.* 2001). Inverse power models fitted the data well ($r^2 = 0.44$–0.89, $P < 0.01$ for all parts). Model curves were significantly different between treatments for both the spread and dispersal of *Rhinanthus*, but were not different for either measure of *Leucanthemum*.

persal and demographic parameters, and to manipulate the models to determine sensitivity to changes in these parameters. There are a few such modelling studies which have analysed such sensitivities to a greater or lesser degree. Lonsdale (1993), using a regression approach, showed that spread of *Mimosa pigra* within a wetland was much faster than his model predictions and concluded that this reflected underestimates of dispersal. Bergelson *et al.* (1993) varied number, size and distribution of gaps, but not dispersal patterns, in a model of *Senecio vulgaris* invading a grassland and found, unsurprisingly, that gap dynamics affected the rate of spread.

More rigorous studies have been done by Parker and Reichard (1997) and Woolcock and Cousens (2000). The former created a spatially explicit cellular model of the spread of the invasive shrub *Cytisus scoparius* into oldfields and *Festuca* grasslands. While they did not carry out a full sensitivity analysis, they found that demographic variation (among six populations) had a larger effect on rates of spread than did variation in dispersal patterns or distances. However, this model exposed an interesting and important effect of the pattern of seed dispersal. Variation in the maximum seed dispersal distance had a much greater effect on the rate of population expansion in the model than did changes in the mean dispersal distance. Woolcock and Cousens (2000) created a continuous-space model of spread of four arable weeds and carried out a sensitivity analysis of key dispersal and demographic parameters. For all four species, variation in dispersal parameters affected the rate of spread much more strongly than did variation in any demographic parameter.

Neubert and Caswell (2000) introduced an approach to modelling spread that combines traditional matrix population growth models with integrodifference dispersal models. This model can use a variety of demographic and dispersal data to calculate the asymptotic rate of spread. An analysis of *Calathea ovandensis* showed that the rate of spread was determined by the maximum measured dispersal distance; the pattern of dispersal at lesser distances had little effect. A second analysis using *Dipsacus sylvestris* produced similar results. Rate of spread increased proportionately with increases in the modelled maximum dispersal distance, but changes in the percentage of seeds dispersed to that maximum had little effect. We used the Neubert and Caswell (2000) approach to model spread in three species:

1 The dwarf ericaceous shrubs *Calluna vulgaris* and *Erica cinerea* invading abandoned arable land. Dispersal for both species was measured using seed traps at distances up to 80 m from the source plants (Fig. 14.6a) (Bullock & Clarke 2000).

2 The annual forb *Rhinanthus minor* invading grassland, managed by either hay-cutting or grazing. Dispersal in each environment was measured using seed traps up to 25 m from the source plants (Fig. 14.6b) (J. M. Bullock, unpubl. data).

Probability density functions (and subsequently moment-generating functions) were derived from the dispersal data using the non-parametric estimator approach of Clark *et al.* (2001), which has the benefit of representing the actual dispersal data rather than a parametric function of them. For all species, demographic data were gathered in the form of stage-classified matrices. This produced estimates of the asymptotic rate of spread (Table 14.1) which are reasonable (e.g. the observed rate of spread of *Rhinanthus* was 0.6 m $year^{-1}$ under grazing and 9 m $year^{-1}$ under a hay-cut; Fig. 14.4).

Methods such as a form of life table response analysis (Caswell 2000) could be used within this modelling framework to assess the relative importance of demographic variation compared with changes in dispersal in affecting rates of spread. However, for the moment we were interested in whether the form of the dispersal data as gathered in the field would affect the predictions made by the model. Thus, for the dwarf shrubs we calculated new dispersal probability density functions on the assumption that seeds had been trapped up to only 10, 30 or 60 m rather than the actual 80 m. Further, we used the fitted models (Fig. 14.6a) to obtain an estimate of seed numbers that might have been trapped at 100 and 200 m and used these to calculate probability density functions for dispersal curves extended to 100 m (with one extra trapping station at 100 m) and 200 m (with two extra trapping stations at 100 and 200 m). We had no dispersal model for the *Rhinanthus* hay-cut data, so we tested only the consequences of reducing the maximum trapping distance, from 25 m to 16, 10, 8, 6 or 4 m.

These changes in the 'dispersal sampling design' had great effects on the calculated rate of spread (Table 14.2). The rates of spread of *Calluna* and *Erica* calculated for a maximum sampling distance of 10 m were <14% and 6%, respectively, of the

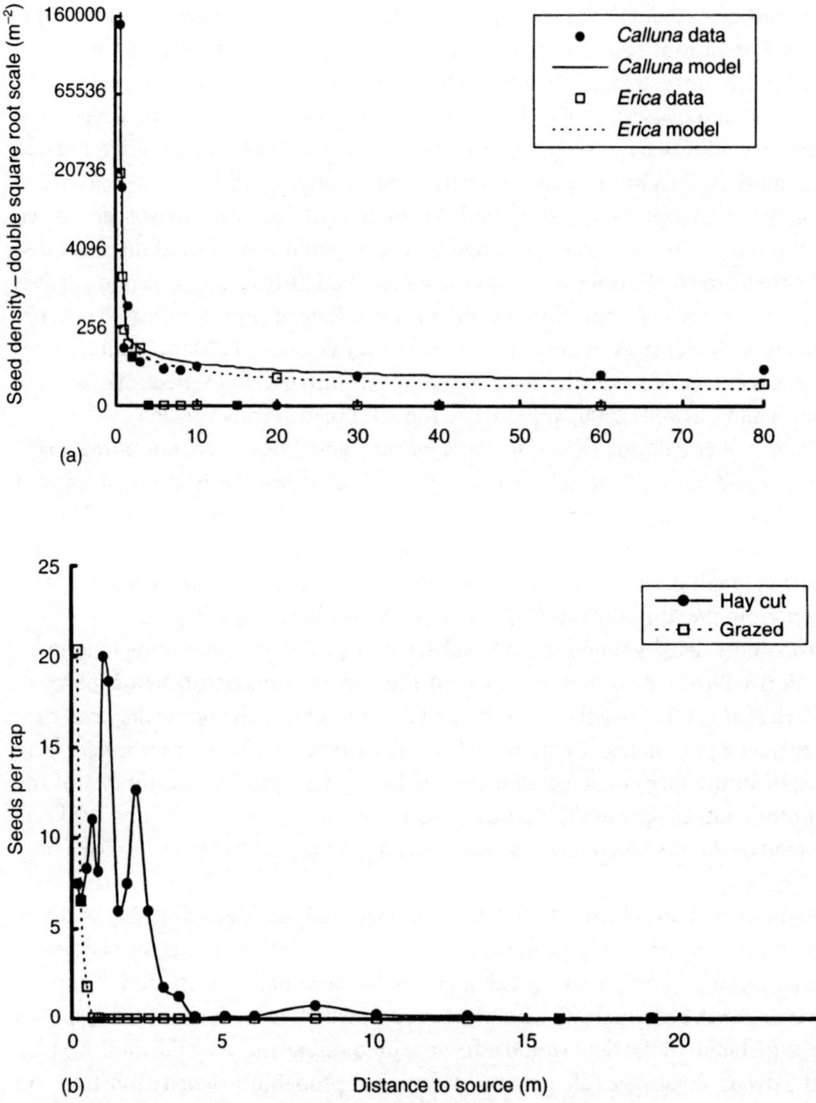

Figure 14.6 Dispersal curves measured over moderately long distances. (a) Data for two ericaceous dwarf shrubs *Calluna vulgaris* and *Erica cinerea.* The fitted models have the form, Seed density = $a.\exp(-b.D) + c.D^{-p}$, where D is the distance from the source and a, b, c and p are fitted parameters (Bullock & Clarke 2000). To allow the relationship between the mode and tail values to be seen easily the y-axis has the scale $y^{0.25}$ (avoiding the zero value problem of log transformation). (b) *Rhinanthus minor* measured under hay-cut and grazed conditions (J. M. Bullock, unpubl. data). The lines connecting the points are for illustrative purposes only.

Table 14.1 Estimates of population growth rate (λ) and asymptotic rate of spread for three species using the Neubert and Caswell (2000) model.

	Calluna	*Erica*	*Rhinanthus* hay-cut	*Rhinanthus* grazed
λ	2.62	2.15	8.02	8.02
Estimated rate of spread (m year^{-1})	7.52	6.27	6.11	0.66

Table 14.2 The effects of different hypothetical dispersal sampling designs on estimates of the asymptotic rate of spread, calculated using the Neubert and Caswell (2000) model. These were defined in terms of the maximum distance of trap data used. The maximum distances measured in the field studies are given in bold. The data for the 100 and 200 m distances for *Calluna* and *Erica* were estimated using models fitted to the dispersal data (see text).

Maximum trap distance (m)	Estimated rate of spread (m year^{-1})
Calluna	
10	1.02
30	2.73
60	5.48
80	7.52
100	8.66
200	16.66
Erica	
10	0.38
30	1.84
60	1.84
80	6.27
100	6.40
200	11.37
Rhinanthus	
4	2.70
6	3.14
8	4.04
10	4.92
16	5.84
25	6.11

rates calculated from the full dispersal data. Similarly, increasing trap distances to 200 m would theoretically increase estimated rates of spread by about 200% for both species. The extrapolations beyond our trap limit of 80 to 100 or 200 m are, of course, statistically invalid. However, they allow an illustration of the consequences of possible rare, long-distance dispersal events. The *Rhinanthus* maximum distances were varied over a smaller range, but a maximum trapping distance of 4 m reduced the model estimate by more than half. These outcomes are despite the fact that the dwarf shrub dispersal curves are extremely leptokurtic and the *Rhinanthus* curve is moderately leptokurtic. Thus for all three species, 99% of seed fell within the limits of our lowest maximum trap distances. The small amount of seed dispersing longer distances and the distance travelled by that seed are of enormous importance in determining rates of spread.

These analyses and the other studies reviewed here suggest that, even at the local scale, dispersal is an important, if not the principal, factor in determining rates of spread of invading species. The limited modelling to date suggests that the important dispersal data may be that for longer distances. Indeed, although it is difficult to get such data (Greene & Calogeropoulos, this volume), there is a suggestion that effort to get such data could be traded off against less informative activities such as gathering dispersal data at shorter distances or precise quantification of the dispersal curve. However, we hesitate to recommend this before more modelling studies are carried out. There is also a need to create such models for applied reasons, such as devising methods for controlling the spread of weeds (Parker & Reichard 1997; Manchester & Bullock 2000) or, conversely, encouraging the spread of reintroduced species (Hodder & Bullock 1997; Coulson *et al.* 2001).

Landscape dynamics

In traditional plant ecology (particularly phytosociology), the prevailing notion was that plant species occupied all suitable habitat within a landscape. Therefore, absence of a species from a community identified a site as having an unsuitable habitat. Straightforward experiments involving the introduction of seed into such unoccupied sites have found this can sometimes allow populations to form (e.g. Primack & Miao 1992; Eriksson 1998; Ehrlen & Eriksson 2000), showing this simple paradigm is not always supported. Dispersal limitation must be invoked as a major factor determining species distributions over a landscape. Here we consider the landscape dynamics of species in terms of their distribution among sites in a landscape. Metapopulation dynamics considers species which are established at sites, but which undergo extinction and recolonization at these sites. The movement of species from occupied to previously unoccupied sites can be considered in terms of landscape succession.

Dynamics of established species (metapopulations)

In this section we shall ask the controversial question 'do plant metapopulations exist?' This may seem a strange question given the enormous popularity of the concept. Many plant ecological studies at the landscape scale cite metapopulation theory. The metapopulation concept is becoming accepted in plant ecology (Husband & Barrett 1996) and is informing conservation strategies (e.g. Spellerberg 1996; Mace *et al.* 1999). However, the metapopulation theory was, and is still being, developed mostly in terms of animals. Plants show differences in their spatial dynamics, particularly in terms of seed banks and limited dispersal, which must be considered in transferring this theory from the animal world (Eriksson 1996).

The fundamental indicators of metapopulation dynamics in a landscape are: (i) the periodic extinction of populations; (ii) recolonization of unoccupied sites (following extinction); and (iii) the exhibition of some degree of independent dynamics by populations (Eriksson 1996; Hanski 1999). Thus, the abundance and distribution of populations in a landscape are controlled by a balance of local dynamics and linked landscape-level colonization and extinction events. The third point is important because populations are maintained by both internal dynamics (births and deaths) and migrations among populations (Thomas & Kunin 1999). The original 'classic' metapopulation concept has been extended to allow a broader family of landscape-level or regional dynamics (Harrison & Taylor 1997; Hanksi 1999). Although some authors would wish to encapsulate all regional dynamics within the metapopulation family, this is dangerous because it assumes that metapopulation theoretical predictions and modelling approaches are valid in all cases. Thus true metapopulations range between classic (a network of sites, with local dynamics occurring at a much faster timescale than metapopulation dynamics; all populations have a significant extinction risk) and source–sink (where some sites have a negative growth rate at low density in the absence of immigration and in others the growth rate at low density is positive; the latter support the former through emigration) (Hanski & Simberloff 1997). Regional dynamics other than metapopulations include: (i) patchy populations, where there is a single population distributed among sites in a region; these show no independent dynamics and dispersal among sites is very high (Harrison & Taylor 1997); and (ii) non-equilibrium or remnant systems, where populations are independent and there is no, or extremely little, migration among sites (Eriksson 1996; Harrison & Taylor 1997). Thomas and Kunin (1999) have also shown that some of these different dynamics can be described along axes relating to the role of internal population dynamics and interpopulation dispersal.

Evidence cited in support of plant metapopulations is almost all of population extinctions and colonizations within a landscape, although there are remarkably few solid examples (Table 14.3). It is interesting that all these examples (except Harrison *et al.* 2000) and the other cases where metapopulation dynamics have been suggested, involve species occupying early-successional habitats. Extinctions are generally due to a site to becoming unsuitable through succession, and colonizations occur at new sites following a disturbance. This pattern is quite different from the standard representation of metapopulations—of suitable sites fixed in space while populations come through dispersal and go through environmental and/or demographic stochasticity. Harrison and Taylor (1997) term this 'habitat tracking' and point out that such metapopulations violate a basic premise of metapopulation theory and models: that extinctions make habitats available for recolonization. The regional dynamics of a species are governed by the turnover of suitable habitat sites as well as migration among sites. Modelling approaches have to take this difference into account (e.g. Johnson 2000).

Table 14.3 Examples of studies which have shown extinctions and colonizations among regional plant populations.

Species	Life form	Habitat	Reference
Aster acuminatus	Perennial	Forest gaps	Hughes *et al.* (1988)
Pedicularis furbishiae	Short-lived perennial	Disturbed river banks	Menges 1990
Primula vulgaris	Perennial	Forest gaps	Valverde & Silvertown (1997)
Senecio jacobeaea and *Cynoglossum oficinale*	Monocarpic perennials	Disturbed areas in dunes	Van der Meijden *et al.* (1992)
Silene alba	Short-lived perennial	Road verges	Antonovics *et al.* (1994)
Silene dioica	Short-lived perennial	Early primary succession on islands	Giles & Goudet (1997)
Five species	Annuals and perennials	Serpentine seeps	Harrison *et al.* (2000)
Several species	Annuals and perennials	Dry grassland	Ouborg (1993)

The major question, however, is whether the (re)colonization we see in these 'metapopulations' is true colonization. Do the first colonists really derive from seed dispersing from other populations or do they come from the seed bank? One must also be sure that a population is truly extinct and that 'recolonization' is not coming from a few small, remnant plants (e.g. Hughes *et al.* 1988). While migration processes are well characterized for metapopulations in other taxa (Okamura & Freeland, this volume; Wilson & Thomas, this volume), we can find only one example (although this is not a complete study of regional dynamics) where plant migration among sites has been measured (for the mistletoe *Phrygilanthus sonoroe*; Overton 1994). It is known that seed banks can be long-lived and that many communities contain species in the seed bank which are not in the vegetation and represent 'extinct' populations (Beatty 1991; Thompson 2000). In truth, an extant seed bank means that a population is not really extinct, but 'sleeping'. Many short-lived, early-successional species, such as those which seem to show metapopulation dynamics, are known to have seed banks which allow them to survive periods of poor or no recruitment (Kalisz & McPeek 1992; van der Meijden *et al.* 1992; Harrison *et al.* 2000). Therefore colonization of a site by dispersal from other populations in a landscape must be distinguished from 'pseudocolonization' which is merely the continuation of an existing population after a lag phase in which the only life stage present was this sleeping seed bank. If the latter is the dominant process, then we are seeing independent, remnant (Eriksson 1996) populations rather than a metapopulation.

It is too early to determine whether true plant metapopulations exist. We do not have the necessary data on migrations and colonizations. A much quoted example of a plant metapopulation, *Pedicularis furbishiae* on disturbed river banks, seems to be valid because it is cited as having no seed survival (Menges 1990). Giles and Goudet's

(1997) example of *Silene dioica* on islands created by land uplift in the Baltic also seems to be a true metapopulation as new islands could not have a seed bank. However, these are extreme situations. The existence of metapopulations in more everyday conditions is suggested by studies which have found isolation effects on colonization rates, a signal which indicates colonists have dispersed from other populations rather than emerged from the seed bank, and on extinction rates, suggesting dispersal among populations is promoting population persistence (although isolation can affect colonization and extinction through other processes, such as a lack of mutualists; e.g. Groom 1998). Thus, Quintana-Ascencio and Menges (1996) found isolation explained occupancy patterns for shrub species in a fire-prone habitat. Ouborg (1993) found isolation effects on extinction and colonization patterns for some species in a landscape of dry grasslands, as did Harrison *et al.* (2000) for five species on serpentine seeps. Counterevidence for the importance of emergence from the seed bank rather than interpopulation dispersal is also rare. The Horvitz and Schemske (1986) study of the understorey herb *Calathea ovandenis* is often quoted as a metapopulation study, although it was not (C. Horvitz, pers. comm.), but it suggested that seed dormancy was much more important than dispersal.

The over-riding importance of the degree of dispersal among sites in determining the type of regional dynamics is illustrated in Fig. 14.7. Metapopulations have an intermediate proportion of recruits which are immigrants. In classic metapopulations this immigrant proportion is approximately equal among populations. The system tends more towards a source–sink metapopulation as the contribution of immigrants to the population becomes more unequal among the sites. Patchy

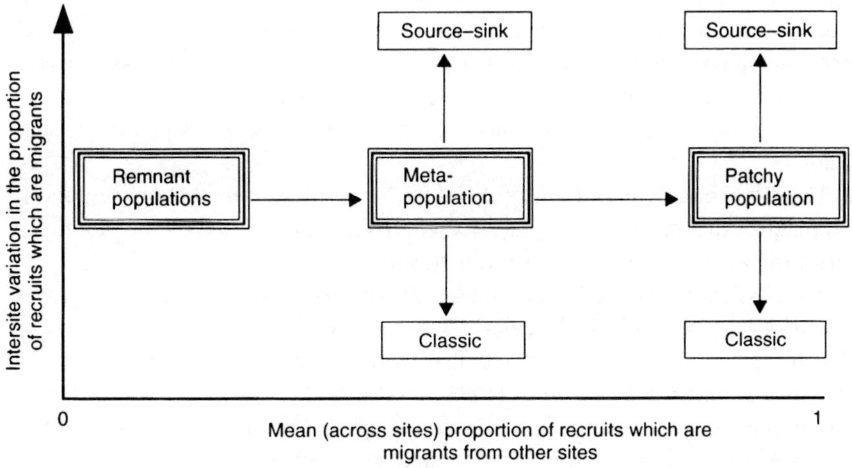

Figure 14.7 A classification of the regional dynamics of plants in terms of the proportion of recruits (in a year or generation) which are from migrant seed derived from other sites. Patchy populations can show source–sink dynamics and the converse of this (i.e. equal proportion of migrants among sites) is termed 'classic' as it is in metapopulations.

populations are distinguished from metapopulations by having a higher degree of dispersal among sites, i.e. the dynamics at each site are less independent and mostly driven by immigrant seed. A surprising number of studies of populations with gap dynamics are cited as metapopulations (e.g. that of Alvarez-Buylla & García-Barrios 1991), but the lack of population dynamics within colonized gaps clearly indicates these are patchy populations. Patchy populations may also show 'source–sink' dynamics (e.g. Kadmon & Shmida 1990). Remnant populations occur where dispersal is near zero (e.g. Wolf *et al.* 1999).

This classification has similarities to Thomas and Kunin's (1999) approach, but is more targeted towards the key process that must be measured in the field. This is the proportion of recruits in patches or populations that are derived by immigration from external sources compared with the internal dynamics of the patch or population. The same classification and research need applies to both the dynamics in occupied sites and the process of (pseudo?) colonization of unoccupied sites. Molecular methods could produce exactly this information (Ouborg *et al.* 1998; Cain *et al.* 2000; Raybould *et al.*, this volume). Figure 14.7 has no boundaries between the three types of regional dynamics because they merge into each other. However, the different types will probably have very different properties, for example in responses to habitat fragmentation. There is therefore a need to gain both field data to determine where different species and systems fall in this classification, and to develop models to understand fully the spatiotemporal properties of these types of regional dynamics.

Arrival of new species (succession)

Directional vegetation change, or succession, within a site involves changes in the abundance of established species, but a major extra factor is invasion by new species. At the landscape scale this involves the movement of species from occupied to unoccupied sites and two aspects of dispersal will govern the identity of these species:

1 The effects of context. Species' source populations will be at different distances from a site. Generally, a more distance source will mean fewer propagules will reach a site.

2 The differential dispersal ability of species. Generally, species that produce more propagules, that can travel longer distances will be less affected by the proximity effect and will get more propagules into the site.

The classic inhibition, tolerance and facilitation theory of succession mechanisms (Connell & Slatyer 1977; Pickett *et al.* 1987) includes dispersal only in the concept that underlies all three mechanisms: that early colonists will be good dispersers (Connell & Slatyer 1977). Tilman (1994) developed these ideas to some degree by describing succession as the slow accretion of species; the arrival of which is constrained by their dispersal abilities. Hovestadt *et al.* (2000) took this further with a model in which the successional trajectory is governed purely by species differences in arrival time through differences in dispersal ability. These developments remain within the classic framework, which is an equilibrium approach to succession dynamics. Thus, succession is a highly deterministic process in which species arrive in a repeatable sequence governed by their dispersal abilities. Causes of intersite variation in succession are due to abiotic differences which directly and indirectly (by affecting species interactions) affect species performances differentially. However, both aspects of dispersal are important and a non-equilibrium approach is valid; i.e. spatial context is vital in determining successional pathways.

One line of empirical evidence for the role of dispersal comes from tests of whether variation in plant performance can explain successional pathways. This repeats our recurring theme of whether variation in demography or dispersal can explain ecological patterns. Several studies of primary succession have shown that intersite variation in successional pathways (i.e. the time trajectory in the abundances of particular species) is not explained by variation in species' performance in terms of seedling establishment (Fastie 1995; Stöcklin & Bäumler 1996; Del Moral 1999). Variation in the abundance of a species at different successional stages may not be related to its establishment performance in these different stages (Whittaker 1993), nor may differences in abundances of species in a succession be related to differences in the establishment of added seed (De Steven 1991).

The studies described above supply negative evidence and do not allow our two dispersal processes (context versus dispersal ability) to be distinguished. More direct evidence comes from studies which have linked intersite differences in successional pathways to the proximity of source populations. We studied the succession of lowland heathland vegetation to scrub and woodland among 116 patches of heathland vegetation in southern England (Nolan *et al.* 1998; Nolan 1999). Virtually all these patches experienced increases in scrub and tree species between 1978 and 1987 (Fig. 14.8) and again between 1987 and 1996, with concomitant loss of the dwarf shrub heath, a vegetation type of conservation importance. The percentage loss in each patch of different types of heathland vegetation, dry humid and wet heath and mire (see Rose *et al.* 2000), was related to a number of spatial variables, and significant regression models ($r^2 = 0.14$–0.48, $P < 0.05$) were derived for six of the eight data sets (four heath vegetation types and two time periods). In all, the most important variables were those describing the relative abundance of scrub and tree species in the vegetation bordering the heath patches, showing that the proximity of source populations was a major factor determining rates of succession.

Del Moral and Bliss (1993) found that, following the Mount St Helens eruption, vegetation developed more slowly on more isolated areas that were more dominated by more 'vagile' species. Matlack (1994) found that planted new woods had slower accumulation of woodland understorey plants the further they were from ancient woodlands. Species showed differences in this isolation effect, with the least affected by isolation being species with ingested seeds; species with adhesive or wind-dispersed seed were more affected, ant-dispersed species more so, and the species most affected were species with no obvious dispersal adaptation. Both Brunet and von Oheimb (1998) and Grashof-Bokdam and Geertsema (1998) found very similar results for understorey species in new forests. Responses to isolation were also linked to dispersal mode, with ant-dispersed species showing larger responses to isolation than species with adhesive or ingested seeds. McClanahan (1986) found

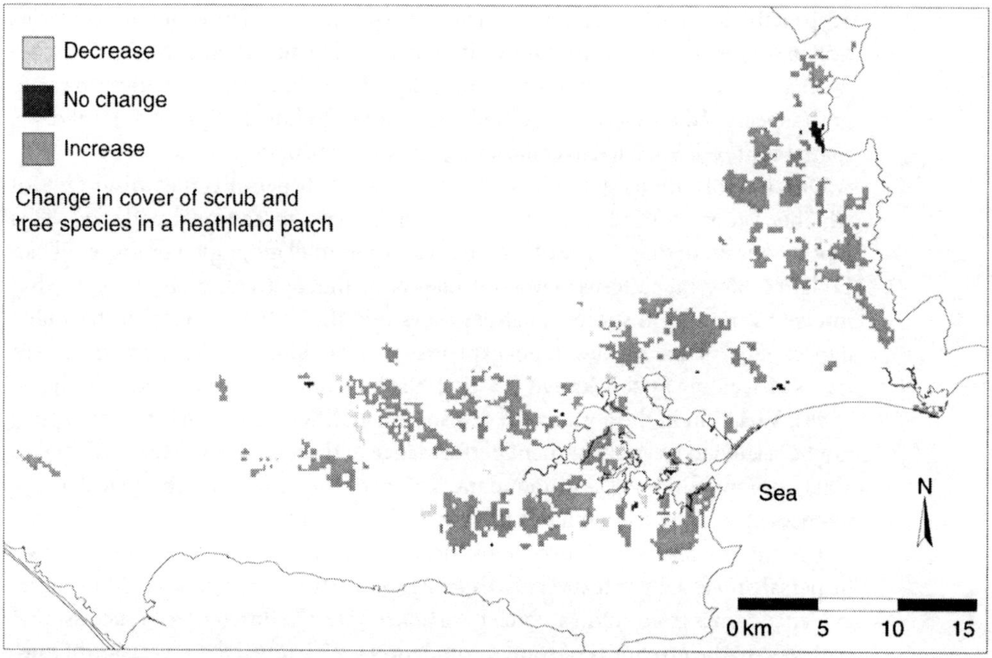

Figure 14.8 Scrub and tree invasion in 116 heathland patches in Dorset, England between 1978 and 1987. Heathland patches are represented by shading; other vegetation types are not mapped. The cover of scrub and trees increased in most patches over the 9 years, but decreased or remained static in a few patches. The coastline (south) and county boundary (east) are shown.

distance to source populations was the best predictor of the presence of late-successional tree species on previously mined sites. Finally, Yao *et al.* (1999) found colonization rates of tree species into oldfields were determined by distance to sources, and bird-dispersed species colonized more rapidly than wind-dispersed species. This isolation effect can be so important that succession can be exceedingly slow or even arrested at sites very far from sources (Duncan & Chapman 1999; Holl 1999; Yao *et al.* 1999). This problem is well known in restoration studies, where dispersal has often to be given a helping hand, by sowing seed, in order to overcome the lack of immigration of seed because sources are too far away and dispersal is too limited in extent (Ash *et al.* 1994; Bakker & Berendse 1999; Pywell *et al.* 2002). Some authors have shown that the provision of bird perches may facilitate the dispersal of bird-dispersed species and thus speed up the development of forest on oldfields (McClanahan & Wolfe 1993; Robinson & Handel 1993; Harvey 2000).

Multiple successional pathways are found in many vegetation types (Miles 1987; Pickett *et al.* 1987; Del Moral 1999). A deterministic understanding of this variation may come from consideration of processes at the landscape level and the joint effects of the proximity of sources and differential dispersal ability of species. This requires studies at the site level of both the seed rain and species' establishment (e.g. Hester *et al.* 1991). At the landscape level there should be measures of seed dispersal from sources both near and far and of how source proximity affects seed rain onto a site. For restoration projects we need to know whether development of the desired vegetation is limited by the availability of seed (dispersal) or the site conditions (establishment), and whether we can aid restoration by facilitating dispersal (e.g. Fischer *et al.* 1996). Conversely, such knowledge is also useful if we want to prevent succession, for example of heathland to scrub. This is generally done by removal of the undesirable species. Our study of the Dorset heathlands suggested that patches subjected to scrub removal (at the level of current management) did not show slower succession (Nolan 1999). It could be more effective to remove sources of colonists and thus induce dispersal limitation.

Conclusions

We have shown here that dispersal and thus dispersal limitation are very important in plant dynamics at local and landscape scales. Dispersal limitation has two aspects. Species differ in dispersal abilities, and varying spatial context means that the effective species pool differs from place to place. The former is another type of life-history variation and is linked to equilibrium explanations of plant dynamics. However, the latter leads to non-equilibrium dynamics reflecting the role of history and spatial structure.

This emphasis on dispersal shows that one must be careful in imposing anthropocentric scales on ecological processes. We separated local and landscape scales, but it is clear that this separation may be arbitrary and false in some cases. Two species with different dispersal patterns and/or demographic attributes may show very different spatial dynamics in a landscape. A forest exhibiting gap turnover may contain a single, patchy population of a gap-dependent tree with good dispersal and a metapopulation of a gap-dependent herb. Further, the tree may exhibit metapopulation dynamics at a larger scale, among forest patches. Thus the 'local' and 'landscape' scales are not the same for all species. We also split the process by which a species invades a new site across two scales: movement between sites and spread within a site. These two processes may be dynamically separate, but an invasion may involve the constant rain of seed from external sources, as well as spread from initial foci (e.g. Debussche *et al.* 1985).

In the section on landscape succession we described increases in the number of sites occupied by a species in a landscape, whereas the idea behind metapopulations is the maintenance of a balance of occupied and unoccupied sites. It is also important to consider how these two forms of landscape dynamics are linked. The successional, or invasion, dynamics may represent an earlier stage in the regional dynamics of a species. Species invading new sites must rely on dispersal, whereas in metapopulations species could recolonize sites by dispersal and/or the seed bank. Subsequent to the expansion of a species in a landscape, it may settle into a form of regional dynamics as described in Fig. 14.7. This argument suggests the analysis of dynamics of species in a landscape must consider the full range of spatiotemporal processes, of occupation of both new and previously occupied sites, the role of the seed bank, whether the population number is increasing, decreasing or is stable, and the effect of dispersal limitation.

There are many studies which have quantified dispersal (see Willson 1993; Nathan & Muller-Landau 2000; Willson & Traveset 2000), but dispersal should be studied with regard to particular ecological questions (see Clark *et al.* 1999). We have shown that certain questions require particular types of dispersal data. Dispersal is poorly studied in all the subject areas discussed. Metapopulation studies, which invoke dispersal in particular, have hardly quantified dispersal at all. Thus, the need over the coming years is to quantify dispersal at different scales. In this way the scale at which particular processes are studied or models are constructed can be determined, as advocated by Harper (1977), by the plant's, not the human's, eye-view.

References

Alvarez-Buylla, E.R. & García-Barrios, R. (1991) Seed and forest dynamics: a theoretical framework and an example from the neotropics. *American Naturalist* **137**, 133–154.

Antonovics, J., Thrall, P., Jarosz, A. & Sratton, D. (1994) Ecological genetics of metapopulations: the *Silene–Ustilago* plant–pathogen system. In: *Ecological Genetics* (ed. L.A. Real), pp. 146–169. Princeton University Press, Princeton, NJ.

Ash, H.J., Gemmell, R.P. & Bradshaw, A.D. (1994) The introduction of native plant species on industrial-waste heaps—a test of immigration and other factors affecting primary succession. *Journal of Applied Ecology* **31**, 74–84.

Bakker, J.P. & Berendse, F. (1999) Constraints in the restoration of ecological diversity in grassland and heathland communities. *Trends in Ecology and Evolution* **14**, 63–68.

Beatty, S.W. (1991) Colonization dynamics in a mosaic landscape—the buried seed pool. *Journal of Biogeography* **18**, 553–563.

Bergelson, J., Newman, J.A. & Floresroux, E.M. (1993) Rate of weed spread in spatially heterogeneous environments. *Ecology* **74**, 999–1011.

Bolker, B.M. & Pacala, S.W. (1999) Spatial moment equations for plant competition: understanding spatial strategies and the advantages of short dispersal. *American Naturalist* **153**, 575–602.

Brokaw, N. & Busing, N.T. (2000) Niche versus chance and tree diversity in forest gaps. *Trends in Ecology and Evolution* **15**, 183–188.

Brunet, J. & von Oheimb, G. (1998) Migration of vascular plants to secondary woodlands in southern Sweden. *Journal of Ecology* **86**, 429–438.

Bullock, J.M. (2000) Gaps and seedling colonization. In: *Seeds: the Ecology of Regeneration in Plant Communities* (ed. M. Fenner), pp. 375–395. CABI, Wallingford, UK.

Bullock, J.M. & Clarke, R.T. (2000) Long distance seed dispersal by wind: measuring and modelling the tail of the curve. *Oecologia* **124**, 506–521.

Bullock, J.M., Clear Hill, B. & Silvertown, J. (1994) Demography of *Cirsium vulgare* in a grazing experiment. *Journal of Ecology* **82**, 101–111.

Bullock, J.M., Clear Hill, B., Silvertown, J. & Sutton, M. (1995) Gap colonization as a source of grassland community change: effects of gap size and grazing on the rate and mode of colonization by different species. *Oikos* **72**, 273–282.

Bullock, J.M., Franklin, J., Stevenson, M.J. *et al.* (2001) A plant trait analysis of responses to grazing in a long-term experiment. *Journal of Applied Ecology* **38**, 253–267.

Cain, M.L., Milligan, B.G. & Strand, A.E. (2000) Long-distance seed dispersal in plant populations. *American Journal of Botany* **87**, 1217–1227.

Caswell, H. (2000) *Matrix Population Models.* Sinauer, Sunderland, MA.

Clark, J.S., Beckage, B., Camill, P. *et al.* (1999) Interpreting recruitment limitation in forests. *American Journal of Botany* **86**, 1–16.

Clark, J., Horvath, L. & Lewis, M. (2001) On the estimation of spread rate for a biological population. *Statistics and Probability Letters* **51**, 225–234.

Connell, J.H. & Slatyer, R.O. (1977) Mechanisms of succession in natural communities and their role in community stability and organisation. *American Naturalist* **111**, 1119–1144.

Coulson, S.J., Bullock, J.M., Stevenson, M.J. & Pywell, R.F. (2001) Colonization of grassland by sown species: dispersal versus microsite limitation in responses to management. *Journal of Applied Ecology* **38**, 204–216.

Cousens, R. & Mortimer, M. (1995) *Dynamics of Weed Populations.* Cambridge University Press, Cambridge.

Dalling, J.W., Swaine, M.D. & Garwood, N.C. (1998) Dispersal patterns and seed bank dynamics of pioneer trees in moist tropical forest. *Ecology* **79**, 564–578.

De Steven, D. (1991) Experiments on mechanisms of tree establishment in old-field succession: seedling emergence. *Ecology* **72**, 1066–1075.

Debussche, M., Lepart, J. & Molina, J. (1985) Seed dispersal by birds, effect of vegetation structure and influence on succession in a mediterranean region. *Acta Oecologica* **6**, 65–80.

Del Moral, R. (1999) Plant succession on pumice at Mount St Helens, Washington. *American Midland Naturalist* **141**, 101–114.

Del Moral, R. & Bliss, L.C. (1993) Mechanisms of primary succession—insights resulting from the eruption of Mount St Helens. *Advances in Ecological Research* **24**, 1–66.

Duncan, R.S. & Chapman, C.A. (1999) Seed dispersal and potential forest succession in abandoned agriculture in tropical Africa. *Ecological Applications* **9**, 998–1008.

Ehrlen, J. & Eriksson, O. (2000) Dispersal limitation and patch occupancy in forest herbs. *Ecology* **81**, 1667–1674.

Eriksson, O. (1996) Regional dynamics of plants: a review of evidence for remnant, source–sink and metapopulations. *Oikos* **77**, 248–258.

Eriksson, O. (1997) Colonization dynamics and relative abundance of three plant species in dry semi-natural grasslands. *Ecography* **20**, 559–568.

Eriksson, O. (1998) Regional distribution of *Thymus serpyllum*: management history and dispersal limitation. *Ecography* **21**, 35–43.

Fastie, C.L. (1995) Causes and ecosystem consequences of multiple pathways of primary succession at Glacier Bay, Alaska. *Ecology* **76**, 1899–1916.

Fischer, S.F., Poschlod, P. & Beinlich, B. (1996) Experimental studies on the dispersal of plants and animals on sheep in calcareous grasslands. *Journal of Applied Ecology* **33**, 1206–1222.

Giles, B.E. & Goudet, J. (1997) A case study of population structure in a plant metapopulation. In: *Metapopulation Biology* (eds I.A. Hanski & M.E. Gilpin), pp. 429–454. Academic Press, San Diego.

Grashof-Bokdam, C.J. & Geertsema, W. (1998) The effect of isolation and history on colonization patterns of plant species in secondary woodland. *Journal of Biogeography* **25**, 837–846.

Groom, M. (1998) Allee effects limit population viability of an annual plant. *American Naturalist* **151**, 487–496.

Gustafsson, E.J., Shirley, S.R., Mladenhoff, D.J., Nimerfro, K.K. & He, H.S. (2000) Spatial simulation of forest succession and timber harvesting. *Canadian Journal of Forest Research* **30**, 32–43.

Hanski, I.A. (1999) *Metapopulation Ecology.* Oxford University Press, Oxford.

Hanski, I.A. & Simberloff, D. (1997) The metapopulation approach, its history, conceptual domain and application to conservation. In: *Metapopulation Biology* (eds I.A. Hanski & M.E. Gilpin), pp. 5–25. Academic Press, San Diego.

Harper, J.L. (1977) *Population Biology of Plants.* Academic Press, London.

Harrison, S. & Taylor, A.D. (1997) Empirical evidence for metapopulations. In: *Metapopulation Biology* (eds I.A. Hanski & M.E. Gilpin), pp. 27–42. Academic Press, San Diego.

Harrison, S., Maron, J. & Huxel, G. (2000) Regional turnover and fluctuation in populations of five plants confined to serpentine seeps. *Conservation Biology* **14**, 769–779.

Harvey, C.A. (2000) Windbreaks enhance seed dispersal into agricultural landscapes in Monteverde, Costa Rica. *Ecological Applications* **10**, 155–173.

He, H.S. & Mladenoff, D.J. (1999) Spatially explicit and stochastic simulation of forest-landscape fire disturbance and succession. *Ecology* **80**, 81–99.

Hester, A.J., Gimingham, C.H. & Miles, J. (1991) Succession from heather moorland to birch woodland. 3. Seed availability, germination and early growth. *Journal of Ecology* **79**, 329–344.

Hobbs, R.J. & Hobbs, V.J. (1987) Gophers and grassland: a model of vegetation response to patchy soil disturbance. *Vegetatio* **69**, 141–146.

Hodder, K.H. & Bullock, J.M. (1997) Translocations of native species in the UK: implications for biodiversity. *Journal of Applied Ecology* **34**, 547–565.

Holl, K.D. (1999) Factors limiting tropical rain forest regeneration in abandoned pasture: seed rain, seed germination, microclimate, and soil. *Biotropica* **31**, 229–242.

Horvitz, C.C. & Schemske, D.W. (1986) Seed dispersal and environmental heterogeneity in a neotropical herb: a model of population and patch dynamics. In: *Frugivores and Seed Dispersal* (eds A. Estrada & T.H. Fleming), pp. 169–186. Dr W. Junk Publishers, Dordrecht.

Hovestadt, T., Poethke, H.J. & Messner, S. (2000) Variability in dispersal distances generates typical successional patterns: a simple simulation model. *Oikos* **90**, 612–619.

Hubbell, S.P., Foster, R.B., O'Brien, S.T. *et al.* (1999) Light-gap disturbances, recruitment limitation, and tree diversity in a neotropical forest. *Science* **283**, 554–557.

Hughes, J.W., Fahey, T.J. & Bormann, F.H. (1988) Population persistence and reproductive ecology of a forest herb *Aster acuminatus. American Journal of Botany* **75**, 1057–1064.

Hurtt, G.C. & Pacala, S.W. (1995) The consequences of recruitment limitation: reconciling chance, history and competitive differences among plants. *Journal of Theoretical Biology* **176**, 1–12.

Husband, B.C. & Barrett, S.C.H. (1996) A metapopulation perspective in plant population biology. *Journal of Ecology* **84**, 461–469.

Johnson, M.P. (2000) The influence of patch demographics on metapopulations, with particular reference to successional landscapes. *Oikos* **88**, 67–74.

Kadmon, R. & Shmida, A. (1990) Spatiotemporal demographic processes in plant populations: an approach and a case study. *American Naturalist* **135**, 382–397.

Kalamees, R. (1999) *Seed bank, seed rain and commmunity regeneration in Estonian calcareous grasslands.* PhD thesis, University of Tartu, Tartu, Estonia.

Kalisz, S. & McPeek, M.A. (1992) Demography of an age-structured annual—resampled projection matrices, elasticity analyses, and seed bank effects. *Ecology* **73**, 1082–1093.

Keddy, P.A. (1981) Experimental demography of a sand-dune annual, *Cakile edentula*, growing along an environmental gradient in Nova Scotia. *Journal of Ecology* **69**, 615–630.

Kinzig, A.P., Levin, S.A., Dushoff, J. & Pacala, S. (1999) Limiting similarity, species packing, and system stability for hierarchical competition–colonization models. *American Naturalist* **153**, 371–383.

Lonsdale, W.M. (1993) Rates of spread of an invading species—*Mimosa pigra* in northern Australia. *Journal of Ecology* **81**, 513–522.

Mace, G.M., Balmford, A. & Ginsberg, J.R. (eds) (1999) *Conservation in a Changing World.* Cambridge University Press, Cambridge.

Malo, J.E., Jimenez, B. & Suarez, F. (1995) Seed bank build-up in small disturbances in a Mediterranean pasture: the contribution of endozoochorous dispersal by rabbits. *Ecography* **18**, 73–82.

Manchester, S.J. & Bullock, J.M. (2000) Non-native species in the UK: impacts on biodiversity and effectiveness of controls. *Journal of Applied Ecology* **37**, 845–864.

Matlack, G.R. (1994) Plant-species migration in a mixed-history forest landscape in eastern North America. *Ecology* **75**, 1491–1502.

McClanahan, T.R. (1986) The effect of a seed source on primary succession in a forest ecosystem. *Vegetatio* **65**, 175–178.

McClanahan, T.R. & Wolfe, R.W. (1993) Accelerating forest succession in a fragmented landscape—the role of birds and perches. *Conservation Biology* **7**, 279–288.

Menges, E.S. (1990) Population viability analysis for an endangered plant. *Conservation Biology* **4** (1), 52–62.

Miles, J. (1987) Vegetation succession: past and present preceptions. In: *Colonization, Succession and Stability* (eds A.J. Gray, M.J. Crawley & P.J. Edwards), pp. 1–29. Blackwell Scientific Publications, Oxford.

Moloney, K.A. & Levin, S.A. (1996) The effects of disturbance architecture on landscape-level population dynamics. *Ecology* **77**, 375–394.

Nathan, R. & Muller-Landau, H.C. (2000) Spatial patterns of seed dispersal, their determinants and consequences for recruitment. *Trends in Ecology and Evolution* **15**, 278–285.

Neubert, M.G. & Caswell, H. (2000) Demography and dispersal: calculation and sensitivity analysis of invasion speed for structured populations. *Ecology* **81**, 1613–1628.

Nolan, A.M. (1999) *Modelling change in the lowland heathlands of Dorset.* PhD thesis, University of Southampton, Southampton.

Nolan, A.M., Atkinson, P.M. & Bullock, J.M. (1998) Modelling change in the lowland heathlands of Dorset, England. In: *Innovations in GIS 5* (ed. S. Carver), pp. 234–243. Taylor & Francis, London.

Ouborg, N.J. (1993) Isolation, population size and extinction: the classical and metapopulation approaches applied to vascular plants along the Dutch Rine-system. *Oikos* **66**, 298–308.

Ouborg, N.J., Piquot, Y. & van Groenedael, J.M. (1998) Population genetics, molecular markers and the study of dispersal in plants. *Journal of Ecology* **87**, 551–568.

Overton, J.M. (1994) Dispersal and infection in mistletoe metapopulations. *Journal of Ecology* **82**, 711–724.

Pacala, S.W. & Levin, S.A. (1997) Biologically generated spatial pattern and the coexistence of competing species. In: *Spatial Ecology: the Role of Space in Population Dynamics and Interspecific Interactions* (eds D. Tilman & P. Karieva), pp. 204–232. Princeton University Press, Princeton, NJ.

Pacala, S.W. & Rees, M. (1998) Models suggesting field experiments to test two hypotheses explaining successional diversity. *American Naturalist* **152**, 729–737.

Pakeman, R.J., Attwood, J.P. & Engelen, J. (1998) Sources of plants colonising experimentally disturbed patches in an acidic grassland, eastern England. *Journal of Ecology* **86**, 1032–1040.

Parker, I.M. (2000) Invasion dynamics of *Cytisus scoparius*: a matrix model approach. *Ecological Applications* **10**, 726–743.

Parker, I.M. & Reichard, S.H. (1997) Critical issues in invasion biology for conservation science. In: *Conservation Biology* (eds P. Fiedler & P. Kareiva), pp. 283–305. Chapman & Hall, New York.

Pickett, S.T.A., Collins, S.L. & Armesto, J.J. (1987) Models, mechanisms and pathways of succession. *Botanical Review* **53**, 335–371.

Primack, R.B. & Miao, S.L. (1992) Dispersal can limit local plant distribution. *Conservation Biology* **6**, 513–519.

Pywell, R.F., Bullock, J.M., Hopkins, A., Walker, K.J. & Burke, M.J.W. (2002) Restoration of species-rich grassland on ex-arable land: a multi-site approach. *Journal of Applied Ecology* (in press).

Quintana-Ascencio, R.F. & Menges, E.S. (1996) Inferring metapopulation dynamics from patch-level incidence of Florida scrub plants. *Conservation Biology* **10**, 1210–1219.

Rees, M. & Paynter, Q. (1997) Biological control of Scotch broom: modelling the determinants of abundance and the potential impact of introduced insect herbivores. *Journal of Applied Ecology* **34**, 1203–1221.

Robinson, G.R. & Handel, S.N. (1993) Forest restoration on a closed landfill—rapid addition of new species by bird dispersal. *Conservation Biology* **7**, 271–278.

Rose, R.J., Webb, N.R., Clarke, R.T. & Traynor, C.H. (2000) Changes on the heathlands in Dorset, England, between 1987 and 1996. *Biological Conservation* **93**, 117–125.

Rusch, G. (1992) Spatial pattern of seedling recruitment at two different scales in a limestone grassland. *Oikos* **65**, 433–442.

Schupp, E.W., Howe, H.F. & Auspurger, C.K. (1989) Arrival and survival in tropical treefall gaps. *Ecology* **70**, 562–564.

Silvertown, J. (2001) Plants stand still, but their genes don't: non-trivial consequences of the obvious. In: *Integrating Ecology and Evolution in a Spatial Context* (eds J. Silvertown & J. Antonovics), pp. 3–20. Blackwell Science, Oxford.

Spellerberg, I.F. (ed.) (1996) *Conservation Biology.* Longman, Harlow, UK.

Stöcklin, J. & Bäumler, E. (1996) Seed rain, seedling establishment and clonal growth strategies on a glacier foreland. *Journal of Vegetation Science* **7**, 45–56.

Thomas, C.D. & Kunin, W.E. (1999) The spatial structure of populations. *Journal of Animal Ecology* **68**, 647–657.

Thompson, K. (2000) The functional ecology of soil seed banks. In: *Seeds: the Ecology of Regeneration in Plant Communities* (ed. M. Fenner), pp. 215–235. CABI, Wallingford, UK.

Tilman, D. (1994) Competition and biodiversity in spatially structured habitats. *Ecology* **75**, 2–16.

Valverde, T. & Silvertown, J. (1997) A metapopulation model for *Primula vulgaris*, a temperate forest understorey herb. *Journal of Ecology* **85**, 193–210.

Van der Meijden, E., Klinkhamer, P.G.L., de Jong, T.J. & van Wijk, C.A.M. (1992) Meta-population dynamics of biennial plants: how to exploit temporary habitats. *Acta Botanica Neerlandica* **41**, 249–270.

Verdú, M. & Garcia Fayos, P. (1996) Nucleation processes in a Mediterranean bird-dispersed plant. *Functional Ecology* **10**, 275–280.

Watkinson, A.R., Lonsdale, W.M. & Andrew, M.H. (1989) Modelling the population dynamics of an annual plant *Sorghum intrans* in the wet dry tropics. *Journal of Ecology* **77**, 162–181.

Whittaker, R.J. (1993) Plant-population patterns in a glacier foreland succession—pioneer herbs and later-colonizing shrubs. *Ecography* **16**, 117–136.

Willson, M.F. (1993) Dispersal mode, seed shadows, and colonization patterns. *Vegetatio* **108**, 261–280.

Willson, M.F. & Traveset, A. (2000) The ecology of seed dispersal. In: *Seeds: the Ecology of Regeneration in Plant Communities* (ed. M. Fenner), pp. 85–107. CABI, Wallingford, UK.

Wolf, A., Brodmann, P.A. & Harrison, S. (1999) Distribution of the rare serpentine sunflower, *Helianthus exilis* (Asteraceae): the roles of habitat availability, dispersal limitation and species interactions. *Oikos* **84**, 69–76.

Woolcock, J.L. & Cousens, R. (2000) A mathematical analysis of factors affecting the rate of spread of patches of annual weeds in an arable field. *Weed Science* **48**, 27–34.

Yao, J., Holt, R.D., Rich, P.M. & Marshall, W.S. (1999) Woody plant colonization in an experimentally fragmented landscape. *Ecography* **22**, 715–728.

Chapter 15
Biogeography and dispersal

Rob Hengeveld and Lia Hemerik

Introduction

This chapter evaluates the role of dispersal in biogeographical processes and their resulting patterns. We consider dispersal as a local process, which comprises the combined movements of individual organisms, but which can dominate processes even at the scale of continents. If this is correct, it is no longer possible to separate local ecological processes from those at broad, geographical scales. However, biogeographical processes differ from those happening in one or a few localities; at the broader scales, there are additional processes occurring which are only evident when examined from this wider perspective.

We integrate biogeography with ecology, explaining broad-scale effects, ranging from processes happening locally as the result of responses of individual organisms to perpetual changes in living conditions in heterogeneous space. The models to be used cannot be those traditional in population dynamics with a dispersal parameter plugged in, but must be spatially explicit. Only a broad-scale perspective of continual redistribution of large groups of individuals or reproductive propagules can give dispersal its biological and biogeographical significance. Our general thesis in this chapter is that adaptation in non-uniform space enables individuals to cope effectively with environmental variation in time.

In our analyses of spatially adaptive processes, we concentrate on principles rather than on details of specific phenomena, such as types of distance distribution. We therefore formulate these principles in terms of simple Poisson processes. In specific cases, these distributions can be replaced by more complex ones which may fit better. Our approach applies both to processes within a range that structure the range, and to external processes that limit it.

This view will be developed in terms of simple stochastic processes in space and time. Eventually, more complicated relations can be formulated to provide greater realism. The present simple approach is intended to set the scene, and assumes, for the time being, biological and spatial independence of individuals and species. Analytical models that would account for various forms of dependence are, for the present, mathematically intractable. In setting up this new framework, we return to data that in earlier publications (e.g. Hengeveld & Haeck 1981, 1982; Hengeveld 1990) were employed in causal explanations of some biogeographical patterns. Here, we have used them to explain functionally the process of continual spatial adaptation by a species.

To evaluate the possible nature and impact of spatial processes, we start by summarizing ideas on ranges with spatially stationary structure, and then look at insights gained from invasion research. We then concentrate on the spatial dynamics of a range, first in the context of biotope accessibility and waiting times to environmental recovery; after this, we briefly introduce a simple model with spatially stochastic movement. The next section discusses the results of a simulation, which highlights the rapidity of spatial adaptation. The fourth section is concerned with consequences of dispersal in the formation of theory in ecology and biogeography as two distinct disciplines; integration of biogeography with ecology has consequences for the development and evaluation of theory in both disciplines. Finally, we consider speciation, which may be viewed as a side effect of ecological adaptation in space, when conspecific organisms lose contact for a certain period of time; effects that can also be considered in the context of broader-scale spatial adaptation. Overall, we begin to model the dynamics of species' ranges, and their consequent structuring, limitation and fragmentation in simple, general terms.

Structure of species' ranges

Generally, a species reaches its highest abundance at the centre of its range. Its abundance, and the number of biotopes occupied, decline towards the range periphery (see Hengeveld & Haeck 1982; Brown 1984; Hengeveld 1990; Brown *et al.* 1995, 1996). However, although the highest abundance tends to be at the centre, abundances across a range often form a multimodal surface, the structure of which can change from one year to the next. Furthermore, species with larger ranges reach higher abundances at the centre of their range than those with smaller ranges (Hengeveld & Hogeweg 1979; Hengeveld & Haeck 1981). Their choice of biotope also differs across the range (see Walter & Walter 1953; Hengeveld 1990), supporting the possibility that preferred habitat conditions remain the same throughout the range.

This implies that the size and structure of the range of a species express the degree to which the species' physiology and ecology match the living conditions in different parts of this range (Hengeveld & Haeck 1981; Hengeveld 1990). Thus, the species' range represented by an optimum response surface across geographical space (Hengeveld & Haeck 1981; Bartlein *et al.* 1986) is comparable to local optimum curves and surfaces relative to environmental gradients (e.g. Gause 1930, 1932; Whittaker 1967; Kessel 1979; Hengeveld 1990). Ranges are spatially structured, as are ecological processes with respect to local gradients (Jerling 1985, 1988).

How can one understand in simple terms both the general structure of a species' range, and its regional and local dynamics? A dynamic form of gradient analysis, taken from agronomy, is helpful in answering this question.

Parry (1978) expressed the energy required by the oat *Avena sativa* to complete its life cycle by the annual number of growing degree days (GDDs). If it is large enough to fulfil the plant's minimum requirement, the plants produce viable seeds, whereas at lower GDDs they cannot. Then the population dies out, unless it has a sufficient seed bank. The risk of crop failure at any location can be determined by the percentage of years with an insufficient GDD. The one-dimensional gradient up a mountain slope of the proportion of years falling short in this requirement represents the species' altitudinal risk distribution and the distribution of the probability of failure across the mountain is the species' risk surface. Conversely, the spatial distribution of crop successes represents the temporal dynamics of the physiological optimum surface relative to some environmental gradient. Over time, this surface can shift (Parry & Carter 1985). In their structure and dynamics, such surfaces can determine the species' range in geographical space (e.g. Hengeveld 1990).

Thus, the annual local risks follow a certain probability distribution, which similarly describes the short-term, local dynamics of the species. Its short-term roughness expresses the spatial heterogeneity of the local conditions. Therefore, single measurements at single dates or locations are inadequate to determine how an organism's requirements match with its environment, and should be replaced by the construction of frequency distributions or time series with regard to the requirement concerned. However, over time these local conditions also vary independently, and statistically form more or less continuous gradients with smooth surfaces.

In this way, the location, size, shape, internal structure, bounds and dynamics of a range all reflect the response of the species to particular environmental variables, rather than to local chance distributions of a large number of variables (e.g. Brown *et al.* 1995). Apart from these few, identified variables, one also needs to understand the dispersal dynamics of species at various spatial scales.

Dynamics of species' ranges

Dispersal and accessibility

Locally, and at a certain frequency, populations die out due to unfavourable conditions, after which the localities vacated can be recolonized as soon as conditions improve. For this to happen, the locality must, of course, be accessible to individuals still living in surrounding biotopes. This means that these individuals should be able to traverse intervening unfavourable biotopes in sufficient numbers and at the right time. A recent invasion model describes this process (Hengeveld & Van den Bosch 1997). This model was derived from an earlier one by Van den Bosch *et al.* (1990, 1992), which assumes that the region which individuals can cross successfully is ecologically uniform. The velocity with which invasion progresses was thought to depend both on the rates of dispersal and of net reproduction. The latter, in turn, depends on age-specific rates of mortality and fertility. The parameter values are not constant and specific for a species, but change when environmental variables change. The resulting invasion velocity depends on responses to particular values of relevant environmental factors. Within the model conditions, the rates of population increase and spatial expansion are exponential.

However, this earlier model proved to be too simple. For example, the rate of progress of the muskrat, *Ondatra zibethicus*, during its invasion across Germany, was greater in areas with many unfavourable biotopes and under unfavourable weather conditions, than where conditions were more suitable (Schröpfer & Engstfeld 1983). Under unfavourable conditions, the dispersing individuals kept moving, thereby increasing the total distance travelled per unit time, and thus increasing the overall invasion velocity. In contrast, individuals tended to stay longer in areas with many favourable biotopes or under favourable conditions, which slowed the progress of the invasion.

This is exactly what the extended version of the model (Hengeveld & Van den Bosch 1997) shows. It takes into account the proportion of unfavourable biotopes within a non-uniform region. The invasion velocity increases with an increasing proportion of unfavourable biotopes. At the same time, though, individuals also suffer a higher mortality in unfavourable biotopes. In fact, with increasing proportions of unsuitable biotopes, the resulting increase in mortality rate eventually overtakes the increase in dispersal rate, implying that at some point the process of invasion will stop. Within this model, these two opposing processes therefore result in a non-linear relationship between the invasion velocity and the amount of favourable biotopes available (Fig. 15.1), irrespective of differences in net reproduction or dispersal risk.

The accessibility of a given biotope from another occupied biotope is given by the rate of movement multiplied by the time available, and divided by the distance separating them. It expresses the ability of individuals to traverse unsuitable biotopes

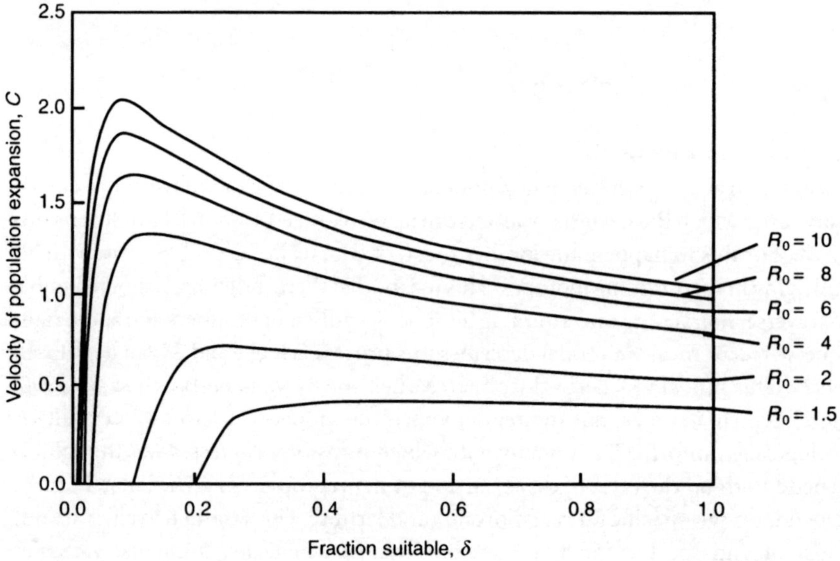

Figure 15.1 Invasion rate C as a function of the fraction of suitable biotope δ in non-uniform conditions for various values of the net reproduction rate R_0. (Adapted from Hengeveld & Van den Bosch 1997.)

and also the possibility that a vacant locality will be reached. This accessibility, therefore, follows a non-linear function, comparable with that mentioned above.

Biotope accessibility across a species' range

The range centre has not only the highest abundance of individuals, but also, as shown, the highest proportion of different biotopes occupied (Hengeveld & Haeck 1981). Therefore, from the range centre outwards, suitable biotopes become increasingly more fragmented, and the fragments of the species' living space become smaller and more widely dispersed. In fact, the average nearest-neighbour distance between biotopes increases exponentially with a decrease in the number of biotopes (e.g. Harris 1984), thus altering their accessibility. Accordingly, this exponential increase in distances and the exponentially distributed time it takes to recolonize a biotope together have a non-linear effect on biotope accessibility. Regions with fewer suitable biotopes and higher dispersal mortality will be recolonized much more slowly, if at all. Thus, biotope accessibility decreases dramatically towards the margin of the range, at some point limiting the range more or less abruptly (for examples, see Davis 1987). The range becomes sharply truncated as a function of rapidly increasing distances between biotope fragments and severity of the unfavourable conditions in the areas between fragments, causing dispersal mortality.

Waiting times to environmental recovery

The time taken for suitable conditions to return in vacated biotopes also plays a role in determining the structure and limitation of a species' range. This time can be considered a waiting time in the terminology of stochastic processes. The Poisson distribution is the simplest description of the number of realizations of particular events over time, and assumes that individual realizations are mutually independent. Often, as for daily temperature variation, this assumption does not apply; it does, however, apply to temperature changes among years within the scale of a few decades. In consequence, waiting times of a return of suitable conditions are exponentially distributed. By the same reasoning, the durations of favourable times are exponentially distributed; they frequently deteriorate, and only rarely remain favourable for a long time.

The means of the different exponential distributions of waiting times for favourable conditions will increase from the range centre towards the periphery; there, unfavourable conditions will last longer. They also become more frequent, since the duration of favourable conditions will be shorter. Within a particular timespan, suitable biotopes will be accessible for fewer and shorter periods. This expectation defines the time available for recolonizing vacant localities, and further constrains the range limit.

Structure and dynamics of range margins

A combination of several processes leads to well-defined abrupt range margins. Some of these processes involve exponential declines or increases. This means that margins may expand at an exponential rate when released through environmental improvement. This effect has actually been observed in the silver spotted skipper, *Hesperia comma*, in southeast England (Thomas *et al.* 2001; Wilson & Thomas, this volume). Thus, a margin which has been very well defined for a long time can become much less definite during times of range expansion (see Davis 1987).

It should be noticed that the structure and dynamics of the margin are determined by the very same processes thought to be operating throughout the range; only the parameter values vary. In some cases, only biotope accessibility limits the range, despite the fact that the local net reproduction rates can be greater than one (Carter & Prince 1985), whereas these are always less than one in models invoking demographic sinks (Pulliam 1988).

We now discuss the relevant, spatially stochastic processes in slightly more detail.

Dispersal movements of individuals

Brownian movements

The invasion models discussed above (Van den Bosch *et al.* 1990, 1992; Hengeveld & Van den Bosch 1997) assume that dispersal is Brownian. This is shown to apply to the expansion wave of the collared dove, *Streptopelia decaocto*, in Europe.

For this species, the locations of both birth and first nesting were mapped (Fisher 1953), after which the distances between each pair of locations were measured and depicted as a frequency distribution (Fig. 15.2) (Hengeveld 1993). Logarithmic transformation of the resulting J-shaped frequency curve—assuming an exponential decay rate with distance (Neubert *et al.* 1995; Kot *et al.* 1996)—results in a more linear curve (Fig. 15.2a). This curve only becomes linear when one takes the square root of the logged distances (Fig. 15.2b). This latter transformation suggests that the dispersal underlying the range expansion is Brownian, because in Brownian movement the distance as the crow flies equals the square root of the sum of distances moved.

These curves show that most birds breed near their parents' nest, whereas only a few engage in long-distance dispersal. Thus, they establish beach-heads which subsequently grow exponentially by short-distance movements and quadratically by long-distance behaviour (for examples, see Hengeveld 1989). These two types of dispersal together constitute hierarchical or stratified diffusion (Hengeveld 1989). Shigesada's models (Shigesada & Kawasaki, this volume) show how the initial population grows exponentially, whereas long-distance dispersers originating from them are at first too few to start a new beach-head because of Allee effects. Only after the initial population reaches a certain critical size can the number of long-distance dispersers become effective, thus initiating the high-velocity expansion phase. The same Allee effects can constrain further range extension at the periphery when populations are too small, are too short lived, or produce too few propagules because of reduced net reproduction. This once more truncates the range limit.

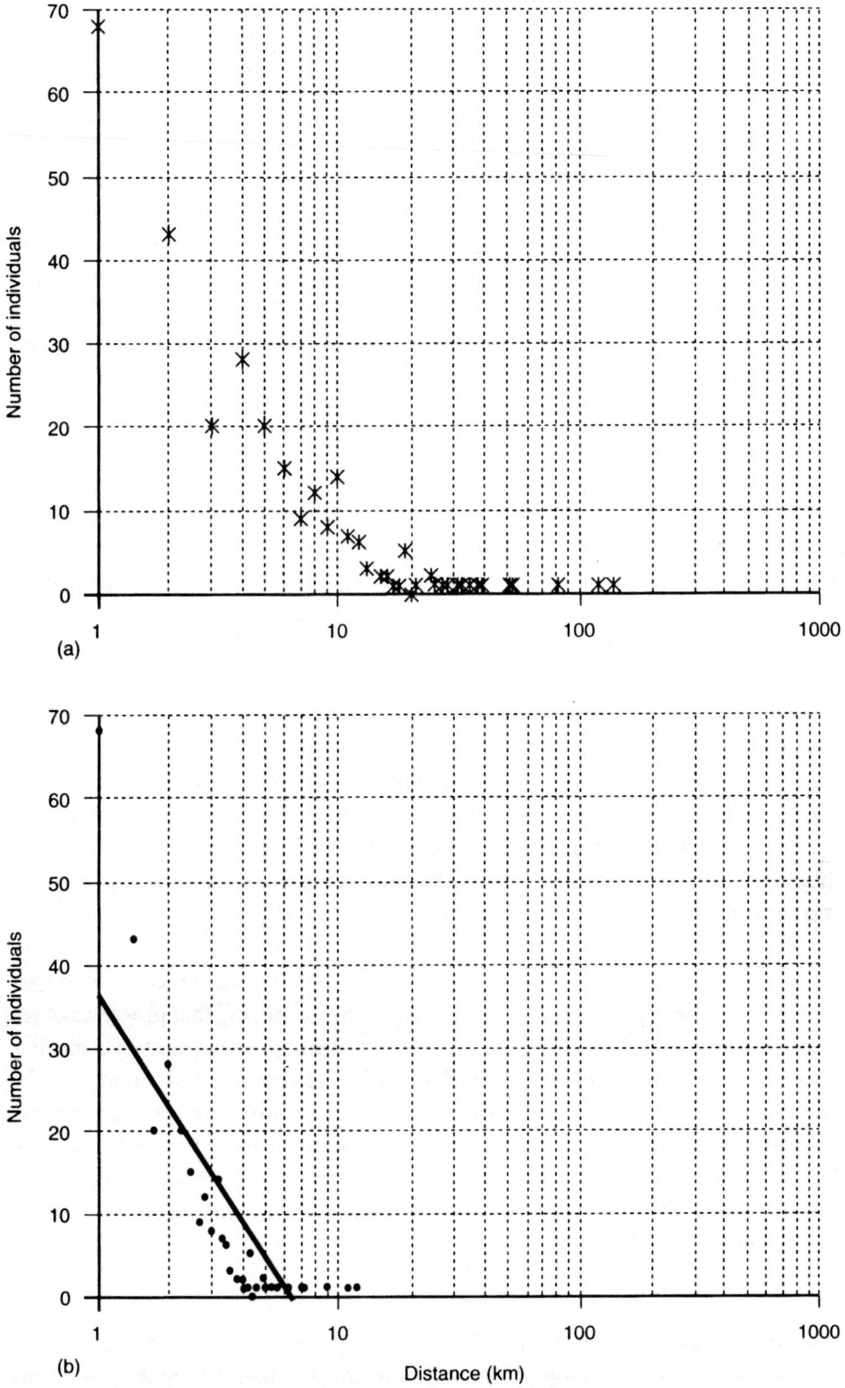

Figure 15.2 Number of breeding birds as a function of the distance from their parent's nest: (a) the logged distances, and (b) the square roots of the logged distances.

Markov chains

Range expansions can be assumed to follow Brownian diffusion, whereas non-expansive spatial movement can be viewed as a Markov process. Such a model describes a dynamic steady state resulting from stochastic movements of individual organisms across a region with ecological conditions that are not uniformly distributed.

For a Markovian process, the region can be represented by a regular lattice, in which each cell is allotted a particular biotope quality with respect to a particular species. Individuals stay longer in a cell of higher biotope quality than in one of poor quality. In the high-quality cells, individuals may still move, but over shorter distances than in cells of unfavourable biotopes (see, for example, Baars 1979). The consequence of the longer residence time in the favourable biotopes is that, according to the Fokker–Planck equation (see Okubo 1980; Kareiva 1982; Turchin 1998), more individuals will be found in these high-quality biotopes. Thus, the size of the congregations of individuals becomes proportional to the local biotope quality.

The resulting distribution of densities remains stationary as long as the configuration of biotope qualities remains the same, despite continual movement of individuals. These continued movements result in a spatially differentiated turnover of individuals across the area as a whole. However, if this spatial configuration changes, the abundances of individuals follow suit, a process known as habitat tracking. Thus, in dynamic environments the spatial distribution of individuals adapts rapidly to the ever-changing living conditions, with more dynamic environments containing more mobile individuals or species (e.g. Lindroth 1949; Southwood 1962). Yet, in contrast to models of (meta)population dynamics based on the logistic equation (e.g. Hanski 1991), the local congregations are open and temporary, and cannot be treated as closed populations with particular properties that have evolved locally. They are open systems, remaining stochastically in a spatially dynamic equilibrium with the ever-changing living conditions.

This spatial view of ecological response also has repercussions for the possibility of species' interactions. In order to have any lasting effect, interactions between species should be in phase, both in time and in space, implying that processes of response to physical factors and to other species should happen at the same scale. For species to remain coherent, this means that mechanisms must operate that tune the individuals to changes in time and space and to each other. An absence of these mechanisms, particularly of spatial tuning in allopatry, is likely to result in speciation. As, in general, no heritable tuning mechanisms exist between individuals of different species, their ecological behaviour becomes individualistic (i.e. species specific and independent).

Simulating range dynamics

Ranges persist under changeable conditions only when favourable and unfavourable biotopes alternate in space. Habitat tracking under such circumstances ideally results when the habitat conditions preferred by individuals remain more or less constant, the more mobile species having the more sharply delimited habitat association (Mayr 1942).

We simulated the process of establishment of a species range by randomly placing 150 individuals in a grid of 128 × 128 cells, using a toroidal grid to avoid edge effects. These individuals thus filled *c.* 1% of their world. Each randomly placed individual had the same probability of 0.2 of moving one cell to the left, right, up or down, or of remaining stationary. After 20 steps, each individual produced two offspring when in a favourable biotope, but in unfavourable biotopes the individual died and no offspring was produced. No upper limit was set to the total number of individuals per biotope. The subsequent 15 generations followed the same process. Next, we defined zones of different favourability, expressed by the proportion of favourable biotopes. The biotopes in the zone near the 'equator' of the torus were 100% favourable, whereas 'polar' biotopes were all unfavourable, with a gradient of partially suitable zones in between. For n zones, the $2n$ bands had an approximate width of $128/2n$; the proportion of randomly placed favourable biotopes increasing from pole to equator with step sizes of $100/(n-1)$%.

How quickly can species with a mean displacement of four grid cells per generation establish in randomly placed favourable biotopes? It appears that the individuals occupy an exponentially increasing proportion of favourable biotopes (Fig. 15.3). Under these non-zonal conditions, the range of a species is established when at least 20% of the biotopes are favourable (Fig. 15.4).

For the zonally arranged biotopes, we simulated 3–6 zones with different proportions of favourable biotope. For six zones, the species' range had built up clearly after

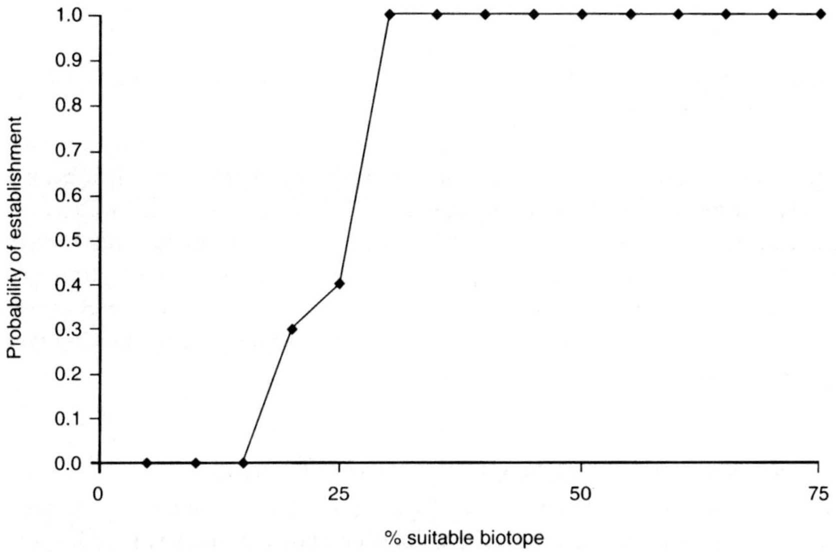

Figure 15.3 Simulated probability of establishment as a function of the percentage of suitable biotope in non-uniform conditions (see text for further details).

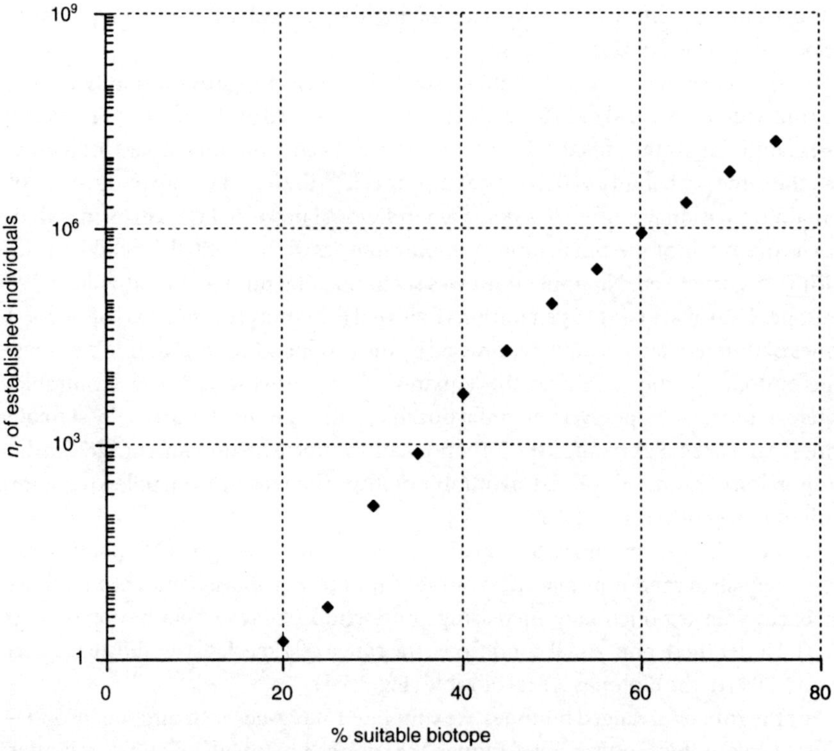

Figure 15.4 Simulated number of established individuals as a percentage of biotope occupied in non-uniform conditions (see text for further details).

as few as 3–6 generations (Fig. 15.5), which also happened in simulations with a smaller number of zones.

So far, we have assumed all animals stay alive during dispersal, an assumption that can be changed in future simulations by incorporating a constant survival probability in each dispersal step. Also, our biotopes were assumed to be constant, which can be relaxed by increasing the turnover of favourable biotopes from the range centre towards the periphery. These assumptions complicate the calculations, although they simplify the spatially adaptive process. However, without these and other complicating assumptions, this simple model serves our purpose by showing the rapidity of spatial adaptation.

Consequences of a spatially dynamic range structure

Surprisingly, for most of their history, both ecology and biogeography have had predominantly spatially static outlooks. In ecology, local populations have mainly been studied as if the processes analysed in one or a few spatially closed (meta)populations or communities were representative of all those found throughout the range of

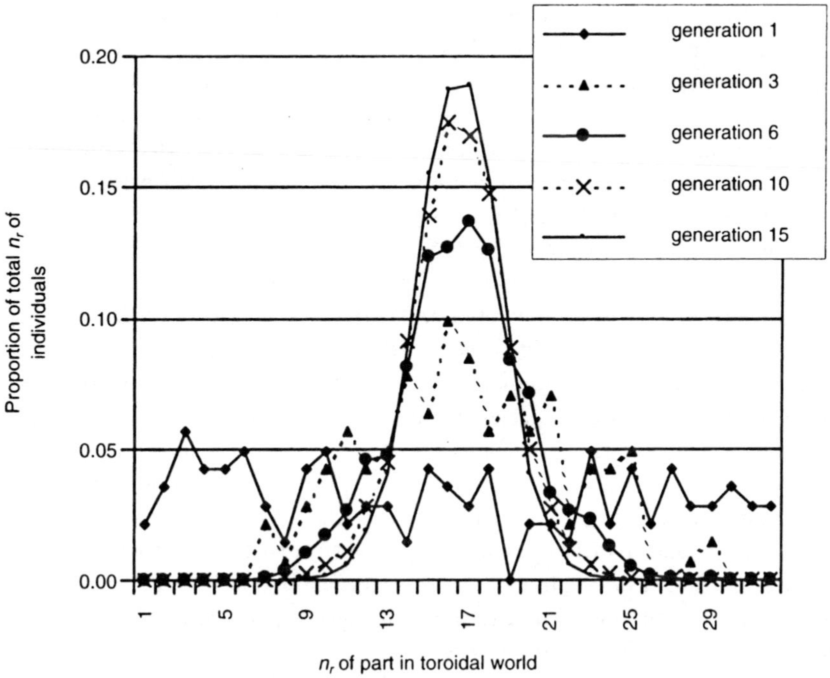

Figure 15.5 Simulated percentage of individuals found across six biotope zones after different numbers of generations (see text for further details).

a species. Similarly, in biogeography, ranges have usually been represented on maps by their outlines only, or as an unstructured shape. According to Ives and Klopfer (1997), this even applies to Brown *et al.* (1995), although they suggested that spatial variation in abundance might require temporal variation, whereas we suggest the opposite direction of dependency. In the next two sections, we look at the impacts the two disciplines have on each other, given the great significance dispersal has for both.

Impact of ecology on biogeography

MacArthur and Wilson's (1967) *The Theory of Island Biogeography* showed that adding dispersal processes permits a more dynamic approach to biogeography. However, vicariance and cladistic biogeography (e.g. Rosen 1978; Nelson & Rosen 1981) still ignore the impact dispersal might have had on the development of geographical patterns, and concentrate on geological (non-biological) explanations (see Nelson 1978).

Five successful approaches

Since MacArthur and Wilson (1967), five developments have taken place: in the anatomy of species ranges, in agronomy, in Quaternary ecology, in research on climatic change and, finally, in invasion research. The first two have already been described and may be summarized thus: (i) larger ranges have higher abundances at the centre than small ranges; the most likely explanation of this phenomenon is based on a physiological optimum response to broad-scale variation in environmental variables (Hengeveld & Haeck 1981, 1982; Hengeveld 1990; Mac Nally 1995); and (ii) there are risks of failure to reproduce, and therefore of local extinction. As shown, these two points imply that a range can be represented by a dynamic distribution of risk in space, continually changing in structure, shape and location.

The third development is in Quaternary ecology. Since the early 1980s, local pollen diagrams have been integrated into geographical maps (e.g. Davis 1981; Huntley & Birks 1983; Bartlein *et al.* 1986). Although pollen diagrams had previously represented the percentage abundance of all pollen taxa, information now showed the distribution of individual species at certain times. Further, the spatial expansion, contraction, fragmentation or coalescence of their ranges was shown in series of maps for successive periods of, for example, 500 or 1000 years, or from plotting successive range margins on a single map (e.g. Davis 1981). Maps like these showed the behaviour of species in various respects: the refugia from which they spread, and the rate and direction of spread. Individual species differ not only in these respects, but also in temporal variation. These results challenged the spatially static picture in biogeography prevailing up to the 1980s.

Modelling shifts in species' ranges due to climate change (see Watkinson & Gill, this volume) was the fourth development in ecology that affected biogeographical thinking. This development relates to those concerning dynamic optimum surfaces and Quaternary range shifts. In order to predict these shifts, information on the ecological preferences of species was used in models, together with assumed changes in climate (e.g. Sykes & Prentice 1995). Monitoring shows that the ranges of several species are already shifting, and not necessarily in concert (e.g. Parmesan *et al.* 1999; Both & Visser 2001). Also, holes within a range can fill up when climatic conditions improve (Eber & Brandl 1994).

Finally, dispersal is the central issue in invasion research (see Hengeveld 1989; Shigesada & Kawasaki 1997; Shigesada & Kawasaki, this volume), in which models are constructed (Mollison 1977; Okubo 1980; Van den Bosch *et al.* 1990) and tested against observed range expansions (Van den Bosch *et al.* 1992; Lensink 1997, 1998). Here, ability to invade and the rate of spatial spread are assumed explicitly to be completely dependent on the values of ecological parameters (e.g. Hengeveld 1994; Hengeveld & Van den Bosch 1997). Thus, not only does the environment change over geographical space, altering local rates of survival and fertility for different species, but so do dispersal rates.

Impact of biogeography on ecology

Despite the large data bases available on diseases like measles (Infantosi 1986; Cliff *et al.* 1981; Cliff & Haggett 1988), influenza (Cliff *et al.* 1986; Patterson 1986) and HIV (Gould 1993), as well as in agricultural epidemiology (Campbell & Madden 1990), the impact of processes at a geographical scale on local disease outbreaks has not yet been fully explored (see, however, Cliff *et al.* 1981). For example, Anderson and May (1992) in their important text, *Infectious Diseases of Humans*, used spatially non-explicit demographic models (for recent spatial models, see Mollison 1995), and hardly discussed spatial processes.

Biotope variation, habitat constancy

A part of metapopulation theory, as derived from MacArthur and Wilson's (1967) biogeographical models, is concerned with the rescue effect (Brown & Kodrick-Brown 1977; Hanski 1982, 1999), rather than with biotope accessibility. The 'rescue effect' assumes that the habitat conditions in all biotope fragments are the same. However, if there are differences in habitat quality, we suggest that an alternative effect, which we dub the 'Santa Claus effect', is more likely to occur. Because individuals congregate in locations with more favourable living conditions, the larger populations become even larger and the smaller, smaller (Verboom 1996). In fact, this effect applies to the Markovian process described above. There have been no empirical studies concerning the construction of distance distributions based on local biotope accessibility, nor concerning changes in accessibility over geographical space towards the margins of species' ranges. In the Netherlands, however, it has been shown that ground beetles occurring in the margins of the species' range are less likely to colonize recently constructed polders than those occurring in the centre (Fig. 15.6).

Gleason (1926), realizing that the locations and shapes of ranges are species specific, long ago introduced the individualistic concept of the ecological behaviour of species (for recent data, see Taper *et al.* 1995). In this, he rejected the idea that species evolve and behave locally as members of a community, to the extent that communities—and, within them, populations as well—might be considered to be evolved entities in their own right (e.g. Emlen 1973; Pianka 1994). In Gleason's formulation, the properties of a species are species-wide, rather than connected with some niche, and locally evolved in the context of a particular community. This formulation is supported, for example, by the facts that a species occupies a different number of biotopes in different parts of its geographical range, and that there is species-wide constancy in habitat preferences (see above).

This habitat constancy implies that dispersal mortality increases towards the range margin, which gives further geographical structure to the demography of a species. Furthermore, there will be mortality due to geographical differences in the recovery time of biotopes, which increases towards the margins, as well as reduced fertility, natality, etc. Local estimates of demographic processes, therefore, lack generality (e.g. Cwynar & MacDonald 1987).

Variable demographic parameters

In fact, even parameters sometimes have to be redefined to include effects of spatially non-uniform living conditions. This applies to the intrinsic rate of natural population growth, r, the value of which varies with the proportion of intervening unfavourable biotopes in an area. Figure 15.7 shows this variation, together with the

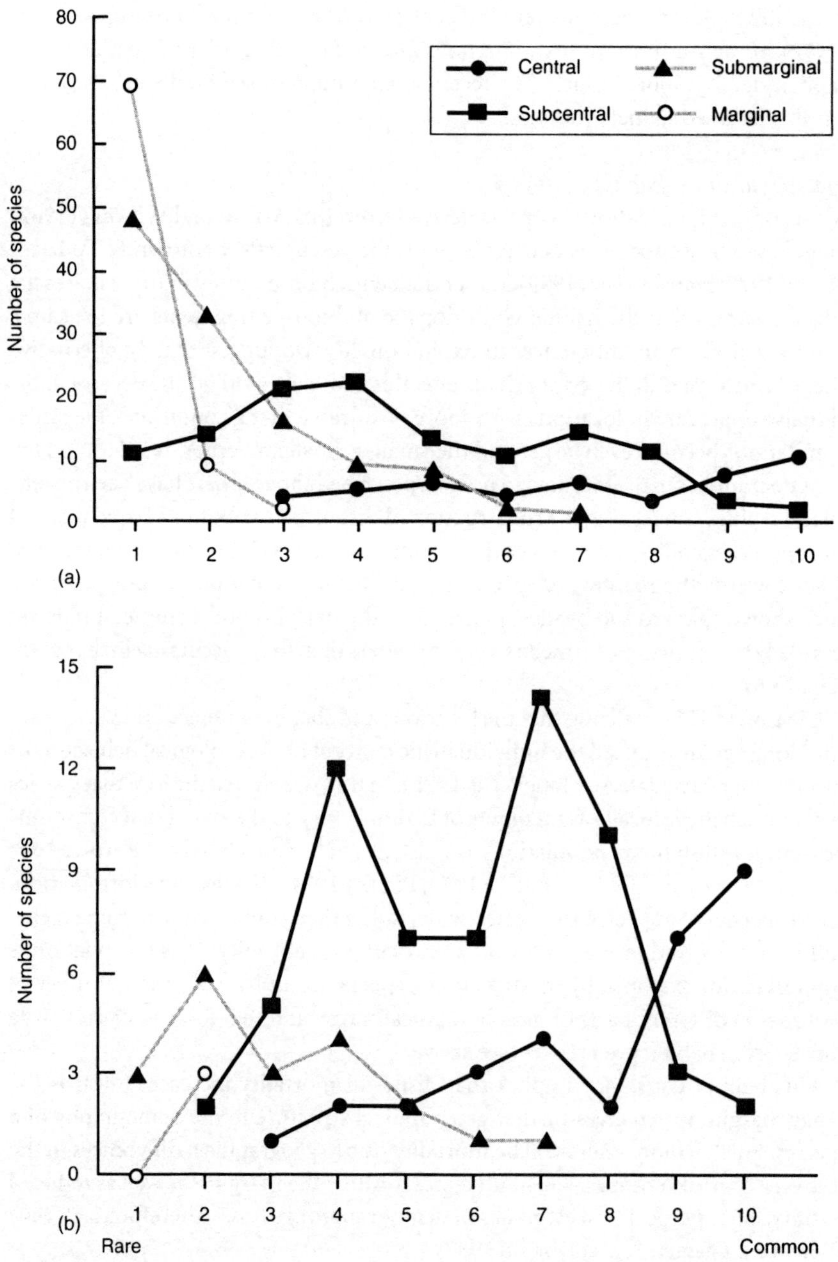

Figure 15.6 The number of ground beetle species of four different range categories in: (a) the Netherlands, and (b) large, newly-reclaimed inland polders within the Netherlands. (From Turin 2000, with permission of the author.)

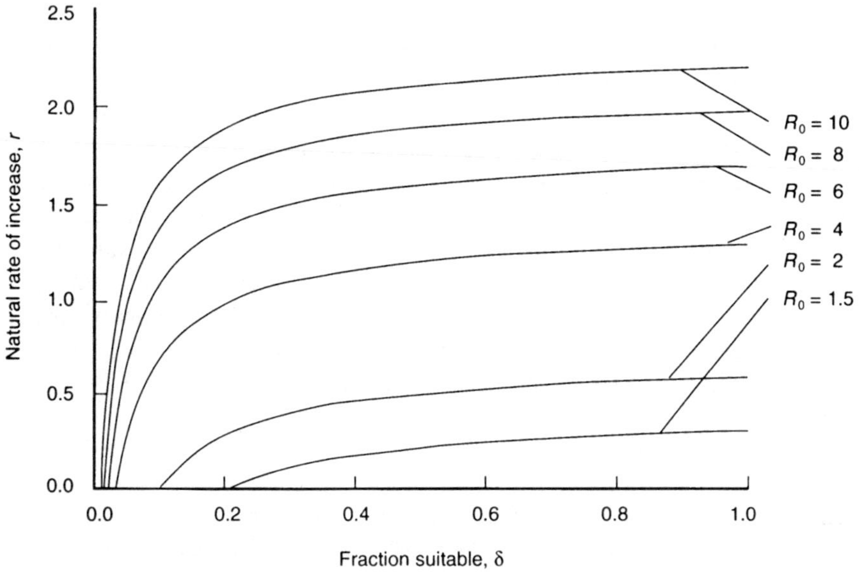

Figure 15.7 Rate of population growth, *r*, as a function of the fraction of suitable biotope δ in non-uniform conditions for various values of net reproduction rate, R_0. (Adapted from Hengeveld & Van den Bosch 1997.)

effect differences in risk during dispersal may have. Similar differences in the level of variation are found in net reproductive rate (Hengeveld & Van den Bosch 1997). Therefore, particular values of *r* have no general relevance, but need to be understood in the context of the spatial heterogeneity of living conditions and of local responses to them. These responses vary with the dynamics and quality of living conditions, which themselves vary in time and across the range of the species.

Geographical differences in abundance across the range cannot easily be interpreted ecologically either, as abundance drives demographic processes among species in communities. Many demographic models assume that processes are density dependent, population numbers being kept within certain specific bounds. These mechanisms become difficult to visualize when the level of numerical fluctuation varies both between different species at the centres of their ranges, as well as across the range of each of them. Therefore, like the number of biotopes occupied in different parts of a range, local estimates of the populations of species cannot be interpreted generally. General models, as well as ecological theory, based on local rather than geographical observations are liable to error.

This has far-reaching theoretical consequences. When a range shifts, for example, one cannot distinguish whether differences observed in local abundance are deviations from the mean, or changes in the mean value itself. This distinction is basic to demographic theory, in which the mean level of fluctuation is assumed to be determined by mechanisms and variables different from those that control the extent of deviations from this mean (Solomon 1949).

Communities as entities

Shifts in the ranges of different species do not take place at equal rates, nor in concert (Davis 1981). Each species responds independently and at its own rate (see Van den Bosch *et al.* 1992) to ecological variation, depending on specific features of its physiology, life history and dispersal rate and direction. The geographical behaviour of a species reflects the ecological responses of individuals to local conditions constituting a Markov chain. Each species thus continually finds itself among other species, all having independent abundances, determined geographically and subject to kaleidoscopic shifts. Therefore, species comprising a community in one locality are replaced by other species at other times and places. Thus, all interactions between species are continually changing, both quantitatively and qualitatively; in consequence, it is not possible for them to coadapt (Davis 1981, 1986; Walter & Paterson 1995).

In fact, because of the individual-based, specific responses of species to continual changes in ecological conditions, communities that may once have seemed well defined gradually change their character, eventually dissolving completely (see Watkinson & Gill, this volume). Huntley (1988) reported that plant communities well established today cannot be found anywhere in the pollen record before *c.* 7000 years ago. This is also seen geographically. Starting from a particular point in space, the vegetation composition gradually changes in all directions (Fig. 15.8); as one species drops out, it is replaced by one or more others (Hengeveld 1997). The temporally non-analogue vegetation types of Huntley and Birks (1983) therefore have their geographical counterpart. Communities are local snapshots of a spatiotemporally fluid system, having no status as distinct biological entities. They are open, dynamic steady-state systems with a certain turnover rate in terms of species, depending on the specific matching of each species individually with the dynamics of its local environment.

Populations as entities

Ecological theory assumes that part of the adaptations of a species to the local environment is demographic, and the remainder is genetic (e.g. Davis & Shaw 2001). Ecological models traditionally exclude spatial adaptation, not recognizing habitat tracking by permanent movement of individuals as an integral part of the response process. However, given the low speed of (population) genetic adaptation relative to the rapid changes in the environment, which are spatially and temporally unpredictable, habitat tracking by random dispersal of individual organisms seems the most likely response to environmental change. This concerns habitat selection by individuals rather than natural selection of individuals. If habitat tracking may take place at a specifically appropriate scale, it is possible for a species to be genetically rigid, with properties static over time (Eldredge and Gould's (1972) periods of stasis) and species-wide in space (Hengeveld 1994). Thus, giving dispersal due weight in the ecological approach, one can envisage a continuous stochastic spatial adaptation to ever-changing conditions.

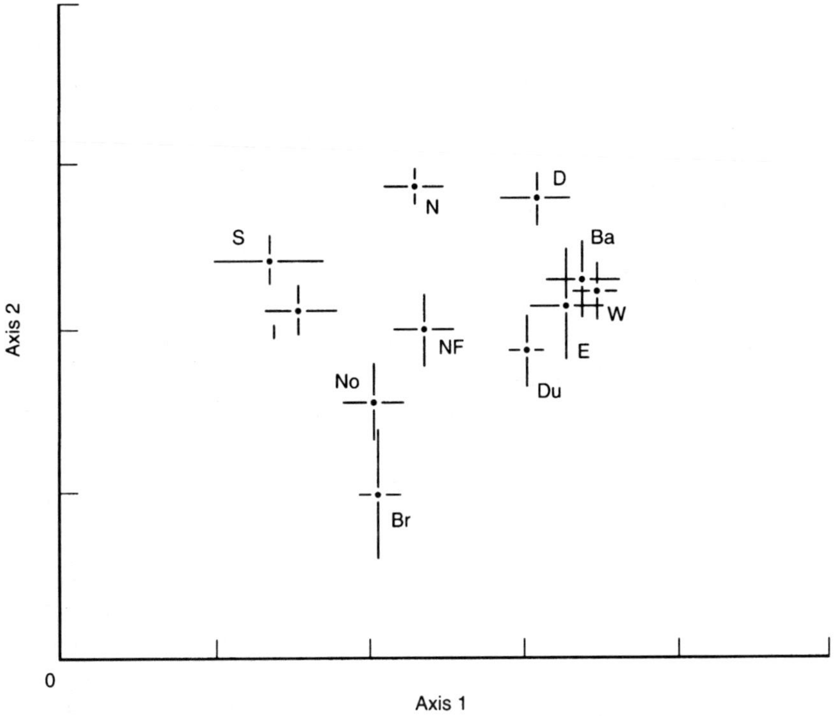

Figure 15.8 Geographical variation in a dune community of dry grasslands throughout northwestern Europe expressed as an ordination of species composition. Ba, Baltic countries; Br, Brittany; D, Denmark; Du, Dune district, western Netherlands; E, England; I, Ireland; N, Normandy; NF, northern France; No, Norway; S, Scotland; W, Wadden district, northern Netherlands. (Adapted from Hengeveld 1994.)

This implies that populations do not exist either, if they are defined as discrete, spatially closed units with a minimum of interchange of individuals, and with specific demographic attributes on which selection can operate. Because of the continual stochastic movements of individuals, congregations of individuals are temporary; but, like communities, they are compositionally steady-state systems with a certain turnover of individuals. The rate of turnover depends on life-history characteristics of individual organisms, such as longevity, fertility and dispersal capacity; and these are influenced or determined by ecological conditions. This turnover rate must match the rate of spatiotemporal change of the environment to prevent the local extinction of the species through spatial rigidity.

In spatially dynamic systems, the individual, rather than the population or the community, is the basic ecological unit. Recognition of this is the direct consequence of extending the spatial and temporal dimensions of observation, and accounting for dispersal at both broad and fine scales of ecological adaptation.

Speciation

How do species persist as identifiable entities with species-wide properties in an extremely loose system of fluid sets of individual organisms, all moving about separately, and mostly unaware of where they are going or of each other's existence? Are they facing the same fate as other supraindividual entities like populations and communities? These questions are particularly acute because the currently accepted model of speciation depends on local partitioning of niches by the erection of various sorts of reproductive barriers against sympatric competitors (Dobzhansky 1937). Accordingly, Mayr (1982) included the occupation of a specific niche as a species characteristic (cf. Hengeveld 1988). As a local phenomenon among individuals and species living sympatrically, though, it would result in allopatry, speciation being an allopatric process (e.g. Mayr 1963). This concept of speciation fits the demographic, non-spatial paradigm of ecology (Hengeveld & Walter 1999; Walter & Hengeveld 2000).

Paterson formulated an alternative model of speciation (e.g. Paterson 1985), which accords with our spatially dynamic approach. Particularly in their reproductive period, the sexes would develop similar habitat preferences, congregating in a specific biotope for mating. Apart from this requirement for reproduction, they also have a fine-tuned life cycle for synchronizing their arrival times at the selected biotope. Finally, the sexes can recognize each other as potential mates with some precision through morphological, behavioural, acoustic, chemical or physiological signals. Because each of these three components results in stabilizing selection, the properties of a species would remain uniform. Habitat tracking by individuals thus leads directly to the stability of species-wide properties and to species identity and rigidity. This, in turn, leads to constraints in variation, and in the structure, limitation and dynamics of species' ranges.

If, however, a number of individuals live for some time under deviating conditions in allopatry—implying that, because they have no genetic interchange, they are lacking the checks of stabilizing selection—they can change and eventually form one or more new species. Thus, species reach and maintain qualitative uniformity and stability, although spatially they are in a dynamic steady state like populations and communities.

This speciation process not only accords with the spatially dynamic view we develop, but it also explains several phenomena of central interest that otherwise remain elusive.

Some new research questions

This approach opens up a vast field of new research. For example, knowledge of risk distributions along transects, or in two-dimensional space at various scales, is badly needed. This may be combined with a study of the local differentiation of ecological processes and its causation. Also, life-history and life-cycle phenomena, together with dynamic biotope association in both their causal and functional aspects, conjoin with other aspects of ecobiogeographical research. Other relevant fields for new research concern the coupling of internal and external processes at the margins of ranges. Spatial risk, expressed by the variation in biotope accessibility or by ecological barriers and their effects, is important in this. Finally, the resulting species-wide, individualistic species behaviour leads to allopatric speciation. Dispersal is central to all these new research questions.

Conclusions

For a long time, ecology and biogeography have stood apart from each other, and have developed independently. However, neither discipline has explicitly adopted dispersal as a significant process for explaining the observed patterns. This remained so even during the integration of biogeography into ecology following MacArthur and Wilson's (1967) groundbreaking study; at best, dispersal represented an additional parameter in the analysis of causes of the assumed stability of local populations. In this approach, dispersal connects islands or locations, but within these the former models and theories still apply (e.g. MacArthur 1972; Brown 1995). The theoretical basis of ecology in which population dynamics were, in the main, spatially static, was amplified by the addition of dispersal, without it being integrated into a new approach. Part of biogeographical theory, in turn, became more dynamic and causative as ecology expanded into its territory, thereby forming ecological biogeography.

Here, we have integrated these two disciplines on the basis of the dispersal of individual organisms. This leads to hypotheses in the form of simple mathematical models of spatially dynamic responses of individuals to ever-changing environmental conditions. Because of spatial heterogeneity, individuals track their preferred conditions, thus dynamically matching their requirements with these conditions in space. Under spatially uniform conditions, the individuals must adapt either physiologically or genetically, or by developing special protective traits. If individuals do not do this quickly enough, they will die out. Habitat tracking in heterogeneous environments is the only way in which to survive under conditions changing rapidly relative to other forms of biological adaptation.

Integration of these two disciplines has significant consequences for the formation of theory in both, as well as in speciation theory. Their separation arises from the different spatial scales considered. Therefore, in this process of integration, with all its consequences for theories and basic concepts, dispersal plays a pivotal role.

References

Anderson, R.M. & May, R.M. (1992) *Infectious Diseases of Humans.* Oxford University Press, Oxford.

Baars, M.A. (1979) Patterns of movement of radioactive carabid beetles. *Oecologia* **44**, 125–140.

Bartlein, P.J., Prentice, I.C. & Webb, T. (1986) Climatic response surfaces for some eastern North American pollen types. *Journal of Biogeography* **13**, 35–57.

Both, C. & Visser, M.E. (2001) Adjustment to climate change is constrained by arrival date in a long-distance migrant bird. *Nature* **411**, 296–298.

Brown, J.H. (1984) On the relationship between abundance and distribution of species. *American Naturalist* **124**, 225–279.

Brown, J.H. (1995) *Macroecology.* University of Chicago Press, Chicago.

Brown, J.H. & Kodrick-Brown, A. (1977) Turnover rates in insular biogeography: effect of immigration and extinction. *Ecology* **58**, 445–449.

Brown, J.H., Mehlman, D.W. & Stevens, G.C. (1995) Spatial variation in abundance. *Ecology* **76**, 2028–2043.

Brown, J.H., Stevens, G.C. & Kaufman, D.M. (1996) The geographical range: size, shape, boundaries, and internal structure. *Annual Review of Ecology and Systematics* **27**, 597–623.

Campbell, C.L. & Madden, L.V. (1990) *Introduction to Plant Disease Epidemiology.* Wiley, New York.

Carter, R.N. & Prince, S.D. (1985) The geographical distribution of prickly lettuce (*Lactuca serriola*) I. A general survey of its habitats and performance in Britain. *Journal of Ecology* **73**, 27–38.

Cliff, A.D. & Haggett, P. (1988) *Atlas of Disease Distributions.* Blackwell, Oxford.

Cliff, A.D., Haggett, P., Ord, J.K. & Versey, G.R. (1981) *Spatial Diffusion.* Cambridge University Press, Cambridge.

Cliff, A.D., Haggett, P. & Ord, J.K. (1986) *Spatial Aspects of Influenza Epidemics.* Pion Ltd, London.

Cwynar, L.C. & MacDonald, G.M. (1987) Geographical variation of lodgepole pine in relation to population history. *American Naturalist* **129**, 463–469.

Davis, M.B. (1981) Quaternary history and the stability of forest communities. In: *Forest Succession* (eds D.C. West, H.H. Shugart & D.B. Botkin), pp. 132–153. Springer Verlag, New York.

Davis, M.B. (1986) Climatic instability, time lags, and community disequilibrium. In: *Community Ecology* (eds J. Diamond & T.J. Case), pp. 269–284. Harper & Row, New York.

Davis, M.B. (1987) Invasions of forest communities during the Holocene: beech and hemlock in the Great Lakes region. In: *Colonization, Succession and Stability* (eds A.J. Gray, M.J. Crawley & P.J. Edwards), pp. 373–393. Blackwell Scientific Publications, Oxford.

Davis, M.B. & Shaw, R.G. (2001) Range shifts and adaptive responses to Quaternary climate change. *Science* **292**, 673–679.

Dobzhansky, Th. (1937) *Genetics and the Origin of Species.* Columbia University Press, New York.

Eber, S. & Brandl, R. (1994) Ecological and spatial patterns of *Urophora cardui* (Diptera: Tephrididae) as evidence for population structure and biogeographical processes. *Journal of Animal Ecology* **63**, 187–199.

Eldredge, N. & Gould, S.J. (1972) Punctuated equilibria: an alternative to phyletic gradualism. In: *Models in Paleobiology* (ed. T.J.M. Schopf), pp. 82–115. Freeman, San Francisco.

Emlen, J.M. (1973) *Ecology: an Evolutionary Approach.* Addison-Wesley, Reading, MA.

Fisher, J. (1953) The collared turtle dove in Europe. *British Birds* **56**, 153–181.

Gause, G.F. (1930) Studies on ecology of the Orthoptera. *Ecology* **11**, 307–325.

Gause, G.F. (1932) Ecology of populations. *Quarterly Review of Biology* **7**, 27–46.

Gleason, H.A. (1926) The individualistic concept of the plant association. *Bulletin of the Torrey Botanical Club* **53**, 1–20.

Gould, P. (1993) *The Slow Plague.* Blackwell, Oxford.

Hanski, I. (1982) Dynamics of regional distribution: the core and satellite species hypothesis. *Oikos* **38**, 210–221.

Hanski, I. (1991) Single-species metapopulation dynamics: concepts, models and observations. In: *Metapopulation Dynamics, Empirical and Theoretical Investigations* (eds M. Gilpin & I. Hanski), pp. 17–38. Academic Press, London.

Hanski, I. (1999) *Metapopulation Ecology.* Oxford University Press, Oxford.

Harris, L.D. (1984) *The Fragmented Forest: Island Biogeography Theory and the Preservation of Biotic Diversity.* University of Chicago Press, Chicago.

Hengeveld, R. (1988) Mayr's ecological species criterion. *Systematic Zoology* **37**, 47–55.

Hengeveld, R. (1989) *The Dynamics of Biological Invasions.* Chapman & Hall, London.

Hengeveld, R. (1990) *Dynamic Biogeography.* Cambridge University Press, Cambridge.

Hengeveld, R. (1993) What to do about the North American invasion by the collared dove? *Journal of Field Ornithology* **64**, 477–489.

Hengeveld, R. (1994) Biogeographical ecology. *Journal of Biogeography* **21**, 341–351.

Hengeveld, R. (1997) Impact of biogeography on a population-biological paradigm shift. *Journal of Biogeography* **24**, 541–547.

Hengeveld, R. & Haeck, J. (1981) The distribution of abundance. II. Models and implications. *Proceedings of the Koninklijke Nederlandse Akademie van Wetenschappen* **C84**, 257–284.

Hengeveld, R. & Haeck, J. (1982) The distribution of abundance. I. Measurements. *Journal of Biogeography* **9**, 303–316.

Hengeveld, R. & Hogeweg, P. (1979) Cluster analysis of the distribution patterns of Dutch carabid species (Col.). In: *Multivariate Methods in Ecological Work* (eds L. Orloci, C.R. Rao & W.M. Stiteler), pp. 65–86. International Co-operative Publishing House, Fairland, MD.

Hengeveld, R. & Van den Bosch, F. (1997) Invading into an ecologically non-uniform area. In: *Past and Future Rapid Environmental Changes: the Spatial and Evolutionary Responses of Terrestrial Biota* (eds B. Huntley, W. Cramer, A.V. Morgan, H.C. Prentice, & J.R.M. Allen), pp. 217–225. Springer Verlag, Berlin.

Hengeveld, R. & Walter, G.H. (1999) The two co-existing ecological paradigms. *Acta Biotheoretica* **47**, 141–170.

Huntley, B. (1988) Europe. In: *Vegetation History* (eds B. Huntley & T. Webb), pp. 341–383. Kluwer, Dordrecht.

Huntley, B. & Birks, H.J.B. (1983) *An Atlas of Past and Present Pollen Maps for Europe: 0–13,000 Years Ago.* Cambridge University Press, Cambridge.

Infantosi, A.F.C. (1986) *Interpretation of case studies in two communicable diseases using pattern analysis techniques.* PhD thesis, Imperial College, London.

Ives, A.R. & Klopfer, E.D. (1997) Spatial variation in abundance created by stochastic temporal variation. *Ecology* **78**, 1907–1913.

Jerling, L. (1985) Population dynamics of *Plantago maritima* belonged a distributional gradient on a Baltic seashore meadow. *Vegetatio* **61**, 155–168.

Jerling, L. (1988) Genetic differentiation in fitness related characters in *Plantago maritima* along a distribution of gradient. *Oikos* **53**, 341–350.

Kareiva, P. (1982) Experimental and mathematical analyses of herbivore movement: quantifying the inference of plant spacing and quality on foraging discrimination. *Ecology* **52**, 261–282.

Kessel, S.R. (1979) *Gradient Modelling.* Springer Verlag, New York.

Kot, M., Lewis, M.A. & Van den Driessche, P. (1996) Dispersal data and the spread of invading organisms. *Ecology* **77**, 2027–2042.

Lensink, R. (1997) Range expansion of raptors in Britain and the Netherlands since the 1960s, testing an individual-based diffusion model. *Journal of Animal Ecology* **66**, 811–826.

Lensink, R. (1998) Temporal and spatial expansion of the Egyptian goose *Alopochen aegyptiacus*, 1967–1994. *Journal of Biogeography* **25**, 251–263.

Lindroth, C.H. (1949) *Die fennoscandischen Carabiden. Vol. 3. Allgemeiner Teil.* Goteborgs Kungliche Vetenskaps- och Vitterhets-sammhalles Handlinger B, Stockholm.

Mac Nally, R.C. (1995) *Ecological Versatility and Community Ecology.* Cambridge University Press, Cambridge.

MacArthur, R.H. (1972) *Geographical Ecology.* Harper & Row, New York.

MacArthur, R.H. & Wilson, E.O. (1967) *The Theory of Island Biogeography.* Princeton University Press, Princeton, NJ.

Mayr, E. (1942) *Systematics and the Origin of Species.* Columbia University Press, New York.

Mayr, E. (1963) *Animal Species and Evolution.* Harvard University Press, Cambridge, MA.

Mayr, E. (1982) *The Growth of Biological Thought.* Harvard University Press, Cambridge, MA.

Mollison, D. (1977) Spatial contact models for ecological and epidemic spread. *Journal of the Royal Statistical Society* **B89**, 283–326.

Mollison, D. (1995) *Epidemic Models.* Cambridge University Press, Cambridge.

Nelson, G. (1978) From Candolle to Croizat: comments on the history of biogeography. *Journal of the History of Biology* **11**, 269–305.

Nelson, G. & Rosen, D.E. (eds) (1981) *Vicariance Biogeography: a Critique.* Columbia University Press, New York.

Neubert, M.G., Kot, M. & Lewis, M.A. (1995) Dispersal and pattern formation in a discrete-time predator–prey model. *Theoretical Population Biology* **48**, 7–43.

Okubo, A. (1980) *Diffusion and Ecological Problems: Mathematical Models.* Springer Verlag, Berlin.

Parmesan, C., Ryrholm, N., Stefanescu, C. *et al.* (1999) Poleward shifts in geographical ranges associated with regional warming. *Nature* **399**, 579.

Parry, M.L. (1978) *Climatic Change, Agriculture and Settlement.* Dawson & Archon, Folkestone, UK.

Parry, M.L. & Carter, T.R. (1985) The effect of climatic variations on agricultural risk. *Climatic Change* **7**, 95–110.

Paterson, H.E.H. (1985) The recognition concept of species. In: *Species and Speciation* (ed. E.S. Vrba), pp. 21–29. Transvaal Museum, Pretoria.

Patterson, K.D. (1986) Pandemic influenza, 1700–1900. Roman & Littlefield, Totowa.

Pianka, E.R. (1994) *Evolutionary Ecology*, 5th edn. Harper & Row, New York.

Pulliam, H.R. (1988) Sources, sinks, and population regulation. *American Naturalist* 132, 652–661.

Rosen, D.E. (1978) Vicariance patterns and historical explanation in biogeography. *Systematic Zoology* **27**, 159–188.

Schröpfer, R. & Engstfeld, C. (1983) Die Ausbreitung des Bisams (*Ondatra zibethicus* Linne, 1977, Rodentia, Arvicolidae) in der Bundesrepublik Deutschland. *Zeitschrift fur Angewandte Zoologie* **70**, 13–37.

Shigesada, N. & Kawasaki, K. (1997) *Biological Invasions: Theory and Practice.* Oxford University Press, Oxford.

Solomon, M.E. (1949) The natural control of animal populations. *Journal of Animal Ecology* **18**, 1–35.

Southwood, T.R.E. (1962) Migration of terrestrial arthropods in relation to habitat. *Biological Reviews* **37**, 171–214.

Sykes, M.T. & Prentice, I.C. (1995) Boreal forest futures: modelling the controls on tree species range limits and transient responses to climate change. *Water, Air and Soil Pollution* **82**, 415–428.

Taper, M.L., Bohning-Goese, K. & Brown, J.H. (1995) Individualistic responses of bird species to environmental change. *Oecologia* **101**, 478–486.

Thomas, C.D., Bodsworth, E.J., Wilson, R.J. *et al.* (2001) Ecological and evolutionary processes at expanding range margins. *Nature* **411**, 577–581.

Turchin, P. (1998) *Quantitative Analysis of Movement.* Sinauer, Sunderland, MA.

Turin, H. (2000) *De Nederlandse Loopkevers, verspreiding en oecologie (Coleoptera: Carabidae). Nederlandse Fauna 3.* Nationaal Natuurhistorisch Museum Naturalis, KNNV Uitgeverij & EIS-Nederland, Leiden.

Van den Bosch, F., Metz, J.A.J. & Diekmann, O. (1990) The velocity of population expansion. *Journal of Mathematical Biology* **28**, 529–565.

Van den Bosch, F., Hengeveld, R. & Metz, J.A.J. (1992) Analysing the velocity of animal range expansion. *Journal of Biogeography* **19**, 135–150.

Verboom, J. (1996) *Modelling fragmented populations: between theory and application in landscape planning.* PhD thesis, Leiden University, Leiden.

Walter, G.H. & Hengeveld, R. (2000) The structure of the two ecological paradigms. *Acta Biotheoretica* **48**, 15–46.

Walter, G.H. & Paterson, H.E.H. (1995) Levels of understanding in ecology: interspecific competition and community ecology. *Australian Journal of Ecology* **20**, 463–466.

Walter, H. & Walter, E. (1953) Einige allgemeine Ergebnisse unserer Forschungsreise nach Sudwestafrika 1952/53: Das Gesetz der relativen Standortskonstanz; das Wesen der Pflanzengesellschaften. *Berichte des Deutschen botanischen Gesellschafts* **66**, 228–236.

Whittaker, R.H. (1967) Gradient analysis of vegetation. *Biological Reviews* **42**, 207–264.

Part 4
Applications of an understanding of dispersal

Chapter 16

Modelling vertebrate dispersal and demography in real landscapes: how does uncertainty regarding dispersal behaviour influence predictions of spatial population dynamics?

Andy B. South, Steve P. Rushton, Robert E. Kenward and David W. Macdonald

Introduction

Humans have increased the rate of change in the spatial distributions of many species through direct effects on demography, by altering habitat availability and by introductions to new areas. Dispersal mediates how species respond to new circumstances since it enables range expansion, recolonization and the rescue of small populations from extinction. Often there is a desire to predict these changes and to determine how they may best be managed. Managers might want to slow the spread of an invading alien species or to reverse the decline of a vulnerable native.

As outlined in Kenward *et al.* (this volume), spatially explicit population models (SEPMs) allow for the explicit representation of complex, heterogeneous landscapes (which is not possible in diffusion and spatial contact models, e.g. Lensink 1997) and are not based upon the repeated extinction and colonization of local populations required by classic metapopulation models (Hanski & Simberloff 1997). SEPMs can be used to assess the impact of habitat management as they can be applied at a similar scale to that at which management takes place. There have been a number of cases where these models have been used to inform management (e.g. McKelvey *et al.* 1993; Liu *et al.* 1995). There has, however, been debate about the reliability of these models, particularly with regard to the paucity of data on which their representation of dispersal is based (Bart 1995; Turner *et al.* 1995; Wennergren *et al.* 1995; Ruckelshaus *et al.* 1997; South 1999).

Here we review the ways that dispersal has been represented within SEPMs and outline how these different approaches may be evaluated. We go on to discuss the potential consequences for model predictions of uncertainty regarding spatial population structure and dispersal mechanisms. We conclude by suggesting ways to address this uncertainty, both for models applied to inform management decisions and for those aimed at furthering the understanding of dispersal itself.

Terminology

Words such as habitat and landscape are used in different ways and this can lead to confusion, particularly when attempting to describe relatively complex models. To avoid this we adopt a very simple classification of the earth's surface. We take 'landscape' to mean the entire study area, 'habitat' to mean areas able to support breeding individuals and 'matrix' to mean areas not able to support breeding individuals (Wiens 1997). Within the matrix there may be a further subdivision into areas through which individuals can disperse and those through which they cannot. We use 'home range' to mean the area covered by an individual in its normal daily activities (Burt 1943), omitting occasional long-distance excursions. We use 'habitat patch' in the population context to mean an area of habitat capable of supporting at least a breeding pair of individuals, separated from other such areas by matrix.

We follow the definition of dispersal adopted in the preface to this volume, namely as 'intergenerational movement'. In the context of modelling vertebrates this means the movement of individuals from one location, where they were born or have bred, to another location where they will breed. The spatial scale at which 'location' is defined will influence which movements are considered as dispersal in a modelling context. Models may represent dispersal as any movement of the home range or may only represent dispersal as movement from one habitat patch to another.

Thus the definitions of dispersal, habitat patch and home range are linked; a habitat patch is partly defined by the fact that home ranges are contained within it and any movement between habitat patches is dispersal. Care should be taken as the behaviour of real animals in real landscapes is inevitably not as simple as this might suggest and any classification will be dependent upon the scale at which the system is observed. Also, for species that do not occupy stable home ranges (Gautestad & Mysterud 1995, Kenward *et al.*, this volume) it is more difficult to differentiate between home ranging and dispersal movements.

SEPMs have been defined as combining a representation of population processes with a map representing the spatial location of individuals and landscape elements (Dunning *et al.* 1995). Thus they differ from spatially implicit models in which spatial separation is represented but with no consideration of the relative positioning of objects or processes (Hanski & Simberloff 1997). Some authors (e.g. Akçakaya & Atwood 1997) describe any model representing local populations linked by dispersal as a metapopulation model. We prefer to reserve this term for models relying on repeated extinctions and colonizations (the 'classical' metapopulation of Hanski & Simberloff 1997), thus differentiating them from the models described here. In addition, we suggest that SEPMs are defined as excluding those models that rely on the assumption of repeated extinctions and colonizations of patches. This would avoid confusion with the incidence function model (Hanski 1994; applied to butterflies: Hanski *et al.* 1996; and to the pika, *Ochotona princeps*: Moilanen *et al.* 1998), which satisfies the other definitions of a SEPM but which is based on classic metapopulation theory.

Representations of the spatial structure of populations and dispersal within SEPMs

A wide variety of SEPMs has been developed. The flexibility within the simulation framework to represent processes in many different ways means that it can be difficult to relate different models to each other. Beissinger and Westphal (1998) provided a general overview of the application of population models to endangered species management. They urged that such models should be used with caution due to the scarcity of demographic and dispersal inputs. To facilitate the evaluation of SEPMs, we provide a detailed review of the way that dispersal has been represented within them.

SEPMs differ principally in two characteristics relating to dispersal. Firstly, their representation of the spatial structure of populations and, secondly, their representation of the dispersal behaviour of individuals (see Table 16.1 for a comprehensive review of published applications to vertebrates and Table 16.2 for generic, off-the-shelf models).

Representation of the spatial structure of populations

Representations of the spatial structure of populations can be put into four categories of increasing spatial resolution: (i) patchy population; (ii) grid cells containing populations; (iii) grid cells containing individual or pair home ranges; and (iv) individual or pair home ranges made up of multiple grid cells (Fig. 16.1).

The patchy population category uses a representation of space similar to that used in metapopulation models (Fig. 16.1a). It is assumed that individuals mix randomly within habitat patches such that any individual can breed with any other, but that individuals may only move between patches by dispersing. This has been the most commonly used approach (Tables 16.1 and 16.2), and has a number of advantages. It is considerably easier to not have to keep track of the spatial location of all individuals within patches. Also population dynamics within patches can be represented using population-scale parameters or by applying individual-scale probabilities using a Monte-Carlo approach (e.g. Burgman *et al.* 1993).

The second category is similar to the first but habitat patches are represented by grid cells (Fig. 16.1b). Population processes are simulated within cells, and dispersal is represented as occurring between cells. This shares the advantages of differentiating between two spatial scales gained by the patchy population approach. However, the division of a contiguous population into a grid of arbitrarily sized subpopulations has less biological meaning.

In the third category, the grid cells are smaller, with each able to represent the home range of a single individual or pair (Fig. 16.1c). This approach does not rely on an explicit differentiation between two spatial scales, as in the within-population and between-population dichotomy described for the previous two approaches. This leads to a difference in what is considered as dispersal. Within the two previous, patch-based approaches, movements within the patch to form pairs are not represented explicitly and are not considered as dispersal. In contrast, in the home-range

Table 16.1 Representations of spatial population structure and dispersal within published SEPMs.

	Dispersal dependent upon				
Representation of space	Habitat	Matrix	Dispersal initiation	Species	Reference
Patchy population	Yes	Partially	Density independent	Spotted owl *Strix occidentalis*	Lahaye *et al.* 1994
	Partially	No	Density dependent	Greater bilby *Macrotis lagotis*	Southgate & Possingham 1995
	Yes	No	Saturation	Red and grey squirrel *Sciurus vulgaris* and *S. carolinensis*	Rushton *et al.* 1997, 2000b; Lurz *et al.* 2001
	Yes	No	Density dependent and independent	California gnatcatcher *Polioptila c. californica*	Akçakaya & Atwood 1997
	Yes	No	Density dependent	Spotted owl	Akçakaya & Raphael 1998
	Yes	Yes	Saturation	Water vole *Arvicola terrestris*	Rushton *et al.* 2000a
	Yes	Yes	Saturation	Beaver *Castor fiber*	South *et al.* 2000
	Yes	No	Density dependent	Greater glider *Petauroides volans*	Lindenmayer *et al.* 2001
Population grid cells	No	No	Density independent	Badgers and bovine tuberculosis *Meles meles* and *Mycobacterium bovis*	White & Harris 1995
	Yes	No	Density independent and saturation	Stephens' kangaroo rat *Dipodomys stephensi*	Price & Gilpin 1996
	No	No	Density independent	Passerines	Baillie *et al.* 2000
Home range grid cells	Yes	Yes	Saturation	Bachman's sparrow *Aimophila aestivalis*	Pulliam *et al.* 1992; Liu *et al.* 1995
	Yes	Yes	Saturation	Spotted owl	McKelvey *et al.* 1993; Bart 1995
Sub-home range grid cells	Yes	Yes	Density independent	Foxes *Vulpes vulpes* and rabies	Smith & Harris 1991
	Yes	Yes	Saturation	Mink *Mustela vison*	Macdonald & Strachan 1999
	Yes	Yes	Saturation	European brown bear *Ursus arctos*	Wiegand *et al.* 1999

Table 16.2 Representations of spatial population structure and dispersal within generic SEPMs.

Representation of space	Model name	Dispersal dependent upon		Dispersal initiation: density independent, density dependent or saturation	Available from
		Habitat	Matrix		
Patchy population					
	Vortex	Yes	Can be	Any	http://pw1.netcom.com/~rlacy/vortex.html
	Ramas	Yes	Can be	Any	http://www.ramas.com/ramas.htm
	Alex	Yes/no	Can be	Any	http://biology.anu.edu.au/research.groups/ecosys/Alex/ALEX.HTM
Sub-home range grid cells	Patch	Yes	Yes	Any	http://www.epa.gov/naaujydh/pages/models/patch/patchmain.htm

grid approach, all movements from one home range to another must be represented and are likely to be considered as dispersal (e.g. Pulliam *et al.* 1992).

One potential drawback of the third approach is that all home ranges are represented as being the same size. To circumvent this, the fourth category of approach builds up home ranges from a variable number of smaller grid cells (Fig. 16.1d). The home ranges of red foxes and brown bears have been represented by 1–18 and 1–9 grid cells, respectively (Smith & Harris 1991; Wiegand *et al.* 1999). The generic spatial demographic model PATCH (Schumaker 1998) combines aspects of the third and fourth approaches. A grid of cells approximating the home range size is imposed upon a grid of smaller cells containing landscape attribute data. This allows for some variation in home range size by allowing home ranges with insufficient habitat cells to include areas from neighbouring home ranges.

The latter two categories can require a more complex representation of breeding. Rather than considering a group of randomly interbreeding individuals (as in the first two categories), it is necessary to represent how spatially separated individuals will come together to breed (South & Kenward 2001). An alternative way of getting around this is to model females only, if it is felt that the distribution of males is unimportant (e.g. Wiegand *et al.* 1999).

Representation of dispersal

Dispersal consists of three phases: starting, moving and stopping (see also Andreassen *et al.*, this volume). These phases have been represented in a diversity of ways within SEPMs. In the starting phase, SEPMs differ in how they determine which individuals disperse (Table 16.1). In some models dispersal is density inde-

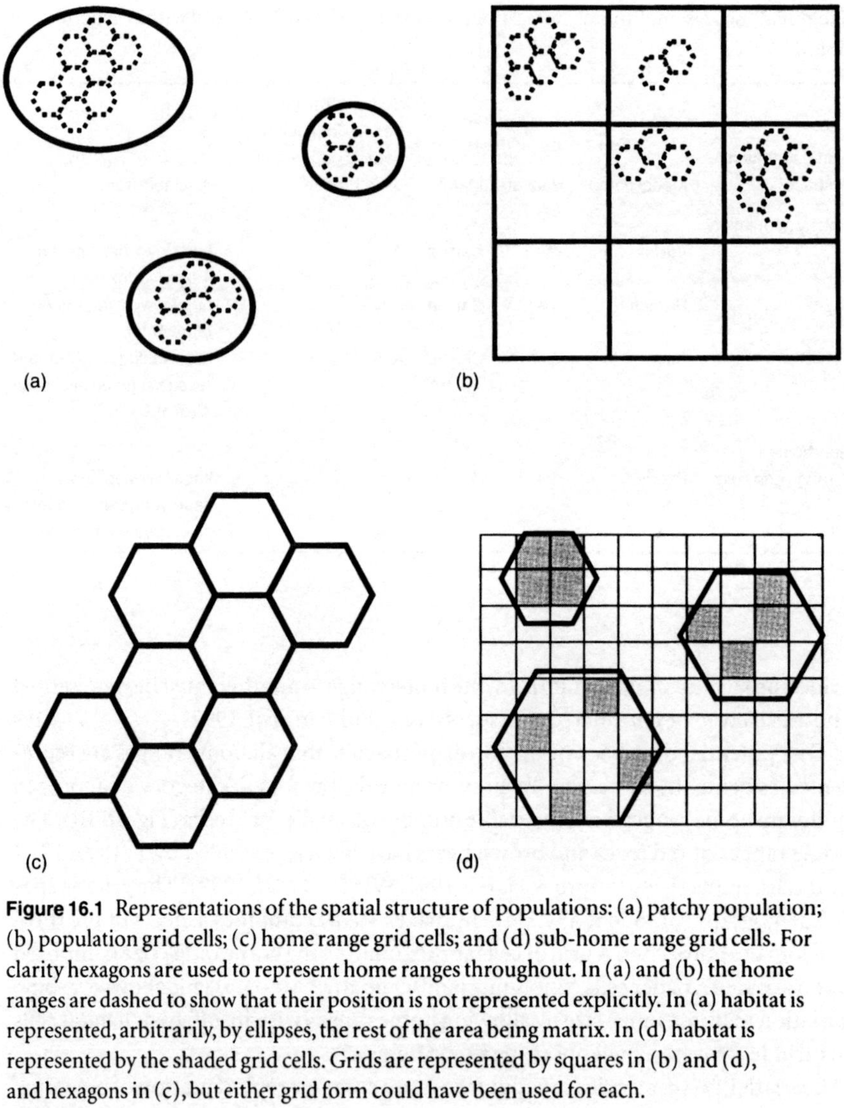

Figure 16.1 Representations of the spatial structure of populations: (a) patchy population; (b) population grid cells; (c) home range grid cells; and (d) sub-home range grid cells. For clarity hexagons are used to represent home ranges throughout. In (a) and (b) the home ranges are dashed to show that their position is not represented explicitly. In (a) habitat is represented, arbitrarily, by ellipses, the rest of the area being matrix. In (d) habitat is represented by the shaded grid cells. Grids are represented by squares in (b) and (d), and hexagons in (c), but either grid form could have been used for each.

pendent whilst in others individuals only disperse once carrying capacity is exceeded (saturation dispersal). For population-based models (Fig. 16.1a,b), carrying capacity is assessed at the level of the patch or grid cell. In home range-based models (Fig. 16.1c,d), carrying capacity has been assessed simply as whether the home range is already occupied by another individual of the same sex (e.g. Pulliam *et al.* 1992). Population-based approaches also allow the representation of density dependence, where dispersal probabilities increase with population density according to a more gradual function. These choices are related to the debate as to whether dispersal occurs principally at saturation or presaturation (Stenseth & Lidicker 1992). Scale effects and the difference between patchy population and home range representations of space also need to be considered here. For example, in patchy population models with saturation dispersal (e.g. Rushton *et al.* 1997), dispersal from a patch only occurs once that patch reaches carrying capacity, but there is an implicit assumption that short-scale dispersal movements occur within the patch when it is below carrying capacity.

Representations of the moving and stopping phases differ in how much dependence there is on the landscape configuration (Table 16.1, Fig. 16.2). Dispersal paths may be simulated without respect to the habitat or matrix (Fig. 16.2a). In this case, the distribution of habitat or properties of the matrix do not influence the simulated path. For example, Baillie *et al.* (2000) simulated dispersers as moving in a random direction for distances determined by randomly drawing numbers from a dispersal distance distribution derived from field data.

Many patchy population models simulate dispersal with similarly little dependence upon the matrix, but with dependence upon the distribution of habitat. These differ in that individuals only disperse to other habitat patches and do not stop in the matrix (Fig. 16.2b), however individuals may still fail to make it to another habitat patch due to the imposition of dispersal mortality. One such approach simulates dispersers as moving from one habitat patch to the nearest other, providing that the latter is within a maximum dispersal distance and not at carrying capacity (Rushton *et al.* 1997). In the generic spatial population viability analysis packages Vortex (Miller & Lacy 1999) and Ramas GIS (Akçakaya 1998), dispersal between habitat patches is represented by a series of dispersal probabilities. For each patch, the probability of a disperser moving to each other patch in the landscape is defined. Such dispersal probabilities can be based upon the straight-line distance between patches (e.g. Akçakaya & Atwood 1997) or can also include information about the matrix between patches such that if two patches are separated by a barrier the probability of dispersal between them is reduced (e.g. Lahaye *et al.* 1994).

These latter models represent an intermediate step between matrix-independent models and models that explicitly simulate the paths of dispersers through the matrix (Fig. 16.2c). Within this category, the matrix is generally divided into areas suitable for dispersal and areas that are not. Dispersal paths are simulated as a series of steps between grid cells with rules governing the choice of route (e.g. Pulliam *et al.* 1992; South *et al.* 2000). Elsewhere these models have been termed individual based (Beissinger & Westphal 1998), but this can be confusing as individual-based models do not necessarily represent the spatial location of individuals (DeAngelis & Gross 1992).

An important part of the dispersal process in terms of its effect on population processes is the associated mortality risk (South 1999). SEPMs have represented either: (i) no additional mortality associated with dispersal (e.g. Akçakaya & Atwood 1997); (ii) a single additional mortality probability per dispersal event (e.g. Rushton *et al.* 1997); (iii) a distance-dependent probability (e.g. Pulliam *et al.* 1992); or (iv) a probability dependent upon both the length of the dispersal path and attributes of the landscape crossed (Wiegand *et al.* 1999). SEPMs often also include additional

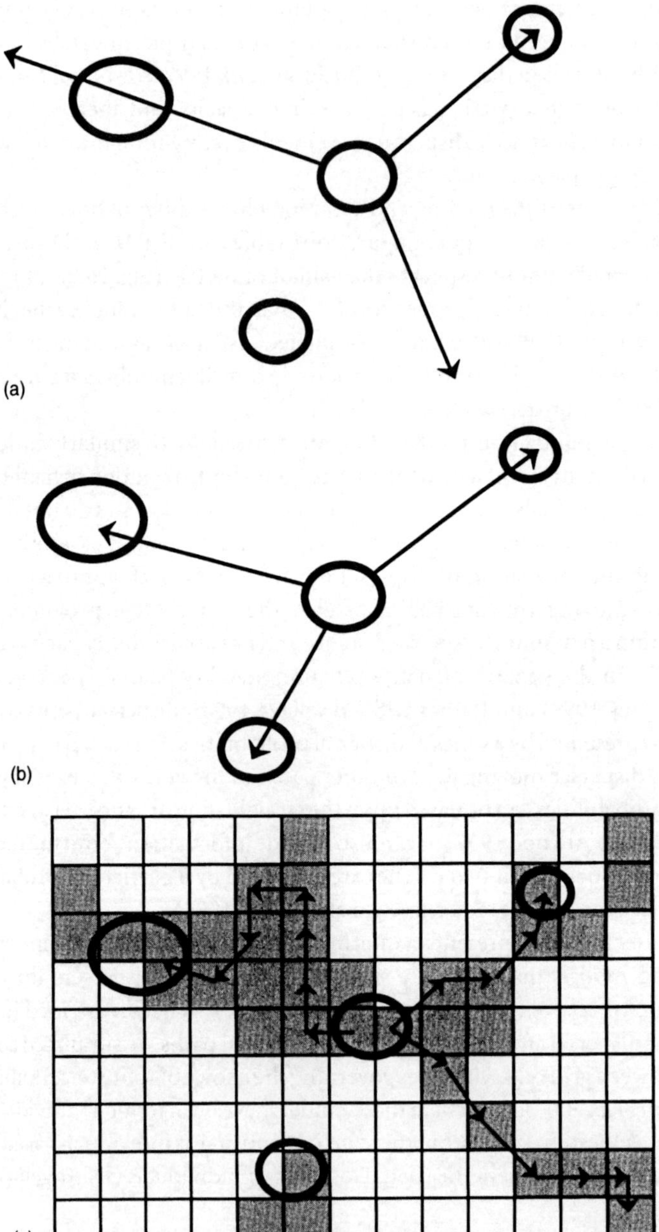

Figure 16.2 Landscape dependency within representations of dispersal: (a) habitat and matrix independent; (b) habitat dependent, matrix independent; and (c) habitat and matrix dependent. Ellipses represent areas of habitat, either supporting individuals or populations, the rest of the area being matrix. Arrows represent a sample of simulated dispersal paths. In (c) the shaded grid cells are suitable for dispersal, and the unshaded cells are not.

mortality risks if simulated dispersers are unable to find a vacant habitat (e.g. Pulliam *et al.* 1992; Rushton *et al.* 1997; South *et al.* 2000).

There is considerable variation in the way in which empirical data are incorporated within the dispersal representations of SEPMs. A recent approach probably using the most empirical data is that of Baillie *et al.* (2000), where simulated dispersal distances were derived by fitting statistical distributions to hundreds of ringing recovery records for the great tit *Parus major*. In contrast, dispersal of beavers *Castor fiber* within the model of South *et al.* (2000) was simulated using rules specifying that beavers moved in relatively straight lines along watercourses, up to the maximum dispersal distance reported in published field studies.

How can the different models be evaluated?

Of course, all of these models are simplifications of reality, and as such none can be considered as true (Oreskes *et al.* 1994), but how can we assess which models are most likely to produce reliable predictions of population behaviour at a chosen scale? Bart (1995) suggests that model reliability be assessed at four levels: structural assumptions, parameter values, secondary predictions and primary predictions.

Evaluating structural assumptions

As outlined in the previous section, the main structural assumptions made by SEPMs are those regarding the representation of the spatial structure of populations and dispersal.

Representation of the spatial structure of populations

The first stage in the representation of the spatial structure of a population is classifying the landscape into habitat and matrix. This will depend upon the habitat specificity of the species, the spatial distribution of habitat relative to home range size and the availability of landscape data. Perhaps the easiest situation to model is when the following conditions are satisfied:

1 Habitat is distributed as clearly defined patches separated by matrix.

2 Home range size is much smaller than habitat patch sizes so that single patches can support populations of multiple individuals.

3 The distance between patches is much greater than home range sizes so that individuals cannot make temporary, non-dispersal movements between patches.

4 Landscape data are available with sufficient detail and resolution to differentiate habitat from matrix.

This is largely the case for woodland species such as the spotted owl and red squirrel in areas where woodland is distributed as distinct blocks within an agricultural matrix (e.g. Lahaye *et al.* 1994; Rushton *et al.* 1997). Differentiation of habitat from matrix can be achieved using a simple rule-based approach and each patch will represent a population isolated from others except for dispersal movements. For these situations a patchy population approach is likely to be the most appropriate.

However, relatively few species–landscape combinations satisfy all of these conditions. When the conditions are violated, representing the spatial structure of populations becomes more complicated. If habitat is not distributed as clearly defined patches it may be appropriate to define patches using more complex methodologies or to use one of the other representations of spatial population structure. Akçakaya and Atwood (1997) applied the former approach to the California gnatcatcher *Polioptila c. californica*. Firstly, they divided the landscape into a grid and calculated a habitat suitability index for each cell based upon logistic regression of presence–absence records against a range of environmental variables. Then they grouped into the same habitat patch those cells that produced an index value above a chosen threshold and were less than a home range diameter apart.

Where the typical block size of habitat is less than the home range size of a species, individuals have to incorporate parts of the matrix within their home range. Species such as the fox *Vulpes vulpes* and buzzard *Buteo buteo* can have home ranges which are many times the size of typical blocks of habitat within the British countryside. This is likely to be the case, particularly for habitat generalists where the landscape comprises a suite of habitats of varying quality. Depending upon the spatial distribution of habitat it may still be possible to represent such situations using a patchy population approach. South *et al.* (2000) classified habitat patches for beaver in Scotland as those groups of neighbouring grid cells that contained greater than a threshold density of wooded riverbanks. This was thought to provide a reasonable approximation for intra- and interpatch processes, as the resultant patches were not too large (less than seven times the maximum home range diameter) and not too small and close together. However, if the distribution of habitat had been different it might have been necessary to use an approach in which home ranges were constructed from smaller grid cells (e.g. Macdonald & Strachan 1999; Wiegand *et al.* 1999).

Representation of dispersal

Perhaps the fundamental problem with modelling dispersal is our lack of understanding of the behavioural mechanisms that cause individuals to disperse particular distances and directions, and to what extent this is a result of responses to the landscape or to the distribution of conspecifics (Lima & Zollner 1996; Wiens 2001; Andreassen *et al.*, this volume). There has been much success in the development of foraging models based upon the optimization of individual behaviour (e.g. Goss-Custard *et al.* 2000), but currently there are simply insufficient data to establish the quantitative costs and benefits of landscape-level dispersal decisions.

Each of the modelling approaches described effectively represents a series of hypotheses as to which are the most important mechanisms driving dispersal. We can differentiate two major hypotheses to explain dispersal distances in vertebrates. The first suggests that dispersal distances can be explained by individuals moving to the nearest vacant home range (Waser 1985). The second suggests that this is insufficient to explain the distances moved and that other factors such as inbreeding avoidance (Wolff 1993) or habitat selection are involved. Waser (1985) produced an analysis which suggested that the first hypothesis is sufficient to explain dispersal distances in some, but not all, vertebrates.

Models in which dispersal is based purely on empirically derived dispersal distance distributions (e.g. Lensink 1997; Baillie *et al.* 2000) are compatible with either hypothesis, in that the mechanisms generating the dispersal distances need not be understood. These models also assume that dispersal distances are characteristic of a species within an area and do not vary according to local or regional landscape factors. This is more compatible with the hypothesis that dispersal distances are an evolved, context-independent behaviour, for example to avoid inbreeding.

Models in which dispersal is represented mechanistically as a search for the nearest vacant space (e.g. Liu *et al.* 1995; Rushton *et al.* 2000a, 2000b; South *et al.* 2000) assume that it is, indeed, competition for space that drives dispersal. Those in which dispersal is habitat dependent but matrix independent, effectively assume that dispersers are able to locate the nearest habitat patches and that the matrix in between is of little consequence. This may be more appropriate for flying vertebrates, particularly birds, which are able to fly high, see long distances and whose movement is less likely to be effected by the characteristics of the matrix than for terrestrial, non-flying vertebrates. In contrast, models that explicitly simulate dispersal paths, step by step, assume less landscape knowledge on the part of the animal and a greater importance of local landscape conditions.

Evaluating parameter values

The parameter values used within a model and the methods used to derive them from field data are usually clearly stated, allowing them to be evaluated. However, the parameter values need to be considered in the context of the structural assumptions made in the model. The accuracy of parameter values will have little meaning if inappropriate structural assumptions have been made.

Evaluating secondary predictions

SEPMs have the capacity to generate secondary predictions, that is predictions other than population distributions or extinction rates (Bart 1995). These include the distribution of individual dispersal distances, the proportion of individuals dispersing or the proportion of individuals settling in different habitat types. Testing secondary predictions offers a means of assessing a model when there are insufficient data to test the primary predictions (Caswell 1976). It can also be used to strengthen model assessment as it is quite possible that within models as complex as SEPMs the right predictions can be generated for the wrong reasons (Caswell 1976; Peters 1991).

The testing of secondary predictions is likely to be particularly useful for the assessment of models that simulate individual responses to the landscape and the distribution of conspecifics. Fine-scale behavioural responses may be tested against patterns in the field data that were recorded at larger spatial and temporal scales. Nevertheless, none of the papers in Table 16.1 stated that their models had been tested in this way. Wiegand *et al.* (1999) have perhaps come the closest to this, including an analysis of the response of dispersal distances within their model to changing landscape structure.

Similarly, models equivalent to the dispersal components of SEPMs have been developed and their behaviour investigated in the absence of demographic processes (e.g. Gustafson & Gardner 1996; Schumaker 1996; Ruckelshaus *et al.* 1997; Zollner & Lima 1999). However, these have also yet to be tested against dispersal data from the field.

Primary predictions

The primary predictions are those generated by the whole model, usually the spatial distribution of the species. It is surprising that only three of the SEPMs listed in Table 16.1 have been tested at this level. In the most comprehensive example, a model of red squirrel decline and grey squirrel expansion was tested against 15 years of data on the changing distribution of the two species (Rushton *et al.* 1997; 2000b). Predictions of grey squirrel spread were found to match the observed data most closely when mortality parameters were set lower and fecundity parameters higher than the averages recorded from field studies (Rushton *et al.* 1997). This enabled the generation of two hypotheses to explain the disparity. The first was that the saturation representation of dispersal used was inappropriate and that individuals dispersing before patches reached carrying capacity lead to higher spread rates. The second was that demographic performance was underestimated by deriving parameters from established populations subject to density-dependent constraints that were not operating in the colonizing populations. Other examples have involved the testing of model predictions against a static picture of population distribution in time (Rushton *et al.* 2000a; Lindenmayer *et al.* 2001), but did not investigate so closely the potential reasons for the differences between predictions and observations.

Consequences of uncertainty in dispersal representations for model predictions

We can examine the uncertainty regarding dispersal behaviour at two levels, firstly at the level of the modelling approach chosen, and secondly at the level of the parameter values used within that approach.

Ideally we might apply all of the different modelling approaches to a series of case studies to demonstrate how the different assumptions affect model predictions. However, the requirements of the different approaches in terms of landscape and species data are different. Thus, there is not a single situation for which, firstly, there is sufficient data to apply all approaches and, secondly, it would be appropriate to do so. For example, we could consider an exchange of approaches between the mechanistic, habitat-dependent, dispersal model applied to red squirrels by Rushton *et al.* (1997) and the empirically derived, habitat-independent, dispersal model applied to passerines by Baillie *et al.* (2000). Application of the Baillie *et al.* (2000) approach to red squirrels is precluded by a lack of an equivalent for red squirrels of the ringing recovery data available for passerines. Application of the Rushton *et al.* (1997) approach to passerines would require partitioning the landscape into habitat patches supporting separate populations. Passerines are able to live in a broad range of habitats, including woodland and farmland, and to move relatively easily between adjoining areas, so the resultant patches would be very large and would probably be an inaccurate representation of the population structure. Indeed, the entire study area could be represented by a single contiguous patch, in which case the model would effectively become aspatial with no dispersal at all.

Table 16.3 summarizes the effect of varying dispersal parameters in those studies that have done so. There are both positive and negative effects on predictions of population size, persistence and spread. There are cases where apparently similar modifications to parameters in the different models produced opposing effects. These can be explained in terms of the different structural assumptions used in the models. For example, Rushton *et al.* (1997) demonstrated an increase in the patches occupied by grey squirrels with increasing dispersal distances, whereas in the models of Akçakaya and Atwood (1997) and Baillie *et al.* (2000) increasing the dispersal distances leads to a decrease in viability of the California gnatcatcher and population sizes of passerines, respectively. For the grey squirrels, individuals only dispersed once their natal patch had reached carrying capacity and they moved to the nearest unoccupied patch if this was within the maximum dispersal distance. Thus, increasing the maximum dispersal distance could only have the positive effect of allowing some dispersers to reach suitable habitat patches that they would not otherwise have reached (Fig. 16.3a). In the case of the gnatcatcher, where interpatch dispersal was represented by distance-dependent probabilities, increasing the maximum dispersal distance meant that some dispersers moved to more distant patches which happened to be smaller and at a greater risk of stochastic extinction (Fig. 16.3b) (Akçakaya & Atwood 1997). This could not have occurred in the model of Rushton *et al.* (1997) where the dispersers would have stopped at the closer, larger, patches if they were not full. The negative effect of increasing dispersal distances on passerine populations modelled by Baillie *et al.* (2000) occurred for a combination of reasons. Firstly, the way that dispersal was modelled meant that dispersal distance and dispersal frequency increased together. Secondly, because dispersal occurred presaturation and dispersal movements were made independently of the landscape, at higher dispersal rates there were a greater number of individuals that left good-quality habitats and ended up in poor ones (Fig. 16.3c).

The probable consequences of using certain inappropriate representations of dispersal are summarized in Table 16.4. Approaches based upon empirically derived dispersal distance distributions but including no responses to the landscape may get the distances right, but may be a poor representation of where individuals choose to settle. Conversely, approaches based upon regional-scale perception (knowing where the nearest habitat patches are) may get individuals into the right habitat types but, because finding the closest vacant site may be insufficient to explain dispersal distances (Waser 1985), may underestimate how far individuals move. Additionally, if a species only has local-level perception (knowing what is immediately

Table 16.3 The effect of varying dispersal parameters within SEPMs.

Representation of space	Effect of changing dispersal parameters	Reference
Patchy population	The risk of population decline was not sensitive to dispersal rates and distances	Lahaye *et al.* 1994
	Increasing dispersal between patches gave no change in population persistence	Southgate & Possingham 1995
	Increasing dispersal distances increased the rate of colonization of the grey squirrel	Rushton *et al.* 1997
	Increasing dispersal distances led to a small decrease in viability as more individuals went to smaller patches (which were slightly further away), and these patches were more vulnerable to stochastic extinction	Akçakaya & Atwood 1997
	Increasing dispersal rates and distances led to an increase in the number of occupied patches but had no effect on population size or viability	Akçakaya & Raphael 1998
	Increasing the maximum dispersal distance increased number of habitat blocks occupied, but not total population size. The effect of maximum dispersal distance was greater when the number of patches was reduced (fragmentation increased). The model was more sensitive to demographic parameters than dispersal ones	Rushton *et al.* 2000a
	Decreasing dispersal mortality and increasing maximum dispersal distances increased rates of population increase at all but the lowest demographic parameter values	South *et al.* 2000
Population grid cells	Increasing dispersal probabilities increased spread and persistence of tuberculosis in the badger population	White & Harris 1995
	Population persistence decreased with increasing dispersal probability for a model with density-independent dispersal, because populations in source cells were reduced. Changing to density-dependent dispersal and adding habitat selection increased population persistence	Price & Gilpin 1996
	Increasing dispersal distances had a negative effect on population sizes as dispersers were more likely to end up in poor habitats	Baillie *et al.* 2000
Home range grid cells	Increasing habitat selectivity of dispersers led to a decrease in population size, because fewer areas were colonized. Adding a memory function to the dispersal mechanism increased population size	Pulliam *et al.* 1992; Liu *et al.* 1995
Sub-home range grid cells	There was no significant effect of maximum dispersal distance upon the rate of colonization of an area	Macdonald & Strachan 1999

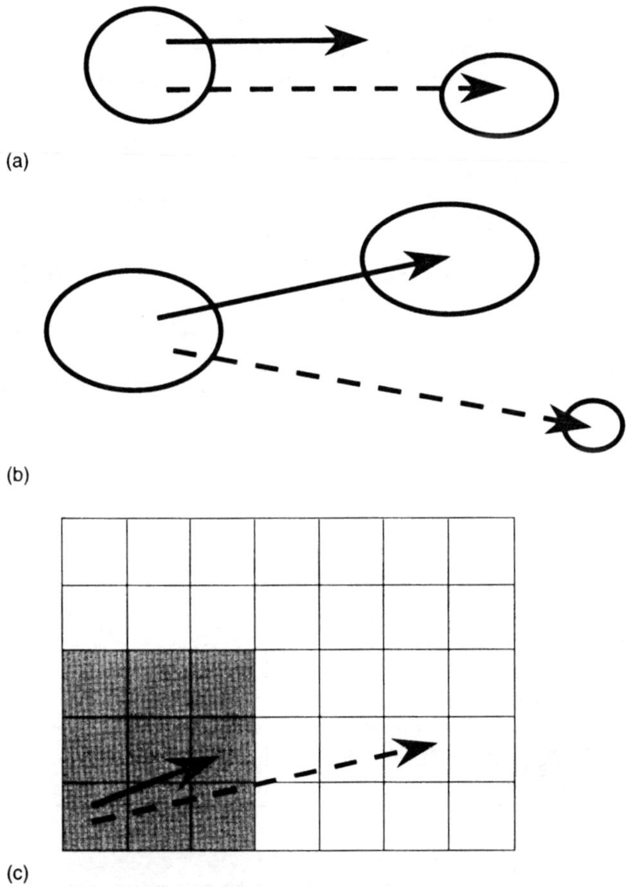

Figure 16.3 Mechanisms explaining the effect of increasing dispersal distances within three SEPMs. The arrows represent dispersal paths, solid for when dispersal distances were short and dashed for when they were longer. In (a) and (b) ellipses represent habitat patches; in (c) filled cells represent good habitat, and empty cells poor habitat. The results of increasing dispersal distance are: (a) colonization of patches that otherwise would not be reached; (b) distribution of the population among more and smaller patches; and (c) individuals are more likely to end up in poor habitat.

around it but not the location of potential habitat), modelling it as having regional-scale perception may overestimate habitat finding.

Whether the initiation of dispersal is represented as being density dependent or independent can also effect model predictions. In presaturation models, increasing the number of habitat patches or the dispersal rates can lead to decreases in population persistence (shown in a spatially implicit model by Lindenmayer and Lacy (1995)) because emigration from patches can make them more vulnerable to extinction. Such an effect would not occur in models based on saturation dispersal as dispersal will not occur when patches are sparsely populated. Thus, for species with

Table 16.4 Potential results of modelling dispersal the wrong way. The columns represent three alternative hypotheses for the mechanisms involved in dispersal for a particular species in a particular landscape. The rows represent the way that dispersal may be modelled based upon one of these three hypotheses. The cells with tick marks represent situations where any appropriate model of dispersal is used and the other cells outline the consequences of using an inappropriate model.

	Way that dispersal is		
Way that dispersal is modelled	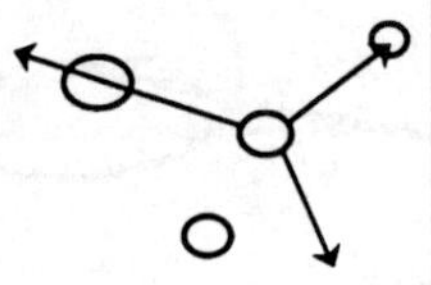	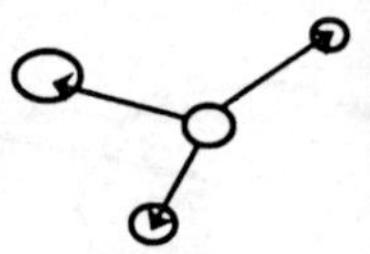	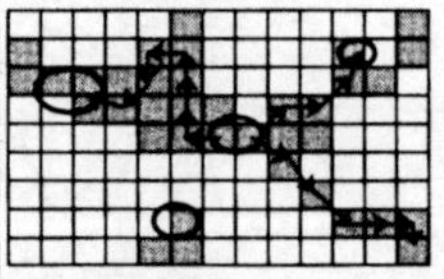
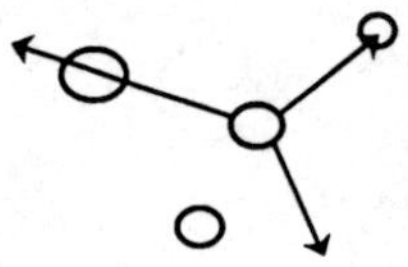	✓	Underpredict habitat finding	Underpredict habitat finding
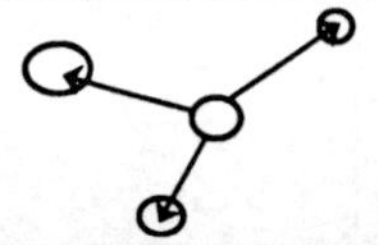	Overpredict habitat finding May underpredict dispersal distances	✓	Overpredict habitat finding May underpredict dispersal distances
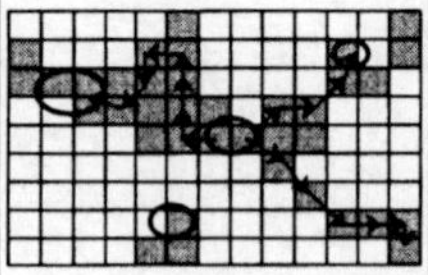	Overpredict habitat finding May underpredict dispersal distances	Underpredict habitat finding May overpredict dispersal distances	✓

poor demographic performance, models based on saturation dispersal are likely to predict higher viability than those based on presaturation dispersal. In contrast, for species with high demographic performance, models based on presaturation dispersal are likely to predict faster rates of population spread.

The consequences of misrepresenting the spatial structure of populations should also be considered. If a series of habitat fragments is represented as separate populations, when in fact relatively frequent foraging movements occur between them (Haila *et al.* 1993), population performance is likely to be underestimated due to an increased exposure to demographic stochasticity. Conversely, if blocks of habitat isolated at the population scale are grouped into the same patch, the spread of a species through an area is likely to be overestimated. In the model, the population would be predicted to colonize all parts of the patch without having to make any movements classed as dispersal.

Dealing with uncertainty regarding dispersal behaviour

The uncertainty regarding dispersal behaviour raises different issues depending upon whether the principal motivation is to make predictions to inform management decisions or to increase understanding of system behaviour. Uncertainty is an inconvenience for making management recommendations and the costs of making the wrong recommendations could be high (Macdonald & Johnson 2001). In contrast, from a research perspective uncertainty generates interesting research questions.

Modelling dispersal uncertainty for management recommendations

The first step is to acknowledge that the uncertainty exists and to communicate this with any model predictions (Bart 1995; Conroy *et al.* 1995). The potential consequences of uncertainty in both parameter values and structural assumptions should be communicated. Whilst some form of sensitivity analysis is often conducted to investigate the effect of changing parameter values, the effect of violations of the structural assumptions are more rarely considered. Some investigation into the effect of structural assumptions has been conducted in the case of the spotted owl where a range of different modelling approaches has been compared (McKelvey *et al.* 1993; Raphael *et al.* 1996) and in the assessment of the European beaver reintroduction to Scotland, where matrix-dependent and -independent dispersal representations were compared (South *et al.* 2000).

Of the studies that do look at the effect of altering parameter values, those that simultaneously investigate each of the parameter values (e.g. White & Harris 1995) provide more useful information than those that alter each parameter in turn, while keeping the others at a constant mean level (e.g. Akçakaya & Atwood 1997). Latin hypercube sampling, can be used when simulating all parameter combinations would be prohibitive, and combined with statistical analysis can show which parameter values have the greatest influence on model outputs (Rushton *et al.* 2000a). One way of presenting model results so that the implications for management may be evaluated easily is to present best, mean and worst case scenarios. For example, South *et al.* (2000) grouped parameters into those relating to demography, habitat quality and dispersal, and displayed the results of using all combinations of high, medium and low values for each. Akçakaya and Raphael (1998) also displayed the predicted effects of management actions in relation to uncertainty in model parameter values.

The effect of uncertainty regarding dispersal should be considered in the context of uncertainty in other model parameters. For a closed population, model predictions of persistence and population size will be sensitive to the values of dispersal parameters across a restricted range of demographic conditions (South 1999). Above this range populations will reach carrying capacity and below it go extinct, both irrespective of dispersal. For species spreading into new areas, sensitivity to dispersal parameters will also be evident when demographic performance is high, with the potential to increase spread rates.

Modelling dispersal uncertainty for research

We believe that SEPMs and other simulation models of dispersal have considerable potential as a research tool to investigate the mechanisms involved in dispersal behaviour. They allow extrapolation from individual-level behavioural rules to population-scale patterns. These rules can be based upon field data, simply be plausible (Zollner & Lima 1999) or include some form of behavioural optimization if expected lifetime reproductive success under different options can be estimated (Baillie *et al.* 2000). To realize this potential it is important that such models are tested against field data for both dispersal distances and spatial population patterns.

Some of the SEPMs developed to date have included relatively complex responses of individuals to the landscape. However, there has been little consideration of how social behaviour and the responses of individuals to conspecifics impacts upon dispersal. Researchers applying analytical models to biological invasions have suggested that rare, long-distance dispersal movements can increase rates of population spread considerably (e.g. Turchin 1998). However, it has also been suggested that the inability of individuals to find mates at the edge of an expanding population slows the rate of spread (Lewis & Kareiva 1993; Veit & Lewis 1996). We have recently developed a model to investigate the interaction between dispersal distances and mate-finding abilities and its effect upon population spread (South & Kenward 2001). This model represents the spatial population structure as a grid of home range-sized cells and draws dispersal distances from a statistical distribution that can be derived from field data. The model predicted that for species exhibiting large mean dispersal distances and poor mate-finding abilities, population growth will be constrained by the inverse density-dependent inability of adults to find mates (an Allee effect). In contrast, for species exhibiting low mean dispersal distances and high mate-finding abilities, growth was predicted to be constrained by the density-dependent inability of dispersers to find vacant home ranges. Population growth and spatial spread were predicted to be highest for species exhibiting mean dispersal distances in between these extremes (Fig. 16.4).

In reality, other behavioural responses of individuals to their conspecifics will also impact upon dispersal and spatial population dynamics. Conspecific attraction has been shown in a range of species (Reed & Dobson 1993). Buzzards have been documented returning to their natal areas after dispersing (Walls & Kenward 1998) and red kites *Milvus milvus* shown to remain close to communal roosts (Newton *et al.* 1994). These behaviours probably reduce the likelihood of new areas being colonized. The use of decoys and recordings of bird calls by conservation managers has been proposed as a means of attracting dispersers into suitable but unoccupied habitats (Reed & Dobson 1993), and has been used in experiments on seabirds (Kress 1983). Alternatively, conspecific attraction could potentially cause dispersers to move further if they are unable to find a mate. This has been suggested to have occurred in the recolonization of Sweden by the European beaver (Hartman 1995).

Recently obtained field data for buzzards suggests that individual dispersal responses to landscape and conspecifics may be even more complex. For example, whether buzzards disperse early in life depends on the number of siblings, the local density of other juveniles and habitat characteristics. Those buzzards that disperse early disperse 3–5 times further than those dispersing later in life, with distances less dependent on habitat than for those that disperse later (Kenward *et al.* 2001). It may

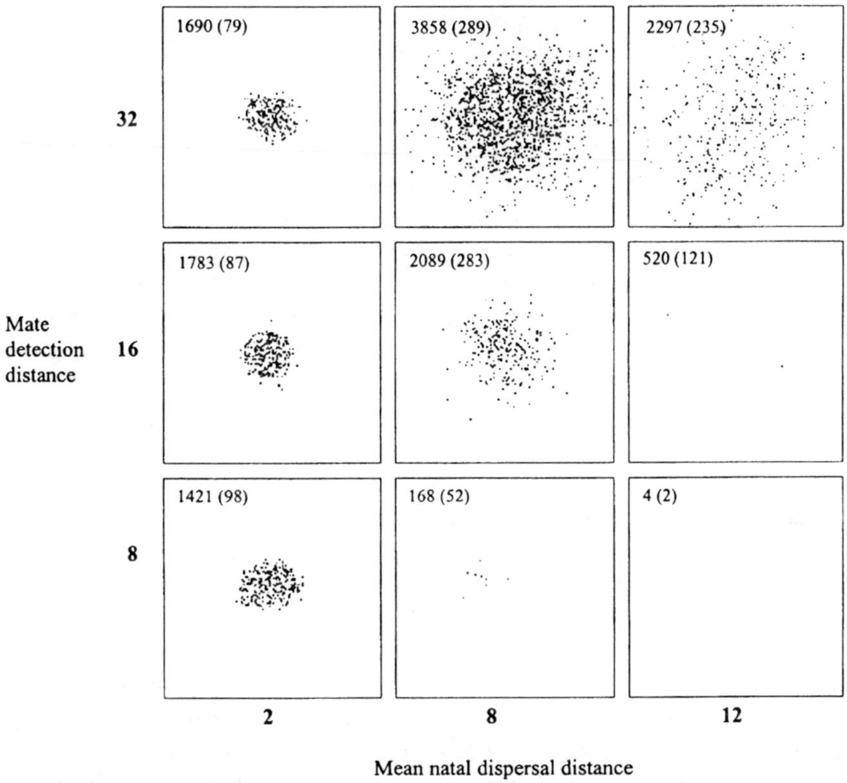

Figure 16.4 Predictions of a generic model incorporating variation in dispersal distances and mate-finding ability. Each of the nine boxes shows the spatial location of adult pairs generated by one model run for 50 years after starting with 11 pairs in the centre. The area displayed is 140 × 140 cells. In the top left of each box the mean population size (and standard error) from 50 model runs is displayed. Mate detection and mean natal dispersal distances are expressed in number of cells. Dispersal distances were sampled from an exponential distribution with the specified mean. (From South & Kenward 2001, with permission).

be necessary to include these mechanisms within models to predict the buzzard's current recolonization of Britain and similarly complex dispersal behaviours may have a bearing on the spatial population dynamics of other species.

Conclusions

We have reviewed the different ways in which dispersal is represented within SEPMs applied to vertebrates. There is a diversity of potential approaches and we suggest that these can most usefully be classified according to the representation of spatial population structure (patchy population; population grid cells; home range grid cells; and sub-home range grid cells) and the dependence of the dispersal representation on different landscape elements (landscape independent; habitat dependent and matrix independent; and habitat and matrix dependent). Whilst there is still the potential for considerable variation between models within these categories, they clarify the main structural assumptions.

We have concentrated on details of the representation of dispersal within SEPMs and only covered more briefly current theories and empirical evidence relating to dispersal in the field. We hope to have provided the information to facilitate the complex task of evaluating whether a particular model formulation provides a useful or reliable representation of the spatial population dynamics of a particular species in a particular landscape. For more information on current theories regarding dispersal behaviour in the field, we point readers to other chapters in this volume, especially those by Sutherland *et al.* and Andreassen *et al.* and to the recent book by Clobert *et al.* (2001) and particularly to chapters on the starting (Lambin *et al.* 2001), moving (Wiens 2001) and stopping (Stamps 2001) phases of dispersal.

Making general recommendations for which approaches are best is difficult and will, of course, depend upon particular aims, the species–landscape combination and data availability. The advantage of patchy population approaches is the relative ease with which they can be developed and the longer history of their application (Table 16.1). Models with finer resolution based upon home range grid cells or sub-home range grid cells, can become considerably more complex and are perhaps sufficiently early in their development to be of most use currently in addressing research questions rather than informing management decisions. There are also advantages in carefully investigating the behaviour of a relatively simple model rather than being able to spend less time in investigating the behaviour of a more complex one. Ideally, models of different levels of detail should be compared to each other and to field data to allow an assessment of how much detail is required (Murdoch *et al.* 1992).

We favour approaches based upon the representation and understanding of behavioural mechanisms rather than the application of purely empirical patterns, but comparing the two can also increase understanding. We do not suggest that either modelling or field work is the more appropriate tool to increase our understanding of dispersal. Either on its own is of little use; they should be advanced together, each being used to generate and test hypotheses for the other.

References

Akçakaya, H.R. (1998) RAMAS GIS: linking landscape data with population viability analysis (ver 3.0). Applied Biomathematics. Setauket, New York.

Akçakaya, H.R. & Atwood, J.L. (1997) A habitat-based metapopulation model of the California gnatcatcher. *Conservation Biology* **11**, 422–434.

Akçakaya, H.R. & Raphael, M.G. (1998) Assessing human impact despite uncertainty: viability of the northern spotted owl metapopulation in the northwestern USA. *Biodiversity and Conservation* **7**, 875–894.

Baillie, S.R., Sutherland, W.J., Freeman, S.N., Gregory, R.D. & Paradis, E. (2000) Consequences of large-scale processes for the conservation of

bird populations. *Journal of Applied Ecology* **37** (Suppl. 1), 88–102.
Bart, J. (1995) Acceptance criteria for using individual-based models to make management decisions. *Ecological Applications* **5**, 411–420.
Beissinger, S.R. & Westphal, M.I. (1998) On the use of demographic models of population viability in endangered species management. *Journal of Wildlife Management* **62**, 821–841.
Burgman, M., Ferson, S. & Akçakaya, H.R. (1993) *Risk Assessment in Conservation Biology.* Chapman & Hall, New York.
Burt, W.H. (1943) Territoriality and home range concepts as applied to mammals. *Journal of Mammology* **24**, 346–352.
Caswell, H. (1976) The validation problem. In: *Systems Analysis and Simulation in Ecology* (ed. B.C. Patten), pp. 313–325. Academic Press, New York.
Clobert, J., Danchin, E., Dhondt, A.A. & Nichols, J.D. (2001) *Dispersal.* Oxford University Press, Oxford.
Conroy, M.J., Cohen, Y., James, F.C., Matsinos, Y.G. & Maurer, B.A. (1995) Parameter-estimation, reliability, and model improvement for spatially explicit models of animal populations. *Ecological Applications* **5**, 17–19.
DeAngelis, D.L. & Gross, L.J. (1992) *Individual-based Models and Approaches in Ecology.* Chapman & Hall, New York.
Dunning, J.B., Stewart, D.J., Danielson, B.J. *et al.* (1995) Spatially explicit population-models—current forms and future uses. *Ecological Applications* **5**, 3–11.
Gautestad, A.O. & Mysterud, I. (1995) The home-range ghost. *Oikos* **74**, 195–204.
Goss-Custard, J.D., Stillman, R.A., West, A.D., McGrorty, S., Durell, S.E.A. le V. dit & Caldow, R.W.C. (2000) Role of behavioural models in predicting the impact of harvesting on populations. In: *Behaviour and Conservation* (eds M. Gosling & W.J. Sutherland), pp. 65–82. Cambridge University Press, Cambridge.
Gustafson, E.J. & Gardner, R.H. (1996) The effect of landscape heterogeneity on the probability of patch colonization. *Ecology* **77**, 94–107.
Haila, Y., Hanski, I.K. & Raivio, S. (1993) Turnover of breeding birds in small forest fragments: the 'sampling' colonization hypothesis corroborated. *Ecology* **74**, 714–725.
Hanski, I. (1994) A practical model of population dynamics. *Journal of Animal Ecology* **63**, 151–162.
Hanski, I. & Simberloff, D. (1997) The metapopulation approach, its history, conceptual domain, and application to conservation. In: *Metapopulation Biology: Ecology, Genetics and Evolution*, Vol. 42 (eds I. Hanski & M. Gilpin), pp. 3–16. Academic Press, San Diego.
Hanski, I., Moilanen, A., Pakkala, T. & Kuussaari, M. (1996) Metapopulation persistence of an endangered butterfly: a test of the quantitative incidence function model. *Conservation Biology* **10**, 578–590.
Hartman, G. (1995) Patterns of spread of a reintroduced beaver *Castor fiber* population in Sweden. *Wildlife Biology* **1**, 97–103.
Kenward, R.E., Walls, S.S. & Hodder, K.H. (2001) Life path analysis: scaling indicates priming effects of social and habitat factors on dispersal distances. *Journal of Animal Ecology* **70**, 1–13.
Kress, S.W. (1983) The use of decoys, sound recordings and gull control for re-establishing a tern colony. *Colonial Waterbirds* **6**, 185–196.
Lahaye, W.S., Gutierrez, R.J. & Akçakaya, H.R. (1994) Spotted owl metapopulation dynamics in southern california. *Journal of Animal Ecology* **63**, 775–785.
Lambin, X., Aars, J. & Piertney, S.B. (2001) Dispersal, intraspecific competition, kin competition and kin facilitation: a review of the empirical evidence. In: *Dispersal* (eds J. Clobert, E. Danchin, A.A. Dhondt & J.D. Nichols), pp 110–122. Oxford University Press, Oxford.
Lensink, R. (1997) Range expansion of raptors in Britain and the Netherlands since the 1960s: testing an individual based diffusion model. *Journal of Animal Ecology* **66**, 811–826.
Lewis, M.A. & Kareiva, P. (1993) Allee dynamics and the spread of invading organisms. *Theoretical Population Biology* **43**, 141–158.
Lima, S.L. & Zollner, P.A. (1996) Towards a behavioral ecology of ecological landscapes. *Trends in Ecology and Evolution* **11**, 131–135.
Lindenmayer, D.B. & Lacy, R.C. (1995) A simulation study of the impacts of population subdivision on the mountain brushtail possum *Trichosurus caninus* Ogilby (Phalangeridae: Marsupialia) in south-eastern Australia. I. Demographic stability

and population persistence. *Biological Conservation* **73**, 119–129.

Lindenmayer, D.B., Ball, I., Possingham, H.P., McCarthy, M.A. & Pope, M.L. (2001) A landscape scale test of the predictive ability of a spatially explicit model for population viability analysis. *Journal of Applied Ecology* **38**, 36–48.

Liu, J., Dunning, J.B. & Pulliam, H.R. (1995) Potential effects of a forest management plan on Bachman's sparrows (*Aimophila aestivalis*): linking a spatially explicit model with GIS. *Conservation Biology* **9**, 62–75.

Lurz, P.W.W., Rushton, S.P., Wauters, L.A. *et al.* (2001) Predicting grey squirrel expansion in north Italy: a spatially explicit modelling approach. *Landscape Ecology* **16** (5), 407–420.

Macdonald, D.W. & Johnson, D.D.P. (2001) Dispersal in theory and practice: consequences for conservation biology. In: *Dispersal* (eds J. Clobert, E. Danchin, A.A. Dhondt & J.D. Nichols), pp. 361–374. Oxford University Press, Oxford.

Macdonald, D.W. & Strachan, R. (1999) *The Mink and the Water Vole: Analyses for Conservation.* WildCRU, Oxford.

McKelvey, K., Noon, B.R. & Lamberson, R.H. (1993) Conservation planning for species occupying fragmented landscapes: the case of the northern spotted owl. In: *Biotic Interactions and Global Change* (eds P.M. Kareiva, J.G. Kingsolver & R.B. Huey), pp. 424–450. Sinauer Associates, Sunderland, MA.

Miller, P.S. & Lacy, R.C. (1999) *Vortex: a Stochastic Simulation of the Extinction Process. Version 8 Users' Manual.* Conservation Breeding Specialist Group (SSC/IUCN), Apple Valley, MN.

Moilanen, A., Smith, A.T. & Hanski, I. (1998) Long-term dynamics in a metapopulation of the American pika. *American Naturalist* **152**, 530–542.

Murdoch, W.W., McCauley, E., Nisbet, R.M., Gurney, W.S.C. & de Roos, A.M. (1992) Individual-based models: combining testability and generality. In: *Individual-based Models and Approaches in Ecology* (eds D.L. DeAngelis & L.J. Gross), pp. 18–35. Chapman & Hall, New York.

Newton, I., Davis, P.E. & Moss, D. (1994) Philopatry and population growth of red kites, *Milvus milvus*, in Wales. *Proceedings of the Royal Society of London, Series B* **257**, 317–323.

Oreskes, N., Shrader-Frechette, K. & Belitz, K. (1994) Verification, validation, and confirmation of numerical models in the earth sciences. *Science* **263**, 641–646.

Peters, R.H. (1991) *A Critique for Ecology.* Cambridge University Press, Cambridge.

Price, M.V. & Gilpin, M. (1996) Modelers, mammalogists and metapopulations. In: *Metapopulations and Wildlife Conservation* (ed. D.R. McCullough), pp. 217–240. Island Press, Washington.

Pulliam, H.R., Dunning, J.B. & Liu, J.G. (1992) Population-dynamics in complex landscapes—a case-study. *Ecological Applications* **2**, 165–177.

Raphael, M.G., Anthony, R.G., DeStefano, S. *et al.* (1996) Use, interpretation, and implications of demographic analyses of northern spotted owl populations. *Studies in Avian Biology* **17**, 102–112.

Reed, J.M. & Dobson, A.P. (1993) Behavioural constraints and conservation biology: conspecific attraction and recruitment. *Trends in Ecology and Evolution* **8**, 253–256.

Ruckelshaus, M., Hartway, C. & Kareiva, P. (1997) Assessing the data requirements of spatially explicit dispersal models. *Conservation Biology* **11**, 1298–1306.

Rushton, S.P., Lurz, P.W.W., Fuller, R. & Garson, P.J. (1997) Modelling the distribution and abundance of the red and grey squirrel at the landscape scale: a combined GIS and population dynamics approach. *Journal of Applied Ecology* **34**, 1137–1154.

Rushton, S.P., Barreto, G.W., Cormack, R.M., Macdonald, D.W. & Fuller, R. (2000a) Modelling the effects of mink and habitat fragmentation on the water vole. *Journal of Applied Ecology* **37**, 475–490.

Rushton, S.P., Lurz, P.W.W., Gurnell, J. & Fuller, R. (2000b) Modelling the spatial dynamics of parapox virus disease in red and grey squirrels: a possible cause of the decline in the red squirrel in the UK? *Journal of Applied Ecology* **37**, 997–1012.

Schumaker, N.H. (1996) Using landscape indices to predict habitat connectivity. *Ecology* **77**, 1210–1255.

Schumaker, N.H. (1998) *A Users Guide to the PATCH Model.* US Environmental Protection Agency publication no. EPA/600/R-98/135. US Environmental Protection Agency, Environmental Research Laboratory, Corvallis, OR.

Smith, G.C. & Harris, S. (1991) Rabies in urban foxes (*Vulpes vulpes*) in Britain: the use of a spatial stochastic simulation model to examine the pattern of spread and evaluate the efficacy of different control regimes. *Philosophical Transactions of the Royal Society London, Series B* **334**, 459–479.

South, A.B. (1999) Dispersal in spatially explicit population models. *Conservation Biology* **13**, 1039–1046.

South, A.B. & Kenward, R.E. (2001) Mate finding, dispersal distances and population growth in invading species: a spatially explicit model. *Oikos* **95** (1), 53–58.

South, A.B., Rushton, S.P. & Macdonald, D.W. (2000) Simulating the proposed reintroduction of the European beaver (*Castor fiber*) to Scotland. *Biological Conservation* **93**, 103–116.

Southgate, R. & Possingham, H. (1995) Modelling the re-introduction of the greater bilby *Macrotis lagotis* using the metapopulation model analysis of the likelihood of extinction (ALEX). *Biological Conservation* **73**, 151–160.

Stamps, J.A. (2001) Habitat selection by dispersers: integrating proximate and ultimate approaches. In: *Dispersal* (eds J. Clobert, E. Danchin, A.A. Dhondt & J.D. Nichols), pp. 230–242. Oxford University Press, Oxford.

Stenseth, N.C. & Lidicker, W.Z.J. (1992) *Animal Dispersal: Small Mammals as a Model.* Chapman & Hall, London.

Turchin, P. (1998) *Quantitative Analysis of Movement.* Sinauer Associates, Sunderland, MA.

Turner, M.G., Arthaud, G.J., Engstrom, R.T. *et al.* (1995) Usefulness of spatially explicit models in land management. *Ecological Applications* **5**, 12–16.

Veit, R.R. & Lewis, M.A. (1996) Dispersal, population growth, and the Allee effect: dynamics of the house finch invasion of eastern North America. *American Naturalist* **148**, 255–274.

Walls, S.S. & Kenward, R.E. (1998) Movements of radio-tagged buzzards *Buteo buteo* in early life. *Ibis* **140**, 561–568.

Waser, P.M. (1985) Does competition drive dispersal? *Ecology* **66**, 1170–1175.

Wennergren, U., Ruckelshaus, M. & Kareiva, P. (1995) The promise and limitations of spatial models in conservation biology. *Oikos* **74**, 349–356.

White, P.C.L. & Harris, S. (1995) Bovine tuberculosis in badger (*Meles meles*) populations in southwest England—the use of a spatial stochastic simulation-model to understand the dynamics of the disease. *Philosophical Transactions of the Royal Society of London, Series B* **349**, 391–413.

Wiegand, T., Moloney, K.A., Naves, J. & Knauer, F. (1999) Finding the missing link between landscape structure and population dynamics: a spatially explicit perspective. *American Naturalist* **154**, 605–627.

Wiens, J.A. (1997) Metapopulation dynamics and landscape ecology. In: *Metapopulation Biology: Ecology, Genetics and Evolution* (eds I. Hanski & M. Gilpin), pp. 3–16. Academic Press, San Diego.

Wiens, J.A. (2001) The landscape context of dispersal. In: *Dispersal* (eds J. Clobert, E. Danchin, A.A. Dhondt & J.D. Nichols), pp. 96–109. Oxford University Press, Oxford.

Wolff, J.O. (1993) What is the role of adults in mammalian juvenile dispersal? *Oikos* **68**, 173–176.

Zollner, P.A. & Lima, S.L. (1999) Search strategies for landscape-level interpatch movements. *Ecology* **80**, 1019–1030.

Chapter 17

Invasion and the range expansion of species: effects of long-distance dispersal

Nanaka Shigesada and Kohkichi Kawasaki

Introduction

The range expansion of an invading species shows a variety of spatiotemporal patterns depending on how the dispersal and reproduction of the organisms proceed under the influence of life-history attributes, disturbance regime and landscape structure (Higgins & Richardson 1999). In particular, the dispersal process is crucial in determining the range frontal pattern and its expansion speed. The dispersal often involves two modes, short-distance dispersal and long-distance dispersal. When offspring depart from their parents, most of them undergo short-distance dispersal to settle near the parents' range, while a minor fraction is infrequently observed at distant locations. Short-distance dispersal generally occurs through an organism's own inherent movement such as walking, swimming or flying, whereas long-distance dispersal is mediated by passive transport on wind, storm, water flow, frugivorous animals, birds or larger flying arthropods, or even artificial transportation (see Davis 1987; Cain *et al.* 1998; Clark *et al.* 1998; Higgins & Richardson 1999; Nathan & Muller-Landau 2000). Since distant colonization by long-distance dispersal is rare and stochastic, the pattern of invasion in its presence may vary from case to case.

Nevertheless, the range expansion of an invading species that is introduced at a local point can be qualitatively classified into three types as schematically illustrated in Fig. 17.1 (Shigesada *et al.* 1995; Shigesada & Kawasaki 1997). If a species undergoes only short-distance dispersal, its range expands continuously from the periphery of the primary population, as illustrated in Fig. 17.1a (type 1). When a species employs both long- and short-distance dispersal, it will form a more complex spatial pattern, namely, a primary colony accompanied by some patchy satellite colonies that are generated by long-distance dispersers. Although the number and location of satellite colonies could be highly variable, we may further distinguish their spatial patterns into two types. In type 2, long-distance dispersers settle fairly close to the primary colony, so that the resultant satellite colonies will remain in isolation only for a short time until they eventually coalesce with their parent population (Fig. 17.1b). On the other hand, when the long-distance dispersers move far away from the parent population, or the range expansion mediated by short-distance dispersal

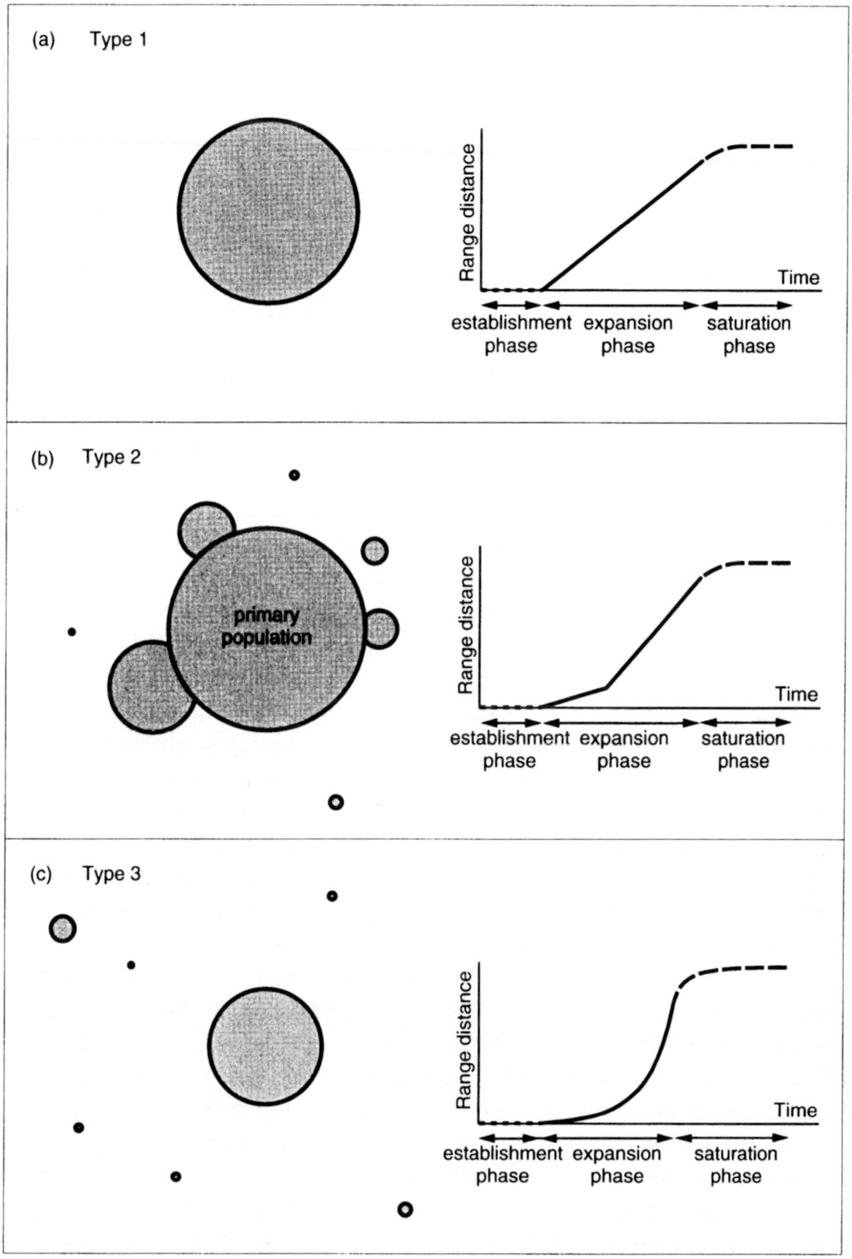

Figure 17.1 Three types of range expansions: (a) type 1, (b) type 2, and (c) type 3. For each type, the spatial pattern and range-versus-time curve are shown on the left and right, respectively. See text for detail. (Adapted from Shigesada & Kawasaki 1997, with permission of Oxford University Press.)

is relatively slow, satellite colonies expand their range independently of the other populations for a prolonged period of time, as seen in Fig. 17.1c (type 3).

The range distance in patchy distributions such as type 2 and type 3 is generally difficult to evaluate, because outliers sporadically arise and coalesce. For instance, if we measure the range distance by the distance from the initial point of introduction to the farthest satellite colony, it would be highly variable and far larger than the distance to the majority of satellite colonies. Here we present two definitions for the range distance: 'radial distance' and 'scattering distance'. The radial distance is the conventional measure as defined by the square root of the total range area, including satellite colonies, divided by $\pi^{1/2}$ (Skellam 1951; Okubo 1980; Shigesada & Kawasaki 1997). Thus it represents the radius of a circle whose area is equivalent to the total range area. The scattering distance is defined as the average distance from the initial point to every satellite colony (excluding the primary colony) weighted by its area. In general, the scattering distance lies between the radial distance and the distance to the farthest colony. As satellite colonies become more widely distributed under the constraint that the total area is kept fixed, the scattering distance grows larger while the radial distance remains constant. Thus the difference between the two distances provides a measure of how remotely the satellite colonies are distributed.

For each type of range expansion, the distance versus time curve often shows a characteristic feature as depicted on the right in Fig. 17.1, where the distances are measured by the radial distance. As described below in the stratified diffusion model section, this feature remains unchanged if the scattering distance is used instead. All three distance versus time curves have in common an 'initial establishment phase' during which little or no expansion takes place, followed by an 'expansion phase', and a final 'saturation phase' if there is a geographical limit for expansion. If we focus on the expansion phase, the spatial spread of type 1 shows a linear increase with time. The corresponding curve for type 2 involves a slow initial spread followed by a linear expansion at a higher rate. In type 3, the rate of spread is continually increasing with time, showing a concave curve. Typical examples for type 1 are muskrats in Europe (Skellam 1951; Andow *et al.* 1990, 1993; Williamson 1996), the California sea otter (Lubina & Levin 1988), epidemics of rabies among wild animals (Macdonald 1980; Murray *et al.* 1986) and bubonic plague that raged in 14th century Europe (Nobel 1974); for type 2, red deer and Himalayan thar in New Zealand (Clarke 1971; Caughley 1970), birds such as the European starling and house finch in North America, and the collared dove in Europe (Kessel 1953; Mundinger & Hope 1982; Hengeveld 1988; Okubo 1988); and for type 3, the rice water weevil in Japan (Andow *et al.* 1993) and cheat grass (*Bromus tectorum*) in western North America (Mack 1981) (for other examples, see Elton 1958; Hengeveld 1989; Metz & Van den Bosch 1995; Williamson 1996; Shigesada & Kawasaki 1997).

To grasp the spatial pattern of invasion and its rate of spread more quantitatively, we need detailed statistical data on dispersal distances together with the life history of the invading species. Ecological and demographic studies could provide life-history data through which the local population dynamics would be delineated (Metz & Diekmann 1986; Veit & Lewis 1996; Williamson 1996; Caswell 2000; Neubert & Caswell 2000). Determining the dispersal distance distribution from field observations is a laborious task in general, although various methods have been devised. Most of these are based on mark–recapture techniques or documenting variations in distance from parents (sources) to landing positions (Cain *et al.* 1998; Clark 1999; Nathan & Muller-Landau 2000). Unfortunately, these methods are difficult to apply to dispersal on a large spatial scale. In fact, most of the field data so far available are limited to short-distance dispersal within a few hundred metres (Clark *et al.* 1999; but see Takasu *et al.* 2000).

The empirical data of dispersal distances have been fitted with various mathematical functions. Kot *et al.* (1996) proposed several functional forms for the dispersal kernel (redistribution kernel) $k(\boldsymbol{x})$ that describes the probability density for a propagule arriving at $\boldsymbol{x}$ with respect to its parent (see also Mollison 1997; Hengeveld 1988; Neubert *et al.* 1995). In particular, when dispersal is two-dimensional and its direction is rotationally symmetrical, the following dispersal kernel (the transect distribution in polar coordinates (r, θ)) can accommodate a variety of dispersal patterns (Clark *et al.* 1999):

$$k(r) = \frac{\nu}{2\pi d^2 \Gamma(2/\nu)} \exp\left[-\left(\frac{r}{d}\right)^{\nu}\right] \tag{17.1}$$

where r is the distance from its source and $\Gamma(s) = \int_0^\infty z^{s-1} e^{-z} dz$ is a gamma function. The mean dispersal distance is given by $<r> = \int_0^\infty 2\pi r^2 k(r) dr = d\Gamma(3/\nu)/\Gamma(2/\nu)$. With decreases in ν, the distribution has larger kurtosis, being more fat-tailed. For example, when $\nu = 2$, $k(r)$ is a Gaussian with mean distance of $\sqrt{\pi} d/2$. When $\nu = 1$, $k(r)$ is an exponential damping function with a mean distance of $2d$, which is longer tailed than a Gaussian. As typical examples fitted to these functions, some herbivorous insect species show Gaussian distributions (Kareiva 1983), some seeds, vertebrates, midges, airborne spores or pollen have exponentially damping kernels (Kettle 1951; Fitt & McCartney 1986; Buechner 1987; Willson 1993; Higgins & Richardson 1999), and *Drosophila pseudoobscura* exhibits a fat-tail with $\nu = 1/2$ (Kettel 1951; Taylor 1978; Kot *et al.* 1996). Various dispersal kernels different from equation (17.1) have also been proposed by many authors (see Wolfenbarger 1975; Okubo 1980; Fitt *et al.* 1987; Okubo & Levin 1989; Portnoy & Willson 1993; Greene & Johnson 1995; Metz *et al.* 2000), among which the modified Bessel function of order zero kind, $\gamma K_0(\sqrt{\gamma / D} r)/(2\pi D)$, is often used when organisms become sedentary at a rate γ while dispersing randomly with diffusion coefficient D (Broadbent & Kendall 1953; Williamson 1961; Shigesada 1980; Metz *et al.* 2000). As noted before, however, since most field data are taken within a range of less than a few hundred metres, these fitted kernel functions are useful only for a limited range, even though formally they may appear to be applicable to an infinite expanse of space (Kot *et al.* 1996).

The dispersal pattern of long-distance dispersers could be qualitatively different from that of short-distance dispersers, since dispersal processes are mediated by different agents. However, because of the rarity of such events and a paucity of information on the maximal possible leap distance, no systematic measurements have been made on distance distributions, though some mammals, insects and seeds have been reported to travel more than several kilometres (Wilkinson 1997; Cain *et al.*

1998; Higgins & Richardson 1999). As extreme cases, airborne fungal, bacterial and viral spores are known to leap hundreds of kilometres (Fitt *et al.* 1987; Kot *et al.* 1996; Brown *et al.*, this volume). Recently, Clark (1998) and Higgins and Richardson (1999) have independently developed statistical methods for parameterizing the dispersal kernels from field data on seed dispersal. They assumed that a dispersal kernel is given by a composite function (or continuous mixture) of short- and long-distance dispersal kernels, $k_S(r)$ and $k_L(r)$, respectively, as:

$$k(r) = (1-p)k_S(r) + pk_L(r) \tag{17.2}$$

where p is the fraction allocated to long-distance dispersal. However, from their spatially explicit individual simulation, Higgins and Richardson (1999) suggested that most existing data sets on dispersal are insufficient to parameterize the long-distance dispersal kernel. Thus we need to develop a mathematical tool for estimating the dispersal kernel from actual data on range expansion (Clark *et al.* 1999).

Long-distance dispersal poses an additional theoretical problem regarding the establishment of satellite colonies. Since long-distance dispersal scatters the organisms sparsely, it may cause an increased risk of extinction due to demographic stochasticity, inbreeding depression or the Allee effect.

In this chapter we will introduce several models that deal with short- and long-distance dispersal as well as the Allee effect and/or demographic stochasticity. In the next section, we describe a classic diffusion model that has successfully explained many cases of range expansion caused by simple diffusion; the model is then extended to more general cases that involve long-distance dispersal or environmental heterogeneity. Subsequent sections present two different mathematical models: (i) the stratified diffusion model; and (ii) the integral kernel-based models—both of which explicitly deal with short-distance and long-distance dispersal. By solving these models either analytically or numerically, we present some formulae for the rate of spread and explore how long-distance dispersal and reproduction interplay to accelerate the speed, or how the Allee effect or demographic stochasticity decelerates it. Finally these models are compared to each other and with other more general approaches.

Diffusion model

When invading organisms undergo random dispersal while multiplying their number, the simplest possible model to describe the spatial spread of the population in two dimensions is the Fisher equation (Fisher 1937; Skellam 1951):

$$\frac{\partial n}{\partial t} = \nabla \cdot D\nabla n + (\varepsilon - \mu n)n \tag{17.3}$$

where $n(\boldsymbol{x},t)$ denotes the population density at time t and spatial coordinate $\boldsymbol{x} = (x, y)$, and the operator ∇ represents the spatial gradient. The left-hand side indicates the change in the population density with time, which is caused by random diffusion and local population growth of the logistic type, expressed respectively by the first and second terms on the right-hand side. ε is the intrinsic rate of increase, μ represents the effect of intraspecific competition on the reproduction rate, and D is the diffusion coefficient, the quadruple of which is a measure of the mean square of dispersal distance per unit time. If the individuals disperse solely by random diffusion without growth, solving equation (17.3) for $\varepsilon = \mu = 0$ under the initial condition $n(\boldsymbol{x},0) = \delta(\boldsymbol{x})$ provides the probability density that an organism initially released at the origin will be found at position $\boldsymbol{x}$ and at time t as $\exp\{-r^2/(4Dt)\}/(4\pi Dt)$ where $r^2 = x^2 + y^2$. Thus the dispersal kernel for random diffusion conforms to equation (17.1) for $v=2$ and $d^2=4Dt$.

Since Skellam (1951) applied equation (17.3) to the range expansion of muskrats, the diffusion model has been widely used as a fundamental equation to describe various biological invasions ranging from bacterial colony growth (Kawasaki *et al.* 1997), population genetics, ecology, epidemiology and even to the spread of human cultures (Okubo 1980; Ammerman & Cavalli-Sforza 1984; Williamson 1986; 1996; Hengeveld 1989; Murray 1989; Renshaw 1991; Banks 1994; Shigesada & Kawasaki 1997; Turchin 1998). For instance, by solving equation (17.3) under the conditions, $\mu>0$ and $\varepsilon>0$, we can show that the initial propagules invading the origin evolve to expand their range in a concentric manner and the range front tends to advance at a constant velocity $2\sqrt{\varepsilon D}$, while the population density within the range is maintained at the carrying capacity, ε/μ. The range front forms a continuously decaying tail of an exponential form, $\exp(-\sqrt{\varepsilon/D}r)$, without showing any patchy distribution, because diffusion model (17.3) is deterministic, neglecting stochastic aspects. Overall, the range expands in concentric circles at a constant speed, conforming to the type 1 pattern shown in Fig. 17.1a. It should be pointed out that essentially the same conclusion is obtained when the logistic growth function is replaced by a more generalized function, but having no Allee effect, $f(n)$, such that $f(n) \geq 0$ for $0 \leq n \leq K$, $f(K) = 0$, and $f'(0)n \geq f(n)$ for $n > 0$. More exactly, the frontal wave advances at a constant velocity $2\sqrt{\varepsilon D}$, where we put $\varepsilon=f'(0)$.

The effect of long-range dispersers in the framework of a one-dimensional diffusion model was first examined by Goldwasser *et al.* (1994) for a population that consists of two classes of individuals, dispersers and non-dispersers, as:

$$\frac{\partial n}{\partial t} = D\frac{\partial^2 n}{\partial x^2} + p\, f(n+s)$$

$$\frac{\partial s}{\partial t} = (1-p) f(n+s) \qquad (17.4)$$

where $n(x,t)$ represents the population density of dispersers and $s(x,t)$ that of non-dispersers. $f(n)$ is the growth rate at density n, and p is the probability of any reproduced individual being a disperser. If we regard the non-dispersers and dispersers as extreme cases of short- and long-distance dispersers, respectively, the model enables us to quantify the effect of composite dispersal on the invasion speed. When the growth function $f(n)$ does not exhibit an Allee effect as in the above-mentioned case, J. Cook (unpublished; see also Goldwasser *et al.* 1994) showed analytically that the solution of equation (17.4) yields a constant wave speed of:

$$V = \sqrt{\varepsilon D}(1 + \sqrt{p}) \qquad (17.5)$$

where $\varepsilon = f'(0)$. When all individuals disperse ($p = 1$), the speed conforms to that of the Fisher model, $2\sqrt{\varepsilon D}$, as expected. However, when p tends to zero, speed is merely reduced to half that of the Fisher model, $\sqrt{\varepsilon D}$, rather than to zero. The discrepancy between this asymptotic speed and the true speed at $p=0$ (i.e. $V=0$) may be explained as follows. Even if p is infinitesimally small, the dispersal kernel (Gaussian in the present case) extends to infinity so that organisms disperse over an infinite expanse instantaneously. Therefore, however small the density of propagules, it would increase exponentially to form a travelling wave after a sufficient lapse of time. Accordingly, the speed at $p \to 0$ stands for the ultimate speed reached after infinite time. Such a paradoxical result would not arise in more realistic situations where an invasion occurs within a finite time, and demographic stochasticity or the Allee effect are also present. In subsequent sections, this issue will be investigated with models of different frameworks.

The diffusion model is further extendable to cases in which the local growth function and the diffusion coefficient involve complicated realistic situations (Okubo 1980; Shigesada *et al.* 1986; Murray 1989; Lewis & Kareiva 1993; Banks 1994; Shigesada & Kawasaki 1997; Keitt *et al.* 2001; Wang & Kot 2001). For instance, environmental heterogeneity may inevitably lead to spatially structured patterns of invasion. When the environment is fragmented by habitat destruction in a regular manner, the diffusion model can sometimes provide analytical solutions for the invasion pattern and its rate of spread. Here we consider a single-species invasion in a heterogeneous environment in which favourable and unfavourable habitats are arranged regularly in a striped pattern, where the habitat widths are assigned l_1 and l_2, respectively (Fig. 17.2). Such heterogeneity may be produced by segmenting an original favourable habitat into alternating belts. Thus, diffusion model (17.3) is modified to incorporate the intrinsic growth rate, ε, and the diffusion coeficient, D, varying with habitats as:

$$D(x) = \begin{cases} D_1 \\ D_2 \end{cases}; \quad \varepsilon(x) = \begin{cases} \varepsilon_1 & \text{(in favourable habitats)} \\ \varepsilon_2 & \text{(in unfavourable habitats)} \end{cases} \qquad (17.6)$$

The intrinsic growth rate in the unfavourable habitats, ε_2, is lower than in the favourable habitats, ε_1 ($\varepsilon_1 > \varepsilon_2$), and can be negative (i.e. the birth rate is lower than the death rate). The relation between D_1 and D_2 varies with species so that no particular condition is imposed on them, except that they are positive. By appropriately choosing the units for the variables, we can put $\varepsilon_1 = D_1 = \mu = 1$ (Shigesada & Kawasaki 1997).

Figure 17.2 shows a snapshot of a numerical solution of equations (17.3) with (17.6), obtained when a few propagules were initially released at the centre. The population grows faster in the favourable habitat than in the unfavourable habitat, resulting in range expansion with a wavy range front. The global shape of the contour map is oval like (but not oval in the mathematical sense). The rate of spread

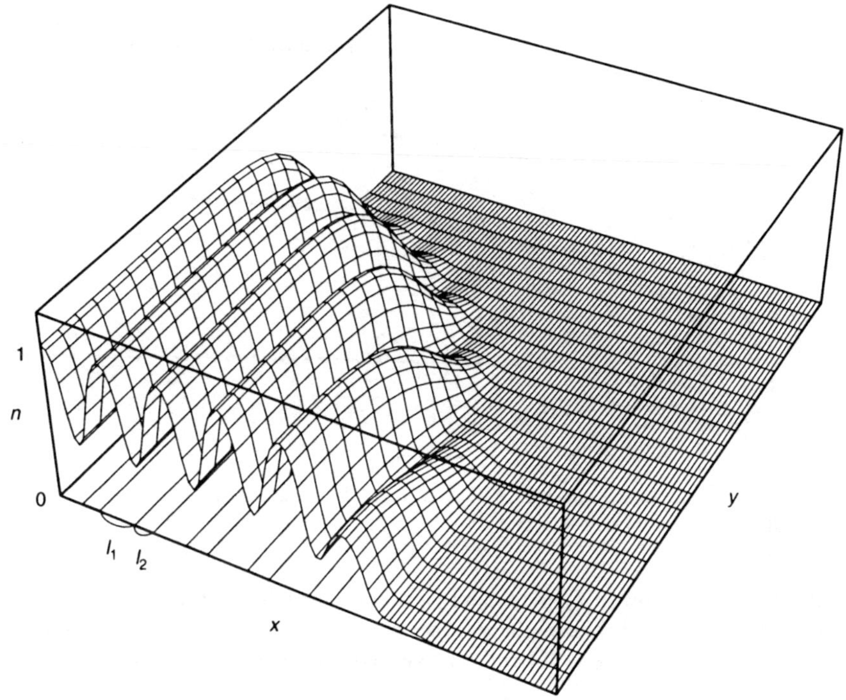

Figure 17.2 Range expansion in a periodically striped environment. Parameters are chosen as $D_1 = D_2 = \varepsilon_1 = \mu = 1, \varepsilon_2 = -0.5, l_1 = 40$ and $l_2 = 20$.

towards any radial direction tends to be periodic in accordance with the spatial period so that the average speed becomes constant, which is analytically determinable (Kinezaki *et al.*, in preparation). Overall, the range expands keeping a similar shape. In Fig. 17.3, the speed parallel to the stripes, c_y, and the speed across the stripes, c_x, are plotted as functions of D_2 for varying values of l_2, with the other parameters kept constant at $\varepsilon_2 = -0.5$ and $l_1 = 1$. The speed parallel to the stripes is always faster than that across the stripes. The difference in speed is most pronounced at $D_2 = 0$, and decreases with increasing D_2 to vanish at a certain value of D_2, beyond which it increases again, when l_2 is smaller than a threshold (two in the present case). Thus the range has an oval-like shape elongated in the direction of the stripes when D_2 is small or large; and at an intermediate D_2 it shows roughly a circular shape as in the homogenous environment (Fig. 17.4, upper row). On the other hand, when l_2 is larger than the threshold of two, the elongated shape becomes more rounded and shrunken with a larger D_2 and ultimately goes to extinction (Fig. 17.4, lower row). An intuitive interpretation of such a dramatic change in the range pattern is as follows. When D_2 is sufficiently small, organisms scarcely cross the adjacent unfavourable habitats, so that the range expands only along the favourable belt. As D_2 increases, they enter unfavourable habitats where some of them die before reaching the

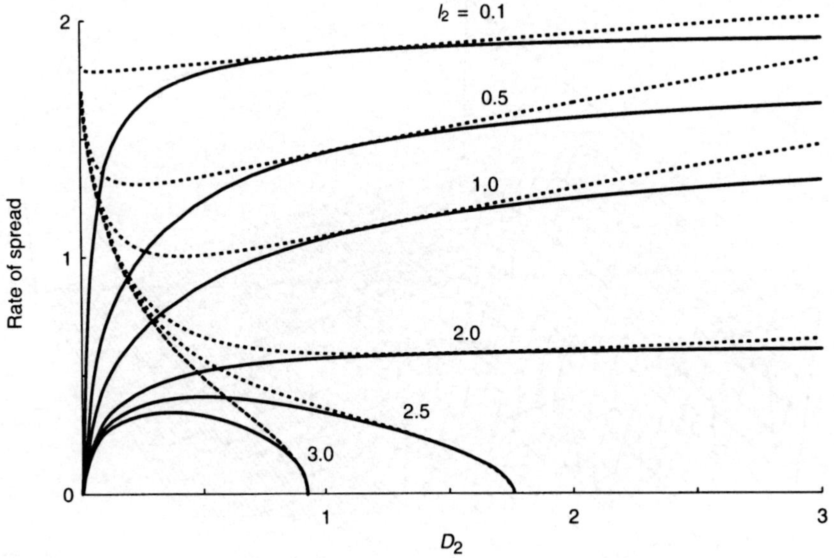

Figure 17.3 Rates of spread in a periodically striped environment as a function of D_2 for varying values of l_2. Solid lines show the speed across the stripes, c_x; and the dotted lines show the speed parallel to the stripes, c_y. The other parameters are fixed as $D_1 = \varepsilon_1 = \mu = l_1 = 1, \varepsilon_2 = -0.5$.

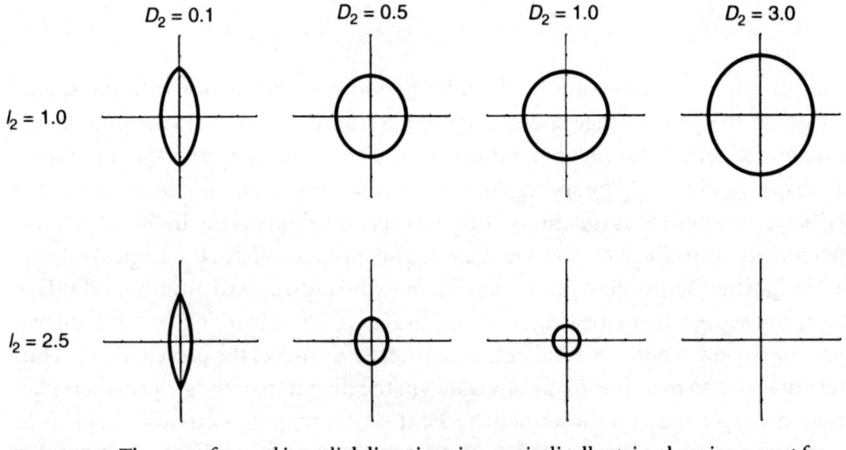

Figure 17.4 The rate of spread in radial directions in a periodically striped environment for the indicated sets of D_2 and l_2. The distance from the origin to a point on the solid curve represents the spread rate in that radial direction. The other parameters are the same as in Fig. 17.3.

favourable habitat lying ahead, thereby resulting in decreased overall speeds. With further increases in D_2, the speed goes up again at small l_2, because organisms quickly pass through unfavourable habitats to reach favourable habitats with no heavy loss. However, the speed monotonically decreases at large l_2, even going to zero, because most organisms die while passing through the wide unfavourable habitats.

Although the diffusion model can further accommodate interspecific interactions such as competition, predation or host–parasite interaction, it has difficulty in incorporating dispersal mechanisms other than random diffusion, and demographic stochasticity. The model in the following section focuses on the stochasticity of long-range dispersal and the Allee effect.

Stratified diffusion model

Although the diffusion model can reasonably explain the range expansion of type 1, it has difficulty in directly describing invasions of types 2 and 3, which involve disjunct patchy distributions due to rare long-distance dispersal events. Here we introduce a stratified diffusion model which hierarchically incorporates distant colonization in addition to the main range expansion by short-distance dispersal (Shigesada *et al.* 1995; Shigesada & Kawasaki 1997).

Let us first consider the simplest case where a few propagules of a single species invade a point in a homogenous environment in two dimensions and create there the nucleus of an isolated colony. As we saw in the previous section, if the invading organisms spread by short-range random diffusion, the occupied area forms a disk shape centred at the initial point of invasion, and its radius tends to increase at a constant rate, say c. At the same time, the population in each colony will produce long-distance migrants, which may create nuclei of new colonies at positions well separated from the parent colony. Since the long-distance dispersal is rare, we assume that the number of nuclei produced per unit time by long-distance migrants from a parent colony per unit area is a random variable chosen from a Poisson distribution with average λ, and their dispersal distances are probabilistically determined by a dispersal kernel $k(\boldsymbol{x})$. Generally, a successful colonization is achieved only when at least a few organisms (including a male–female pair in animals) settle in the same region and succeed in producing their offspring against the risk of extinction due to demographic stochasticity or the Allee effect. Thus, we assume that a new colony can establish only if its nearest neighbour colony is located within a distance r_A. The parameter, r_A, is referred to as the Allee radius, hereafter. In extreme cases, when $r_A = 0$, the range expansion occurs only by short-distance dispersal so that only the primary colony exists, while at $r_A = \infty$, no Allee effect is involved in colony establishment. Once a new nucleus is successfully established, it will expand its range by short-distance dispersal at a rate of c, while simultaneously producing long-distance migrants, just as the parent population does. Thus repeating these processes over again, the satellite colonies will show a patch-wise distribution surrounding the primary population, as schematically illustrated in Figs 17.1b and c.

The stratified diffusion model as introduced above constitutes the dispersal

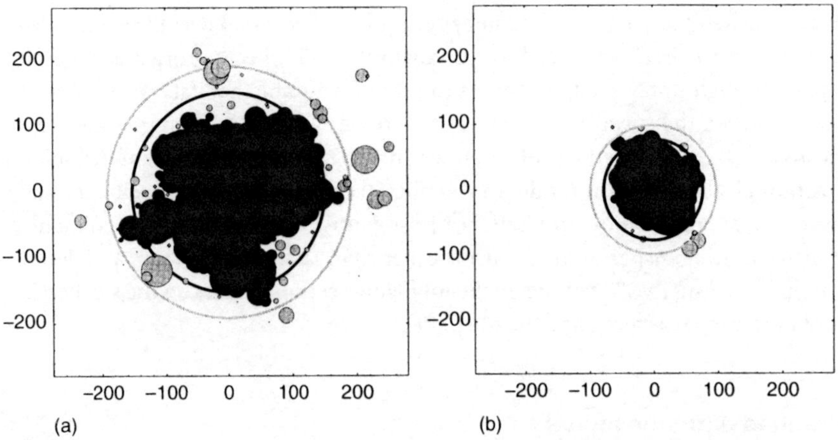

Figure 17.5 Snapshots of range expansion in the stratified diffusion model. The black and grey areas represent the primary and satellite colonies, respectively. The black and grey circles indicate the radial and scattering distances, respectively. (a) $r_A = \infty$; (b) $r_A = 20$. The other parameters are $\lambda = 0.001$, $d_L = 25$, $c = 1$ and $t = 60$.

kernel, $k(\mathbf{x})$, and three parameters: (i) the rate of spread by short-distance dispersal, c; (ii) the colonization rate, λ; and (iii) the Allee radius, r_A. Of these, c can be reduced to one if we take c as the spatial unit length, so that λc^2, $\mathbf{x}/c$ and r_A/c can be replaced by λ, $\mathbf{x}$ and r_A, respectively. Note that the actual radial distance and rate of spread are obtained by multiplying those obtained in the scaled unit by c.

In the following, we focus on the case where the dispersal kernel is an exponentially bounded function $k(r) = 1/(2\pi d_L^2)\exp(-r/d_L)$ (i.e. eqn (17.1) for $v = 1$), for which the mean dispersal distance is given by $2d_L$. d_L is referred to as the 'long-dispersal distance' hereafter. Figures 17.5a and b are snapshots of computer simulations for Allee radii, $r_A = \infty$ and $r_A = 20$, respectively, with fixed $\lambda = 0.001$, $d_L = 25$ and $c = 1$. The radial distance and scattering distance as defined in the introduction are indicated by the black and grey circles, respectively. The range expansion for the case of $r_A = 20$ is clearly restrained by the Allee effect, because the magnitude of Allee radius $r_A = 20$ is small compared with the mean dispersal distance $2d_L = 50$.

Figure 17.6 shows the radial and scattering distances plotted as functions of time, each for a single run, in the black and grey curves, respectively. The radial distance gradually increases at first and then approaches the later phase where it increases linearly with time, conforming to type 2 expansion as shown Fig. 17.1b. On the other hand, the scattering distance shows discontinuous changes, but its average for many runs gives a smoothed curve with a similar feature to that of the radial distance. In fact, we confirmed that an average of 50 simulations for each case gives a linear increase with the same slope in the later phase, where the standard deviations are less than 15 and 20 for the radial and scattering distances, respectively. Accordingly, the difference between the two distances in the later phase remains constant

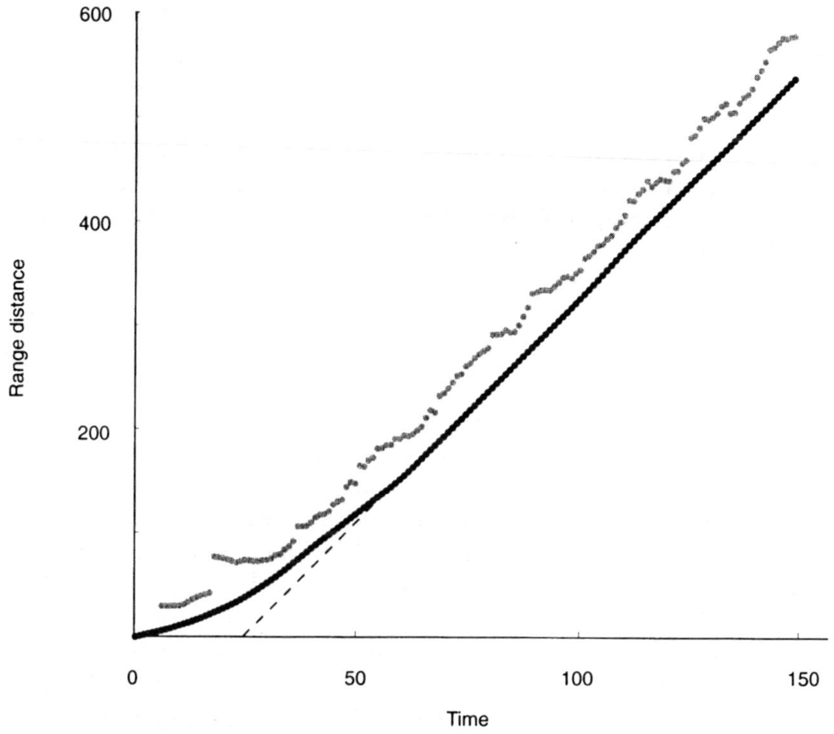

Figure 17.6 Time developments of the radial and scattering distances in the stratified diffusion model. The black and grey lines indicate the radial and scattering distances obtained from a single run of simulation, respectively. The thin dashed line represents a slope fitted to the later linear phase of the radial distance. The parameters are the same as in Fig. 17.5a.

with time. This difference, which reflects the degree of scattering of satellite colonies, becomes greater as d_L or r_A increases, but it does not depend so much on the value of λ.

We can also see from the time sequence of simulated spatial patterns that in the initial accelerating expansion phase, most colonies are small enough to grow independently without coalescing with each other; whereas in the later phase, colonies often collide to form new combined colonies. Thus we may expect that when dispersal distance d_L is sufficiently large, the initial accelerating expansion phase continues for a long time, resulting in type 3 expansion as illustrated in Fig. 17.1c. For example, when $d_L = 25$ and $\lambda = 0.001$, most colonies are isolated until $t = 30$. For the extreme case where neither coalescence nor the Allee effect exists, we can analytically derive the radial distance as a function of time (see Shigesada & Kawasaki 1997). By fitting that analytical solution to the patch-wise range expansion of cheat grass, *Bromus tectorum*, observed in the northwestern United States in early 1990s (Mack 1981), the colonization rate is estimated to be $\lambda = 1.1 \times 10^{-5}\,\text{km}^{-2}\,\text{year}^{-1}$.

We now compare the speeds in the later phase for various values of d_L, λ and r_A. As seen in Fig. 17.7a, the speed increases almost linearly with d_L for given values of r_A. On the other hand, the speed sharply rises from $c\,(=1)$, as the colonization rate λ increases from zero, particularly when r_A is large (Fig. 17.7b). However, as r_A decreases (the Allee effect is intensified), the initial rise becomes less pronounced. Furthermore, the Allee effect acts to reduce the speed for any value of λ. This is because the density at the leading edge is always low enough to invoke the Allee effect, irrespective of the magnitude of λ.

Particularly in the case of $r_A = \infty$ (no Allee effect), we obtain a heuristic formula for the speed, V, as:

$$V = 4.2cd_L\sqrt{\lambda} + c \tag{17.7}$$

Here we reasoned that, since the speed increases almost linearly with d_L (see Fig. 17.7a), it could also be proportional to $c\sqrt{\lambda}$ from dimension analysis. The coefficient 4.2 is chosen by regression. In fact, this function, as plotted by the dotted lines in Figs 17.7a and b, agrees well with the simulated results provided $\lambda < 0.005$. Unlike speed obtained from the diffusion model (eqn 17.5), speed in the present model (eqn 17.7) continuously increases from c as λ increases from zero. In comparing equations (17.5) and (17.7), it should be noted that p, the fraction of long-range dispersers as seen in equation (17.5), is proportional to λ, the number of long-range dispersers generated per unit area. This difference in the expression for speed comes from the fact that the stratified diffusion model deals with long-range dispersal as a stochastic process and the speed is obtained by averaging a finite number of sample processes for a finite duration of time. Thus we can conclude that the demographic stochasticity dramatically decreases the rate of spread when the number of long-range dispersers is very small. In addition, the Allee effect, if it exists, further decelerated the rate of spread for any value of λ.

In some cases, the dispersal kernel has a unimodal distribution with a peak at a certain distance from the origin (Beer & Swaine 1977; Neubert *et al.* 1995). For such a case, we have previously obtained an approximate formula for the rate of spread, which was applied to the spreads of the house finch and the European starling in North America (Shigesada & Kawasaki 1997; Takasu *et al.* 1997).

Integral kernel-based models

In the stratified diffusion model, we dealt with the local rate of spread, c, as a parameter directly obtainable from observed local expansion processes, while disregarding the detailed life history of invading organisms or their interaction with resident species. More mechanistic models that incorporate the fecundity schedule and dispersal have been presented in various mathematical frameworks such as renewal equations (Diekmann 1978, 1979; Thieme 1979; van den Bosch *et al.* 1990, 1999; Metz *et al.* 2000), integral differential equations (Mollison 1977) and integrodifference equations (Kot *et al.* 1996; Veit & Lewis 1996; Clark 1998; Clark *et al.* 1999; Neubert & Caswell 2000). If dispersal and population spread occur on a one-

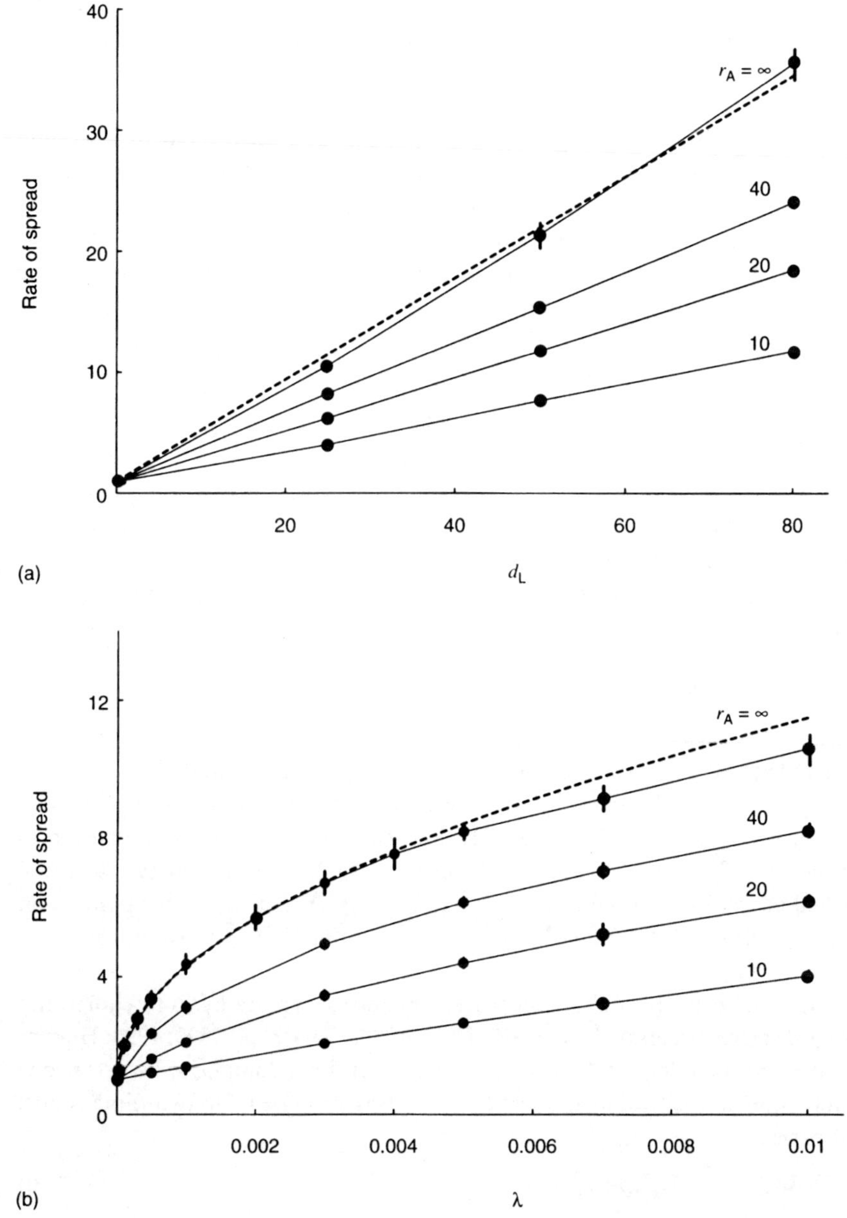

Figure 17.7 The rate of spread measured by the radial distance in the later linear phase for various values of Allee radius, r_A. The solid dots are simulated results: (a) plots as a function of d_L for $\lambda = 0.01$ and $c = 1$; and (b) plots are a function of λ for $d_L = 25$ and $c = 1$. The dashed lines are spread rates for $r_A = \infty$ calculated from the heuristic equation (17.7).

dimensional domain (or if, in a two-dimensional expansion, the range has already expanded enough so that its leading front is approximated by a straight line), these models are analytically tractable for the asymptotic invasion speed, provided that a travelling wave solution exists (Metz *et al.* 2000).

To capture the effect of long-distance dispersal concisely, here we focus on the unstructured integrodifference model presented by Kot *et al.* (1996). This model deals with the range expansion of a single species which grows and disperses in synchrony once a year on a continuous one-dimensional habitat. Let $n_t(y)$ denote the population density at location y and at time t. Then the population density at x in the next year is given by summing the contribution from all location y as:

$$n_{t+1}(x) = \int_{-\infty}^{\infty} k(x-y)F(n_t(y))dy \tag{17.8}$$

where $F(n_t(y))$ is the growth function at location y and $k(x)$ is a dispersal kernel defined in one-dimensional space. Kot *et al.* (1996) (see also Weinberger 1978, 1982; Lui 1982a, 1982b, 1983) demonstrated that when $F(n_t) \le F'(0)n_t$ (i.e. there is no Allee effect) and the dispersal kernel $k(x)$ has a tail not longer than an exponential bounded distribution, the solution of equation (17.8) starting from any localized initial distribution evolves to a travelling wave and its asymptotic speed, V, is given by the following formula:

$$V = \min_s \frac{\log R_0 M(s)}{s} \tag{17.9}$$

where $M(s) = \int_{-\infty}^{\infty} k(x)\exp(sx)dx$, and R_0 is the net reproductive rate defined by $F'(0) = R_0$. On the other hand, when the dispersal kernel is more fat-tailed than exponential, the speed increases, accelerating with time, showing type 3 range expansion as depicted in Fig. 17.1c (Mollison 1977; Kot *et al.* 1996). In reality, however, since the actual tail should be bounded by a maximum dispersal distance, the population wave speed ultimately tends to be a constant constrained by this maximum dispersal distance (Kot *et al.* 1996).

We now focus on a situation where organisms migrate by both short- and long-distance dispersal (Lewis 1997; Clark 1998; Clark *et al.* 1998, 1999; Higgins & Richardson 1999), so that the kernel is given by a composite kernel in one dimension formally similar to (17.2) that was proposed for two-dimensional dispersal:

$$k(x) = (1-p)k_S(x) + pk_L(x) \tag{17.10}$$

Let us first consider the case that k_S and k_L are given by the following exponential functions:

$$k_S(x) = \frac{1}{2d_S}\exp\left(-\left|\frac{x}{d_S}\right|\right) \quad \text{and} \quad k_L(x) = \frac{1}{2d_L}\exp\left(-\left|\frac{x}{d_L}\right|\right) \tag{17.11}$$

where the long-dispersal distance d_L is assumed to be much larger than that of the short-dispersal distance d_S ($d_S/d_L \ll 1$). Substituting equation (17.10) with equation

(17.11) into equation (17.9), we numerically obtain the wave speed for varying p, R_0, d_S and d_L.

The solid lines in Fig. 17.8a and b show the speed, V, as functions of the long- and short-dispersal distances, d_S and d_L, respectively for $R_0 = 3, 5, 10$ and $p = 0.01$. We can see that both the reproductive rate R_0 and the long-dispersal distance d_L crucially influence the speed, while the short-dispersal distance d_S scarcely does so. The dependency of the speed on the fraction allocated to long-distance dispersal p is most intriguing (Fig. 17.8c). In the absence of long-distance dispersal, the speed is calculated from formula (17.9) for $p = 0$, which is approximated by $V_S \approx 2d_S\sqrt{\log R_0}$ for small values of $\log R_0$ and $V_S \approx d_S \log R_0$ for large values of $\log R_0$. However, with the slightest shift of p above zero, it abruptly increases to $d_L \log R_0$ and then further increases gradually with increases in p.

Although equation (17.9) cannot be described in an explicit functional form, the following approximate expression for speed is derived by using perturbation expansion, when $d_S/d_L \ll 1$ and $p \ll 1$:

$$V = V_S \quad (\text{for } p = 0)$$

$$V \approx d_L\{\log R_0 + \sqrt{2p\log R_0}\} \quad \left(\frac{p}{2} \ll \log R_0 \ll \frac{1}{2p}\right) \qquad (17.12)$$

In Fig. 17.8, the approximate speed (eqn 17.12) is also plotted by dashed lines, which agree remarkably well with the theoretical speeds for a wide range of parameter values.

We also performed similar analyses for the case that the dispersal kernel of the short-distance dispersal is a Gaussian, $k_S(x) = 1/(\sqrt{\pi}\, d_S)\exp[-(x/d_S)^2]$, while the dispersal kernel of the long-range dispersal follows the same exponential function as in equation (17.11). Surprisingly, the wave speeds numerically solved from equation (17.9) show almost the same curves as the solid curves in Fig. 17.8a–c. In fact, perturbation expansion of the present case gives the same approximate expression as equation (17.12). Thus it seems that the invasion speed is almost completely determined by the long-distance dispersal events, even if they are rare. More exactly, the speed is mainly defined by the product of the long-dispersal distance, d_L, and reproductive rate, $\log R_0$ (i.e. the total reproductive rate including short- and long-distance dispersers), but it is indifferent to short-distance dispersal regardless of what dispersal kernel is involved. This unexpected property may arise for the same reason as the counterintuitive result obtained from the diffusion model (eqn 17.5): one major factor could be the lack of demographic stochasticity.

Recently, Lewis and Pacala (2000) examined the effect of demographic stochasticity in the integrodifference model (eqn 17.8) by adopting a moment expansion method and evaluated patchiness in terms of a spatial covariance function. Furthermore, Lewis (2000) demonstrated that when non-linear interactions are acting locally over small neighbourhoods, the spread rate becomes lower than that for the equivalent non-linear deterministic model. This feature is somewhat similar to that derived from the stratified diffusion model as described in the previous section.

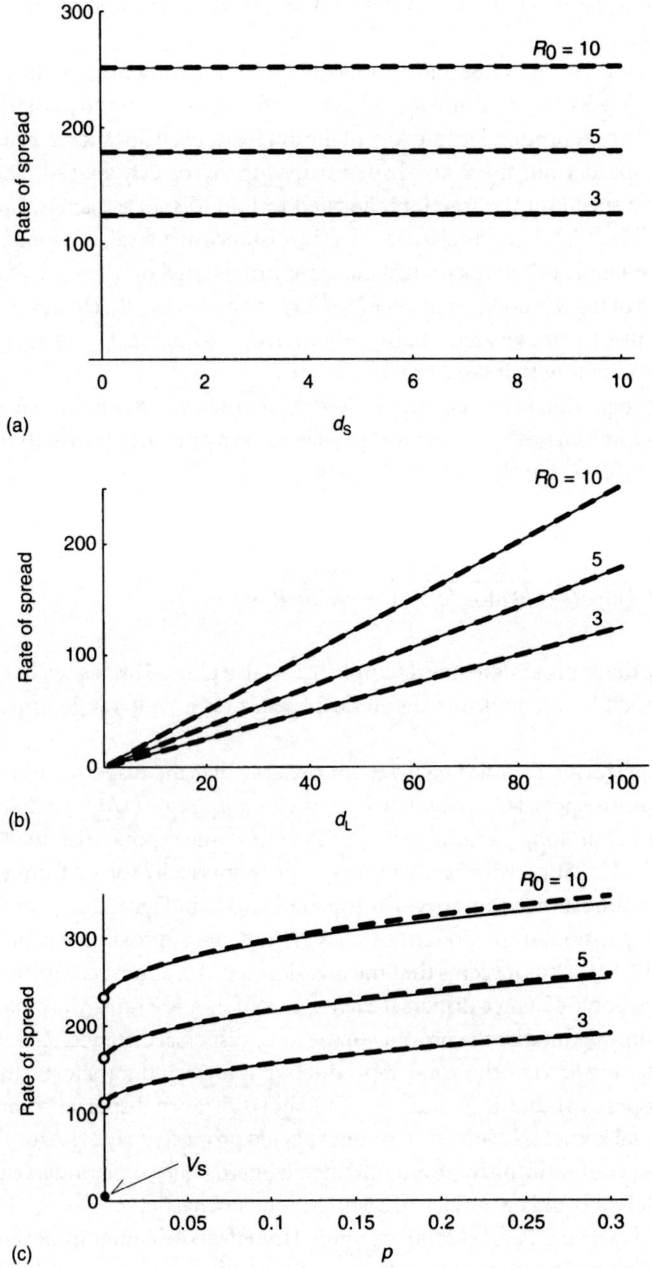

300
200
100
Rate of spread
$R_0 = 10$
5
3
0
2
4
6
8
10
(a)
d_S
200
100
0
Rate of spread
$R_0 = 10$
5
3
20
40
60
80
100
(b)
d_L
300
200
100
0
Rate of spread
$R_0 = 10$
5
3
V_S
0.05
0.1
0.15
0.2
0.25
0.3
(c)
p

However, Lewis has not addressed how the speed changes as the fraction of the long-range disperser, p, tends to zero. Here it should be noted that the stratified diffusion model automatically includes both demographic stochasticity and non-linear interactions in the contexts of long-distance dispersal events and a saturated density within the expanded range, respectively.

The model equation (17.8) can also allow us to investigate the Allee effect by appropriate modification of the growth function $F(n_t)$. In this case, however, no general formula is available for the wave speed as equation (17.9) (but for a special case see Kot *et al.* 1996). Thus we numerically solve equation (17.8) in which the Allee effect is superimposed on the growth function of Ricker type as:

$$F(n) = \frac{n\exp\left\{r\left(1-\frac{n}{K}\right)\right\}}{1+\exp\{-(n-n_A)/a\}} \qquad (17.13)$$

where n_A represents a threshold density (the Allee threshold) below which the population size decreases next year and $1/a$ is a measure of how sharply the growth function $F(n)$ changes around $n = n_A$ (Fig. 17.9). At the extreme of $a = 0$, $F(n)$ stays zero for $n < n_A$, while it converges to the growth function without the Allee effect for $n > n_A$.

The numerical results of equation (17.8) for various values of n_A with $r = \log 3$, $K = 4$ and $a = 10^{-8}$ are illustrated in Fig. 17.10. In the presence of the Allee effect, even if both n_A and a are very small, the speed continuously drops to V_S when p approaches zero. Thus the presence of the Allee effect helps to resolve the apparent enigmatic discontinuous behaviour of model (17.8). Furthermore, the Allee effect causes decreased speeds for any value of p. This feature is attributable to sparse population densities at the leading edge, as explained for Fig. 17.7b in the stratified diffusion model.

More mechanistic models incorporating the Allee effect are explored by Veit and Lewis (1996) for the spread of the house finch in eastern North America, and by Takasu *et al.* (2000) for the spread of pine wilt disease in Japan. In particular, Takasu *et al.* estimated the proportion of long-range dispersers in the population of the disease vector (the pine sawyer beetle) on the basis of field data.

Figure 17.8 (*opposite*) The rate of spread in the integrodifference model with a composite dispersal kernel as given by equation (17.10) with (17.11). The solid curves and the dashed curves represent numerical results from equation (17.9) and approximated results from equation (17.12), respectively. (a) Spread rates as a function of d_S for $R_0 = 3, 5$ and 10. Other parameters are fixed as $d_L = 100$ and $p = 0.01$. (b) Spread rates as a function of d_L for $R_0 = 3, 5$ and 10. Other parameters are fixed as $d_S = 1$ and $p = 0.01$. (c) Spread rates as a function of p for $R_0 = 3, 5$ and 10. Other parameters are fixed as $d_S = 1$ and $d_L = 100$. The solid dot indicates the spread rate in the absence of long-distance dispersal, V_S, whose value is 2.5, 3.2 and 4.2 for $R_0 = 3, 5$ and 10, respectively.

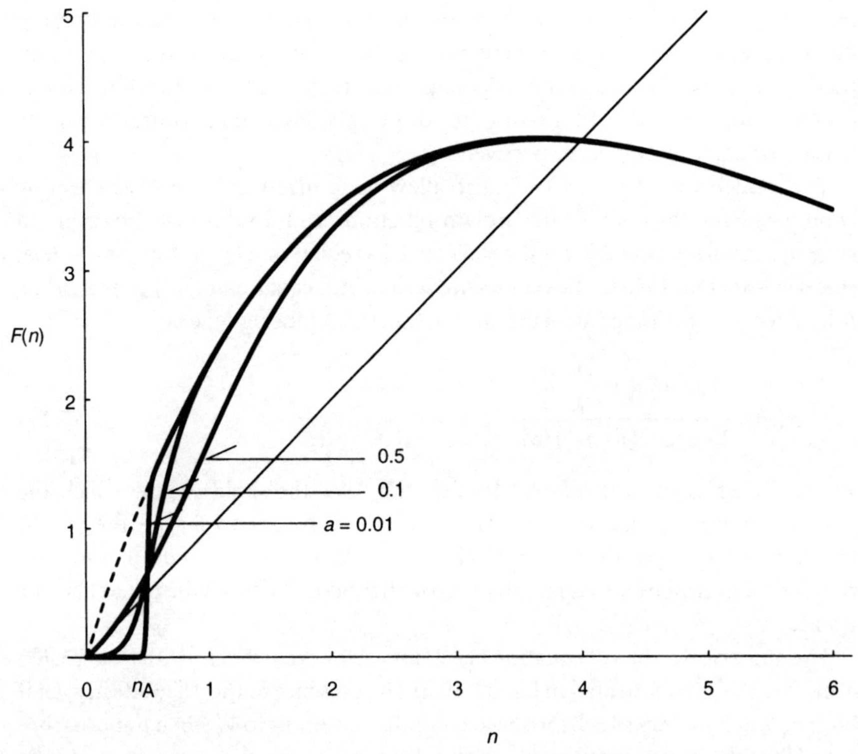

Figure 17.9 Properties of the proposed growth function of Ricker type with an Allee effect as defined by equation (17.13). The growth functions are plotted as a function of n for $a = 0.01$, 0.1 and 0.5 with fixed $n_A = 0.5$. The dashed line indicates the Ricker growth function with no Allee effect.

Conclusions

Three different models—the diffusion model, stratified diffusion model and integrodifference model—have been analysed to derive the rate of spread of a single unstructured population that expands their range by both short- and long-range dispersal. Particularly, in the absence of an Allee effect, we have derived analytical or approximate equations for the speed (eqns 17.5, 17.7 and 17.12) for the three models, respectively. The mathematical frameworks of these models are essentially different from each other. The diffusion model solely deals with continuous time and its dispersal kernel is restricted to a Gaussian distribution. By contrast, the other two models are adaptable to both continuous time and discrete time, and also can accommodate any functional form for the dispersal kernel, though in the present analyses, time is taken as discrete and the dispersal kernel is an exponentially damping function. Note here that the one-dimensional exponential kernel used in the integrodifference model does not correspond to the marginal kernel derived from the two-dimensional exponential kernel used in the stratified diffusion model.

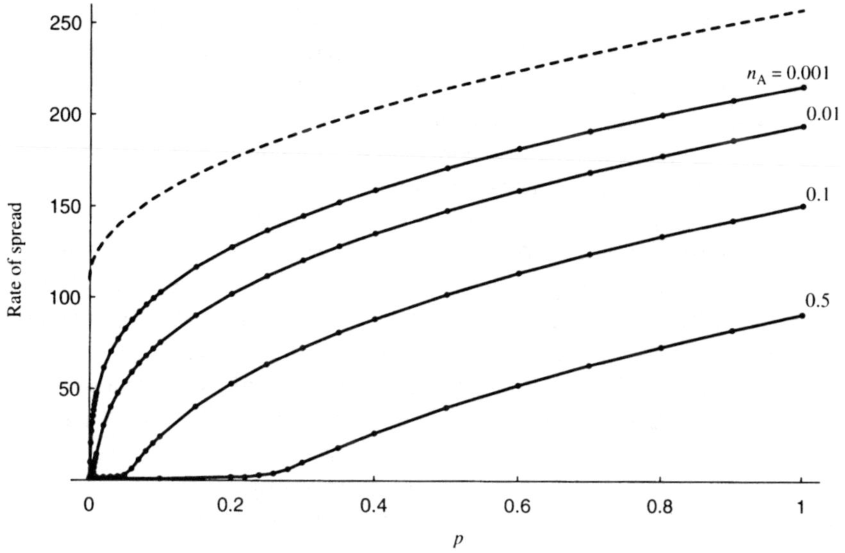

Figure 17.10 Influence of the Allee effect defined by equation (17.13) on the rate of spread in the integrodifference model. The spread rates are plotted as a function of p for varying values of n_A. Other parameters are fixed as $d_S = 1$, $d_L = 100$, $r = \log 3$, $a = 10^{-8}$ and $K = 4$. The dashed line represents the spread rate in the absence of the Allee effect obtained from equation (17.9).

Instead the former is derived from the two-dimensional kernel of the modified Bessel function as described in the introduction section (Metz *et al.* 2000).

In spite of such numerous differences among the three models, equations for the rate of spread (eqns 17.5, 17.7 and 17.12) show formally similar features to each other when long-range dispersers are relatively rare. Those functions are roughly proportional to the product of average long-dispersal distances ($\sqrt{D}$, d_L and d_L) and terms related to reproduction ($\sqrt{\varepsilon}$, $c\sqrt{\lambda}$ and $\log R_0$): namely $\sqrt{D\varepsilon}$, $d_L c\sqrt{\lambda}$ and $d_L \log R_0$ for equations (17.5, 17.7 and 17.12), respectively. It should be pointed out, however, that ε and $\log R_0$ stand for regeneration rates of both long- and short-distance dispersers, while λ represents that of long-distance dispersers alone. Therefore, when the number of long-distance dispersers approaches zero, the speed in the stratified diffusion model tends to zero, whereas the speeds in either the diffusion–reaction model or the integrodifference model does not vanish because ε and $\log R_0$ have positive values owing to regeneration of short-distance dispersers. As we have already explained in the diffusion model section, the discontinuity in speed at $p = 0$ in the first and third models may arise because the rate of spread in these models is defined by the wave speed of the travelling wave, which is established after infinite time. In contrast, the speed in the stratified diffusion model exhibits no such anomaly because it is estimated from patchy colony distributions that are generated stochastically in a finite time. Coincidentally, Lewis (2000) recently analysed the effect of demographic stochasticity in the integrodifference model when individuals interact locally over small neighbourhoods, and found that the spread in the stochastic model is slower than in the corresponding deterministic model.

However, even in the deterministic integrodifference model, the abrupt increase in speed at $p=0$ disappears if we incorporate any of the following factors: (i) the Allee effect; (ii) a long-distance dispersal kernel truncated within a certain maximum distance; or (iii) a long-range dispersal kernel with a Gaussian distribution. We have already demonstrated how the Allee effect lowers the spread rate for the stratified diffusion model and the integrodifference model. As for the latter two cases (ii and iii), we can verify from equation (17.9) that the speed increases with p in a manner similar to those in the presence of the Allee effect as shown in Fig. 17.10. This outcome probably occurs because the tail of the dispersal kernel is absent or too thin to support a travelling wave when $p \to 0$. Note, however, that in the diffusion model a Gaussian kernel yields the discontinuous increase in speed as given in equation (17.5). This discrepancy may be ascribed to the fact that dispersal takes place in continuous time in the diffusion model, but in discrete time in the integrodifference model.

Although we have restricted our analyses to the simplest models neglecting demographic structures, there have been a number of more generalized models that incorporate age- or stage-specific data on demography and dispersal. Continuous-time models including age structures have been developed by Thieme (1977, 1979), van den Bosch *et al.* (1990) and Metz *et al.* (2000). A discrete-time model (matrix model) based on stage-specific data was extensively examined for plants by Neubert and Caswell (2000; see also Bullock *et al.*, this volume). The recent review by Metz *et al.* (2000) gives elaborate comparisons between the different models so far presented. There they pointed out that in continuous-time integral differential models, the speed again shows an abrupt increase when the long-distance dispersal kernel is a Gaussian, as in the diffusion model described above. Therefore, we have to be cautious when choosing models to deal with actual ecological situations. Generally speaking, the stratified diffusion model is advantageous when data are available on the local rate of spread of patchy colonies but not on the demography of local populations, while the integral kernel-based models are more suitable if detailed demographic data are known and long-distance dispersal is not extremely rare.

References

Ammerman, A.J. & Cavalli-Sforza, L.L. (1984) *The Neolithic Transition and the Genetics of Populations in Europe.* Princeton University Press, Princeton, NJ.

Andow, D., Kareiva, P., Levin, S. & Okubo, A. (1990) Spread of invading organisms. *Landscape Ecology* **4**, 177–188.

Andow, D., Kareiva, P., Levin, S. & Okubo, A. (1993) Spread of invading organisms: patterns of spread. In: *Evolution of Insect Pests: The Pattern of Variations* (ed. K.C. Kim), pp. 219–242. John Wiley & Sons, New York.

Banks, R.B. (1994) *Growth and Diffusion Phenomena.* Springer Verlag, Berlin.

Beer, T. & Swaine, M.D. (1977) On the theory of explosively dispersed seeds. *New Phytologist* **78**, 681–694.

Broadbent, S.R. & Kendall, D.G. (1953) The random

walk of *Trichostrongylus Retorateformis. Biometrics* **9**, 460–466.
Buechner, M. (1987) A geometric model of vertebrate dispersal: tests and implications. *Ecology* **68**, 310–318.
Cain, M.L., Damman, H. & Muir, A. (1998) Seed dispersal and the Holocene migration of woodland herbs. *Ecological Monographs* **68**, 325–347.
Caswell, H. (2000) *Matrix Population Models*, 2nd edn. Sinauer Associates, Sunderland, MA.
Caughley, G. (1970) Liberation, dispersal and distribution of himalayan thar (*Hemitragus jemlahicus*) in New Zealand. *New Zealand Journal of Forestry Science* **13**, 200–239.
Clark, J.S. (1998) Why trees migrate so fast: confronting theory with dispersal biology and the paleorecord. *American Naturalist* **152**, 204–224.
Clark, J.S., Fastie, C., Hurtt, G. *et al.* (1998) Reid's paradox of rapid plant migration. *BioScience* **48**, 13–24.
Clark, J.S., Silman, M., Kern, R., Macklin, E. & HilleRisLambers, J. (1999) Seed dispersal near and far: patterns across temperate and tropical forest. *Ecology* **80**, 1475–1494.
Clarke, C.M.H. (1971) Liberations and dispersal of red deer in northern South Island districts. *New Zealand Journal of Forestry Science* **1**, 194–207.
Davis, M.B. (1987) Invasion of forest communities during the Holocene: beech and hemlock in the Great Lakes region. In: *Colonization, Succession, and Stability* (eds A.J. Gray, M.J. Crawley & P.J. Edwards), pp. 373–393. Blackwell Scientific Publications, Oxford.
Diekmann, O. (1978) Thresholds and travelling waves for the geographical spread of infection. *Journal of Mathematical Biology* **6**, 109–130.
Diekmann, O. (1979) Run for your life. A note on the asymptotic speed of propagation of an epidemic. *Journal of Differential Equations* **33**, 109–130.
Elton, C.S. (1958) *The Ecology of Invasion by Animals and Plants.* Methuen, London.
Fisher, R.A. (1937) The wave of advance of advantageous genes. *Annals of Eugenics* **7**, 255–369.
Fitt, B.D. & McCartney, H.A. (1986) Spore dispersal in splash droplets. In: *Water, Fungi and Plants* (eds P.G. Ayres & L. Boddy), pp. 7–104. Cambridge University Press, Cambridge.
Fitt, B.D., Gregory, P.H., Todd, A.D., McCartney, H.A. & Macdonald, O.C. (1987) Spore dispersal and plant disease gradients; a comparison between two empirical models. *Journal of Phytopathology* **118**, 227–242.
Goldwasser, L., Cook, J. & Silverman, E.D. (1994) The effects of variability on metapopulation dynamics and rates of invasions. *Ecology* **75**, 40–47.
Greene, D.F. & Johnson, E.A. (1995) Long-distance wind dispersal of tree seeds. *Canadian Journal of Botany* **73**, 1036–1045.
Hengeveld, R. (1988) Mechanisms of biological invasions. *Journal of Biogeography* **15**, 819–828.
Hengeveld, R. (1989) *Dynamics of Biological Invasions.* Chapman & Hall, London.
Higgins, S.K. & Richardson, M. (1999) Predicting plant migration rates in a changing world: the role of long-distance dispersal. *American Naturalist* **153**, 464–475.
Kareiva, P.M. (1983) Local movement in herbivorous insects: applying a passive diffusion model to mark–recapture field experiments. *Oecologia* **57**, 322–327.
Kawasaki, K., Mochizuki, A., Matsushita, M., Umeda, T. & Shigesada, N. (1997) Modeling spatio-temporal patterns generated by *Bacillus subtili. Journal of Theoretical Biology* **188**, 177–185.
Keitt, T.H., Lewis, M.A. & Holt, R.D. (2001) Allee effects, invasion pinning, and species' borders. *American Naturalist* **157**, 203–216.
Kessel, B. (1953) Distribution and migration of the European starling in North America. *Condor* **55**, 49–67.
Kettle, D.S. (1951) The spatial distribution of *Culicoides impunctatus* Goet. Under woodland and moorland conditions and its flight range through woodland. *Bulletin of Entomological Research* **42**, 239–291.
Kinezaki, N., Kawasaki, K., Takasu, F. & Shigesada N. Modelling biological invasions into fragmented environments. (In preparation.)
Kot, M., Mark, A.L. & van den Driessche, P. (1996) Dispersal data and the spread of invading organisms. *Ecology* **77**, 2027–2042.
Lewis, M.A. (1997) Variability, patchiness and jump dispersal in the spread of an invading population. In: *Spatial Ecology* (eds D. Tilman & P. Kareiva), pp. 46–49. Princeton University Press, Princeton. NJ.
Lewis, M.A. (2000) Spread rate for a nonlinear sto-

chastic invasion. *Journal of Mathematical Biology* **41**, 430–454.

Lewis, M.A. & Kareiva, P. (1993) Allee dynamics and the spread of invading organisms. *Theoretical Population Biology* **43**, 141–158.

Lewis, M.A. & Pacala, S. (2000) Modeling and analysis of stochastic invasion processes. *Journal of Mathematical Biology* **41**, 387–429.

Lubina, J.A. & Levin, S.A. (1988) The spread of a reinvading species: range expansion in the California sea otter. *American Naturalist* **131**, 526–543.

Lui, R. (1982a) A nonlinear integral operator arising from a model in population genetics. I. Monotone initial data. *SIAM Journal of Mathematical Analysis* **13**, 913–937.

Lui, R. (1982b) A nonlinear integral operator arising from a model in population genetics. II. Initial data with compact support. *SIAM Journal of Mathematical Analysis* **13**, 938–953.

Lui, R. (1983) Existence and stability of travelling wave solutions of a nonlinear integral operator. *Journal of Mathematical Biology* **16**, 199–220.

Macdonald, D.W. (1980) *Rabies and Wildlife.* Oxford University Press, Oxford.

Mack, R.N. (1981) Invasion of *Bromus tectorum* L. into western North America: an ecological chronicle. *Agro-Ecosystems* **7**, 145–165.

Metz, J.A.J. & Diekmann, O. (1986) *The Dynamics of Physiologically Structured Populations.* Springer Verlag, New York.

Metz, J.A.J. & van den Bosch, F. (1995) Velocities of epidemic spread. In: *Epidemic Models: their Structure and Relation to Data* (ed. D. Mollison), pp. 150–186. Cambridge University Press, Cambridge.

Metz, J.A.J., Mollison, D. & van den Bosch, F. (2000) The geometry of ecological interactions simplifying spatial complexity. In: *The Dynamics of Invasion Waves. Cambridge Studies in Adaptive Dynamics* (eds U. Diekmann & J.A.J. Metz), pp. 482–512. Cambridge University Press, Cambridge.

Mollison, D. (1977) Spatial contact models for ecological and epidemic spread. *Journal of the Royal Statistical Society, Series B* **39**, 283–326.

Mundinger, P.C. & Hope, S (1982) Expansion of the winter range of the house finch: 1947–79. *American Birds* **36**, 347–353.

Murray, J.D. (1989) *Mathematical Biology.* Springer Verlag, Berlin.

Murray, J.D., Stanley, E.A. & Brown, D.L. (1986) On the spatial spread of rabies among foxes. *Proceedings of the Royal Society of London, Series B* **229**, 111–150.

Nathan, R. & Muller-Landau, H.C. (2000) Spatial patterns of seed dispersal, their determinants and consequences for recruitment. *Trends in Ecology and Evolution* **15**, 278–285.

Neubert, M.G. & Caswell, H. (2000) Demography and dispersal: calculation and sensitivity analysis of invasion speed for structured populations. *Ecology* **81**, 1613–1628.

Neubert, M.G., Kot, M. & Lewis, M.A. (1995) Dispersal and pattern formation in a discrete-time predator–prey model. *Theoretical Population Biology* **48**, 7–43.

Nobel, J.V. (1974) Geographic and temporal development of plagues. *Nature* **250**, 726–728.

Okubo, A. (1980) *Diffusion and Ecological Problems: Mathematical Models.* Springer Verlag, New York.

Okubo, A. (1988) Diffusion-type models for avian range expansion. In: *Acta XIX Congress Internationalis Ornithologici, Vol. 1* (ed. H. Quellet), pp. 1038–1049. University of Ottawa Press, Ottawa.

Okubo, A. & Levin, S.A. (1989) A theoretical framework for data analysis of wind dispersal of seeds and pollen. *Ecology* **70**, 329–338.

Portnoy, S. & Willson, M.F. (1993) Seed dispersal curves: behavior of the tail of the distribution. *Evolutionary Ecology* **7**, 25–44.

Renshaw, E. (1991) *Modelling Biological Populations in Space and Time.* Cambridge University Press, Cambridge.

Shigesada, N. (1980) Spatial distribution of dispersing animals. *Journal of Mathematical Biology* **9**, 85–96.

Shigesada, N. & Kawasaki, K. (1997) *Biological Invasions: Theory and Practice.* Oxford Series in Ecology and Evolution. Oxford University Press, Oxford.

Shigesada, N., Kawasaki, K. & Teramoto, E. (1986) Traveling periodic waves in heterogeneous environments. *Theoretical Population Biology* **30**, 143–160.

Shigesada, N., Kawasaki, K. & Takeda, Y. (1995) Modeling stratified diffusion in biological invasions. *American Naturalist* **146**, 229–251.

Skellam, J.G. (1951) Random dispersal in theoretical populations. *Biometrika* **38**, 196–218.

Takasu, F., Kawasaki, K. & Shigesada, N. (1997) Simulation study of stratified diffusion model. *Forma* **12**, 167–175.

Takasu, F., Yamamoto, N., Kawasaki, K., Togashi, K. & Shigesada, N. (2000) Modeling the range expansion of an introduced tree disease. *Biological Invasion* **2**, 141–150.

Taylor, R.A.J. (1978) The relationship between density and distance of dispersing insects. *Ecological Entomology* **3**, 63–70.

Thieme, H.R. (1977) A model for the spread of an epidemic. *Journal of Mathematical Biology* **4**, 337–351.

Thieme, H.R. (1979) Density-dependent regulation of spatially distributed populations and their asymptotic speed of spread. *Journal of Mathematical Biology* **8**, 173–187.

Turchin, P. (1998) *Quantitative Analysis of Movement: Measuring and Modeling Population Redistribution in Animals and Plants.* Sinauer Associates, Sunderland, MA.

Van den Bosch, F., Metz, J.A.J. & Diekmann, O. (1990) The velocity of spatial population expansion. *Journal of Mathematical Biology* **28**, 529–565.

Van den Bosch, F., Metz, J.A.J. & Zadoks, J.C. (1999) Pandemics of focal plant disease, a model. *Phytopathology* **89**, 495–505.

Veit, R.R. & Lewis, M.A. (1996) Dispersal, population growth and the Allee effect: dynamics of the house finch invasion of North America. *American Naturalist* **148**, 255–274.

Wang, M. & Kot, M. (2001) Speeds of invasion in a model with strong or weak Allee effects. *Mathematical Biosciences* **171**, 83–97.

Weinberger, H.F. (1978) Asymptotic behavior of a model of population genetics. In: *Nonlinear Partial Differential Equations and Applications* (ed. J. Chadam), Lecture Notes in Mathematics No. 684, pp. 47–95. Springer Verlag, Berlin.

Weinberger, H.F. (1982) Long-time behavior of a class of biological models. *SIAM Journal of Mathematical Analysis* **13**, 353–396.

Wilkinson, D.M. (1997) Plant colonization: are wind dispersed seeds really dispersed by birds at larger spatial and temporal scales? *Journal of Biogeography* **24**, 61–65.

Williamson, E.J. (1961) The distribution of larvae of randomly moving insects. *Australian Journal of Biological Sciences* **14**, 598–604.

Williamson, M.H. (1996) *Biological Invasions.* Chapman & Hall, London.

Williamson, M.H. & Brown, K.C. (1986) The analysis and modelling of British invasion. *Philosophical Transactions of the Royal Society of London, Series B* **314**, 505–522.

Willson, M.F. (1993) Dispersal mode, seed shadows, and colonization patterns. *Vegetatio* **107/108**, 261–280.

Wolfenbarger, D.O. (1975) *Factors Affecting Dispersal Distances of Small Organisms.* Exposition Press, Hickswill, NY.

Chapter 18
Success factors in the establishment of human-dispersed organisms

Andy N. Cohen

Introduction

As recently as a decade ago, exotic organisms—species dispersed beyond their native ranges by human activities—rarely ranked high among recognized threats to biodiversity. Habitat loss and habitat alteration, including global climate change, usually headed lists of such threats. Pollution was also placed high, along with hunting and fishing, including the hunting of whales and fur seals, the harvesting of animal tusks and organs for ornament or medicine, the taking of sea turtles and coral reef fish, and the death of dolphins entangled in tuna nets. Exotic species were generally placed near the bottom, if indeed they were on the list at all. But perceptions are changing.

In 10 recent asssessments of threats to different groups of organisms, most studies concluded that exotic species are a leading threat to biodiversity (Table 18.1). For example, Wilcove *et al.* (1998) found that competition with, or predation by, exotic species constituted the second greatest threat to imperiled plants and animals in the United States, second to habitat loss or degradation, but affecting more than twice as many imperiled species as pollution, and nearly three times as many species as overexploitation. In 1998, the World Conservation Union ranked exotic species as the second greatest threat to biodiversity, after habitat loss (Raver 1999); and in 2000, the Director of the US Geological Survey predicted that the spread of invasive organisms would be the second most serious ecological problem facing the United States in the 21st century (Groat 2000).

Besides affecting biodiversity, the introduction of exotic organisms can have substantial economic, social and public health impacts—and we live in a world today in which those impacts are likely to grow. Unless countermeasures are taken, a massive increase in the rapid transport and release of organisms around the world is a predictable consequence of the current expansion in international trade. As new global markets are developed, organisms are moved both intentionally and accidentally along a growing number of pathways, between a changing array of source regions and destinations, via an alarming diversity of mechanisms. Among marine species, for example, organisms are transported between and across oceans as plankton in

Table 18.1 Ranking of factors in the imperilment, endangerment or extinction of species. 'ESA-listed' taxa are species, subspecies or vertebrate populations listed as threatened or endangered under the US Endangered Species Act.

Assessment	Ranking of exotic species	Reference
Factors contributing to the loss of the 40 recently extinct North American fish species and subspecies	2nd most frequent of 5 factors	Miller *et al.* 1989
Threats to the 364 imperiled North American fish species and subspecies. Exotic species included in one category in combination with hybridization, predation and competition	2nd most frequent of 4 categories of threats	Williams *et al.* 1989
Primary causes of endangerment for 98 ESA-listed plant species	Tied for 8th most frequent of 14 causes	Schemske *et al.* 1994
Primary threats to 1111 imperiled bird species worldwide	4th most frequent threat	Collar *et al.* 1994
General factors adversely affecting 667 ESA-listed species. Exotic species included in an 'interspecific interactions' factor, defined to include disease, predation and competition, particularly as associated with exotic species	2nd most frequent of 5 factors	Flather *et al.* 1994
Specific reasons contributing to the endangerment of 667 ESA-listed species (among reasons affecting at least 15% of species)	2nd most frequent of 18 reasons	Flather *et al.* 1994
Factors cited in listings of 68 ESA-listed fish	2nd most frequent of 3 factors	Lassuy 1995
Sources of stressors cited by biologists as causing historic declines in 135 imperiled species in the USA	4th most common of 20 sources	Richter *et al.* 1997
Sources of stressors cited by biologists as limiting recovery in 135 imperiled species in the USA	3rd most common of 20 sources	Richter *et al.* 1997
Factors contributing to the imperilment of 1880 taxa of plants and animals in the USA	2nd most frequent of 5 factors	Wilcove *et al.* 1998

the ballast tanks and seawater piping systems of cargo vessels, as sedentary organisms attached to the hulls of boats and ships, as breeding stock and food for aquaculture, as goods in the saltwater aquarium, live seafood and live bait trades, and as endozoic, epizoic or otherwise associated biota inadvertantly shipped with any of these. Terrestrial and freshwater species are transported by a similarly diverse set of vectors. These globe-trotting species will inevitably include some organisms that substantially alter native ecosystems on land and in the sea; pests of forests, crops and livestock; and human parasites and diseases, including emergent diseases and antibiotic-resistant strains of diseases that had been thought to be under control.

There has thus been much interest in understanding which traits of organisms make them more or less successful as invaders, and which characteristics of habitats make them more or less vulnerable to invasions. Charles Elton, for example, in *The Ecology of Invasions by Animals and Plants* (1958), noted that 'invasions most often come to cultivated land, or land much modified by human practice,' or to more natural areas that none the less 'have also suffered the results of human occupation.' He also suggested that islands were especially heavily invaded, devoting an entire chapter to island invasions. In a 1964 symposium on the genetics of colonizing species, organized by Herbert Baker and Ledyard Stebbins, the genetic, reproductive and life-history traits of successful invaders were frequent points of discussion. In his own presentation, Baker (1965a) compared congeneric pairs of plant species, one of which had spread widely and the other of which had not; and in an appendix to his talk provided an oft-cited list of the traits of the 'ideal weed'.

In the 1980s, the Scientific Committee on Problems of the Environment (SCOPE) of the International Council of Scientific Unions supported an extensive programme of enquiry into biological invasions, which focused on the questions of what makes a species a successful invader and what makes a site prone to invasion. The programme sponsored several international symposia that resulted in a series of publications described by Mooney and Drake (1989) and Williamson (1996). Contributions to these publications, and other papers published in the scientific literature, have suggested a bewilderingly large number of traits as being typical of successfully invading species, including such possibly contradictory traits as plant seeds being, on the one hand, large (Baker 1965b; Mayr 1965b), or, on the other hand, small (Rejmanek & Richardson 1996) and numerous (Mulligan 1965; Groves 1986; Pimental 1986); a lifespan that is short (Orians 1986; Di Castri 1990) or long (Crawley 1986; Möller 1996; Townsend 1996); and a body size that is small (Crawley 1986, 1987; Di Castri 1990) or large (Ehrlich 1986, 1989; Möller 1996; Townsend 1996).

Compelling data that either support or refute the various characteristics proposed for successful invaders or vulnerable environments have been difficult to come by, leading some workers to suggest that the search for broad, predictive characteristics may be futile (Simberloff 1986, 1989; Williamson 1996). Nevertheless, that certain traits do generally contribute to the success of invasions has been widely accepted by many researchers and resource managers, and used in some risk assessment protocols for invasions (Ruesink *et al.* 1995). Some of the most frequently cited generalizations are that organisms that invade successfully are opportunistic, '*r*-selected' species with a high reproductive output and wide environmental tolerances, are generalists in resource use, and have escaped from their natural predators or parasites in their new environment; and that islands, disturbed habitats and species-poor communities are particularly vulnerable to invasions. These propositions generally have their roots in the idea that the success or failure of invasions is governed by biotic interactions between the invading and resident species—that is, that species with these traits are in some broad sense better competitors than species lacking them, and that the more vulnerable types of environments host fewer or weaker competitors and predators.

In this chapter, I will examine the evidence for four of these propositions: (i) that islands are particularly vulnerable to invasions; (ii) that disturbed habitats are particularly vulnerable to invasions; (iii) that organisms that produce large numbers of young are more successful as invaders; and (iv) that organisms frequently succeed as invaders because they have left their native parasites behind. Specifically, I will ask whether there is evidence to support a conclusion that these characteristics increase

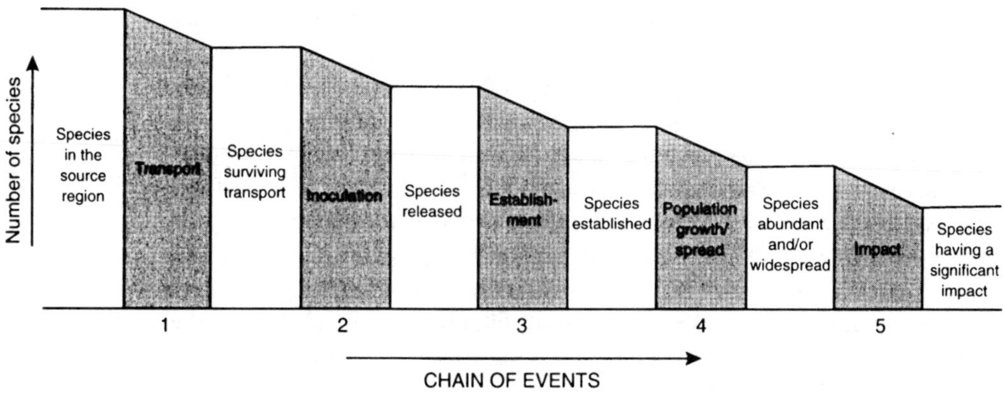

Figure 18.1 An invasion conceived as stages separated by steps or 'filters'.

the probability that a species released into a novel environment will become established there. First, however, I will refer to a conceptual model of invasions to define some of the terms of this enquiry, and discuss the types of evidence and analyses that have been offered in regard to these propositions.

A conceptual model of invasions

Figure 18.1 provides a visual model of invasions, similar to models proposed by other workers. An invasion is shown as a series of stages characterized by progressively shrinking sets of species—the species present in the source region, the smaller set of species that survive transport to the new region, and so on. Between these stages are steps or processes, distinguished by a grey background and numbered for discussion, labelled 'transport', 'inoculation', etc. It is helpful to think of these steps as filters, and our questions may then be framed as asking what species' traits make it more likely that a species will pass through one or other filter or group of filters, and what environmental characteristics make it more likely that a larger number of species will pass through.

Different researchers have been concerned in their investigations with different filters, and unfortunately, due to ambiguities in terminology or in the way they have presented their results, it has not always been clear to which filters their work refers. The main value of a model like this is that it allows us to discuss this without ambiguity, or at least with less ambiguity, even where different terminology has been used. Thus Williamson and Fitter (1996a, 1996b) and Williamson (1996) discuss patterns in the percentage of organisms that pass the second filter (here called 'inoculation', but which they call 'escaping'), the third filter (as here, called 'establishing') and the fourth and fifth filters ('becoming a pest'). Some research on invasion success has addressed other filters or combinations, such as which established organisms will spread in a new environment (the fourth filter), or which of the organisms released into an environment are likely to have an impact there (the third through fifth filters), or which of the organisms present in a region are likely to spread to other regions (the first through third filters).

In the general view, particular characteristics of species or environments have usually been perceived as either promoting or inhibiting the invasion process as a whole, without considering the potential for different effects at different stages of invasion. Characteristics associated with passage through one filter may differ from those associated with passage through another, and a characteristic that assists passage through one filter may even retard passage through another. For example, Crawley (1986) found that among insects released to control weeds, egg dispersal and a long lifespan generally correlated with the likelihood of establishment (the third filter), while egg aggregation and a short lifespan generally correlated with the degree of control (the fourth and fifth filters).

In a recent review, Kolar and Lodge (2001) considered studies published in the scientific literature between 1986 and 1999 with reference to an invasion model with three steps or 'transitions': transport and introduction (corresponding to the first two filters), establishment (the third filter) and invasion (defined as becoming widespread, the fourth filter). They looked for studies that included at least 20 species and that quantitatively analysed whether species' traits other than taxonomic identity were associated with success or failure in passing through these transitions in the field. They found only one study that addressed the transport and introduction transition, and 14 studies that addressed the establishment or invasion transitions. Among these, analyses of the establishment transition were primarily conducted on birds, and analyses of the invasion transition were primarily conducted on plants.

In this chapter I will be focusing on the third filter, seeking quantitative analyses of success and failure in the field that indicate whether certain species' traits and habitat characteristics assist or prevent the establishment of species released into novel environments. Consistent with Kolar and Lodge (2001), the relevant analyses that I found were limited to animal data.

Approaches to analysing the effect of species and environmental traits on invasion success

Species' traits said to contribute to the success of invasions have often been identified on the basis of those traits being common in species that had passed through the filter or filters of interest, such as traits common in species that had spread to new regions, or that had caused problems in new regions, etc. But, as others have pointed out (e.g. Crawley 1986; Simberloff 1986, 1989, 1995), drawing meaningful conclusions about the effect of these traits requires information about both the successful and the unsuccessful invaders (i.e. what was poured into the filters), not just about the successful invaders (what passed through the filters). Similarly, determining the role played by environmental characteristics requires knowledge of the rates of success and failure in environments displaying and lacking the characteristics of interest. In a few, but very few, cases, this sort of knowledge has been compiled and analysed. With regard to the ability of exotic organisms inoculated into an environment to become established there, such analyses have been primarily conducted on biocontrol agents (primarily arthropods; e.g. Hall & Ehler 1979; Crawley 1986, 1987; Simberloff 1986), released and escaped birds (e.g. Moulton & Pimm 1986; Newsome & Noble 1986; Pimm 1989; Simberloff & Boecklen 1991; Lockwood *et al.* 1993; Moulton 1993; Lockwood & Moulton 1994; Brooke *et al.* 1995; Case 1996; Veltman *et al.* 1996; Duncan 1997; Green 1997) and intentionally released fish (Cohen 1996). But in most cases other than intentional releases, and in many circumstances where intentional releases were poorly documented, we know little about which organisms were released into the environment but failed to establish.

Accordingly, some researchers have turned to proxy data sets of species, on the sometimes unstated assumption that the distribution of traits in the proxy set is the same as the distribution of traits in the set of species released. A frequent practice has been to use the native species in a region as a proxy for the species released into the region (e.g. Mulligan 1965 using Canadian plants; Crawley 1986 using British birds and mammals; Crawley 1987 using northeastern USA plants). But without any *a priori* reason for believing that the traits of the proxy set and the traits of the set of released species are the same, such analyses are suspect.

That certain traits contribute generally to invasion success has also been argued on theoretical grounds (e.g. Pimm 1989, based on community assembly and food web models). Of course, however useful such arguments are for framing hypotheses, they cannot inform us about what is really going on in the world. Physical models consisting of artificial ecosystems (primarily aquatic microcosms, e.g. Robinson & Dickerson 1984) and field experiments (usually manipulating plots of terrestrial plants, e.g. Tilman 1997; other studies referenced in Hobbs & Huenneke 1992; and Stohlgren *et al.* 1999) have also been used to test for species' traits or environmental characteristics that influence invasion success. Some observers commend the field experimental approach as the best hope for progress toward understanding what affects the potential for successful invasions (Kareiva 1996; Mack 1996). But unless corroborated by analyses of invasions in real ecosystems, the relevance of studies in micocosms or manipulated plots remains uncertain.

Are islands especially vulnerable to invasions?

The large number of exotic species established on islands is impressive, in many cases accounting for a substantial portion of the island biota (e.g. Sailer 1978; Simberloff 1986; Atkinson 1989). This has encouraged a long-held belief that islands are more easily invaded than continents, which is also partly based on a perception that island species are in general weaker competitors, less aggressive predators and more poorly defended than mainland species, and that island communities as a whole are more fragile and characterized by less intensive competition than mainland communities (Elton 1958; Carlquist 1965; Mayr 1965a; Simberloff 1986, 1989, 1995; Loope & Mueller-Dombois 1989; Bowen & van Vuren 1997). This is summed up in the concept of 'biotic resistance', a term invoking the tendency to exclude

Table 18.2 Establishment of biocontrol arthropods on islands and continents. (Adapted from Hall & Ehler 1979; based on worldwide data from Clausen 1978.)

Habitat	n	Rate of establishment	χ^2	P
Continents	1468	0.30	21.96	<0.0005
Islands	827	0.40		

organisms either by competition or by attack from predators or parasites (Simberloff 1985, 1989, 1995). Islands are said to offer less biotic resistance than continents.

A somewhat different explanation for the presumed greater invasibility of islands is that they contain what is generally described as numerous vacant niches (however unpalatable that term is to some researchers) (Herbold & Moyle 1986) relative to mainland ecosystems, and thus more opportunities for exotic species to establish without encountering any direct competition (Loope & Mueller-Dombois 1989; Simberloff 1995). Simberloff (1986) argued further that because of the smaller number of native species on islands, the filling of each new niche would represent a proportionally greater habitat change on islands than on continents, and thus a proportionally greater creation of new niches for yet additional species to invade. Thus, the establishment of a few exotic plants on an island with an impoverished flora would provide a substantially greater proportion of new habitat for phytophagous insects than would the same level of exotic plant establishment on a continent, and the establishment of a few species of exotic vertebrates on an island with a poor vertebrate fauna would create relatively more new niches for vertebrate parasites (Simberloff 1986).

These ideas have been much debated, as reviewed by Simberloff (1986, 1989, 1995). Several researchers have drawn conclusions about the relative invasibility of islands and continents based on the numbers of exotic species or the proportion of the biota consisting of exotic species in these habitats (e.g. Imms 1931; DeBach 1965; Sailer 1978). However, there have been few analyses comparing the rates of successful establishment of organisms released on islands and continents. In one, Hall and Ehler (1979) analysed global data from Clausen (1978) on arthropods used to control insects and arachnids, and found that the released organisms were more successful at establishing on islands than on continents (Table 18.2).

Simberloff (1986) argued that the organisms used for analysis should be related, to avoid possible confounding factors. He selected for analysis from the records in Clausen (1978) the six insect genera with the largest number of species that had been moved among islands and continents, each with about 20 species recorded. His analysis addressed together the two hypotheses that: (A) continents are more resistant to invasions than islands, and (B) continental species are more successful at invading than island species, by looking at the relative rates of establishment in four types of releases:

1 Continental species released on islands (should have the highest rate of establishment if hypotheses A and B are true).
2 Continental species released on continents (should have an intermediate rate of establishment).
3 Island species released on islands (should also have an intermediate rate of establishment).
4 Island species released on continents (should have the lowest rate of establishment).

Of the four genera for which there were adequate data to conduct an analysis, two (*Aphytis* and *Bracon*) generally fitted the above pattern and two did not (Table 18.3). When data for the six genera were pooled, the data fitted the pattern but the differences were not significant. Reviewing this and other evidence (though apparently not including Hall and Ehler's (1979) analysis), Simberloff (1989) found that although the data were insufficient to draw strong conclusions, it appeared likely to him that islands are more vulnerable to invasion than continents.

The data used by Simberloff can also be assessed directly for differences in rates of establishment on islands and continents. When releases of the continental and island species in a genus are considered separately (to avoid the possibility of confounding differences between island and continental species in their ability to establish), there is a higher rate of establishment on islands than on continents in six of the seven cases where there are data (Table 18.3). However, in only one of the six cases of greater establishment on islands is the difference significant (the release of continental species of *Opius*: $\chi^2 = 6.277$, $P = 0.0122$).

Newsome and Noble (1986) analysed the birds released in Australia relative to establishment in island and mainland regions. They included releases of both foreign birds and of translocated Australian birds and included Tasmania as part of the mainland, and found a non-significant higher rate of establishment on the islands. When they removed Kangaroo Island, the largest island, from the analysis, the difference became significant (Table 18.4).

Case (1996) assembled data on the numbers of successful and failed bird introductions in 22 islands and regions within continents. For analysis, he took the residuals from a regression of the number of successful introductions on the number of failed introductions at each site, which he described as a measure of the relative success rate, and used them as the dependent variable in a stepwise regression with various candidate independent variables, including the areas of the regions. The areas were not significantly correlated with relative success rate. Table 18.5 lists Case's (1996) data on the rates of establishment of exotic birds, along with data from a few other sources. Areas are included to allow sorting into islands and continents for analysis. A pattern of increasing rate of establishment with decreasing area seems clear, but as these data were assembled by different researchers for different time periods using different methods or definitions to determine the number of species introduced and established, it seems unwise to attempt a statistical analysis. Instead I analysed Case's data by sorting the regions into continent and island groups, counting all successful and unsuccessful introductions of a species into a region, and

Table 18.3 Rates of establishment in insect genera widely deployed for biocontrol. (Adapted from Simberloff 1986.)

	Continental species released on				Island species released on			
	Islands		Continents		Islands		Continents	
	n	Rate of establishment	*n*	Rate of establishment	*n*	Rate of establishment	*n*	Rate of establishment
Opius	24	0.46	37	0.14	6	0.67	6	0.50
Aphytis	16	0.69	48	0.67	0	No data	2	0.50
Bracon	11	0.36	28	0.14	6	0.33	2	0.00
Apanteles	11	0.73	27	0.59	5	0.40	1	1.00
Six genera of insects	71	0.49	176	0.43	19	0.47	15	0.33

Table 18.4 Rates of establishment of birds in Australia, including both foreign species and translocated native species. (Adapted from Newsome & Noble 1986.)

Habitat	*n*	Rate of establishment	χ^2	*P*
Including Kangaroo Island				
Mainland (Australia and Tasmania)	72	0.47	2.52	ns
Islands	65	0.63		
Excluding Kangaroo Island				
Mainland (Australia and Tasmania)	72	0.47	6.45	<0.05
Islands	52	0.73		

ns, not significant.

Table 18.5 Rates of establishment of birds. For the larger regions the area given is for the entire continent or archipelago (i.e. British Isles, Australia and North America), though the establishment rates refer to parts of these.

Invaded area	Rate of establishment	Area of island or continent (km^2)	Reference
Lord Howe	0.69–0.75	13	Mayr 1965a; Case 1996
Norfolk	0.92	40	Case 1996
Bermuda	0.41–0.88	54	Mayr 1965a; Case 1996
Chagos Archipelago	0.58	65	Case 1996
Ascension	0.29	90	Case 1996
Rodriguez	0.64	109	Case 1996
Saint Helena	0.31	125	Case 1996
Seychelles (granitic)	0.67	233	Case 1996
Tahiti	0.20	1041	Case 1996
Hawaii (non-mongoose)	0.52	1422	Case 1996
Mauritius	0.43	1865	Case 1996
Comoros	0.50	1958	Case 1996
Fiji (non-mongoose)	0.39	2375	Case 1996
Reunion	0.53	2512	Case 1996
Kangaroo Island	0.58	3890	Case 1996
Hawaii (mongoose)	0.46	12 136	Case 1996
Hawaii (all)	0.48	16 708	Mayr 1965a
Fiji (mongoose)	0.33	15 921	Case 1996
Tasmania	0.81	67 900	Case 1996
New Zealand	0.20–0.30	266 800	Mayr 1965a; Case 1996; Veltman *et al.* 1996; Green 1997
Great Britain	0.30	312 900	Case 1996
Australia (Victoria)	*c.*0.20–0.33	7 642 000	Mayr 1965a; Case 1996
Australia (Sydney County)	≤0.30	7 642 000	Mayr 1965a
Australia (including islands)	0.44	7 642 000	Newsome & Noble 1986
Europe	0.15	10 036 000	Mayr 1965a
Continental United States	0.13	24 326 000	Case 1996

Table 18.6 Rates of establishment of birds on continents and islands. (Based on data from Case 1996.)

	Continents		Islands			
Continents are defined to include	*n*	Rate of establishment	*n*	Rate of establishment	χ^2	*P*
North America	98	0.13	743	0.41	27.02	<0.001
North America, Australia	146	0.20	695	0.41	23.01	<0.001
North America, Australia, British Isles	176	0.22	665	0.42	23.71	<0.001
North America, Australia, British Isles, New Zealand	325	0.24	516	0.46	39.49	<0.001
North America, Australia, British Isles, New Zealand, Tasmania	341	0.27	500	0.45	27.27	<0.001

comparing these for the two groups using a 2 × 2 contingency test. I defined the lower limit of continent size at, successively, the area of North America, of Australia, of the British Isles, of New Zealand, and of Tasmania, so that there are five tests (Table 18.6). The differences were highly significant for all definitions of continent.

There may be some pseudoreplication problems with these data. For example, Case divided the Hawaiian and Fijian island groups by the presence/absence of the mongoose for a different analysis, but Simberloff and Boecklen (1991) argued that for the present sort of analysis observations on different islands within an archipelago are not truly independent observations since birds established on one island may then colonize another. Case also counted established exotic birds for which he had no record of introduction as introduced and established. These probably represent a combination of self-colonizations, cage escapes or intentional but unrecorded releases. Failures from these types of introductions would rarely or never be recorded, so the rates of establishment are probably inflated. If self-colonizations of species from continents to nearby islands are more common than the reverse (which seems likely given that continents have many species that are absent from nearby islands and thus a large pool of potential colonists, while islands rarely have species that are absent from nearby continents), then rates of establishment in these data are probably inflated for islands relative to continents.

Despite these problems, the highly significant results from this analysis of birds and from Hall and Ehler's (1979) analysis of biocontrol releases, and the weak pattern in Simberloff's (1986) analysis of biocontrol releases and Newsome and Noble's (1986) analysis of birds, when taken together strongly suggest that islands are in fact more vulnerable to the establishment of exotic species than continents. Simberloff (1986, 1989, 1995) notes that biocontrol releases are primarily into agricultural communities, which host few native organisms and are more similar to islands and continents than are native communities. Thus any patterns derived from differences in the characteristics of island and mainland species or the resistance of natural communities would be muted or missing from biocontrol data. Simberloff (1995) argues that the same issue arises with bird data, at least on the Hawaiian and Mascarene islands (but not in Bermuda; Lockwood & Moulton 1994), where exotic birds primarily occupy highly altered, anthropogenous habitat and have limited interactions with native birds. That statistically strong differences in establishment rates between islands and continents can be found in spite of this is striking. If interactions with native biota or native communities on islands and continents are not the cause of these differences, then what is?

Are disturbed habitats more easily invaded?

The idea that invasions are more successful in disturbed than in undisturbed habitats has a long pedigree (e.g. Elton 1958; Mooney & Drake 1989; Hobbs & Huenneke 1992), and is most often attributed to a reduction in the level of competition in disturbed habitats (Ehrlich 1989; Luzon & MacIsaac 1997). Lozon and MacIssac (1997), in a review of 133 studies of invasions published in 10 journals from 1993 to 1995, found that disturbance was associated with the establishment of exotic species by 68% of the papers discussing exotic plants and by 28% of the papers discussing exotic animals. In these papers, 86% of the exotic plants and 12% of the exotic animals studied were reported to be dependent on disturbance for establishment. However, disturbance means different things to different investigators. Hobbs and Huenneke (1992) reviewed definitions ranging from the relatively specific ('a process that removes or damages biomass'—Grime 1979) to the perhaps uselessly broad ('any process that alters the birth and death rates of individuals present in the patch'—Petraitis *et al.* 1989). Others have considered any change from past conditions to be a disturbance. Examples of disturbance cited in these studies include fire, the suppression of fire, flood, drought, irrigation, soil disturbance by animals or mechanical activity, grazing, the removal of grazing, and the release of exotic species. In different studies, a site may be considered disturbed if it is subject to repeated or continuous perturbation; has undergone a change from previous conditions; differs from natural conditions; has a reduced biota, in terms of species or individuals; has reduced plant cover; or has disturbed soils. The flexible meaning of disturbance, combined with the various definitions of invasion discussed above, have given this proposition a particularly amorphous character.

It is none the less apparent that exotic organisms are often common in areas of substantial human activity or alterations, which would be considered to be disturbed environments by several definitions. Exotic organisms are generally more common in areas altered by urban or agricultural development than in pristine areas, more common along roadsides than away from roads, more common in rivers that have been altered by dams than in those that have not, and more common in harbours and estuaries than in open coast or open ocean areas. However, this is not necessarily because areas of human activity and alteration are more vulnerable to invasion. There are at least three other possible explanations.

First, exotic organisms are primarily transported to, and released into, areas with substantial human activity (Simberloff 1986, 1989; Williamson 1996; Cohen & Carlton 1998; Mack *et al.* 2000). Exotic crop and livestock species are purposefully introduced into agricultural regions, bringing with them accidental inoculations of exotic weeds, pests and parasites, which sometimes in turn lead to releases of additional exotic organisms to control the weeds, pests and parasites. People and goods from distant lands, and associated organisms, arrive in urban areas in great numbers. Exotic fish are frequently released in the reservoirs impounded by dams. Harbours and estuaries, as centres for shipping and aquaculture, have received far larger inputs of exotic marine organisms than have open coast and ocean areas.

Second, human-associated mechanisms that transport organisms around the world, either intentionally or accidentally, primarily take those organisms from areas of substantial human activity. Thus we mainly move organisms that are adapted to, or are tolerant of, substantial human-caused disturbance (Williamson 1996; Cohen & Carlton 1998). Crop and livestock species and their associates are moved from areas of agriculturally related disturbance; people and goods travelling long distances generally begin their journeys in areas subjected to urban disturbance; and ships load goods and ballast water (containing aquatic organisms) from harbour areas, which are often highly altered and heavily polluted environments.

Third, areas disturbed by human activity, such as agricultural lands and lands associated with human habitation, are of substantial importance to humanity and are generally nearer to research institutes than are pristine areas. If, either because of importance or proximity, human-disturbed habitats are more intensively studied than pristine ones, then exotic species are more likely to be detected in them (Simberloff 1986, 1989).

The problem of acceptably sorting environments into more disturbed and less disturbed categories, combined with the difficulty of finding adequate data on successful and unsuccessful inoculations of exotic species into these environments, has made analysis of this question particularly challenging. In the only pertinent study I could find, Hall and Ehler (1979) analysed data from Clausen (1978) on arthropods used to control insects and arachnids, sorting the data into releases in three habitat types: (i) the most unstable or frequently disturbed sites, consisting of annual or short-cycle crops including most vegetable and field crops; (ii) sites with an intermediate level of disturbance, including orchards and other perennial crop systems; and (iii) sites with the least disturbance, including forest and rangeland. They found that the rate of establishment increased as the level of disturbance decreased, with a highly significant increase between the greatest and least levels of disturbance, contrary to conventional wisdom (Table 18.7). Analysing Canadian data in a similar manner, they found that the released organisms established most successfully at intermediate levels of disturbance, and least successfully at the greatest levels of disturbance (Table 18.7). As noted above, there may be problems with using results derived from biocontrol releases to draw conclusions about natural communities. At a minimum, however, there appears to be no published statistical support for the idea that habitats that have been disturbed are more easily invaded.

Table 18.7 Establishment of biocontrol agents at different levels of disturbance. (Adapted from Hall & Ehler 1979.)

Based on worldwide data from Clausen (1978)		
Habitat type:	*n*	Rate of establishment
Greatest level of disturbance	640	0.28
Intermediate level of disturbance	916	0.32
Least level of disturbance	535	0.36
Habitats compared:	χ^2	*P*
Greatest vs. intermediate disturbance	3.19	>0.05
Intermediate vs. least disturbance	2.72	>0.05
Greatest vs. least disturbance	9.66	<0.005
Based on Canadian data from Beirne (1975)		
Habitat type:		Rate of establishment
Greatest level of disturbance (annual crops)		0.16
Intermediate level of disturbance (orchards and ornamental shrubs)		0.43
Least level of disturbance (forests)		0.23

Are prolific organisms more successful as invaders?

It has often been proposed that successful invaders are characterized by a large reproductive potential or intrinsic rate of increase (e.g. MacArthur & Wilson 1967; Crawley 1986, 1987; Pimm 1989). Crawley (1986) described the components of the intrinsic rate of increase as fecundity, survivorship and developmental rate, and annual fecundity is a function of both clutch size and the number of clutches per year. Several studies have analysed some of these factors relative to the rate of successful establishment, with variable but usually non-significant results (Table 18.8). Some limitations of these studies should be noted. Analyses based on colonizations effected across natural dispersal distances (O'Connor 1986) may not apply to longer-distance, human-mediated inoculations. Analyses based on translocated species (Newsome & Noble 1986; Griffith *et al.* 1989), which include both species released within their historic ranges and species released outside those ranges, may not apply to introductions of exotic species, which by definition are invasions outside their native ranges. Analyses based on biocontrol agents (Crawley 1986, 1987), which are typically released into habitats with diminished native biota and with the targeted pest providing abundant food resources for the released agent, may have limited relevance to invasions in natural communities. Green's (1997) different analyses of exotic birds in New Zealand yielded both positive and negative relationships (before Bonferroni correction) of clutch size with success, but these became non-significant when analysed for within-family variation, suggesting caution when interpreting analyses that have not accounted for between-family variation. Overall these analyses provide little or no support for hypotheses that organisms exhibiting one or another component of a high reproductive potential are more likely to become established when released into a new environment. On the other hand, Crawley's (1996, 1997) analyses of data on biocontrol insects suggest the possibility

Table 18.8 Relationship of fecundity variables to success in the establishment of species. The effect of an increase in the variable (greater, faster) is reported as significantly increased success (+), significantly decreased success (−) or no significant difference (ns). O'Connor (1986) provided data but no statistical analysis on the greater success of multibrooded bird species as colonists; the effect, as reported here, is not significant ($n=48$, $df=1$, $\chi^2=2.82$, $P>0.05$). Griffith *et al.* (1989) reported a significantly higher success rate in translocated animals with large clutches and early breeding based on a value of $0.05 \le P \le 0.1$, but this is reported more conventionally here as not significant.

Variable analysed	*n*	Effect	Reference
Seasonal egg production in birds colonizing Britain	48	+	O'Connor 1986
Broods per season in birds colonizing Britain	48	ns	After O'Connor 1986
Clutch size of foreign bird species released in Australia	61	ns	Newsome & Noble 1986
Clutch size of translocated bird species released in Australia	47	ns	Newsome & Noble 1986
Fecundity of biocontrol insects:			
on *Opuntia*	≤22	ns	Crawley 1986
on *Lantana*	≤30	ns	
on other weeds	≤173	+	
Developmental rate of biocontrol insects:			
on *Opuntia*	≤22	ns	Crawley 1986
on *Lantana*	≤30	+	
on other weeds	≤173	ns	
Intrinsic rate of increase (indirectly calculated) of biocontrol insects:			
on *Opuntia*	≤22	+	Crawley 1987
on *Lantana*	≤30	+	
on other weeds	≤173	+	
Clutch size and early/late breeding in translocated mammals and birds in Australia, Canada, Hawaii, New Zealand and the USA	198	ns	Griffith *et al.* 1989
Clutch size of exotic bird species released in New Zealand	79	ns	Veltman *et al.* 1996
Broods per season in exotic bird species released in New Zealand	79	ns	Veltman *et al.* 1996
Maximum reported egg production of exotic fish species released into the San Francisco Bay watershed	42	ns	Cohen 1996
Mean clutch size of exotic bird species released in New Zealand:			
variable treated on its own	47	−/ns	Green 1997
in selected regression model	47	−/ns	(with / without
in selected regression model, analysis of within-family variation	36	ns/ns	Bonferroni adjustment)
in selected regression model including only British birds	27	+/ns	

that studies addressing the intrinsic rate of increase, rather than components of the rate of increase, may yet reveal a consistent relationship.

Do invading organisms benefit by leaving their parasites behind?

It has been proposed that invading species gain an advantage by losing parasites when they are transported from their native region, so that in the invaded region

Table 18.9 Parasitization rates in native and exotic fish in northern California. (Based on data from Haderlie 1953.)

	Native fish		Exotic fish			
	n	Fraction parasitized	*n*	Fraction parasitized	χ^2	*P*
Centrarchidae	18	0.44	485	0.70	7.7	<0.0055
Cyprinidae	415	0.66	71	0.68	0.1	<0.7644
Salmonidae	141	0.56	38	0.87	12.1	<0.0005
All fish	960	0.63	1038	0.77	45.0	<0.0001

they host fewer parasite species and smaller numbers of parasites than do their potential competitors (Dobson & May 1986). Parasites may be lost through the chance absence of a parasite species from the founding population, through the mortality of infected hosts during transport, and through the absence of suitable intermediate hosts in the invaded region (Dobson & May 1986; Guégan & Kennedy 1993). This 'lost parasites' hypothesis has provided one of the main rationales for classic biological control, which has frequently sought to identify and import the parasites lost by invading species (Simmonds *et al.* 1976; Van den Bosch *et al.* 1982).

One prediction of the lost parasites hypothesis is that successful invading species will host fewer species of parasites and have a smaller parasite burden than do competing native species. To test this, I analysed data from three studies of fish parasites in California that examined both native and exotic fish, comparing both parasite species richness and the host parasitization rate (the fraction of fish in the population that carry some parasites) (Cohen 1996). In these studies, overall and within families that included both native and exotic fish, parasitization rates were usually higher for exotic species, sometimes very significantly so, contrary to the hypothesis (Table 18.9). The mean number of parasite species per fish species was also more often higher for exotic than for native fish.

One possible explanation is that California is a special case. Most fish introduced into the state came from regions with richer fish faunas. These regions may also have richer fish parasite faunas, so that even after losing some parasites, invading fish could still host more parasites than do native fish. If true, however, and if in general having fewer parasites confers a competitive advantage, then California's native fish communities should have been especially resistant to invasion. Instead they have been extensively invaded (Moyle 1976).

Another possibility is that the invading fish may have benefited by retaining their parasites. These may have prevented native California fish parasites from attacking the invading fish (by competitive exclusion); or parasites that arrived with the invading fish may have attacked competing native fish. Since parasite strains or species that are new to a host are often more harmful than parasites to which the host has a history of exposure (Esch & Fernàndez 1993), retaining its natural parasites could provide substantial benefits to an invading host both by reducing its vulnerability to novel parasites and by increasing its competitors' exposure to novel parasites. The value to invaders of carrying parasites or diseases with them has been documented most extensively for human invasions—historical studies have shown that smallpox and other diseases carried by European soldiers and settlers, rather than superior military technology, led to the rapid destruction and successful invasion of aboriginal human populations in many regions of the globe (e.g. Crosby 1986).

Conclusions

It has been often stated in the scientific literature that testing for the characteristics that either enable invading organisms to become established or that make environments more vulnerable to invasion, requires data on the rates of success and failure in becoming established. Surprisingly, given the level of interest and the volume of published material on this subject, few such analyses have been done. Data on intentional releases of biocontrol agents, birds, fish, and possibly mammals and shellfish, must surely exist for many regions, though compiling the data in an appropriate form for analysis is time consuming. There are often significant problems with such data—including uncertainties due to the vagaries of reporting and possible confounding factors in the manner and size of the releases—but as it is the best data we have, we should make more thorough use of it.

From the data and analyses reviewed for this chapter, there appears to be little evidence that disturbed habitats are especially vulnerable to the establishment of exotic species and some evidence to the contrary; no clear evidence that prolific organisms are more successful at establishing than less prolific ones; and some counterevidence to the hypothesis that species improve their ability to establish by shedding their native parasites. There does, however, appear to be some good quantitative evidence that islands are more vulnerable than continents are to invasions. The studies reviewed here all involved data on animal invasions, as, surprisingly, there appear to be no published statistical analyses of the success in establishment of introduced plants relative to these four characteristics. Most analyses of the success of plant invasions focus instead on the postestablishment spread of plants. Further analyses of data on establishment rates could only improve the current situation, wherein many researchers and resource managers believe it has been proven that certain characteristics of organisms or environments increase the probability of exotic species becoming established, when in fact these propositions have only rarely been tested, and when tested have mainly produced ambiguous or contradictory results.

References

Atkinson, I. (1989) Introduced animals and extinctions. In: *Conservation for the Twenty-first Century* (eds D. Western & M. Pearl), pp. 54–69. Oxford University Press, New York.

Baker, H.G. (1965a) Characteristics and modes of origin of weeds. In: *The Genetics of Colonizing Species* (eds H.G. Baker & G.L. Stebbins), pp. 147–172. Academic Press, New York.

Baker, H.G. (1965b) In 'Discussion of paper by Dr Stebbins'. In: *The Genetics of Colonizing Species* (eds H.G. Baker & G.L. Stebbins), p. 194. Academic Press, New York.

Beirne, B.P. (1975) Biological control attempts by introductions against pest insects in the field in Canada. *Canadian Entomologist* **107**, 225–236.

Bowen, L. & van Vuren, D. (1997) Insular endemic plants lack defenses against herbivores. *Conservation Biology* **11**, 1249–1254.

Brooke, R.K., Lockwood, J.L. & Moulton, M.P. (1995) Patterns of success in passeriform introductions on Saint Helena. *Oecologia* **103**, 337–342.

Carlquist, S. (1965) *Island Life: a Natural History of the Islands of the World.* Natural History Press, Garden City, NY.

Case, T.J. (1996) Global patterns in the establishment and distribution of exotic birds. *Biological Conservation* **78**, 69–96.

Clausen, C.P. (1978) *Introduced Parasites and Predators of Arthropod Pests and Weeds: a World Review.* Agriculture Handbook No. 480. US Department of Agriculture, Washington, DC.

Cohen, A.N. (1996) *Biological invasions in the San Francisco Estuary: a comprehensive regional analysis.* PhD thesis, University of California, Berkeley, CA.

Cohen, A.N. & Carlton, J.T. (1998) Accelerating invasion rate in a highly invaded estuary. *Science* **279**, 555–558.

Collar, N.J., Crosby, M.J. & Stattersfield, A.J. (1994) *Birds to Watch 2. The World List of Threatened Birds.* Birdlife International, Cambridge.

Crawley, M.J. (1986) The population biology of invaders. *Philosophical Transactions of the Royal Society of London, Series B* **314**, 711–731.

Crawley, M.J. (1987) What makes a community invasible? In: *Colonization, Succession and Stability* (eds A.J Gray, M.J. Crawley & P.J. Edwards), pp. 429–453. Blackwell Scientific Publications, London.

Crosby, A.W. (1986) *Ecological Imperialism: the Biological Expansion of Europe, 900–1900.* Cambridge University Press, Cambridge.

DeBach, P. (1965) Some biological and ecological phenomena associated with colonizing entomophagous insects. In: *The Genetics of Colonizing Species* (eds H.G. Baker & G.L. Stebbins), pp. 287–303. Academic Press, New York.

Di Castri, F. (1990) On invading species and invaded ecosystems: the interplay of historical chance and biological necessity. In: *Biological Invasions in Europe and the Mediterranean Basin* (eds F. Di Castri, A.J. Hansen & M. Debussche), pp. 3–16. Kluwer Academic Publishers, Dordrecht.

Dobson, A.P. & May, R.M. (1986) Patterns of invasion by pathogens and parasites. In: *Ecology of Biological Invasions of North America and Hawaii* (eds H.A. Mooney & J.A. Drake), pp. 58–77. Springer Verlag, New York.

Duncan, R.P. (1997) The role of competition and introduction effort in the success of passeriform birds introduced to New Zealand. *American Naturalist* **149**, 903–915.

Ehrlich, P.R. (1986) Which animal will invade? In: *Ecology of Biological Invasions of North America and Hawaii* (eds H.A. Mooney & J.A. Drake), pp. 79–95. Springer Verlag, New York.

Ehrlich, P.R. (1989) Attributes of invaders and the invading processes: vertebrates. In: *Biological Invasions: a Global Perspective* (eds J.A. Drake, H.A. Mooney, F. Di Castri, R.H. Groves, F.J. Kruger, M. Rejmanek & M. Williamson), pp. 315–328. John Wiley & Sons, New York.

Elton, C.S. (1958) *The Ecology of Invasions by Animals and Plants.* Methuen, London.

Esch, G.W. & Fernàndez, J.C. (1993) *A Functional Biology of Parasitism.* Chapman & Hall, London.

Flather, C.H., Joyce, L.A. & Bloomgarden, C.A. (1994) *Species Endangerment Patterns in the United States.* General Technical Report No. RM–241. US Department of Agriculture Forest Service, Rocky Mountain Forest and Range Experiment Station, Fort Collins, CO.

Green, R.E. (1997) The influence of numbers released on the outcome of attempts to introduce exotic bird species to New Zealand. *Journal of Animal Ecology* **66**, 25–35.

Griffith, B., Scott, J.M., Carpenter, J.W. & Reed, C. (1989) Translocation as a species conservation tool: status and strategy. *Science* **245**, 477–480.

Grime, J.P. (1979) *Plant Strategies and Vegetation Processes.* Wiley, New York.

Groat, C. (2000) Statement of Charles Groat, Director US Geological Survey, from *People, Land and Water* (US Department of the Interior) July/August 2000, at http://www.usgs.gov/invasive_species/plw/usgsdirector01.html, accessed 26 January 2001.

Groves, R.H. (1986) Plant invasions of Australia: an overview. In: *Ecology of Biological Invasions* (eds R.H. Groves & J.J. Burdon), pp. 137–149. Cambridge University Press, Cambridge.

Guégan, J.F. & Kennedy, C.R. (1993) Maximum local helminth parasite community richness in British freshwater fish: a test of the colonization time hypothesis. *Parasitology* **106**, 91–100.

Haderlie, E.C. (1953) Parasites of the fresh-water fishes of northern California. *University of California Publications in Zoology* **57**, 303–440.

Hall, R.W. & Ehler, L.E. (1979) Rate of establishment of natural enemies in classical biological control. *Bulletin of the Entomological Society of America* **25**, 280–282.

Herbold, B. & Moyle, P.B. (1986) Introduced species and vacant niches. *American Naturalist* **128**, 751–760.

Hobbs, R.J. & Huenneke, L.F. (1992) Disturbance, diversity, and invasion: implications for conservation. *Conservation Biology* **6**, 324–337.

Imms, A.D. (1931) *Recent Advances in Entomology*. Churchill, London.

Kareiva, P. (1996) Developing a predictive ecology for non-indigenous species and ecological invasions. *Ecology* **77**, 1651–1652.

Kolar, S.K. & Lodge, D.M. (2001) Progress in invasion biology: predicting invaders. *Trends in Evolution and Ecology* **16**, 199–204.

Lassuy, D.R. (1995) Introduced species as a factor in extinction and endangerment of native fish species. *American Fisheries Society Symposium* **15**, 391–396.

Lockwood, J.L. & Moulton, M.P. (1994) Ecomorphological pattern in Bermuda birds: the influence of competition and implications for nature preserves. *Evolutionary Ecology* **8**, 53–60.

Lockwood, J.L., Moulton, M.P. & Anderson, S.K. (1993) Morphological assortment and the assembly of communities of introduced passeriforms on oceanic islands: Tahiti versus Oahu. *American Naturalist* **141**, 398–408.

Loope, L.L. & Mueller-Dombois, D. (1989) Characteristics of invaded islands, with special reference to Hawaii. In: *Biological Invasions: a Global Perspective* (eds J.A. Drake, H.A. Mooney, F. Di Castri, R.H. Groves, F.J. Kruger, M. Rejmanek & M. Williamson), pp. 257–280. John Wiley & Sons, New York.

Luzon, J.D. & MacIsaac, H.J. (1997) Biological invasions: are they dependent on disturbance? *Environmental Review* **5**, 131–144.

MacArthur, R.H. & Wilson, E.O. (1967) *The Theory of Island Biogeography*. Princeton University Press, Princeton, NJ.

Mack, R.N. (1996) Predicting the identity and fate of plant invaders: emergent and emerging approaches. *Biological Conservation* **78**, 107–121.

Mack, R.N, Simberloff, D., Lonsdale, M., Evans, H., Clout, M. & Bazzaz, F. (2000) *Biotic Invasions: Causes, Epidemiology, Global Consequences and Control*. Issues in Ecology No. 5. Ecological Society of America, Washington, DC.

Mayr, E. (1965a) The nature of colonizations in birds. In: *The Genetics of Colonizing Species* (eds H.G. Baker & G.L. Stebbins), pp. 29–43. Academic Press, New York.

Mayr, E. (1965b) Summary. In: *The Genetics of Colonizing Species* (eds H.G. Baker & G.L. Stebbins), pp. 553–562. Academic Press, New York.

Miller, R.R., Williams, J.D. & Williams, J.E. (1989) Extinctions of North American fishes during the past century. *Fisheries* **14** (6), 22–38.

Möller, H. (1996) Lessons for invasion theory from social insects. *Biological Conservation* **78**, 125–142.

Mooney, H.A. & Drake, J.A. (1989) Biological invasions: a SCOPE program overview. In: *Biological Invasions: a Global Perspective* (eds J.A. Drake, H.A. Mooney, F. Di Castri, R.H. Groves, F.J. Kruger, M. Rejmanek & M. Williamson), pp. 491–508. John Wiley & Sons, New York.

Moulton, M.P. (1993) The all-or-none pattern in introduced Hawaiian passeriforms: the role of competition sustained. *American Naturalist* **141**, 105–199.

Moulton, M.P. & Pimm, S.L. (1986) Species introductions to Hawaii. In: *Ecology of Biological Invasions of North America and Hawaii* (eds H.A. Mooney & J.A. Drake), pp. 231–249. Springer Verlag, New York.

Moyle, P.B. (1976) Fish introductions in California: history and impact on native fishes. *Biological Conservation* **9**, 101–118.

Mulligan, G.A. (1965) Recent colonization by herbaceous plants in Canada. In: *The Genetics of*

Colonizing Species (eds H.G. Baker & G.L. Stebbins), pp. 127–146. Academic Press, New York.

Newsome, A.E. & Noble, I.R. (1986) Ecological and physiological characters of invading species. In: *Ecology of Biological Invasions* (eds R.H. Groves & J.J. Burdon), pp. 1–20. Cambridge University Press, Cambridge.

O'Connor, R.J. (1986) Biological characteristics of invaders among bird species in Britain. *Philosophical Transactions of the Royal Society of London, Series B* **314**, 583–598.

Orians, G.H. (1986) Site characteristics favoring invasions. In: *Ecology of Biological Invasions of North America and Hawaii* (eds H.A. Mooney & J.A. Drake), pp. 133–148. Springer Verlag, New York.

Petraitis, P.S., Latham, R.E. & Niesenbaum, R.A. (1989) The maintenance of species diversity by disturbance. *Quarterly Review of Biology* **64**, 393–418.

Pimental, D. (1986) Biological invasions of plants and animals in agriculture and forestry. In: *Ecology of Biological Invasions of North America and Hawaii* (eds H.A. Mooney & J.A. Drake), pp. 149–162. Springer Verlag, New York.

Pimm, S.L. (1989) Theories of predicting success and impact of introduced species. In: *Biological Invasions: a Global Perspective* (eds J.A. Drake, H.A. Mooney, F. Di Castri, R.H. Groves, F.J. Kruger, M. Rejmanek & M. Williamson), pp. 351–367. John Wiley & Sons, New York.

Raver, A. (1999) What's eating America? Weeds. *New York Times* 18 October.

Rejmanek, M. & Richardson, D.M. (1996) What attributes make some plant species more invasive? *Ecology* **77**, 1655–1661.

Richter, B.D., Braun, D.P., Mendelson, M.A. & Master, L.L. (1997) Threats to imperiled freshwater fauna. *Conservation Biology* **11**, 1081–1093.

Robinson, J.V. & Dickerson Jr, J.E. (1984) Testing the invulnerability of laboratory island communities to invasion. *Oecologia (Berlin)* **61**, 169–174.

Ruesink, J.L., Parker, I.M., Groom, M.J. & Kareiva, P.M. (1995) Reducing the risks of nonindigenous species introductions: guilty until proven innocent. *BioScience* **45**, 465–477.

Sailer, R.I. (1978) Our immigrant insect fauna. *Bulletin of the Entomological Society of America* **24**, 3–11.

Schemske, D.W., Husband, B.C., Ruckelshaus, M.H., Goodwillie, C., Parker, I.M. & Bishop, J.G. (1994) Evaluating approaches to the conservation of rare and endangered plants. *Ecology* **75**, 584–606.

Simberloff, D. (1986) Introduced insects: a biogeographic and systematic perspective. In: *Ecology of Biological Invasions of North America and Hawaii* (eds H.A. Mooney & J.A. Drake), pp. 3–26. Springer Verlag, New York.

Simberloff, D. (1989) Which insect introductions succeed and which fail? In: *Biological Invasions: a Global Perspective* (eds J.A. Drake, H.A. Mooney, F. Di Castri, R.H. Groves, F.J. Kruger, M. Rejmanek & M. Williamson), pp. 61–75. John Wiley & Sons, New York.

Simberloff, D. (1995) Why do introduced species appear to devastate islands more than mainland areas? *Pacific Science* **49**, 87–97.

Simberloff, D. & Boecklen, W. (1991) Patterns of extinction in the introduced Hawaiian avifauna: a reexamination of the role of competition. *American Naturalist* **138**, 300–327.

Simmonds, F.J., Franz, J.M. & Sailer, R.I. (1976) History of biological control. In: *Theory and Practice of Biological Control* (eds C.B. Huffaker & P.S. Messenger), pp. 17–39. Academic Press, New York.

Stohlgren, T.J., Binkley, D., Chong, G.W. *et al.* (1999) Exotic plant species invade hot spots of native plant diversity. *Ecological Monographs* **69**, 25–46.

Tilman, D. (1997) Community invasibility, recruitment limitation, and grassland biodiversity. *Ecology* **78**, 81–92.

Townsend, C.R. (1996) Invasion biology and ecological impacts of brown trout *Salmo trutta* in New Zealand. *Biological Conservation* **78**, 13–22.

Van den Bosch, R., Messenger, P.S. & Gutierrez, A.P. (1982) *An Introduction to Biological Control.* Plenum, New York.

Veltman, C.J., Nee, S. & Crawley, M.J. (1996) Correlates of introduction success in exotic New Zealand birds. *American Naturalist* **147**, 542–557.

Wilcove, D.S., Rothstein, D., Dubrow, J., Phillips, A. & Losos, E. (1998) Quantifying threats to imperiled species in the United States. *BioScience* **48**, 607–615.

Williams, J.E., Johnson, J.E., Hendrickson, D.A. *et al.* (1989) Fishes of North America endangered, threatened, or of special concern: 1989. *Fisheries* **14**, 2–20.

Williamson, M. (1996) *Biological Invasions.* Chapman & Hall, London.

Williamson, M.H. & Fitter, A. (1996a) The varying success of invaders. *Ecology* **77**, 1661–1666.

Williamson, M.H. & Fitter, A. (1996b) The characters of successful invaders. *Biological Conservation* **78**, 163–170.

Chapter 19

Oases in the desert: dispersal and host specialization of biotrophic fungal pathogens of plants

James K. M. Brown, Mogens S. Hovmøller, Rebecca A. Wyand and Dazhao Yu

The importance of wind dispersal for biotrophic fungi

Obligately biotrophic fungi are among the most significant pathogens of plants, causing a range of diseases, notably rusts, powdery mildews and downy mildews (Agrios 1997). Many of these are economically important as they may cause substantial losses in yield or quality of field crops unless they are controlled by the use of fungicides or disease-resistant cultivars. Rust diseases are caused by fungi in the order Uredinales of the Basidiomycotina and powdery mildews caused by Erysiphales among the Ascomycotina. Downy mildews are caused by Peronosporaceae, a family of Oomycetes, which are not true fungi, being classified within the Stramenopiles (Chromophyta) (Bhattacharya *et al.* 1992; Sogin & Silberman 1998).

Three features of these fungi are relevant to the discussion in this chapter. The first is the simple fact that they are obligate biotrophs, because they can only grow and reproduce on living host tissue. So, unlike the great majority of fungi, they are unable to colonize dead matter such as leaf litter, cut stumps and roots of trees, animal dung and so forth. It has proved possible to culture several rust and downy mildew fungi in the laboratory (Maclean 1982) but the conditions required for success are so restrictive that it is most unlikely that these fungi grow saprotrophically in nature, by feeding on non-living substrates, either within a plant or in the soil.

Secondly, biotrophic fungi show a high level of host specificity. This operates at two levels. At the species level, several obligate biotrophic fungi are subdivided into special forms, each of which, strictly speaking, can only infect a limited range of hosts, typically those of a single genus or a few, closely related genera. For example, *Blumeria graminis* has eight special forms, four of which—*avenae, hordei, secalis* and *tritici*—are parasitic on oats, barley, rye and wheat, respectively (Marchal 1902; Oku *et al.* 1985). The system of special forms tends to oversimplify a complex situation, however, because specialization to the host genus or genera may not be as absolute as the nomenclature suggests (Hardison 1944; Eshed & Wahl 1970; Sheng *et al.* 1993).

Another level of specificity operates at the genotypic level, with a system of specific interactions between host plants and parasitic fungi determined by gene-for-gene relationships (Fig. 19.1). The gene-for-gene paradigm is that each resistance

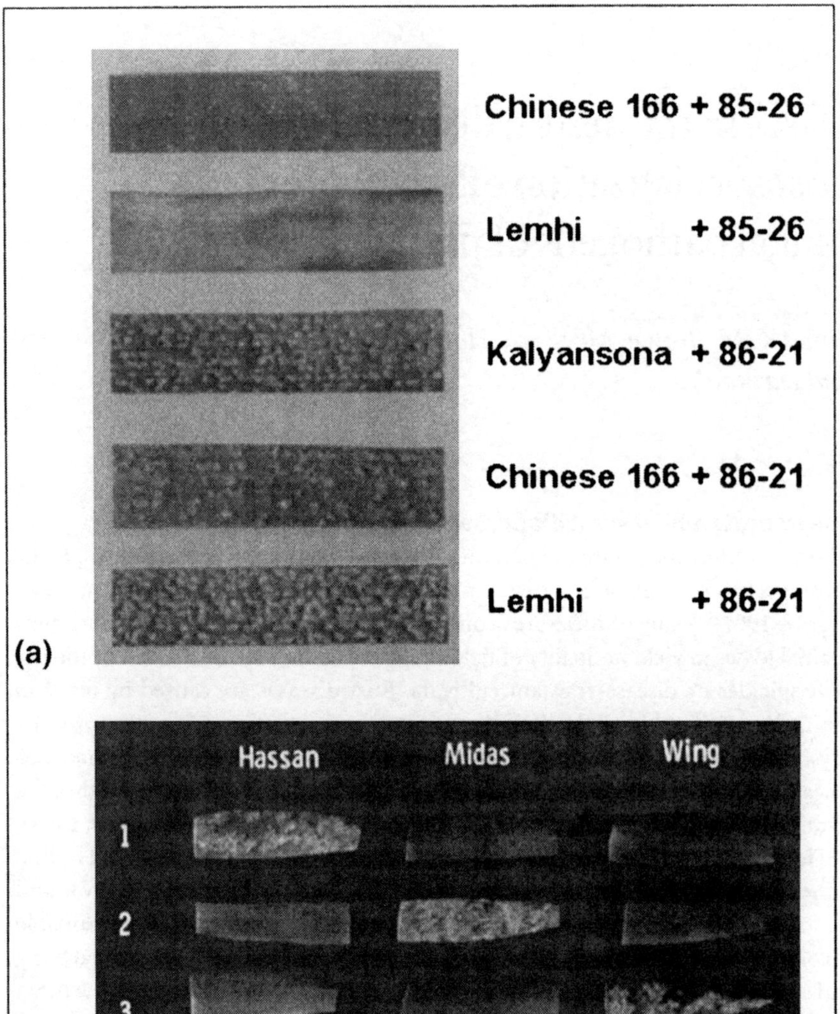

Figure 19.1 Specific interactions between plant host and fungal pathogen genotypes. (a) Wheat and the yellow rust (stripe rust) fungus, *Puccinia striiformis* f.sp. *tritici*. The resistance of cultivar Chinese 166 to *P. striiformis* isolate 85-26 is controlled by the resistance gene *Yr1*; Chinese 166 is one of the set of standard differential varieties used by rust workers (e.g. Hovmøller *et al.* 2002; Justesen *et al.* 2002). Lemhi is resistant to 85-26, but note that its resistance involves the development of extensive necrosis in the leaf, unlike Chinese 166. Kalyansona (carrying the *Yr2* resistance gene), Chinese 166 and Lemhi are all susceptible to isolate 86-21, so the rust fungus forms pustules which release uredospores. (b) Barley and the powdery mildew fungus, *Blumeria graminis* (syn. *Erysiphe graminis*) f.sp. *hordei*. The cultivars Hassan, Midas and Wing have the resistance genes *Mla12*, *Mla6* and *Mla7* + *Mlk1*, respectively. Each cultivar is susceptible to one of the isolates, Hassan to 1, Midas to 2 and Wing to 3, but resistant to the other two. On susceptible cultivars, the mildew fungus produces large numbers of conidia from colonies on the leaf surface.

gene in the host confers an incompatible interaction, with no disease or a reduced amount of disease, when the pathogen carries the matching avirulence gene (Flor 1971). (Note that plant pathologists use the term 'virulence' in a different sense to that used in medical or veterinary pathology. In plant pathogens, 'virulence' usually describes a qualitative ability to cause disease on a plant genotype rather than a quantitative ability to cause an increased amount of disease.) Typically, many gene-for-gene resistances are involved in any one plant disease, so there need only be one pair of matching resistance and avirulence genes for an interaction to be incompatible. However, if the host has no resistance gene matching any of the pathogen's avirulence genes, there is a compatible interaction between the plant and the fungus. Once again, the basic gene-for-gene model oversimplifies what is often a complex situation. Indeed, in the barley powdery mildew fungus, *Blumeria graminis* (synonym *Erysiphe graminis*) f.sp. *hordei*, which is one of the best studied plant pathogens, few of the avirulence phenotypes that have been studied in depth actually follow the simple gene-for-gene model (Brown 2002). Nevertheless, the gene-for-gene relationship is a useful foundation underpinning research on host specificity in biotrophic fungi.

A third, distinctive feature of these pathogens bears on the subject of this book—their asexually produced spores, which are responsible for multiplication of the pathogens during epidemics, are air-borne and therefore wind dispersed. The conidia of powdery mildew fungi and the uredospores of rust fungi are dispersed only in this way, while the zoospores of downy mildews are dispersed both by wind and by water (Agrios 1997). Air-borne spores may travel over very long distances, as shown by the wind dispersal of spores of rust and powdery mildew pathogens of cereals for hundreds of kilometres between the UK and Denmark (Hermansen *et al.* 1978); the transport of spores for thousands of kilometres may even be possible. In the light of the previous two points—biotrophy and host specialization—it is not hard to understand why wind dispersal should be so closely associated with biotrophy. For an individual fungus to multiply, its daughter spores must be able to infect healthy plant tissue, and that plant must be of an appropriate species and a compatible genotype. If a fungus were to reproduce only as a mycelium, its daughters would be in direct competition with itself. Splash-dispersed spores might impact upon a neighbouring plant as well as the plant infected by their parent, but in a natural situation of diverse plant species, few neighbours are likely to be of an appropriate host species. Production of air-borne spores, with their potential for long-distance dispersal to other plants of the same species, maximizes the opportunity for growth and reproduction on fresh host tissue while minimizing parent–offspring competition.

Farmland: oases in a desert

A fourth factor to be considered in a discussion of dispersal in relation to host specificity is that of the distribution of the host populations of these fungi. To human eyes, the English landscape in summer may seem idyllic, a patchwork of lush pastures and burgeoning crops. For an obligately biotrophic pathogen, however, this is an almost entirely hostile environment since any individual fungus is unable to infect the great majority of fields, and therefore unable to grow and reproduce on them. Yet if a pathogen successfully infects one host plant, it will almost always be in a field in which every other plant is also susceptible, because many modern crops are grown as inbred cultivars in monoculture and are, therefore, genetically highly uniform. A field of an appropriate species and compatible cultivar therefore appears to a rust or mildew fungus as an oasis in the desert, a patch of ideal habitat in the midst of what might as well be a barren waste.

When an air-borne spore of an obligate pathogen lands on a plant in a particular field, it must surmount a series of obstacles to the growth, reproduction and further dispersal of its own daughter spores. Are there plants in the field? If so, is it an appropriate host species? If so, is it of a compatible cultivar? If the farmer has used a fungicide, is the fungus resistant to that chemical? So, for a wind-dispersed pathogen to reproduce, its spores must be transported from one 'oasis' to another. This is obviously true within a growing season. However, a field which can support a particular pathogen genotype one year will, much more often than not, be sown with a different crop species or cultivar the next year. Dispersal of air-borne spores may therefore be crucial to a pathogen's chances of surviving from one year to the next. In particular, long-distance dispersal enables a pathogen clone to infect crops at a considerable distance from its point of origin.

This chapter explores the relationship between strong host specialization, the 'oasis' distribution of the host population and long-distance dispersal of spores of biotrophic plant pathogens. We show that long-distance dispersal is a crucial factor in the population genetics and epidemiology of rust and powdery mildew fungi, and may have had a critical role in the evolution of these pathogens during the history of agriculture. In relation to the pathogens' population genetics, different crop varieties are susceptible to different pathogen genotypes. In terms of epidemiology, susceptible crops are present in different places at different times. Finally, and more speculatively, recent data suggest that, as people chose to grow different species of cereals, special forms of biotrophic, pathogenic fungi became adapted for infection and reproduction on different host crops.

Population genetics: wheat yellow rust in northern Europe

The effects of long-distance dispersal on pathogen population genetics are illustrated for the yellow rust (or stripe rust) disease of wheat, caused by *Puccinia striiformis* f.sp. *tritici.* Until the advent of modern plant breeding, this was the most serious disease of wheat in temperate regions. Although it is generally well controlled nowadays by the use of fungicides and resistant cultivars, it still has the potential to cause disastrous crop losses if these measures fail. New disease-resistance genes are periodically introduced into wheat-breeding programmes, often from wild grasses related to wheat. Cytogenetic methods can be used to introgress chromosomes or parts of chromosomes carrying resistance genes from wild species into modern wheat cultivars (Gale & Miller 1987). As this process is long and expensive, a resistance gene which is effective against a target disease tends to be used by many wheat breeders so, sometimes, control of a disease relies excessively on a single resistance gene. Such was the case for yellow rust in Europe in the mid-1990s, when many varieties had good resistance conferred by the *Yr17* gene (Hovmøller 2001), introgressed from a wild grass related to wheat, *Triticum ventricosum* (Bariana & McIntosh 1993). This wild grass was originally used in plant breeding as a source for resistance to the fungus causing eyespot of wheat, *Tapesia yallundae*.

A resistance which follows a gene-for-gene relationship is inherently vulnerable to being overcome by virulent pathogens, since, in principle, just one mutation is needed for the fungus to change from avirulence to virulence matching the resistance gene. In practice, more than just one gene may be involved, as discussed for barley powdery mildew by Brown (2001). Nevertheless, the principle that the effectiveness of a host resistance may be lost through the emergence of pathogen genotypes with the corresponding virulence holds good. This process is often described colloquially as 'breakdown' of the resistance gene, although the resistance gene, strictly speaking, has not been changed. A recent study has shown that the long-distance dispersal of *P. striiformis* uredospores has striking effects on the population genetics of the rust fungus (Hovmøller *et al.* 2002).

Samples of *P. striiformis* were obtained from wheat fields and trial sites between 1988 and 1998. The majority of isolates were from Denmark, and these were compared with isolates from the UK, France and Germany, covering a geographical range of 1700 km (Fig. 19.2). Each isolate was characterized for virulence on a differential set of wheat varieties which, between them, have the yellow rust resistance genes used in European wheat breeding. All isolates had virulence to two of the resistances and avirulence to another five, but polymorphism was detected in the virulence phenotypes matching seven resistances. Isolates were also screened for amplified fragment length polymorphism (AFLP) (Vos *et al.* 1995) using 21 primer pairs. Although the overall level of polymorphism was very low, it was possible to generate an unrooted evolutionary tree of the isolates from data on the presence or absence of AFLP bands, since it is reasonable to presume that this type of molecular variation is selectively neutral (Ridout & Donini 1999). *P. striiformis* has no known sexual stage (Manners 1988) and statistical analysis of the evolutionary tree confirmed that the population structure was consistent with the fungus reproducing entirely by asexual means.

Yr17 first broke down—i.e. virulent pathogens first caused disease on cultivars with this resistance—in the UK in 1994. Yellow rust was observed for the first time on *Yr17*-resistant varieties in Denmark in 1997. Remarkably, the *P. striiformis* clones that caused these two breakdown events were not only virulent to *Yr17* but were identical in all the other virulence characters tested and in all 28 polymorphic AFLP bands examined (type K, Table 19.1). Shortly afterwards, identical isolates of *P. striiformis* were also found in France and Germany in 1997 and 1998. In the light of the diversity in the *P. striiformis* isolates as a whole, it was concluded that these isolates were, in fact, probably members of a single clonal lineage. This lineage appears to have originated from a lineage of *P. striiformis* that was virulent on wheat varieties

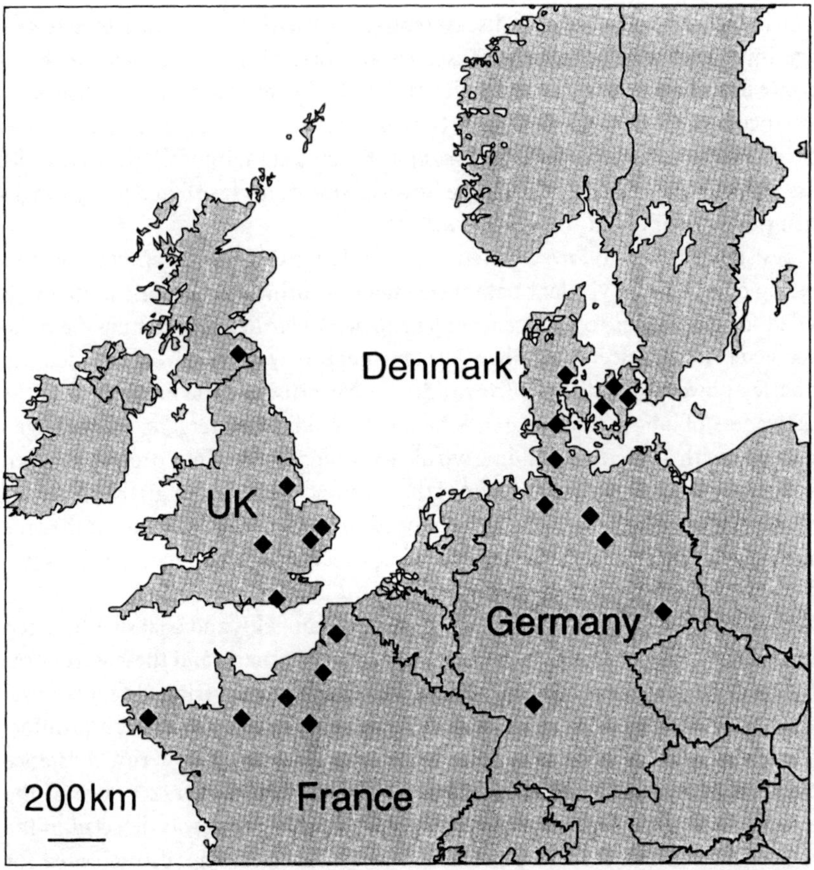

Figure 19.2 Map of northwest Europe, showing sites where isolates of the wheat yellow rust fungus, *Puccinia striiformis* f.sp. *tritici*, used in the research of Justesen *et al.* (2002) and Hovmøller *et al.* (2002), were obtained. The sample area covers a maximum distance of 1700 km.

with another widely used resistance gene, *Yr9*, and to have first mutated to *Yr17* virulence in England, and then to have spread to other countries in northern Europe where wheat cultivars with *Yr17* were grown.

At least one virulence mutation occurred subsequently in this lineage, as isolates with virulence to *Yr17*, *Yr9* and another gene, *Yr6*, were found in England and Germany (type K′, Table 19.1). Whether this was the result of a single mutation event in one place, followed by dispersal over northern Europe, or of separate mutations in the two countries, is not known. However, these isolates had identical AFLP patterns to that of the original *Yr17*/*Yr9*-virulent lineage. This implies that the average rate of evolution of AFLP bands is slow relative to that of virulences.

It was striking, therefore, that a second, quite different *Yr17*-virulent clonal lineage was detected. This differed from the first *Yr17*-virulent clone only in one

Table 19.1 Instances in which isolates of the wheat yellow rust (stripe rust) fungus, *Puccinia striiformis* f.sp. *tritici* with indistinguishable virulence characters and AFLP types (after Justesen *et al.* 2002) were found in more than one country in northwest Europe (details in Hovmøller *et al.* 2002).

AFLP type	Virulences (corresponding resistances listed)	Countries (and years)
A	*Yr2, Yr3b + Yr4b, Yr6, Yr9*	UK (1989), Denmark (1994–95)
E	*Yr2, Yr3b + Yr4b, Yr9*	UK (1988 and 1990), Denmark (1993–95)
J	*Yr1, Yr2,* Carstens V	UK (1990), Denmark (1993)
K	*Yr1, Yr2, Yr9, Yr17*	UK (1994), Denmark (1997–98), France (1997–98), Germany (1998)
K′	*Yr1, Yr2, Yr6, Yr9, Yr17*	UK (1996), Germany (1998)
O	*Yr1, Yr2, Yr3b + Yr4b, Yr9, Yr17*	Denmark (1995), UK (1997–98), France (1997), Germany (1998)

virulence phenotype, as isolates of the second lineage were virulent on cultivars with a complex resistance controlled by the genes *Yr3b+ Yr4b*, whereas isolates of the first lineage were not. However, it differed from the first lineage in 11 AFLP bands. Again, this lineage was detected in Denmark, France, Germany and the UK (type O, Table 19.1). It is likely that, like the first *Yr17*-virulent lineage, it too had a single origin and spread rapidly over northern Europe.

At least three other similar events, involving other host-resistance genes, occurred in the 1980s and 1990s (types A, E and J, Table 19.1). This indicates that transport of *P. striiformis* spores over many hundred kilometres, and the development of yellow rust epidemics as a result of long-distance spore dispersal, may occur frequently in northern Europe.

What are the factors that cause this to happen so often? When a new resistance gene is used in a wheat cultivar, it imposes strong selection on the pathogen for the corresponding virulence. As fields are genetically uniform, each being sown with a single, highly inbred cultivar, the new, virulent clone may multiply rapidly once it is established within a field. When one field of a hitherto resistant cultivar has succumbed to a virulent race of *P. striiformis*, uredospores may be rapidly wind dispersed to other fields of the same cultivar or of other cultivars with the same resistance gene. Then, since the resistance gene is used in wheat cultivars throughout Europe, epidemics of rust develop on a continental scale because of long-distance dispersal of spores and the selection of virulent clones by cultivars with the same resistance gene. It was previously proposed that a similar process of long-distance dispersal of a few virulent clones occurred in the barley powdery mildew fungus, *B. graminis* f.sp. *hordei* (Brown 1994). The process may be characteristic of obligate biotrophic fungal pathogens of plants in general, as it appears to be a simple consequence of the factors discussed here: strict specialization to host genotypes, wind transport of asexual spores and 'oases' of susceptible host tissue.

Epidemiology: wheat powdery mildew in central China

Wind dispersal of spores is crucial for the propagation of an obligately biotrophic pathogen when the host plant is not available in a particular area throughout the year, because it allows annual re-establishment of the disease from external sources. Here, we illustrate this with the example of the epidemiology of wheat powdery mildew in central China (Yu 2000). Once again, this links strong host specialization, wind-borne dispersal and 'oases' of susceptible hosts as determining factors in the population biology of biotrophic fungi.

Hubei province, where this investigation was carried out, lies in the valley of the Yangtze River (Fig. 19.3). It is approximately 700 km from east to west and 450 km from north to south and, at its eastern extremity, is approximately 500 km from the sea. Much of its area consists of fertile agricultural land, although mountain ranges rise to nearly 3000 m in the west of the province. Wheat is the second most important crop in Hubei, by area and production, after rice. Although Hubei is not as large a

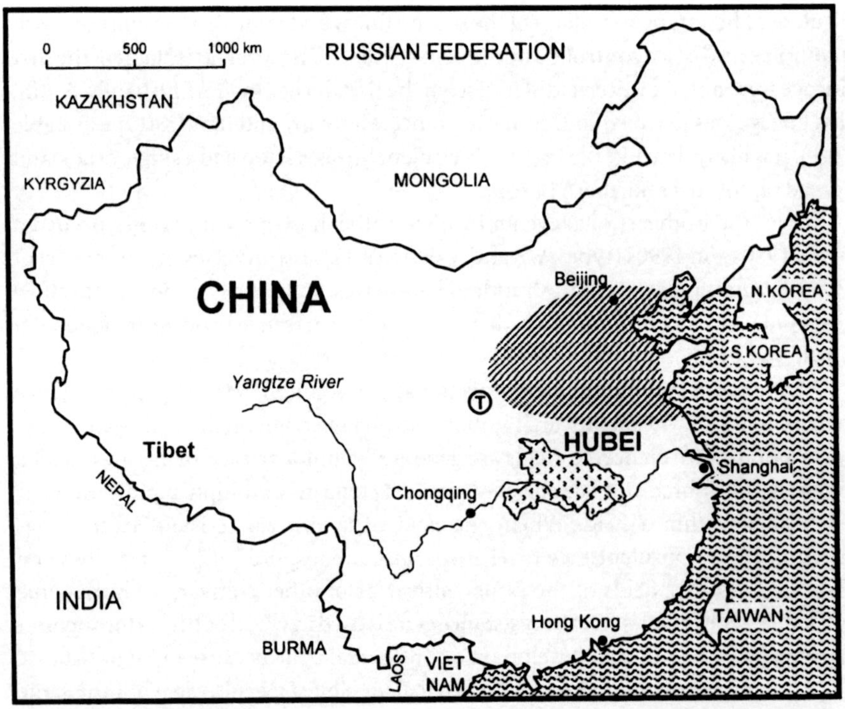

Figure 19.3 Map of China, showing the location of Hubei province. The Yangtze River and major cities are shown for reference; not all neighbouring countries are marked. The major wheat-growing region of northern China is diagonally shaded. The encircled 'T' marks the location of Tianshui prefecture in Gansu province, believed to be an annual source of spores of *Puccinia striiformis* f.sp. *tritici*, the yellow rust (stripe rust) fungus, which infect wheat crops throughout northern China (Shan *et al.* 1998).

wheat producer as northern provinces such as Shaanxi, Shanxi, Hebei, Henan or Shandong (Fig. 19.3), it may have a crucial position in the epidemiology of wheat mildew in China. In northern parts of the country, winter temperatures are too low for the powdery mildew fungus, *Blumeria graminis* f.sp. *tritici*, to survive. Mildew must therefore be re-established in these regions each year from elsewhere. In southwestern China, Guizhou province and the eastern half of Sichuan province have mild winter climates, so mildew persists there throughout the year. Wheat crops in the Yangtze valley provinces, including Hubei, may act as a bridge for *B. graminis* spores, facilitating disease transmission from southwestern China to the north of the country. Understanding the epidemiology of wheat mildew in the Yangtze provinces may therefore help in devising disease control measures, not only in those provinces but in China as a whole.

A model of the epidemiology of powdery mildew in central China was developed by Yu (2000) on the basis of observations of the incidence of mildew on crops and volunteer crops in highland and lowland regions, experiments on the survival of volunteer plants in summer, and research on the conditions under which ascospores of *B. graminis* are released from cleistothecia (Fig. 19.4). In the lowland areas of Hubei, in the basin of the Yangtze and Han Rivers, wheat is harvested in early May and sown in late October. Summer temperatures are too high for self-sown wheat volunteers to survive and, as a highly specialized fungus, *B. graminis* f.sp. *tritici* cannot survive on alternative hosts which are present in the summer, such as rice. *B. graminis*, therefore, cannot oversummer in the asexual phase in lowland Hubei, as no host plants are available (Yu 2000). In areas where the summer weather is dry, cleistothecia, the sexual reproductive structures of *B. graminis*, allow the pathogen to oversummer (Koltin & Kenneth 1970) but this is almost certainly not possible in the lowlands of Hubei, because the high humidity caused by the monsoon climate is likely to cause all ascospores to be released from the cleistothecia before the seedlings in new fields of wheat emerge in late autumn (Yu 2000).

Mildew must therefore be re-established each year on wheat crops in the lowlands of Hubei. The most likely source of infection is conidiospores from wheat crops in the highland areas of Hubei and, possibly, neighbouring provinces. In the highland parts of the province, the growing season of wheat is longer because temperatures are cooler. Above 1000 m, wheat is harvested in early July and sown in early September. Volunteers are common in summer, because temperatures are sufficiently cool for them to survive, and they may act as a bridge host which allows *B. graminis* to persist between seasons. In the model proposed by Yu (2000), volunteer seedlings are infected by ascospores of *B. graminis* after the harvest and then support populations of the fungus through the summer. In the autumn, conidia of *B. graminis* are dispersed from volunteers onto germinating wheat crops. In turn, in the following spring, conidia are dispersed from wheat fields in the highlands to the lowlands.

Wind-borne spore dispersal between wheat-growing regions within the province, therefore, appears to be a critical factor in the epidemiology of powdery mildew in Hubei. This is a consequence of the pathogen being an obligate biotroph and being highly specialized to its host species, so that it must be re-established in

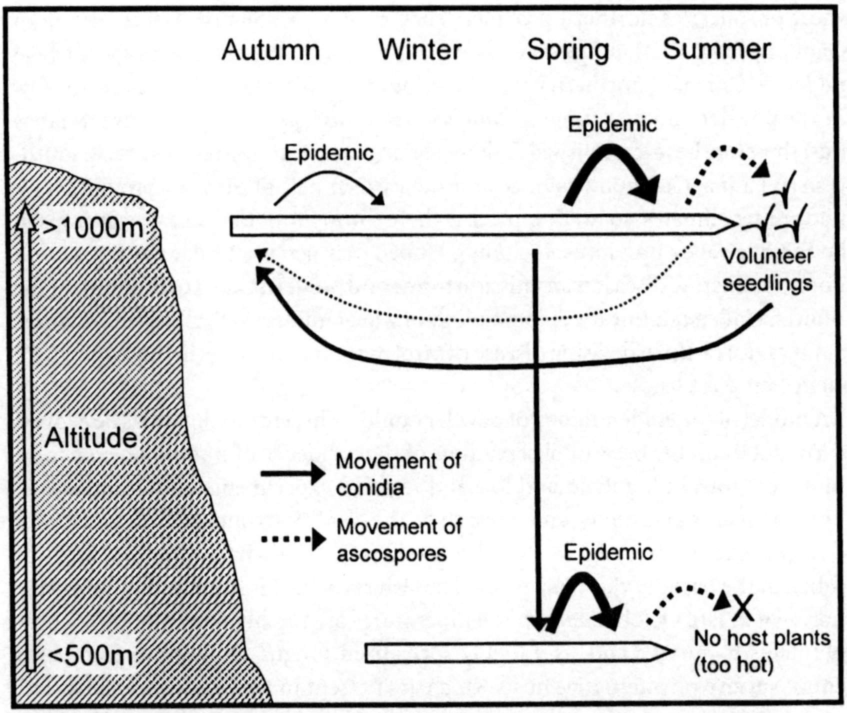

Figure 19.4 Model of the epidemiology of wheat powdery mildew in Hubei province, central China, after Yu (2000). At high altitudes, mildew is present throughout the growing season of wheat because green plant tissue is available and temperatures are favourable. Cleistothecia of *Blumeria graminis* f.sp. *tritici* are produced as a result of sexual reproduction at the end of the growing season and release ascospores onto volunteer wheat seedlings in the summer. The resulting mildew colonies generate conidia by asexual reproduction, which then infect crops in the autumn. Some cleistothecia of *B. graminis* may survive the summer to release ascospores directly onto wheat plants in the autumn. At low altitudes, all ascospores are released from cleistothecia during the summer because of continuous warm, wet conditions, so no ascospores are present by the time the autumn crop has emerged. Moreover, summer temperatures are too hot for volunteer seedlings to survive. Lowland crops do not emerge until late autumn, so they are infected by conidia of *B. graminis* from higher altitudes in the spring.

lowland wheat-growing areas each year. Dispersal of *B. graminis* spores on an even larger scale, between provinces (i.e. over distances equivalent to several European countries), may be significant for the epidemiology of wheat mildew in China as a whole. Firstly, Yu (2000) suggested that conidia may be dispersed from Hubei to neighbouring provinces and vice versa, but the impact of *B. graminis* populations in different provinces in central China on one another is not at all well understood. Secondly, the extent to which mildew epidemics in the major wheat-growing area of northern China (Fig. 19.3) are caused by dispersal of *B. graminis* conidia from central and southwestern China is not yet known. These questions about very long-distance dispersal, as well as the model of Yu (2000) (Fig. 19.4), may be tested by means of markers such as virulences and DNA polymorphisms.

The dispersal of spores over very long distances is a well-known feature of the epidemiology of diseases caused by obligate biotrophic fungi, because living plants of appropriate hosts are present at different times of the year in different places. Very long-distance dispersal is believed to be a critical factor in the epidemiology of wheat yellow rust in northern China (Shan *et al.* 1998, and references therein). *P. striiformis* cannot survive the summer in the main wheat-growing provinces (Fig. 19.3), because of the absence of wheat crops and the hot, humid climate, so yellow rust must be re-established afresh in that region each year. Tianshui prefecture of southern Gansu province, to the west of the main wheat-growing region, may be a yellow rust 'hot-spot' that is cool enough for the fungus to survive during summer and to be dispersed to other provinces in autumn.

On a similar geographical scale and for very similar reasons, long-distance dispersal is a critical factor in the survival of another cereal rust pathogen, *Puccinia graminis*, the causal agent of stem rust (or black rust) in both North America and India. In the former case, the fungus cannot survive the winter in the northern prairie states of the USA or the western provinces of Canada, so the disease must be re-established each spring by uredospores blown northwards from Mexico and Texas along the so-called '*Puccinia* pathway'. In the latter case, host wheat plants are absent in the kharif (hot and rainy) season in the northern plains, but uredospores of *P. graminis* are blown northwards in the autumn from the Nilgiri Hills in South India, which have a relatively mild climate (Hau & de Vallavieille-Pope 1998).

Coevolution of cereals and grasses with their powdery mildew parasite

The two examples above consider pathogen population biology in modern, intensive agriculture. However, wind dispersal of obligately biotrophic fungi may have been crucial in the evolution of host specialization itself, one of the most significant features of these organisms as they now exist on agricultural crops. Marchal (1902) described four special forms of *Blumeria graminis* infecting the cereal crop species barley, oats, rye and wheat and three forms specialized to the wild grass genera *Agropyron*, *Bromus* and *Poa*, while an eighth form, specialized to the genus *Dactylis*, was added by Oku *et al.* (1985). Several researchers have shown that the host range of *B. graminis* isolates sampled from wild grasses may extend well beyond the genus from which they were sampled (Hardison 1944; Eshed & Wahl 1970; Sheng *et al.* 1993), but this work has not caused the view that there are four separate, cereal-infecting forms of *B. graminis* in modern agriculture to be challenged.

To understand the evolution of agricultural pathogens to their present state, it is important to bear in mind that the cultivation of uniform crop species is quite a recent invention. For most of the evolutionary history of *B. graminis*, it was a parasite of wild grasses, which presumably grew in diverse ecosystems then as now. Grasses were first harvested for food in the Upper Paleolithic, at least 20 000 years ago, and cultivation of what are now our temperate cereal crops began in West Asia about 9000 BC, marking the start of the Neolithic period. Wheat and barley were cultivated as separate species from about 7000 BC onwards (Zohary & Hopf 1988), while oats were first grown as a separate crop from the late Bronze Age, about 2000 BC (Helbaek 1959). Rye is a more recent crop still, as it seems to have been a weed of wheat, or at least grown together with wheat, until Roman times, when it was first cultivated separately (Zohary & Hopf 1988; Evans 1995).

A paradigm for host–parasite coevolution is Fahrenholz's rule (Eichler 1948), that parasite phylogeny mirrors host phylogeny. A striking illustration of this rule is provided by the work of Hafner and Nadler (1988) on pocket gophers and chewing lice, where the branch points of the phylogenies of the host and parasite taxa are congruent, with two exceptions. The louse is intimately associated with its host throughout its life cycle, so louse populations are likely to diverge and eventually speciate as gopher populations diverge. The two exceptions to Fahrenholz's rule may represent host-switching events; one may imagine that a burrow made by one gopher species and containing eggs of the corresponding louse species, was taken over by a gopher of a different species. At any rate, the almost perfect fit to the prediction of Fahrenholz's rule is almost certainly the result of the very close association between the host and the parasite.

A very different situation is presented in the comparison of the phylogeny of *B. graminis* on the one hand and that of cereals and grasses on the other (Wyand 2001). The *B. graminis* phylogeny was based on sequencing of the intervening transcribed spacers (ITSs) of the ribosomal DNA cistron. Three features of these data are relevant to the discussion in this chapter. Firstly, *B. graminis* from cultivated cereals formed a single clade, well separated and distinct from most clades of *B. graminis* from wild grasses. Secondly, host and parasite phylogenies were not congruent; for example, oat is a member of the tribe Aveneae, but *B. graminis* f.sp. *avenae* is most closely related to *B. graminis* from wheat and barley, which are members of the tribe Triticeae, than to isolates from Aveneae grasses such as *Alopecurus*. Thirdly, *B. graminis* f.sp. *secalis*, from rye, has an identical ITS sequence to that of *B. graminis* f.sp. *tritici*.

These observations can be interpreted in the light of the propagation of *B. graminis* by wind-dispersed spores, which contrasts with the close association between chewing lice and their pocket gopher hosts. In nature, different grasses, i.e. potential host species of *B. graminis*, grow together. The pathogen is therefore regularly dispersed between different host species and there is likely to be selection for a wide host range. In agriculture, by contrast, fields of uniform crop species have been grown from the mid-Neolithic period onwards (Zohary & Hopf 1988), although uniform crop cultivars are a much more recent innovation. The fungus would, therefore, usually have been dispersed to the same host species, which would have caused selection for highly specialized forms of the pathogen. The clade of *B. graminis* which includes the cereal-infecting forms may have been a local type of the fungus which became associated with early arable crops and subsequently became specialized to wheat, barley and oats. The wheat-infecting form diverged thereafter, giving rise to a separate, rye-infecting form. This argument assumes that specialization involves a cost, in that the putative generalist, preagricultural genotypes of *B. graminis* would have reproduced more slowly on any particular host species than the corresponding special form of the fungus found nowadays in agriculture.

Dispersal and sex as survival strategies in agriculture

Host specialization and wind dispersal are characteristic features of obligate biotrophic fungal pathogens. Clearly, wind dispersal is an ancient adaptation, as it allows spores to be dispersed between susceptible host plants in natural ecosystems, whereas agriculture is a comparatively very recent invention. Research on *Blumeria graminis* (Wyand 2001) suggests that, in this pathogen at least, host specialization evolved as agriculture developed, probably because the average contiguous area occupied by a single host genotype increased enormously, from perhaps no more than the area covered by one plant to tens or even hundreds of hectares on the most intensive modern farms.

In modern agriculture, each field of an inbred cultivar of a single crop species is an ideal environment for an appropriately specialized genotype of a biotrophic fungus but a hostile one for inappropriate genotypes. This highly simplified landscape has made long-distance, mass dispersal of spores even more crucial for the multiplication of these pathogens in agriculture. In turn, the dispersal of pathogen spores between 'oases' of host plants causes widespread epidemics of rust and mildew diseases, so that resistance genes deployed in widely separated places may lose their effectiveness against the pathogen population almost simultaneously (Table 19.1) (Brown 1994; Hovmøller *et al.* 2002) and crops may be reinfected from distant sources in areas where the pathogen cannot survive harsh summer or winter conditions (Fig. 19.4) (Hau & de Vallavieille-Pope 1998; Shan *et al.* 1998; Yu 2000).

A striking feature of the asexual spores of biotrophic fungal pathogens is that they are not adapted for long-term survival. Indeed, many live only a few hours. By contrast, many of these fungi have long-lived sexual structures: cleistothecia of powdery mildews, telia of rusts and oospores of oomycetes (Agrios 1997). The sexually produced spores—ascospores, aeciospores and sporangia of the three groups, respectively—are also wind dispersed (although sporangia of oomycetes, like zoospores, may also be dispersed by water such as rain splash). However, while sexual spores may initiate epidemics of some diseases (Fig. 19.4) (Agrios 1997), pathogen multiplication during epidemics is almost always by means of asexual spores. The contrast between the roles of sexual and asexual spores reflects the role of recombination in pathogen adaptation. As host genes for resistance are recombined during sex and as seeds are dispersed, a pathogen population in a natural ecosystem is likely to be confronted with a different array of host genotypes (or even host species) from one year to the next. Sex in the pathogen is an adaptation to host diversity because it generates diverse pathogen genotypes with recombined virulence genes, at least some of which match host genotypes with novel combinations of resistance genes

(Brown *et al.* 1993; Brown 1999). The longevity of the sexual structures is also adaptive because it allows fungi to survive until susceptible hosts are available (Chamberlain & Ingram 1997). However, once plants have germinated, the pathogen may pass through several generations until the plant sets seed and becomes senescent. This inequality between host and pathogen generation times favours asexual reproduction (Hamilton *et al.* 1990) because daughter spores may infect the same plant as that on which they were produced or nearby plants of similar genotypes.

The simplification of the distribution of plant genotypes in agriculture has tilted the balance in favour of asexual reproduction. Within a growing season, asexual reproduction allows rapid multiplication of pathogen genotypes which are virulent on the cultivars that are currently grown and this is facilitated by long-distance spore dispersal between fields of susceptible host plants, as discussed above (Table 19.1) (Brown 1994; Hovmøller *et al.* 2002). The advantage provided by long-distance dispersal of asexual spores is therefore even greater in agriculture than in nature. Since relatively few genes for resistance to a disease are used in plant-breeding programmes at any one time (Wolfe 1984; Brown 1994; McIntosh *et al.* 1995), a pathogen which oversummers or overwinters in the asexual phase has a substantially higher chance of its asexual spores encountering a susceptible host genotype than would be the case in a natural ecosystem. This means that clones of a pathogen may persist from year to year (indeed, *Puccinia striiformis* has no known sexual stage and only exists in the asexual phase; Manners 1988), provided that the asexually produced spores are dispersed by the wind from one location to another, where the host is present and the climate is favourable (Hau & de Vallavieille-Pope 1998; Shan *et al.* 1998; Yu 2000).

References

Agrios, G.N. (1997) *Plant Pathology*, 4th edn. Academic Press, San Diego.

Bariana, H.S. & McIntosh, R.A. (1993) Cytogenetic studies in wheat. XV. Location of rust resistance genes in VPM1 and their genetic linkage with other disease resistance genes in chromosome 2A. *Genome* **36**, 476–482.

Bhattacharya, D., Medlin, L., Wainright, P.O. *et al.* (1992) Algae containing chlorophylls a + c are paraphyletic—molecular evolutionary analysis of the Chromophyta. *Evolution* **46**, 1801–1817.

Brown, J.K.M. (1994) Chance and selection in the evolution of barley mildew. *Trends in Microbiology* **2**, 470–475.

Brown, J.K.M. (1999) The evolution of sex and recombination in fungi. In: *Structure and Dynamics of Fungal Populations* (ed J.J. Worrall), pp. 73–95. Kluwer Academic Publishers, Dordrecht.

Brown, J.K.M. (2002) Comparative genetics of avirulence and fungicide resistance in the powdery mildew fungi. In: *The Powdery Mildews: a Comprehensive Treatise* (eds R.R. Belanger, A.J. Dik & W.R. Bushnell). APS Press, St Paul, MN (in press).

Brown, J.K.M., Simpson C.G. & Wolfe, M.S. (1993) Adaptation of barley powdery mildew populations in England to varieties with two resistance genes. *Plant Pathology* **42**, 108–115.

Chamberlain, M. & Ingram, D.S. (1997) The balance and interplay between asexual and sexual reproduction in fungi. *Advances in Botanical Research incorporating Advances in Plant Pathology* **24**, 71–87.

Eichler, W. (1948) Some rules in ectoparasitism. *Annals and Magazine of Natural History* **12**, 588–598.

Eshed, N. & Wahl, I. (1970) Host ranges and interrelations of *Erysiphe graminis hordei*, *Erysiphe graminis tritici*, and *Erysiphe graminis avenae*. *Phytopathology* **60**, 628–634.

Evans, G.M. (1995) Rye. In: *Evolution of Crop Plants* (eds J. Smartt & N.W. Simmonds), pp. 166–170. Longman, Harlow.

Flor, H.H. (1971) Current status of the gene for gene hypothesis. *Annual Review of Phytopathology* **9**, 275–296.

Gale, M.D. & Miller, T.E. (1987) The introduction of alien genetic variation into wheat. In: *Wheat Breeding: its Scientific Basis* (ed. F.G.H. Lupton), pp. 173–210. Chapman & Hall, London.

Hafner, M.S. & Nadler, S.A. (1988) Phylogenetic trees support the coevolution of parasites and their hosts. *Nature* **332**, 258–259.

Hamilton, W.D., Axelrod, R. & Tanese, R. (1990) Sexual reproduction as an adaptation to resist parasites (a review). *Proceedings of the National Academy of Sciences of the USA* **87**, 3566–3573.

Hardison, J.R. (1944) Specialization of pathogenicity in *Erysiphe graminis* on wild and cultivated grasses. *Phytopathology* **34**, 1–20.

Hau, B. & de Vallavieille-Pope, C. (1998) Wind-dispersed diseases. In: *The Epidemiology of Plant Diseases* (ed. D.G. Jones), pp. 323–347. Kluwer Academic Publishers, Dordrecht.

Helbaek, H. (1959) Domestication of food plants in the old world. *Science* **130**, 365–372.

Hermansen, J.E., Torp, U. & Prahm, L.P. (1978) Studies of transport of live spores of cereal mildew and rust fungi across the North Sea. *Grana* **17**, 41–46.

Hovmøller, M.S. (2001) Disease severity and pathotype dynamics of *Puccinia striiformis* f.sp. *tritici* in Denmark. *Plant Pathology* **50**, 181–189.

Hovmøller, M.S., Justesen, A.F. & Brown, J.K.M. (2002) Clonality and long-distance migration of *Puccinia striiformis* f.sp. *tritici* in NW-Europe. *Plant Pathology* **51**, 24–32.

Justesen, A.F., Ridout, C.J. & Hovmøller, M.S. (2002) The recent history of *Puccinia striiformis* f.sp. *tritici* in Denmark as revealed by disease incidence and AFLP markers. *Plant Pathology* **51**, 13–23.

Koltin, Y. & Kenneth, R. (1970) The role of the sexual stage in the over-summering of *Erysiphe graminis* DC. f. sp. *hordei* Marchal under semi-arid conditions. *Annals of Applied Biology* **65**, 263–268.

Maclean, D.J. (1982) Axenic culture and metabolism of rust fungi. In: *The Rust Fungi* (eds K.J. Scott & A.K. Chakravarti), pp. 37–120. Academic Press, London.

Manners, J.G. (1988) *Puccinia striiformis*, yellow rust (stripe rust) of cereals and grasses. *Advances in Plant Pathology* **6**, 373–387.

Marchal, E. (1902) De la spécialisation du parasitisme chez l'*Erysiphe graminis*. *Comptes Rendus de l'Académie des Sciences* **135**, 210–212 [in French].

McIntosh, R.A., Wellings, C.R. & Park, R.F. (1995) *Wheat Rusts: an Atlas of Resistance Genes*. Kluwer Academic Publishers, Dordrecht.

Oku, T., Yamashita, S., Doi, Y. & Nishihara, N. (1985) Host range and forma specialis of cocksfoot powdery mildew fungus (*Erysiphe graminis* DC.) found in Japan. *Annals of the Phytopathological Society of Japan* **51**, 613–615.

Ridout, C.J. & Donini, P. (1999) Use of AFLP in cereals research. *Trends in Plant Science* **4**, 76–79.

Shan, W.X., Chen, S.Y., Kang, Z.S., Wu, L.R. & Li, Z.Q. (1998) Genetic diversity in *Puccinia striiformis* Westend. f.sp. *tritici* revealed by pathogen genome-specific repetitive sequence. *Canadian Journal of Botany* **76**, 587–595.

Sheng, B.Q., Zhou, Y.L., Duan, X.Y. & Xiang, Q.J. (1993) The host range of wheat powdery mildew (*Erysiphe graminis* f.sp. *tritici*) in Shanxi province. *Acta Phytophylacica Sinica* **20**, 105–112.

Sogin, M.L. & Silberman, J.D. (1998) Evolution of the protists and protistan parasites from the perspective of molecular systematics. *International Journal for Parasitology* **28**, 11–20.

Vos, P., Hogers, R., Bleeker, M. *et al.* (1995) AFLP: a new concept for DNA fingerprinting. *Nucleic Acids Research* **23**, 4407–4414.

Wolfe, M.S. (1984) Trying to understand and control powdery mildew. *Plant Pathology* **33**, 451–466.

Wyand, R.A. (2001) *Molecular evolution of Blumeria graminis*. PhD thesis, University of East Anglia, Norwich.

Yu, D. (2000) *Wheat powdery mildew in central China: pathogen population structure and host resistance*. PhD thesis, Wageningen University and Research Centre, Wageningen.

Zohary, D. & Hopf, M. (1988) *Domestication of Plants in the Old World*. Clarendon Press, Oxford.

Chapter 20
Climate change and dispersal

Andrew R. Watkinson and Jenny A. Gill

Introduction

While we are particularly concerned at the present time about anthropogenic-driven change in the environment, study of the past indicates that change is an integral feature of the world we live in (Tallis 1991; Bennett 1997). This change has been documented in terms of a wide range of factors including plate tectonics, geomorphology, climate and biodiversity. Over the course of the 21st century, Sala *et al.* (2000) have estimated the potential impacts of a number of different drivers of environmental change on biodiversity in the following rank order: land use, climate, nitrogen deposition, biotic exchange and atmospheric carbon dioxide. All of these drivers will impact on the demographic processes of species through the numbers of births, deaths, immigrants and emigrants in populations. Predicting the impact of drivers on the distribution and abundance of species therefore depends upon quantifying dispersal along with numbers of births and deaths; the critical role of density dependence in determining patterns of abundance through its impact on demographic parameters is discussed elsewhere in this volume by Sutherland *et al.*

The movement of individuals through dispersal is particularly important in the case of climate change. While dispersal undoubtedly impacts on conservation in a landscape that is becoming increasingly fragmented (Macdonald & Johnson 2001; Bullock *et al.*, this volume; South *et al.*, this volume), long-distance dispersal is of major significance if species are to track climate change at the rate predicted over the course of the current century. Elsewhere in this book we have seen that our understanding of long-distance dispersal through monitoring is poor (Bullock *et al.*, this volume; Greene & Calogeropoulos, this volume). An examination of changes in the distribution of species in relation to past climate change therefore presents the opportunity to quantify past rates of long-distance dispersal and predict how species and consequently communities may respond to future climate change (Adams & Woodward 1992). For much of geological time, however, the resolution of the relationship between climate change and the distribution of species is poor and it is only since the last glaciation that we can clearly resolve the impact of climate change on the dispersal of a range of species.

Here we focus our attention on the past and predicted effects of climate change in northern temperate latitudes. The aim is to review: (i) recent and predicted future climate change; (ii) evidence for the impact of climate change on the routes and rates of dispersal; (iii) the potential consequences of climate change for community structure as a result of dispersal; and (iv) the issues that dispersal raises for the conservation and management of species.

Climate change

Over the last 65 million years, the earth's climate system has experienced a continuous shift from warm, ice-free periods to glacial periods of extreme cold (Zachos *et al.* 2001). The most recent glacial stage lasted from approximately 115 000 BP until about 10 000 BP. Warming began from about 16 000 BP and continued until about 10 000 BP, since when it has varied much less. However, the average long-term rate of warming between 16 000 and 7000 BP is at least an order of magnitude less than that predicted over the next century (Newman 2000). This gives concern if one wants to try and extrapolate from historical distribution changes in response to climate change how species might respond to climate change in the future. In some localities at least, though, there were shorter periods where warming is likely to have been at rather faster rates; for example, Anklin *et al.* (1993) report warming by 7°C over a few decades in Greenland. However, Briffa and Atkinson (1997) warn that in general there are considerable problems in separating the climatic signal from non-climatic noise in reconstructing past climates and that it is consequently often difficult to be precise about rates of temperature change at well-resolved temporal or spatial scales.

The last millennium has seen considerable variations in temperature that have recently been reviewed for the northern hemisphere (Briffa *et al.* 2001; Jones *et al.* 2001b). Following a relatively warm period during the early part of the last millennium, there was a gradual cooling over the 15th and 16th centuries to a minimum in the 17th century. There was then a period of strong warming through the early 1700s, relatively stable temperatures for the remainder of the 18th century, followed by abrupt cooling in the early 19th century. The 20th century saw a rapid warming with mean global temperatures increasing by 0.15°C per decade over the latter part, so that temperatures are now warmer than at any other period during the last millennium and indeed possibly since the last interglacial period, over 120 000 years ago.

During the latter part of the 20th century, there have been increasing concerns that human activity and, in particular, the emission of carbon dioxide and other greenhouse gases are responsible for the observed increases in mean temperature and other aspects of climate change (Houghton *et al.* 2001). The 1990s climate in the UK was, for example, 0.5°C warmer than the 1961–1990 average, and there was a decrease in the frequency of cold days in winter coupled with a less perceptible increase in the frequency of hot days, a systematic change in the seasonality of precipitation with winters becoming wetter and summers becoming drier, and an earlier arrival of spring (Hulme *et al.* 2001). Overall, the evidence for climate warming now appears overwhelming, with evidence not only from climate observations, but also from the physical and biological indicators of environmental change such as retreating glaciers and longer growing seasons (Briffa *et al.* 2001; Hulme *et al.* 2001; Jones *et al.* 2001b). But what of the future?

Climate futures

Fundamental to the prediction of future climates are estimates of future greenhouse gas emissions, and in particular those of carbon dioxide, the greenhouse gas that causes about 60% of the human-induced greenhouse effect. It is predicted that the 2001 concentrations of about 370 ppmv may rise to somewhere between 550 and 830 ppmv by 2100 (Wigley 1999). There are considerable uncertainties in the impact that these changes may have on global climate as a result of uncertainties of how sensitive the earth's climate is to rising greenhouse gas concentrations. The recent Intergovernmental Panel on Climate Change (IPCC) projections (Houghton *et al.* 2001) are that the annual global mean surface air temperature will rise from 14.0°C (1961–1990 average) to between 15.5 and 19.9°C by 2100. By comparison, maximum temperatures during the last interglacial about 125 000 years ago are estimated to have reached between 15.0 and 15.5°C. The latest projections represent rates of change between 0.1 and 0.5°C per decade, which compare with 0.15°C per decade since the 1970s and a warming rate of about 0.05°C per decade since the 19th century. While such large decadal rates of change have previously been observed over the past two millennia, what obviously distinguishes the predicted rates of change is that they are consistently positive, whereas in the past periods of warming have been followed by periods of cooling.

Mean temperature is, of course, only one aspect of climate change; there are also predicted changes in seasonal and diurnal temperatures together with precipitation and the frequency of extreme events at both a global and regional scale (e.g. Hulme & Jenkins 1998; NAST 2000; Houghton *et al.* 2001). These will undoubtedly have important consequences for the dispersal and distribution of species. In looking to the future, as to the past, we are, however, much less certain about such aspects of climate change. We are also much less certain about how they impact on dispersal and distribution.

Dispersal and climate

In the face of climate change species have five potential responses: no response, persistence, adaptation, extinction and dispersal. Many species will potentially be able to remain in much of their present range as predicted future climates fall within their tolerance range (Sykes 1997; Dockerty 1998). Moreover, many species have the potential to adapt genetically to differences in climate, as shown by the fact that relevant ecotypic variation occurs within species (Norton *et al.* 1995, 1999). However, if climate change results in organisms experiencing conditions that are outside their physiological and ecological tolerances, then local extinction will be the net result. The rapid warming at the beginning of the Holocene (13 000–10 000 BP) resulted in the loss of many cold climate species of plants and animals from western Europe, but most appear to have survived at either higher latitudes or altitudes (Adams & Woodward 1992; Coope & Wilkins 1994). This contrasts with the extinction of many large mammals at the same time (Sher 1997). Climate change is not, however, the only potential cause of these extinctions; hunting by people was also probably involved. This nevertheless perhaps serves to illustrate that human influences on natural populations may make them less able to survive future climate changes (Adams & Woodward 1992). Climate change is one of a number of factors that are currently being implicated in the falling numbers of amphibians in many parts of the world as a result of increased exposure to UV-B (Kiesecker *et al.* 2001).

Dispersal has in the past resulted in major shifts in the distribution of species as climate has changed. In this section we will review how species responded to climate warming after the end of the last glaciation, and the evidence for recent changes in the distribution of organisms as a result of dispersal and climate warming over the last few decades. Here we define dispersal as the movement of individuals (following the definition in the preface to this volume) and migration as the change in population distribution resulting from dispersal.

Dispersal since the last glaciation

Routes of migration

Information on the migration routes of species has primarily come from two sources: the fossil/pollen record and gene sequencing. Pollen sequences from a range of sites across North America and western Europe, in particular, coupled with radiocarbon dating have allowed the migration routes of a large number of individual plant species to be ascertained (Davis 1981; Huntley & Birks 1983), while fossils have allowed the vast distributional changes that have occurred in beetle faunas during the Quaternary climatic changes to be documented (Coope 1995). Modern DNA technology has allowed the migration routes of a range of species to be calculated (Hewitt 1999, 2000). Polymerase chain reaction (PCR), in particular, has provided an invaluable tool for showing that the distributions of many species are subdivided by hybrid zones that have resulted from the meeting of two diverged genomes that have expanded from different glacial refugia. Twelve cases have so far been collated which allow postglacial colonization routes from their refugia across Europe to be deduced. Each has its own distinctive pattern but three broad patterns are evident—the grasshopper, the hedgehog and the bear—that Hewitt considers may serve as paradigms (Fig. 20.1).

With climate warming after the last glaciation, the Balkans would appear to have been the source for all species in the east and for many species in the west. In contrast, Italian genomes expanded less frequently into northern Europe with the ice-capped Alps presenting a barrier to dispersal. Similarly, the Pyrenees acted as a barrier for organisms such as the meadow grasshopper and beech. With different organisms dispersing into western Europe from different refugia, the flora and fauna of Britain is consequently cosmopolitan in its origin with oaks, shrews, hedgehogs and bears (now extinct) coming from Spain and the meadow grasshopper, alder, beech and newts from the Balkans (Hewitt 2000).

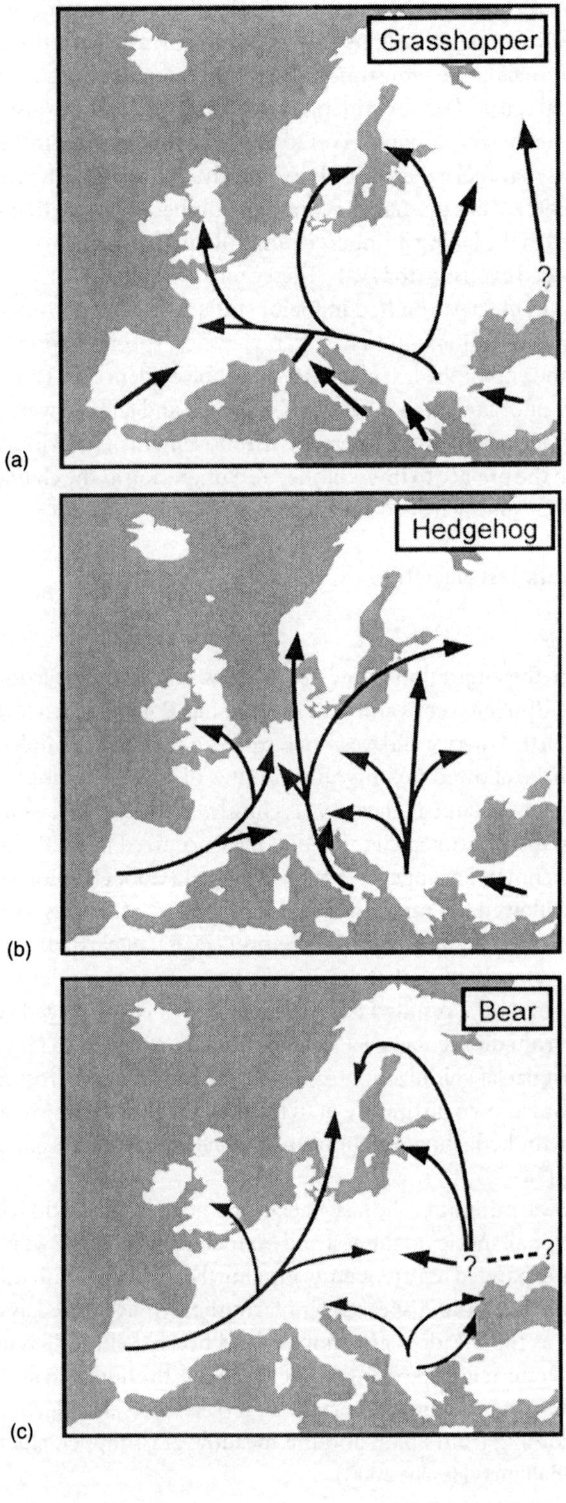
Grasshopper
?
(a)
Hedgehog
(b)
Bear
?
?
(c)

The data from gene sequencing do not allow us currently to say anything about the rate of dispersal across Europe, but they do provide a fascinating insight into the barriers to dispersal for different organisms and also into the movement of genes once they meet. The mountain ranges of the Pyrenees and the Alps clearly provided major barriers to the dispersal of some organisms, but not others. Where expanding genomes met, sometimes at major barriers to dispersal, they formed hybrid zones (Hewitt 2000); there appear to be clusters of such zones in the Alps and central Europe, the north Balkans, Pyrenees and central Sweden. Analysis of the structure of hybrid zones indicates the width of the zone will gradually expand at a rate dependent on gene dispersion (Barton & Hewitt 1985).

Rates of migration

During the last glaciation we know that the northern limit of European forests was in the Mediterranean region, while in North America it was near the Gulf of Mexico. From the glacial maximum at about 18 000 BP to 6000 BP when the ice caps had retreated to approximately their current position, most of the forest trees had reached their current limits (Williamson 1996). It is only for a relatively small number of taxa, mostly trees with readily identifiable pollen, that we are able to monitor spread and thus infer rates of dispersal during this period. In Europe, Huntley and Birks (1983) used intervals of 500 years on contoured maps, while for North America, Delcourt and Delcourt (1987) used 2000-year intervals. While there remains considerable uncertainty over the details of the spread at a fine spatial and temporal resolution, Williamson (1996) argues that the average speed for such long periods and large areas must be about right.

Bringing together the European and American data for the average and maximum rates of tree migration (Williamson 1996) indicates that 200 m $year^{-1}$ is a typical average speed, while maximal speeds are an order of magnitude greater at 2 km $year^{-1}$, although the typical maximum is slightly less than this at a figure of more like 600 m $year^{-1}$ (Fig. 20.2). The American averages are rather lower than the British ones. This may in part be because they are taken over a longer period (Williamson 1996), but the northwards flow of the rivers in northern Europe as opposed to the southerly flow in North America may be implicated in the faster European dispersal (Huntley & Birks 1983).

How were these rates of movement achieved, especially given that it is not only the distance per year that has to be taken into account but also the number of years from one generation to another? The answer is not at all clear. The species for which migration rates have been quantified since the last glaciation include wind-dispersed

---

Figure 20.1 (*opposite*) Three paradigm colonizations of northern Europe from southern refugia deduced from DNA sequencing. (a) The grasshopper *Chorthippus parallelus*, crested newt *Triturus cristatus*, alder *Alnus glutinosa* and beech *Fagus sylvatica* show similar patterns. (b) The hedgehog *Erinaceus europeus*, oaks *Quercus* spp. and silver fir *Abies alba* show similar patterns. (c) The bear *Ursos arctos*, shrews and water voles *Arvicola terrestris* show similar patterns. (From Hewitt 2000, with permission by *Nature*.)

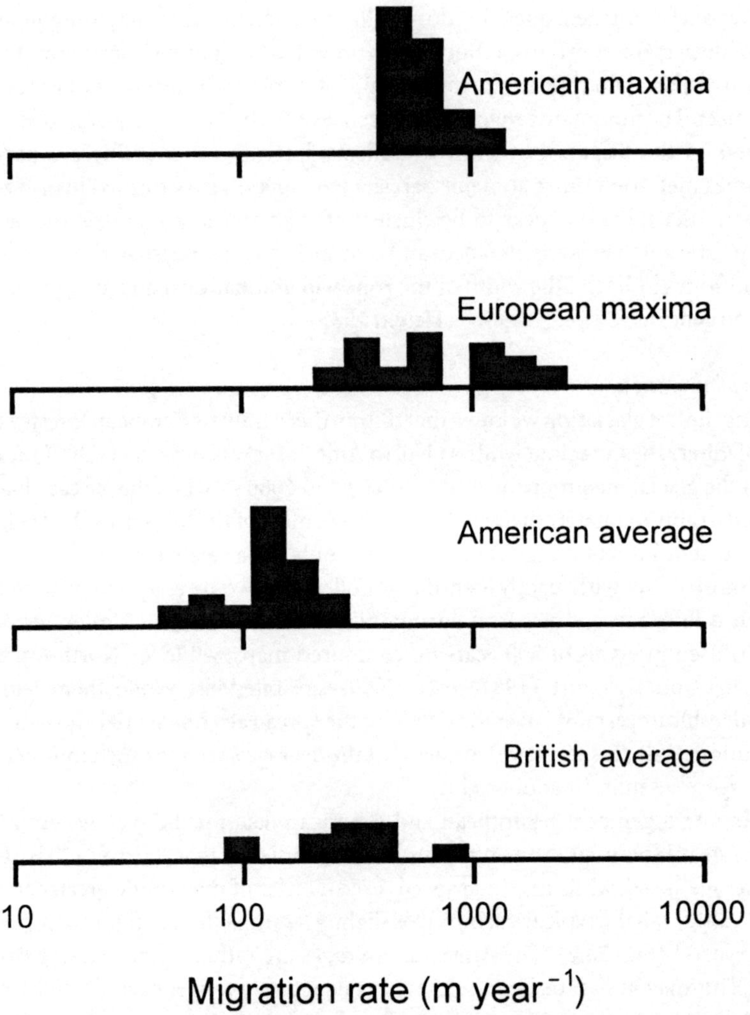

Figure 20.2 The rates of spread of trees in both Europe and North America following the last glaciation. (From Williamson 1996, with permission.)

species with relatively light seeds and others with heavy nuts, e.g. *Corylus*. Explaining the high rates of dispersal is especially problematic for these species (Huntley 1993). Greene and Calogeropoulos (this volume) have shown that we know a great deal about the shape of the dispersal curve for plants except for the tail of the distribution. Given that we know the shape of most of the dispersal curve and the rate of spread from the pollen record, it is possible to estimate what the incidence of long-distance dispersal must have been during the early Holocene migration. In the case of *Asarum canadense* (Aristolochiaceae) which travelled over 200 km in 16 000 years, Cain *et al.* (1998) estimated that one seed per thousand would have to disperse more than 1 km to explain the northward expansion of this species after the Holocene; the observed dispersal distance of seeds by ants is up to 35 m! Similarly, Clark *et al.* (1998) estimated for several animal- and wind-dispersed tree species that 2–10% of seeds must be dispersed distances of 1–10 km.

Long-distance dispersal is most common in animal-dispersed species and wind-dispersed species that have very light seeds (Willson 1993). Birds, in particular, have been implicated in the high rates of dispersal of some of the larger-seeded trees (Johnson & Webb 1989; Clark *et al.* 1998). That observed dispersal curves fail to capture the potential of species to colonize over large distances is shown by *Vulpia fasciculata*, an annual grass found on fixed and yellow dunes. *V. fasciculata* has a dispersal distance that is typically recorded in centimetres and no seeds have been observed to travel further than 1 m despite the large majority of seeds being accounted for (Watkinson 1978). Nevertheless, this species colonized newly formed dunes approximately 400 m from the nearest known colony in just a few years, equivalent to a dispersal rate two orders of magnitude greater than the observed distance of dispersal from direct observation of seeds. There are many other recent examples of high rates of spread in the invasions literature (Williamson 1996). It is clear that while the detailed monitoring of seed dispersal from plants tells us a lot about the immediate fate of seeds close to parent plants, where most of the seed falls, we are obviously struggling to measure the shape of the dispersal tail. Observed rates of spread would appear to be at least an order of magnitude greater than the observed seed-dispersal curves would indicate. In contrast to the passive dispersal of plants, animal species with active dispersal might be expected to disperse much faster. Beetles, for example, appear to have responded quickly to climate change in the past (Coope & Wilkins 1994; Coope 1995), with populations becoming established up to 1000 years more rapidly than populations of trees following a climate warming.

At mid-latitude in the northern hemisphere, the annual temperature gradient is about 0.7°C per 1° of latitude (Tallis 1991). A projected 2.5°C rise in temperature between 1990 and 2050 would, therefore, represent a northward shift of climatic isotherms by about 300–400 km. To track this rate of climatic change, species would have to show an overall rate of movement of 5–6 km $year^{-1}$ (Adams & Woodward 1992). We have seen that such rates are possible for some species, but it appears that most tree species at least would be left far behind, although they might catch up if warming stabilized on a timescale of centuries or perhaps millennia. In a more recent and detailed analysis, Malcolm and Markham (2000) concluded that required migration rates under global warming are 10 times higher than the rapid rates during the early Holocene. Moreover, in large parts of the northern hemisphere (Canada, Russia, Fennoscandia) required migration rates in excess of 1 km $year^{-1}$ are common, with the highest required rates being in the taiga/tundra, temperate evergreen forest, temperate mixed forest and boreal coniferous forest. In general, though, we know too little at a sufficiently fine scale of resolution of the past rates of climatic change and plant movements to be confident about future responses to climate change. What are the indications of changes in the range of organisms as a result of dispersal in response to recent climate warming?

Recent responses to climate change

There is growing evidence to suggest that species have responded to recent climate change through both changes in phenology and distribution, although the direct response to climate is often difficult to ascertain. Advanced phenology in response to milder winters and, especially, springs (Sparks & Carey 1995) has been seen in a range of organisms with earlier flowering in a range of plants (Walkovsky 1998; Cayan *et al.* 2001; Post *et al.* 2001), egg laying in birds (Crick & Sparks 1999; Dunn & Winkler 1999), flight times in insects (Roy & Sparks 2000), and breeding in rabbits (Bell & Webb 1991) and amphibians (Beebee 1995). Some species of birds with annual migration may be able to respond to climate change by altering their arrival dates on the breeding or wintering grounds. However, long-distance annual migrants may be unable to alter annual migration schedules in response to climate change and may suffer as a result (Both & Visser 2001).

It is much less certain what the impact of climate has been on the overall distribution of species, but there are indications that a number of species have moved polewards and to increased altitudes (IUCN 2001). These include the arctic fox (Hersteinsson & MacDonald 1992), mountain plants (Grabherr *et al.* 1994) and temperate butterflies (Parmesan *et al.* 1999). In addition, Thomas and Lennon (1999) analysed the distribution of British breeding birds between the 1970s and the 1990s and showed that the northern margins of many species had moved northwards by, on average, 18.9 km during that time.

While some remain sceptical, others are arguing that we are already seeing the biological consequences of global warming (Hughes 2000). To what extent will species be able to track changes in climate space in the future? This will depend on the magnitude of the change in mean climate, the rate of climate change, the changes in extremes, the timing of the seasons, the dispersal ability of individual species and the availability of suitable habitat to be colonized. As species move in relation to climate change, what will be the consequence for the communities of which they are a part?

Biodiversity consequences

Evidence from the movement of species in the past in response to climate change indicates that populations of different taxa have migrated independently of each other and not as communities (Tallis 1991). Forest development in each of the preceding interglacials followed a unique course (Bennett 1997) and it would appear from the pollen record, at least, that plant communities are impermanent (1000–10 000 years) assemblages resulting from the individualistic behaviour of taxa in response to environmental changes (see also Hengeveld & Hemerik, this volume). The 'unit of response' in relation to climate change is clearly the population, not the community. Consequently, patterns of biodiversity can be expected to change in response to future climate change both within sites and communities.

The potential consequences of climate change for community structure within nature reserves in Great Britain has recently been explored (Dockerty 1998). The analysis used was 'climate envelope analysis', which takes as its starting point the assumption that the eventual response of a species to a change in climate is an adjustment in range, as the organism realigns its distribution in accordance with its climatic tolerances (Huntley & Webb 1989). While other studies have used this approach to examine the potential for changes in the distribution of species (Huntley *et al.* 1995; Sykes *et al.* 1996), Dockerty focuses on the impact of a shift in climatic suitability at a particular site for the suite of species that exist there. Analyses of a sample of 241 plant species from 86 nature reserves in Great Britain, showed that future climates could be favourable for a large proportion of plants on Scottish reserves, with the exception of montane species, and less favourable for many plants on the reserves in the south of England, with no change in the majority of Welsh reserves and those further north in England (Fig. 20.3). For example, at Loch Lomond National Nature Reserve in Scotland, in excess of 60% of the species modelled are likely to benefit from higher temperatures, while in the southeast of England at Pashford Poors Fen, 56% of the species modelled are likely to experience a decline in climatic suitability. The results reaffirm the individualistic response of species to climate change at a site and indicate how communities are potentially more likely to disassemble in the south of Great Britain than in the north.

The validity of this approach depends critically on the relation between species distribution and climate (Dockerty 1998). Essentially, the models reflect the realized climatic niche of species, which reflect not only the impact of climate on species distributions but also a range of other factors such as species interactions. A major criticism of the climate envelope approach is that it ignores how climate change may modify the interactions between species (Davis *et al.* 1998a, 1998b). In ignoring species interactions, climate envelope modelling may consequently be misleading in predicting changes in species distribution. At present, though, we know too little about how species interactions impact on distribution patterns to include them in our models. That species interactions may be altered in different climates is, however, well established (Davis *et al.* 1998a; Buse *et al.* 1999; Visser & Holleman 2001) and the analysis of Dockerty (1998) also clearly indicates that the species composition and consequently competitive interactions in plant communities are likely to be radically modified, but with as yet unknown consequences for community structure.

Those that use the climate envelope approach recognize that they are assuming that the distribution of species is determined primarily by climatic tolerances and argue that other factors such as human land use, plant migration history, succession and other factors act on a smaller scale, causing minor deviations from the larger-scale climatic trends (Davis *et al.* 2000). Clearly the scale at which predictions are made is critical, but the analyses of both Dockerty (1998) for nature reserves within the UK and Davis *et al.* (2000) for the National Parks of the Western Great Lakes indicate the extent to which conservation areas may be impacted by climate change. This has many implications for how we approach the issue of biodiversity conservation in a world with a changing climate. This is not the place to examine all these issues. Rather we concentrate in the next section on how the question of dispersal impacts on the question of conservation in the face of climate change.

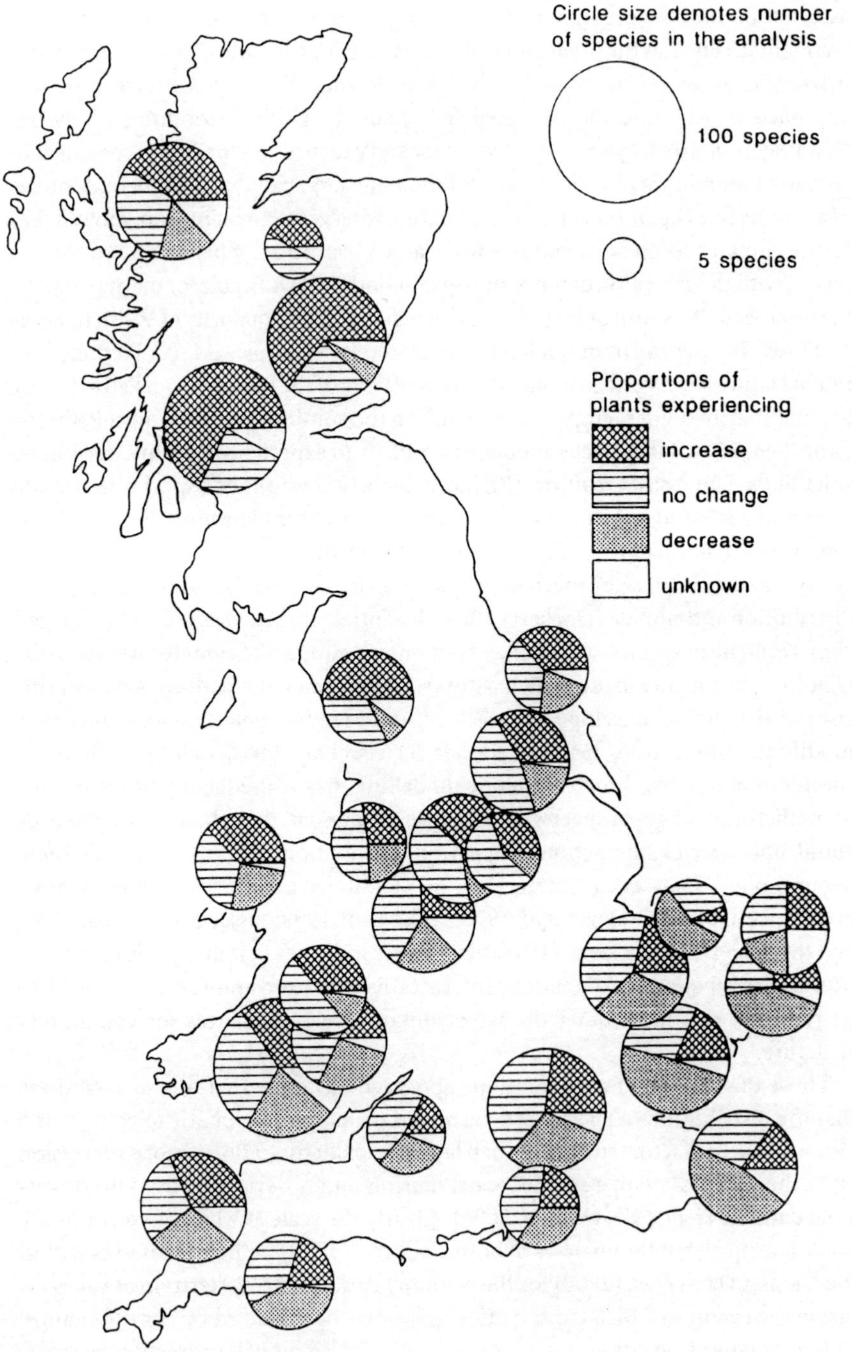

Figure 20.3 A climate change trend analysis for the UKHI scenario for climate change over the 21st century, illustrating the potential change in climatic suitability for species at selected nature reserves. (From Dockerty 1998, with permission.)

Issues for biodiversity conservation

Present approaches to conservation with their emphasis on stasis are likely to prove inadequate in the face of climate change, which will in turn drive the movement of species, the disassociation of communities and changes in habitat structure. Future conservation policies must take into account the fact that dispersal and changes in species distribution are likely to become increasingly significant.

The requirements of a future conservation strategy in the advent of climate change, with an increased emphasis on dispersal, have been outlined by Huntley *et al.* (1997). To facilitate dispersal where large-scale range changes are projected, they advocate the creation of a network of habitats and habitat corridors. There are two elements to such a strategy: (i) habitat corridors and patches will potentially facilitate the movement of species from one region to another; and (ii) the creation of numerous and diverse protected areas, that individually and together include the widest possible variety of habitats, will potentially allow species to find microclimates within their existing range that will require only short-distance dispersal. However, two other issues must also be considered that relate to the question of dispersal, climate change and biodiversity conservation: translocations and invasions.

Corridors

In an increasingly fragmented landscape, habitat corridors are frequently seen as a method of encouraging dispersal between habitat patches (Beier & Noss 1998; Boudjemadi *et al.* 1999; Diaz *et al.* 2000). Population persistence has been shown in many cases to be positively correlated with physical proximity of other populations (Sjögren 1991; Thomas & Jones 1993), and this has been widely interpreted as evidence that dispersal between patches can have a 'rescue effect' for, particularly small, populations. This has resulted in a great deal of interest in landscape connectivity, particularly in relation to habitat 'corridors' linking larger patches and potentially allowing dispersal through otherwise inhospitable habitat.

If organisms are to disperse in the face of climate change then we need to improve the connections that allow dispersal between habitat patches. Either we increase the number of habitat patches to allow jump dispersal between them or we provide corridors along which dispersal can take place. However, despite many studies, the conservation value of corridors is still a contentious issue (Simberloff *et al.* 1992). In many cases it is difficult to envisage that suitable habitat will be available within corridors to allow the dispersal of more than a subset of species.

A review by Beier and Noss (1998) found only limited evidence for corridors enhancing population viability, largely as a result of poor experimental design. As well as a lack of empirical evidence that corridors are used by target species, there are concerns over negative impacts such as increased disease transmission rates (Hess 1994), increased distribution of non-native species (Downes *et al.* 1997) and increased interspecific competition (Clinchy 1997). Such concerns, however, reflect a static view of nature conservation largely based on the protection of separate sites and do not take into account the dispersal over the landscape that will have to be encouraged with climate change.

Habitat patchworks

If, for example, an ectothermic organism is absolutely dependent upon a particular habitat structure and the microclimate within it, then a rise in temperature without any change in habitat structure will make that habitat unsuitable. As a consequence, the organism will have to move to a latitude or altitude where the appropriate habitat structure and temperature again coincide. If, however, only the microclimate is critical then the organism might only have to disperse a short distance to a habitat structure that provides the appropriate microclimate. The key, in the latter case, to reducing the distance that an organism will have to disperse is to provide a habitat patchwork.

The interaction between habitat structure and climate change for four ectothermic heathland animals (the silver-studded blue butterfly *Plebejus argus*, a red ant *Myrmica sabuleti*, the heath grasshopper *Chorthippus vagans* and the sand lizard *Lacerata agilis*) was examined by Thomas *et al.* (1999), by comparing the biotope available to them at the northern edge of their range in southern Britain with that available 300–400 km further south where summer temperatures are 2–3°C warmer. For example, *P. argus*, at the northern edge of its range, is found only on south-facing slopes in early successional heathland. In contrast, at the centre of its range it is found in both early and later successional heaths on slopes of any aspect. Modelling the impact of climate change on the distribution of the butterfly at the northern edge of its range showed that there would be an increase in the area of habitat available to it, if the structure of the heathland vegetation at the current range edge remained the same.

Many invertebrates occupy narrower niches at the edges of their range because of the need for their basic resources to coexist with substantially warmer microclimates than are typical for that latitude (Thomas 1993). However, the niche not only narrows towards the edge of the range as in the case of *P. argus*, but may also alter. For *M. sabuleti* the south-facing slopes that are suitable under cool conditions at the northern edge of the range are typically unsuitable under warm ones (Thomas *et al.* 1999). With an increase in temperature, *M. sabuleti* will therefore have to change habitat to maintain populations at a particular location.

The above argument has been put forward in terms of ectothermic animals, but there is no reason why the same should not apply to plants (Thomas *et al.* 1999). While the key for maintaining ground-dwelling ectotherms at what is now the northern edge of their range may be to maintain more vegetation in later successional stages (Webb & Thomas 1994; Thomas *et al.* 1999), the key for others may be just to ensure a patchwork of habitat diversity.

Translocations

If the rate of dispersal of a species is too slow to match the rate of future climate change, or if the landscape is too fragmented relative to the distance of dispersal, then one option for at least some species is deliberate translocation by humans. Many plant species and especially trees have already been dispersed outside of their natural range. *Quercus ilex*, a native of the Mediterranean and southwest Europe, has been transplanted extensively to the north of its natural range in Europe, including eastern England, where it has established in at least a few localities (James *et al.* 1981). With global warming, the issue in this case would be one of local expansion from translocated populations rather than long-distance dispersal from the natural range.

There is much debate about the value of translocations. Williamson (1996) states that deliberate releases of any sort should be discouraged because of their often considerable impacts on the communities to which they have been introduced. Concerns are typically expressed over the establishment of populations outside of their range and in terms of genetic conservation within the range. Agricultural and forestry plants have always been moved about extensively but there is growing concern in some quarters, especially with the restoration of wildflower meadows, that introducing alien genes into local populations may result in poor adaptation to local conditions and may have destabilizing effects on the genetic integrity of native populations (Keller *et al.* 2000). A recent study of hawthorn *Crataegus monogyna* by Jones *et al.* (2001a) has highlighted the fact that local provenances generally perform better under current climatic conditions. However, there has been widespread importation of continental hawthorns into the British Isles. While these provenances may not perform as well at present there is the possibility that they will match the climate better in the future.

Other translocation projects have met with much greater support from conservation bodies. These have typically involved the movement of species to areas in which they have become extinct. Examples from the UK alone include the white tailed sea eagle *Haliaeetus albicilla* (Green *et al.* 1996), red kite *Milvus milvus* (Carter *et al.* 1999) and large blue butterfly *Maculinea arion* (Elmes & Thomas 1992). White tailed sea eagles were reintroduced to Scotland from northern Norway, while the large blue was translocated from Sweden to the southwest of England and red kites were reintroduced to Scotland and England from Wales, Sweden, Germany and Spain. Does it matter where individuals are moved from in translocation programmes? The answer is probably in some cases yes and in others no. Norton (1996) transplanted *Vulpia fasciculata* from three sites on the coast of Great Britain (eastern England, southwest England, north Wales) to 14 sites across the UK, including a number outside its current range in northern England and Scotland. The site of origin was found to have little impact on performance as measured by the finite rate of population increase. Conservation organizations will consequently have to confront the question of whether to translocate species in the face of global warming, as well as from where within the current range of the species individuals should be taken.

Invasions

Another major challenge that will have to be met relates to the question of invasions. We have shown above that many species have already been dispersed by humans, either deliberately or accidentally, to areas outside their recent range limits. The proportion of the biota in each country that consists of aliens varies considerably; the ratio of natives to aliens in the flora of a range of countries has recently been shown by Dalmazzone (2000) to vary from 0.02 to 1.84. Given that large numbers of species have already been dispersed to areas where they are not naturally found and that global warming will change the conditions and species interactions in a wide range of habitats, it seems highly probable that changing climate will cause a new wave of invasions. Can we predict which of the alien species will be most likely to become invasive? Prediction is not the same as the statistical explanation of past invasions (Lawton 1996; Williamson 1999) and in general predicting invasions has met with little success (Williamson 1999; Cohen, this volume). The reasons for this lack of success remain unclear, but the general lack of species-specific detail with too few estimates of demographic parameters undoubtedly hinders prediction. Invasion is essentially a demographic process and unless one understands the details of the factors affecting the births, deaths, immigrants and emigrants in a population, then the prediction of which species will invade and what impact they will have on an ecosystem will remain elusive. It is likely, however, that as communities change there will be an over-representation of species that are capable of rapid migration (i.e. species with high fecundity and long-distance dispersal).

Conclusions

While climate change poses a major challenge for biodiversity conservation over the course of this century, changes in land use are predicted to have the greatest effect on biodiversity (Sala *et al.* 2000). It is, however, primarily climate change that brings the issue of long-distance dispersal to the fore as an issue for biodiversity conservation. The extent to which dispersal will allow species to track changes in climate remains unclear. We know only about the dispersal rates of a relatively small number of species from studies of the past and the landscape over which species now have to disperse is highly fragmented and heavily modified by human activity. Moreover, the potential for the deliberate movement of species by humans in the future, the presence of large numbers of aliens already outside their natural range and the fact that the warming is of an extent that the world has not seen for thousands of years, means that the pattern of dispersal and population expansion during the course of this century will be very different from that we have seen in the past.

References

Adams, J.M. & Woodward, F.I. (1992) The past as a key to the future: the use of palaeoenvironmental understanding to predict the effects of man on the biosphere. In: *Global Climate Change: the Ecological Consequences* (ed. F.I. Woodward), pp. 257–314. Academic Press, London.

Anklin, M., Barnola, J.M., Beer, J. *et al.* (1993) Climate instability during the last interglacial period recorded in the GRIP ice core. *Nature* **364**, 203–207.

Barton, N.H. & Hewitt, G.M. (1985) Analysis of hybrid zones. *Annual Review of Ecology and Systematics* **16**, 113–148.

Beebee, T.J.C. (1995) Amphibian breeding and climate. *Nature* **374**, 219–220.

Beier, P. & Noss, R. (1998) Do habitat corridors provide connectivity? *Conservation Biology* **12**, 1241–1252.

Bell, D.J. & Webb, N.J. (1991) Effects of climate on reproduction in European wild rabbit (*Oryctolagus cuniculus*). *Journal of Zoology* **224**, 639–648.

Bennett, K.D. (1997) *Evolution and Ecology.* Cambridge University Press, Cambridge.

Both, C. & Visser, M.E. (2001) Adjustment to climate change is constrained by arrival date in a long-distance migrant bird. *Nature* **411**, 296–298.

Boudjemadi, K., Lecomte, J. & Clobert, J. (1999) Influence of connectivity on demography and dispersal in two contrasting habitats: an experimental approach. *Journal of Animal Ecology* **68**, 1207–1224.

Briffa, K. & Atkinson, T. (1997) Reconstructing late-glacial and Holocene climates. In: *Climates of the British Isles* (eds M. Hulme & E. Barrow), pp. 84–111. Routledge, London.

Briffa, K.R., Osborn, T.J., Schweingruber, F.H. *et al.* (2001) Low-frequency temperature variations from a northern tree ring density network. *Journal of Geographical Research* **106**, 2929–2941.

Buse, A., Dury, S.J., Woodburn, R.J.W., Perrins, C.M. & Good, J.E.G. (1999) Effects of elevated temperature on multi-species interactions: the case of pedunculate oak, winter moth and tits. *Functional Ecology* **13** (Suppl. 1), S74–S82.

Cain, M.L., Damman, H. & Muir, A. (1998) Seed dispersal and the Holocene migration of woodland herbs. *Ecological Monographs* **68**, 325–347.

Carter, I., McQuaid, M., Snell, N. & Stevens, P. (1999) The red kite (*Milvus milvus*) reintroduction project: modeling the impact of translocating kite young within England. *Journal of Raptor Research* **33**, 251–254.

Cayan, D.R., Kammerdiener, S.A., Dettinger, M.D., Caprio, J.M. & Peterson, D.H. (2001) Changes in the onset of spring in the western United States. *Bulletin of the American Meteorological Society* **82**, 399–415.

Clark, J.S., Fastie, C., Hurtt, G. *et al.* (1998) Reid's paradox of rapid plant migration. *BioScience* **48**, 13–24.

Clinchy, M. (1997) Does immigration 'rescue' populations from extinction? Implications regarding movement corridors and the conservation of mammals. *Oikos* **80**, 618–622.

Coope, G.R. (1995) Insect faunas in ice age environments: why so little extinction? In: *Extinction Rates* (eds J.H. Lawton & R.M. May), pp. 55–74. Oxford University Press, Oxford.

Coope, G.R. & Wilkins, A.S. (1994) The response of insect faunas to glacial–interglacial climatic fluctuations. *Philosophical Transactions of the Royal Society, Series B* **344**, 19–26.

Crick, H.Q.P. & Sparks, T.H. (1999) Climate change related to egg-laying trends. *Nature* **399**, 423–424.

Dalmazzone, S. (2000) Economic factors affecting the vulnerability to biological invasions. In: *The Economics of Biological Invasions* (eds C. Perrings, M. Williamson & S. Dalmazzone), pp. 17–30. Edward Elgar, Cheltenham, UK.

Davis, A.J., Jenkinson, L.S., Lawton, J.H., Shorrocks, B. & Wood, S. (1998a) Making mistakes when predicting shifts in species range in response to global warming. *Nature* **391**, 783–786.

Davis, A.J., Lawton, J.H., Shorrocks, B. & Jenkinson, L.S. (1998b) Individualistic species responses invalidate simple physiological models of community dynamics under global environmental change. *Journal of Animal Ecology* **67**, 600–612.

Davis, M.B. (1981) Quaternary history and the stability of forest communities. In: *Forest Succession* (eds D.C. West, H.H. Shugart & D.B. Botkin), pp. 132–153. Springer, New York.

Davis, M., Douglas, C., Calcote, R., Cole, K.L., Winkler, M.G. & Flakne, R. (2000) Holocene climate in the Western Great Lakes National Parks and lakeshores: implications for future climate change. *Conservation Biology* **14**, 968–983.

Delcourt, P.A. & Delcourt, H.R. (1987) Long-term forest dynamics of the temperate zone. Springer Verlag, New York.

Diaz, J.A., Carbonell, R., Virgos, E., Santos, T. & Telleria, J.L. (2000) Effects of forest fragmentation on the distribution of the lizard *Psammodromus algirus. Animal Conservation* **3**, 235–240.

Dockerty, T.L. (1998) *Developing a climate–space modelling approach using a GIS to estimate the impacts of climate change on nature reserves in Great Britain.* PhD thesis, University of East Anglia, Norwich, UK.

Downes, S.J., Handasyde, K.A. & Elgar, M.A. (1997) Variation in the use of corridors by introduced and native rodents in south-eastern Australia. *Biological Conservation* **82**, 379–383.

Dunn, P.O. & Winkler, D.W. (1999) Climate change has affected the breeding date of tree swallows throughout North America. *Proceedings of the Royal Society of London, Series B* **266**, 2487–2490.

Elmes, G.W. & Thomas, J.A. (1992) Complexity of species conservation in managed habitats—interaction between *Maculinea* butterflies and their ant hosts. *Biodiversity and Conservation* **1**, 155–169.

Grabherr, G., Gottfried, M. & Pauli, H. (1994) Climate effects on mountain plants. *Nature* **369**, 448.

Green, R.E., Pienkowski, M.W. & Love, J.A. (1996) Long-term viability of the reintroduced population of the white-tailed eagle *Haliaeetus albicilla* in Scotland. *Journal of Applied Ecology* **33**, 357–368.

Hersteinsson, P. & MacDonald, D.W. (1992) Interspecific competition and geographical distribution of red and arctic foxes *Vulpes vulpes* and *Alopex lagopus*. *Oikos* **64**, 505–515.

Hess, G. (1994) Conservation corridors and contagious disease: a cautionary note. *Conservation Biology* **8**, 256–262.

Hewitt, G.M. (1999) Post-glacial recolonization of the European biota. *Biological Journal of the Linnean Society* **68**, 87–112.

Hewitt, G. (2000) The genetic legacy of the Quaternary ice ages. *Nature* **405**, 907–913.

Houghton, J.T., Ding, Y., Griggs, D.J. *et al.* (2001) *Climate Change: the Scientific Basis*. Cambridge University Press, Cambridge.

Hughes, L. (2000) Biological consequences of global warming: is the signal already apparent? *Trends in Ecology and Evolution* **15**, 56–61.

Hulme, M. & Jenkins, J. (1998) *Climate Change Scenarios for the United Kingdom*. UKCIP Technical Note No. 1. Climatic Research Unit, Norwich, UK.

Hulme, M., Jenkins, G., Brooks, N. *et al.* (2001) What is happening to global climate and why? In: *Health Effects of Climate Change in the UK*, pp. 21–56. Department of Health, London

Huntley, B. (1993) Rapid early-Holocene migration and high abundance of hazel (*Corylus avallana* L.): alternative hypotheses. In: *Climate Change and Human Impact on the Landscape* (ed. F.M. Chambers), pp. 205–215. Chapman & Hall, London.

Huntley, B. & Birks, H.J.B. (1983) *An Atlas of Past and Present Pollen Maps for Europe: 0–13000 Years Ago*. Cambridge University Press, Cambridge.

Huntley, B. & Webb, T. (1989) Migration: species' response to climate variations caused by changes in the earth's orbit. *Journal of Biogeography* **16**, 5–19.

Huntley, B., Berry, P.M., Cramer, W. & McDonald, A.P. (1995) Modelling present and potential future ranges of some European higher plants using climate response surfaces. *Journal of Biogeography* **22**, 967–1001.

Huntley, B., Cramer, W., Morgan, A.V., Prentice, H.C. & Allen, J.R.M. (1997) *Past and Future Rapid Environmental Changes: the Spatial and Evolutionary Responses of Terrestrial Biota*. NATO ASI Series No. 1, Global Environmental Change, Vol. 47. Springer Verlag, Berlin.

IUCN (2001) *Climate Change and Species Survival: Implications for Conservation Strategies*. IUCN, Gland, Switzerland.

James, R., Mitchell, S.C., Kett, J. & Leaton, R. (1981) The natural history of *Quercus ilex* in Norfolk. *Watsonia* **13**, 271–286.

Johnson, W.C. & Webb, T.I. (1989) The role of blue jays (*Cyanocitta cristata* L.) in the postglacial dispersal of fagaceous trees in eastern North America. *Journal of Biogeography* **16**, 561–571.

Jones, A.T., Hayes, M.J. & Sackville Hamilton, N.R. (2001a) The effect of provenance on the performance of *Crataegus monogyna* in hedges. *Journal of Applied Ecology* **38**, 952–962.

Jones, P.D., Osborn, T.J. & Briffa, K.R. (2001b) The evolution of climate over the last millennium. *Science* **292**, 662–667.

Keller, M., Kollmann, J. & Edwards, P.J. (2000) Genetic introgression from distant provenances reduces fitness in local weed populations. *Journal of Applied Ecology* **37**, 647–659.

Kiesecker, J.M., Blaustein, A.R. & Belden, L.K. (2001) Complex causes of amphibian population declines. *Nature* **410**, 681–683.

Lawton, J. (1996) Corncrake pie and prediction in ecology. *Oikos* 76, 3–4.

Macdonald, D.W. & Johnson, D.D.P. (2001) Dispersal in theory and practice: consequences for conservation biology. In: *Dispersal* (eds J. Clobert, E. Danchin, A.A. Dhondt & J.D. Nichols), pp. 358–372. Oxford University Press, Oxford.

Malcolm, J.R. & Markham, A. (2000) *Global Warming and Terrestrial Biodiversity Decline*. WWF World Wide Fund for Nature, Gland, Switzerland.

NAST (National Assessment Synthesis Team) (2000) *Climate Change Impacts on the United States: the Potential Consequences of Climate Variability and Change*. US Global Change Research Programme, Washington.

Newman, E.I. (2000) *Applied Ecology and Environmental Management.* Blackwell Science, Oxford.

Norton, L.R. (1996) *The responses of plant populations to climate change.* PhD thesis, University of East Anglia, Norwich, UK.

Norton, L.R., Firbank, L.G. & Watkinson, A.R. (1995) Ecotypic differentiation of response to enhanced CO_2 and temperature levels in *Arabidopsis thaliana. Oecologia* **104**, 394–396.

Norton, L.R., Firbank, L.G., Gray, A.J. & Watkinson, A.R. (1999) Responses to elevated temperature and carbon dioxide in the perennial grass *Agrostis curtisii* in relation to population origin. *Functional Ecology* **13**, S29–S37.

Parmesan, C., Ryrholm, N., Stefanescu, C. *et al.* (1999) Poleward shifts in geographical ranges of butterfly species associated with regional warming. *Nature* **399**, 579–583.

Post, E., Forchhammer, M.C., Stenseth, N.C. & Callaghan, T.V. (2001) The timing of life-history events in a changing climate. *Proceedings of the Royal Society of London, Series B* **268**, 15–23.

Roy, D.B. & Sparks, T.H. (2000) Phenology of British butterflies and climate change. *Global Change Biology* **6**, 407–416.

Sala, O.E., Chapin, F.S., Armesto, J.J. *et al.* (2000) Global biodiversity scenarios for the year 2100. *Science* **287**, 1770–1774.

Sher, A.V. (1997) Late-Quaternary extinction of large mammals in northern Eurasia: a new look at the Siberian contribution. In: *Past and Future Rapid Environmental Changes* (ed. B. Huntley), pp. 319–339. Springer Verlag, Berlin.

Simberloff, D., Farr, J.A., Cox, J. & Mehlmann, D.W. (1992) Movement corridors—conservation bargains or poor investments. *Conservation Biology* **6**, 493–504.

Sjögren, J. (1991) Extinction and isolation gradients in metapopulations: the case of the pool frog (*Rana lessonae*). *Biological Journal of the Linnaean Society* **42**, 135–147.

Sparks, T.H. & Carey, P.D. (1995) The responses of species to climate over two centuries: an analysis of the Marsham phenological record, 1736–1947. *Journal of Ecology* **83**, 321–329.

Sykes, M.T. (1997) The biogeographical consequences of forecast changes in the global environment: individual species' potential range changes. In: *Past and Future Rapid Environmental Changes* (ed. B. Huntley), pp. 427–440. Springer Verlag, Berlin.

Sykes, M.T., Prentice, I.C. & Cramer, W. (1996) A bioclimatic model for the potential distributions of north European tree species under present and future climates. *Journal of Biogeography* **23**, 203–233.

Tallis, J.H. (1991) *Plant Community History.* Chapman & Hall, London.

Thomas, C.D. & Jones, T.M. (1993) Partial recovery of a skipper butterfly (*Hesperia comma*) from population refuges: lessons for conservation in a fragmented landscape. *Journal of Animal Ecology* **62**, 472–481.

Thomas, C.D. & Lennon, J.J. (1999) Birds extend their ranges northwards. *Nature* **399**, 213.

Thomas, J.A. (1993) Holocene climate change and warm man-made refugia may explain why a sixth of British butterflies inhabit early-successional habitats. *Ecography* **16**, 278–284.

Thomas, J.A., Rose, R.J., Clarke, R.T., Thomas, C.D. & Webb, N.R. (1999) Intraspecific variation in habitat availability among ectothermic animals near their climatic limits and their centres of range. *Functional Ecology* **13**, S55–S64.

Visser, M.E. & Holleman, L.J.M. (2001) Warmer springs disrupt the synchrony of oak and winter moth phenology. *Proceedings of the Royal Society of London, Series B* **268**, 289–294.

Walkovsky, A. (1998) Changes in phenology of the locust tree (*Robinia pseudoacacia* L.) in Hungary. *Journal of Biometeorology* **41**, 155–160.

Watkinson, A.R. (1978) The demography of a sand dune annual: *Vulpia fasciculata.* III. The dispersal of seeds. *Journal of Ecology* **66**, 483–498.

Webb, N.R. & Thomas, J.A. (1994) Conserving insect habitats in heathland biotopes: a question of scale. In: *Large Scale Ecology and Conservation Biology* (eds P. Edwards, R.M. May & N.R. Webb), pp. 131–153. Blackwell Scientific Publications, Oxford.

Wigley, T.M.L. (1999) *The Science of Climate Change: Global and US Perspectives.* Pew Center on Global Climate Change, Arlington, VA.

Williamson, M. (1996) *Biological Invasions.* Chapman & Hall, London.

Williamson, M. (1999) Invasions. *Ecography* **22**, 5–12.

Willson, M.F. (1993) Dispersal mode, seed shadows, and colonization patterns. *Vegetatio* **107/108**, 261–280.

Zachos, J., Pagani, M., Sloan, L., Thomas, E. & Billups, K. (2001) Trends, rhythms and aberrations in global climate 65 Ma to present. *Science* **292**, 686–693.

Part 5
Overview

Chapter 21

Overview and synthesis: the tale of the tail

Mark Williamson

Introduction

'Dispersal is one of the most important, yet least understood, features of ecology, population biology, and evolution' (Wiens 2001). It is also a very large subject. One way to see that is to compare this volume with another (Clobert *et al.* 2001) from another conference on dispersal held 2 years earlier. There is no overlap of authorship and no overlap in the chapter titles. Doing a capture–recapture calculation on this would indicate the subject is infinite. Of course, this is an exaggeration—there is quite a lot in common at a lower level. The reason for much of the distinctness is that Clobert *et al.* concentrate on the genetic and evolutionary aspects, whereas this volume is based more on ecology.

Here I will attempt to sketch a map of the study of dispersal and indicate where the various chapters of this volume fit into that. It is in no sense a summary of what the chapters say.

Landscape type: the setting for dispersal

Dispersal takes place across a set of environments and habitats. For the terrestrial studies that make up the bulk of this volume it would be usual to describe that as dispersal across a landscape. Different studies take very different views of the landscape; the landscape type considered is fundamental for comparing different dispersal studies. The variation in landscape types is more or less continuous. Here, for convenience, I use an arbitrary classification, not unlike that used by Keeling (1999).

Homogeneous landscapes

The simplest landscape type is the uniform, infinite, homogeneous one in which there is no variation, or more accurately no variation considered, in all directions. This is the natural starting point for studying the dispersal from a fixed point such as seeds from a tree (Greene & Calogeropoulos, this volume) or insects flying freely (Osborne *et al.*, this volume). This is also the landscape in which dispersal curves are most easily conceptualized. Dispersal curves, the frequency distribution of the number of individuals by distance, are the fundamental and basic mathematical ingredient of dispersal theories and are considered below.

Patchy populations and metapopulations

The natural first extension is to consider dispersal in and between homogeneous patches. Apart from Hengeveld and Hemerik, none of the chapters here attempt to go beyond that to the analysis of dispersal in continuously varying habitats. The simplest types of patches are those all of the same type set in a matrix of a different type: a binary distinction. Depending on details and definitions, these are either metapopulations or patchy populations or both. An impressive body of theory relating to metapopulations has been built up in the last couple of decades, but the theoreticians have shifted their definition, which is a source of mild confusion. Hanski and Gilpin (1997, p. 2) explain: 'In . . . 1991 . . . metapopulation . . . was seen to imply significant turnover of local populations, local extinctions, and colonizations . . . now . . . any assemblage of discrete local populations with migration among them is . . . a metapopulation, regardless of . . . population turnover.' And on p. 3 they say: 'The metapopulation approach is based on the notion that space is not only discrete but that there is a binary distinction between suitable and unsuitable habitat types.' The binary distinction is clear there, but it is clear in this volume that the rest of the approach has not been universally accepted.

In this volume, Wilson and Thomas, Andreassen *et al.*, Bullock *et al.* and Okamura and Freeland all discuss dispersal in metapopulations. Wilson and Thomas talk of a population of populations and also of a metapopulation of metapopulations. Bullock *et al.* offer a scheme for distinguishing remnant populations, metapopulations and patchy populations and look, in their definition of metapopulations, for independent dynamics in the patches and turnover, i.e. extinction and recolonization. They also reiterate how hard it is to be sure of extinction in a patch for most plant species if they have a seed bank. Yet another scheme is offered by Thomas and Kunin (1999).

Patchy habitats are undoubtedly common, especially in such a strongly managed landscape as occurs in Britain, and the binary distinction is a useful simplification. Classic metapopulations, with equilibrium from turnover, are apparently uncommon. Continent/island, otherwise source/sink, populations (discussed by Sutherland *et al.*, this volume) are somewhat commoner, but most metapopulations studied as such seem to fit the 1997 definition rather than the 1991 one. How often do species have patchy, relatively independent, populations? An answer to this is indicated by Asher *et al.* (2001), where British butterflies are divided into habitat specialists and wider countryside species. The distinction is somewhat arbitrary and may not correspond exactly to metapopulation species and others. For instance, the two species considered at most length by Wilson and Thomas (this volume) are *Hesperia comma*, the silver spotted skipper, and *Aricia agestis*, the brown argus. The first is a habitat specialist, the second is not, though the very closely related *A. artaxerxes*, the northen brown argus, is. The boundary between *A. agestis* and *A. artaxerxes* has been redrawn recently (Asher *et al.* 2001). There are 29 habitat specialists, including all the rarer species and 25 wider countryside species. So perhaps somewhat more than half of British species, and much fewer than half the individuals, occur as metapopulations or patchy populations where the populations are reasonably independent. Thomas (1984), using a more sophisticated argument, concludes that 60% of extant British butterfly species have reasonably independent populations, which agrees with this.

Other models with homogeneous habitat

There are other ways to go to more complicated landscapes, starting from the uniform homogeneous one. A popular one with theoretical spatial ecologists is the one step system. In this the landscape is gridded and individuals move on the grid, most commonly one step at a time. If the grid is square, then a move is confined either to one of the four edge squares or to one of the eight surrounding squares; on a hexagonal grid the move must be to one of the six edge squares. Or the move can be a correlated random walk, with random distance and random angle (Wilson & Thomas, this volume). These are individual-based, spatially explicit models, but in their simplest form (scarcely mentioned in this volume) the patches are still homogeneous. Such systems are often studied by simulation, but analytical methods for studying them are being developed (Rand 1999).

Geneticists commonly use two other models with homogeneous patches, Wright's (1931) island model and Kimura and Weiss's (1964) stepping stone model. Wright's model underlies the equating of his F_{ST}, a standardized measure of genetic variance among populations, with $1/(4Nm+1)$, where N is the effective population size and m the number of migrants. Very many geneticists have used that equivalence to estimate gene dispersal, and its use by Raybould *et al.* and Okamura and Freeland in this volume is discussed below. The island model has an infinite number of exactly equal populations all exchanging genes equally with each other or, equivalently, through a central pool. Most ecologists regard such a model as wildly unrealistic, yet Andreassen *et al.* (this volume) find an example which is not unlike it. They also have an example similar to the stepping stone model, where populations are most likely to exchange dispersers with neighbouring ones.

Heterogeneous landscape models

The most complicated model system in common use at present is the quilted landscape of patches, contiguous or not, with numerous types of patches, each homogeneous in itself (see the figures in South *et al.*, this volume). Such a system can also be used in individual-based single-step models (Travis 2001). This system is both the most realistic and the most difficult mathematically. It has usually been studied by simulation as in Boyd, Kenward *et al.* and South *et al.* in this volume. Simulation can give beautiful and fascinating results but it is always difficult to know how general they are or to generalize from them. That can only be done by mathematical analysis.

There has been a recent development that allows, to some extent, the mathematical analysis of a version of an individually based, spatially explicit, model. It is possible to get a deterministic approximation to a stochastic continuous spatial process by using the method of spatial moments. The first moment is the spatial average density of organisms. The second moment is the average density of pairs of individuals a known distance from another individual. The mathematics of this are set out, reasonably simply, by Law *et al.* (2001). A major difficulty is that the differential equations for the first moment include a term involving the second, and so on; the system is not closed. The mathematics has been developed independently and simultaneously by two groups and their work, and their views how to close the system, are conveniently both summarized in one book (Dieckmann *et al.* 2000). Bolker *et al.* (2000) assume that the third moment is small enough to be neglected. Dieckmann and Law (2000) express the third central moment as several alternative functions of the first and second moments; one of these is equivalent to the Bolker *et al.* method. As Law *et al.* (2001) say 'there is still much to learn about appropriate moment closures'.

Single-step individual-based models give a dispersal curve only by implication and that is a negative exponential one (in any particular direction). That is the function used in spatial models by Lewis and Pacala (2000) and Murrell and Law (2000). The latter use different parameters in different habitat patches. Below I note the extent to which papers in this volume support the negative exponential.

The landscape pattern considered is crucial for the type of study and the sort of conclusions that can be drawn about dispersal. That leads to the consideration of three major conceptual questions about dispersal: how (empirical dispersal), why (the causes of dispersal) and what (the consequences of dispersal).

How? Empirical dispersal

Several authors in this book, notably, but not only, Bailey and Lilley, Boyd, Greene and Calogeropoulos, Kenward *et al.* and Osborne *et al.*, discuss some of the large and heterogeneous range of techniques whereby dispersal is studied. Generalizations appear both when we come to results and to a classification of types of dispersal.

The shape of the dispersal curve

The basic quantitative aspect of dispersal is the shape of the dispersal curve. As noted above, this is most easily conceptualized for the homogeneous limitless landscape. In the simplest case, samples are taken at various distances from a source and the results compared to some mathematical distribution. Porter and Dooley (1993) stressed the importance of correcting for sampling effort at various distances, and the failure to do so in many earlier studies. Bullock and Clarke (2000) and Wilson and Thomas (this volume) show how to do this for plants and butterflies, respectively.

The simplest curve conceptually, one that, as noted above, arises naturally from single-step models but also in other ways, is the negative exponential with the form:

$$y = \exp(-x^1) \qquad (21.1)$$

Like all dispersal curves, this is high near the origin and tapers off into a long tail at large distances. It gives a linear plot if the ordinate is on a logarithmic scale and the abscissa untransformed. It can be regarded as having the standard tail; the exponent one is there in the equation above to emphasize this. Another standard distribution, one that arises naturally from random dispersal (Williamson 1996), is the Gaussian or normal with the form:

$$y = \exp(-x^2) \tag{21.2}$$

which has a thinner tail. A fatter tail than the negative exponential can be produced with functions:

$$y = \exp(-x^a) \tag{21.3}$$

with $0 < a < 1$ ($a = 0$ gives a horizontal line) or with a power function of the form:

$$y = ax^b \quad \text{or} \quad \log y = \log a + b \log x \tag{21.4}$$

which is linear if both axes are on logarithmic scales, or in many other ways. Greene and Calogeropoulos (this volume) discuss six dispersal functions, including the flexible but oddly named 2Dt of Clark *et al.* (1999) (derived by a two-dimensional argument but no more or less two-dimensional than the other functions). 2Dt can range, by tuning parameters, from the Gaussian to the standard thick-tailed distribution of statistical texts, the so-called Cauchy distribution (invented long before Cauchy):

$$y = 1/(1 + x^2) \tag{21.5}$$

which has an infinite variance. There are some standard generalizations of the Cauchy (Johnson *et al.* 1994) that have not been used in dispersal studies. The Cauchy arises in several natural ways and, as Wilson and Thomas (this volume) note, Shaw (1995) used it to model wind-borne spore dispersal (cf. Brown *et al.*, this volume).

Empirically, thick tails seem to be common but it is less clear what happens near the origin. That much is clear, for instance in Nathan *et al.* (2000) who found their empirically predicted dispersal distances for pine seeds followed, with scatter, a distribution between a negative exponential and a power curve. Paradis *et al.* (1998), for bird recoveries, found that so many were found near the origin that the median was in the first bin of their histograms, while the rest of the plot was of almost constant thinness. Regrettably, they did not try transforming their data. Both these studies seem free of the distance-related biasses discussed above. The two parts of dispersal curves that are important for ecologists are the shapes at the origin and in the tails. In between many functions will give an adequate fit and there are usually no important ecological consequences of using an arbitrary function.

The fit near the origin is the less studied of the two, but that fit is an important justification by Clark *et al.* (1999) for their 2Dt function. Greene and Calogeropoulos (this volume) nevertheless find that none of the functions they studied fitted all data. Wilson and Thomas (this volume) concur. One reason for this is that there is much variation in the shape of the curve near the origin. Going towards the origin the dispersal curve may either turn upwards or downwards. That may be important ecologically as Bullock *et al.* (this volume) also point out.

The tail is generally more important, but where it starts is of course arbitrary, where it appears to end depends on the study method. In many cases the end of the tail is measured only in metres or hundreds of metres, though Brown *et al.* (this volume) go out to thousands of metres. But the fat tails discussed above undoubtedly exist and have important consequences for ecology and evolution. With molecular genetic techniques, it can be shown that populations far apart are nevertheless identical genetically at the loci studied and so were connected recently by dispersal. Genomics will undoubtedly improve the scope of such studies. Here are two examples.

The first concerns oceanic planktonic foraminifera, which are protists, short-lived with spiral skeletons that do not encyst. Individuals are only about 100 μm in diameter. The example, or rather set of examples, involves movement across the tropics by subpolar taxa. Darling *et al.* (2000) show that no less than four or five pairs of taxa in three morphospecies have identical SSU rDNA (small subunit ribosomal DNA) in arctic and antarctic population pairs (two pairs in *Globigerina bulloides*, one or two in *Turborotalita quinqueloba* and one in right coiling *Neogloboquadrina pachyderma*). Even allowing for the shrinkage of the extent of tropical waters in glacial times, the jumps must be at least 5000 km against the prevailing oceanic currents and across unsuitable waters. All cases seem to involve movement from south to north. There is an ocean current that does that, but it takes decades to cover the distance (Wells *et al.* 1996), mostly at too high a temperature. These dispersals must have involved something much more unusual in surface waters recurring only in times measured in hundreds of thousand of years, which is recent to a geologist.

The second example is *Metrosideros* (Myrtaceae), an important genus of trees in the Pacific. It is from a better known class of example and involves an even longer distance. Wright *et al.* (2000) show that *M. collina* s.l. in New Zealand, the Society Islands and Hawaii have identical ITS regions (the 5.8S gene and two flanking internally transcribed spacers) in their ribosomal DNA sequences or have sequences differing by only 1 base pair (bp). The divergence rate for this region seems to be about 1 bp in 1.5 million years, so the populations have presumably dispersed in a period less than that. The dispersal involves jumps of *c.* 10 000 km from New Zealand to the Cook Islands and/or the Society Islands and another jump of the same size from there to Hawaii. *Metrosideros* has very light seeds which are lofted by wind speeds of only 10 km h^{-1} or so, and thus the dispersal probably involves rare passage through the upper atmosphere, possibly helped by Pleistocene wind patterns.

Species limits in *Metrosideros* are not clear and no doubt depend partly on what is meant by a species. The relevant New Zealand form is called *M. excelsa*; there are other *Metrosideros* taxa there and the subgenus *Metrosideros* seems to have originated there. The Hawaiian populations are prominent in the native vegetation and variable in habit and habitat, genotypically and phenotypically (Carlquist 1980; Cordell *et al.* 1998). They have all been called *M. polymorpha* or split into five or six species. Other cases of long-distance dispersal in plants are often ascribed to birds and other animals (Macphail 1997; Wilkinson 1997; Cain *et al.* 2000), though

Greene and Calogeropoulos (this volume) are sceptical. There are many other cases involving oceanic islands that must have involved rare long-distance dispersal (Williamson 1981), sometimes apparently between oceans (Williamson 1991).

In this volume, Greene and Calogeropoulos and Watkinson and Gill consider long-distance dispersal in plants, while Compton discusses the remarkable distances covered by aphids and fig wasps to reach their targets. In neither case has the shape of the tail of the dispersal distribution been determined. But both for the insects and for trees re-expanding their ranges after the lasat glaciation, long-distance dispersal is essential to produce the situation we see. Trees with all sorts of seeds managed to travel between 100 and 1000 m each year, even though their generation times are measured in tens of years. The rate seems a touch faster in Europe than in America but that is probably from taking averages over shorter periods (Williamson 1996).

Although Greene and Calogeropoulos (this volume) claim a consensus that the rates were constrained, it seems to me that the evidence is against constraint. Different tree species advanced at different speeds along the same tracks (Williamson 1996) so, simplistically, not more than one was constrained on a given track at a given time. It has also been known for some decades from many sites (Coope & Brophy 1972; Buckland & Coope 1991; Walker *et al.* 1993; Coope 2000) that beetles frequently indicate a temperate climate shortly after an arctic one even though no trees are found in the deposits. Coleopterists use scavenging and carnivorous species in these studies, to avoid species dependent on particular plants. Beetles move faster than trees, so trees are presumably not constrained. Again there is no quantitative information on the dispersal curve but a thick tail seems almost certain.

Types of dispersal

Along with the importance of the shape of the dispersal curve and the landscape type, the biological type of the organism dispersing is also basic. It is common with vertebrates to distinguish natal and breeding dispersal. In plants, pollen and seeds behave differently. In many organisms the sexes differ in their dispersal behaviour. Even within one morphological type, there may be differences with age. There are many variants on these themes in this volume in the chapters by Andreassen *et al.*, Boyd, Kenward *et al.* and Sutherland *et al.*

Variations in genotype can affect dispersal. Eukaryotic geneticists, such as Raybould *et al.* (this volume), often assume that differences between genotypes are selectively neutral, while prokaryotic geneticists, such as Bailey and Lilley (this volume), take selection for granted. The latter seems more cautious, though harder to handle mathematically.

The question here with all these types, is whether the type affects dispersal and the dispersal curve. Shigesada and Kawasaki (this volume) point to the differences due to insects walking or flying; Wilson and Thomas indicate that butterflies fly in a different manner within and between their metapopulation patches. In the strand plant *Cakile* (Brassicaceae), the fruit structure, with two seeds, leads to one breaking off and floating away, while the other remains attached to the plant, and the habit of the plant has an effect on the dispersal curves of these two types (Donohue 1998). Other authors seem to have assumed two types of dispersal curve to explain long-distance dispersal (Clark 1998; Higgins & Richardson 1999). Undoubtedly something is needed to get the thick tails implied by extreme dispersal, but fitting two curves to achieve this seems to offer less than using a single suitable function. No doubt this was one reason for Clark developing the 2Dt function (Clark *et al.* 1999).

Why? The causes of dispersal

Naively, if an organism has found a good place, it would seem rash to disperse. Any other place would probably be worse. The reddened spectrum of environmental heterogeneity in space and time makes that a bad strategy. In a famous paper, Hamilton and May (1977) showed that there was an advantage to dispersal even in a uniform habitat, and there has been much development and variation on their theme since (Travis & Dytham 1999; Travis 2001). It is natural to divide the causes, ultimate and proximate, of dispersal behaviour into biological and physical.

Biological causes

Biological causes can be subdivided into three main classes: intraspecific competition, interspecific interactions and the effects associated with mating systems. Intraspecific competition can lead either to many variants of density dependence (Sutherland *et al.*, this volume) or particularly to sibling competition (Andreassen *et al.*, this volume). Interspecific effects are even more variable, including interspecific competition (in the widest sense), predation, parasitism and disease (Boyd; Compton; Dwyer & Hails, all this volume) and mutualism (Compton, this volume). In symbols: −−, +− and ++ interactions.

In mating systems, the avoidance of inbreeding is popular with geneticists (cf. Clobert *et al.* 2001), often ignored by ecologists, though all (including Boyd, this volume and South *et al.*, this volume) recognize the potential importance of dispersing to seek a mate.

Physical causes

Physical causes are both obvious and variable, and include all those processes leading to the different landscape types, particularly variation in patchiness, discussed above. But one aspect not mentioned there is important: the transience of habitats. Transience varies greatly in scale and intensity, though generally following a reddened spectrum. Its importance is shown, most evidently, in this volume in the chapters by Boyd, Osborne *et al.* and Hengeveld and Hemerik.

What? The effects of dispersal

Dispersal has effects in almost all aspects of ecology and evolution. Here I discuss a few themes that are prominent in this volume.

Population structure

Population structure greatly affects dispersal; dispersal greatly affects population structure. A major theme of this volume is the effects of and the affects on patchiness and metapopulations, particularly in the chapters of Bullock *et al.* (plants), Andreassen *et al.* (vertebrates) and Wilson and Thomas (butterflies). These give empirical details that highlight the theoretical findings mentioned above, that spatial structure induces new and interesting phenomena compared with classic (mean field to mathematicians) population dynamics. The chapter by Dwyer and Hails makes the same point in a continuous system. Metapopulations are, as noted above, usually defined by the type of dispersal but Wilson and Thomas (this volume) show how the tail of the dispersal distribution is critical for the analysis of dynamics beyond the local patch.

Density dependence is a fundamental part of population dynamics. Density-dependent dispersal can have many and varied causes and effects, explored here particularly by Sutherland *et al.* largely for birds. Their source–sinks, ideal free pattern and other classifications, though, apply to most organisms.

Spatial population dynamics affect the persistence and stability of individual populations, the synchrony or otherwise of sets of them, and the importance or otherwise of Allee effects, themes explored here by Andreassen *et al.*, Dwyer and Hails and Shigesada and Kawasaki. There can be waves of population density crossing the landscape as a result of dispersion (Lambin *et al.* 1998). We are but starting to characterize and analyse the population ecological phenomena arising from dispersion, but the advances reported here are impressive and important.

Dispersal is important too in genetic and evolutionary phenomena as shown by the chapters in this volume by Raybould *et al.*, Andreassen *et al.* and Okumura and Freeland. As Whitlock and McCauley (1998) argued, and these chapters show, F_{ST} is a useful parameter of the genetic structure of populations. These chapters also show that it is not a useful measure of dispersion; dispersion of genes must be measured in other ways. Others have noticed this before. 'In metapopulations, with frequent extinction and colonization, the relationship between genetic differentiation and gene flow is not straightforward' (Ouborg *et al.* 1999).

F_{ST} is a measure of inbreeding. Inbreeding is generally deleterious genetically for individuals, and, as noted above, geneticists regard its avoidance as an important reason for dispersal. Nieminen *et al.* (2001) show that it can have consequences at the population level too. The butterfly *Melitaea cinxa* has a classic metapopulation structure with extinction and recolonization of patches. Inbred colonies are more prone to extinction and these will be found on the smaller and more isolated patches.

Range

Range is a more biogeographical topic; range expansion follows from colonization and leads to invasion. The study of range flows naturally from the study of population dynamics with dispersal and vice versa. Invasion is a topic considered in this volume by Bullock *et al.* with field observations illuminating theory, by Shigesada and

Kawasaki with theory illuminating field observations, and by Watkinson and Gill who consider the range expansion that may be expected under climate change in the light of what is known of postglacial range expansions by trees.

It has been known for a long time that the postglacial rate of spread of trees was remarkably rapid (Williamson 1996), a phenomenon sometimes known as Reid's paradox. Ordinary reaction–diffusion equations, which produce a Gaussian tail to the dispersal curve, can not account for what is seen. Such equations produce a wavefront expanding at constant speed (Williamson 1996). A thicker tail allows more interesting phenomena (Kot *et al.* 1996), such as an accelerating wavefront or, with a Cauchy form (Shaw 1995), no wave front at all, with new patches arising irregularly in front of the population. Shaw suggested his model for the wind dispersal of pathogen spores. Brown *et al.* (this volume) show the large dispersal distances that can be observed with such organisms. The wind-borne seeds of some trees are much heavier; those dispersed by birds and mammals heavier still. We still lack observations of extreme dispersal in such species. That is, we do not know what the shape of the tail is and still less how it gets that shape. All that can be said is that the observed rates summarized by Watkinson and Gill (this volume) and Williamson (1996) imply that the tail must be thick.

Adaptation

In a sense, the extreme dispersal of trees is a necessary adaptation to rapidly changing climates. The reason why the tree flora of western Europe lacks the characteristic trees of western North America, such as redwoods and other conifers, is that such species went extinct in Europe after the Pliocene, apparently from failing to follow the Pleistocene climatic changes. On a much shorter timescale, Boyd (this volume) shows how marine vertebrates use dispersal to track habitat changes.

Adaptation is a matter of evolution, and the evolution of life histories is an important topic nowadays. One aspect of that is the role of trade-offs. Compton (this volume) discusses the trade-offs between being dispersed far and fast and being dispersed shorter distances more slowly with more control in small, vulnerable short-lived insects that have to travel remarkably far to find their host plants. Boyd (this volume) poses related trade-offs for marine birds and mammals. Thompson *et al.* (this volume) consider three trade-offs that have been suggested in plant dispersal. They distinguish between inescapable trade-offs, such as seed size and number, which inevitably work and those that need not, like the three examples they consider. On their evidence, those that need not work do not, or at least not in any clear-cut way. Trade-offs, like much in ecology, are more complex than they at first appear.

Conclusions

Dispersal is a large subject with effects that ramify through all of population and community ecology. Studies of population dynamics that fail to consider dispersal can now be seen to be inadequate. The basic variable in dispersal is the shape of the dispersal curve, particularly the tail. Having any sort of dispersal affects dynamics, but a thick-tailed dispersal curve makes a critical difference at large scales. But at all scales, the type of landscape, homogeneous or patchy, continuously or discontinuously variable, affects everything else.

Whither dispersal? What will be and what should be the pattern of dispersal research in the next decade or so? Much work will, as usual, be filling in the gaps and building straightforwardly on the structure of present knowledge. A better collaboration and exchange of ideas between geneticists and ecologists is clearly desirable, and has been so for at least half a century in population biology. The increasing importance of genetic techniques in studying ecological problems may well help such collaboration.

Innovative work seems most urgently needed, as my title implies, in learning more about the tail of the dispersal distribution. Considerable ingenuity and technical skill will be necessary to get better measurements of the shape of the tail. Genetic techniques will probably be useful, but direct physical measurements are also needed. The reason for wanting to know the shape of the tail lies in theory. Reciprocally, theory needs to be developed to explore the consequences of variation in tail shape. Predictions of the rate of spread of species and the behaviour of metapopulations, to mention just two major areas, are much affected by the shape of the tail.

In dispersal, it's the tail that wags the dog.

References

Asher, J., Waren, M., Fox, R., Harding, P., Jeffcoate, G. & Jeffcoate, S. (2001) *The Millennium Atlas of Butterflies in Britain and Ireland.* Oxford University Press, Oxford.

Bolker, B.M., Pacala, S.W. & Levin, S.A. (2000) Moment methods for ecological processes in continuous space. In: *The Geometry of Ecological Interactions* (eds U. Dieckmann, R. Law & J.A.J. Metz), pp. 388–411. Cambridge University Press, Cambridge.

Buckland, P.C. & Coope, G.R. (1991) *A Bibliography and Literature Review of Quaternary Entomology.* J.R. Collis Publications, Department of Archaeology and Prehistory, University of Sheffield, Sheffield.

Bullock, J.M. & Clarke, R.T. (2000) Long distance seed dispersal: measuring and modelling the tail of the curve. *Oecologia* **124**, 506–521.

Cain, M.L., Milligan, B.G. & Strand, A.E. (2000) Long-distance seed dispersal in plant populations. *American Journal of Botany* **87**, 1217–1227.

Carlquist, S. (1980) *Hawaii: a Natural History.* Pacific Tropical Botanical Garden, Lawai, Hawaii.

Clark, J.S. (1998) Why trees migrate so fast: confronting theory with dispersal biology and the paleorecord. *American Naturalist* **152**, 204–224.

Clark, J.S., Silman, M., Kern, R., Macklin, E. & HilleRisLambers, J. (1999) Seed dispersal near and far: patterns across temperate and tropical forests. *Ecology* **80**, 1475–1494.

Clobert, J., Danchin, E., Dhondt, A.A. & Nichols, J.D. (eds) (2001) *Dispersal.* Oxford University Press, Oxford.

Coope, G.R. (2000) Middle Devensian (Weichselian) coleopteran assemblages from Earith, Cambridgeshire (UK) and their bearing on the interpretation of 'Full glacial' floras and faunas. *Journal of Quaternary Science* **15**, 779–788.

Coope, G.R. & Brophy, J.A. (1972) Late glacial environmental changes indicated by a coleopteran succession from North Wales. *Boreas* **1**, 97–142.

Cordell, S., Goldstein, G., Mueller-Dumbois, D., Webb, D. & Vitousek, P. (1998) Physiological and morphological variation in *Metrosideros polymorpha*, a dominant Hawaiian tree species, along an altitudinal gradient. The role of phenotypic plasticity. *Oecologia* **113**, 188–196.

Darling, K.E., Wade, C.M., Stewart, I.A., Kroon, D., Dingle, R., & Leigh Brown, A.J. (2000) Molecular evidence for genetic mixing of Arctic and Antarctic subpolar populations of planktonic foraminifers. *Nature* **405**, 43–47.

Dieckmann, U. & Law, R. (2000) Relaxation properties of the method of moments. In: *The Geometry of Ecological Interactions* (eds U. Dieckmann, R. Law & J.A.J. Metz), pp. 412–455. Cambridge University Press, Cambridge.

Dieckmann, U., Law, R. & Metz, J.A.J. (eds) (2000) *The Geometry of Ecological Interactions.* Cambridge University Press, Cambridge.

Donohue, K. (1998) Maternal determinants of seed dispersal in *Cakile edentula*: fruit plant and site traits. *Ecology* **79**, 2771–2788.

Hamilton, W.D. & May, R.M. (1977) Dispersal in stable habitats. *Nature* **269**, 578–581.

Hanski, I.A. & Gilpin, M.E. (eds) (1997) *Metapopulation Biology.* Academic Press, San Diego.

Higgins, S.I. & Richardson, D.M. (1999) Predicting plant migration rates in a changing world: the role of long-distance dispersal. *American Naturalist* **153**, 464–475.

Johnson, N.L., Kotz, S. & Balakrishnan, N. (1994) *Continuous Univariate Distributions—1.* Wiley, New York.

Keeling, M. (1999) Spatial models of interacting populations. In: *Advanced Ecological Theory: Principles and Applications* (ed. J. McGlade), pp. 64–99. Blackwell Science, Oxford.

Kimura, M. & Weiss, G.H. (1964) The stepping stone model of population structure and the decrease of genetic correlation with distance. *Genetics* **49**, 561–576.

Kot, M., Lewis, M.A. & van den Driessche, P. (1996) Dispersal data and the spread of invading organisms. *Ecology* **77**, 2027–2042.

Lambin, X., Elston, D.A., Petty, S.J. & MacKinnon, J. (1998) Spatial asynchrony and periodic travelling waves in cyclic field vole populations. *Proceedings of the Royal Society, Series B* **265**, 1491–1496.

Law, R., Purves, D.W., Murrell, D.M. & Dieckmann, U. (2001) Causes and effects of small-scale spatial structure in plant populations. In: *Integrating Ecology and Evolution in a Spatial Context* (eds J. Silvertown & J. Antonovics), pp. 21–44. Blackwell Science, Oxford.

Lewis, M.A. & Pacala, S. (2000) Modeling and analysis of stochastic invasion processes. *Journal of Mathematical Biology* **41**, 387–429.

Macphail, M.K. (1997) Comment on M. Pole (1994): 'The New Zealand flora—entirely long-distance dispersal?'. *Journal of Biogeography* **24**, 113–117.

Murrell, D.J. & Law, R. (2000) Beetles in fragmented woodlands: a formal framework for dynamics of movement in ecological landscapes. *Journal of Animal Ecology* **69**, 471–483.

Nathan, R., Safriel, U.N., Noy-Meir, I. & Schiller, G. (2000) Spatiotemporal variation in seed dispersal and recruitment near and far from *Pinus halapensis* trees. *Ecology* **81**, 2156–2169.

Nieminen, M., Singer, M.C., Fortelius, W., Schöps, K. & Hanski, I. (2001) Experimental confirmation that inbreeding depression increases extinction risk in butterfly populations. *American Naturalist* **157**, 237–244.

Ouborg, N.J., Piquot, Y. & van Groenendael, J.M. (1999) Population genetics, molecular markers and the study of dispersal in plants. *Journal of Ecology* **87**, 551–568.

Paradis, E., Baillie, S.R., Sutherland, W.J. & Gregory, R.D. (1998) Patterns of natal and breeding dispersal in birds. *Journal of Animal Ecology* **67**, 518–536.

Porter, J.H. & Dooley Jr, J.L. (1993) Animal dispersal patterns: a reassessment of simple mathematical models. *Ecology* **74**, 2436–2443.

Rand, D.A. (1999) Correlation equations and pair approximations for spatial ecologies. In: *Advanced Ecological Theory: Principles and Applications* (ed. J. McGlade), pp. 100–142. Blackwell Science, Oxford.

Shaw, M.W. (1995) Simulation of population expansion and spatial pattern when individual dispersal distributions do not decline exponentially with distance. *Proceedings of the Royal Society, Series B* **259**, 243–248.

Thomas, C.D. & Kunin, W.E. (1999) The spatial structure of populations. *Journal of Animal Ecology* **68**, 647–657.

Thomas, J. (1984) The conservation of butterflies in temperate countries. *Symposia of the Royal Entomological Society* **11**, 333–353.

Travis, J.M.J. (2001) The color of noise and the evolution of dispersal. *Ecological Research* **16**, 157–163.

Travis, J.M.J. & Dytham, C. (1999) Habitat persistence, habitat availability and the evolution of dispersal. *Proceedings of the Royal Society, Series B* **265**, 17–23.

Walker, M.J.C., Coope, G.R. & Lowe, J.J. (1993) The Devensian (Weichselian) lateglacial palaeoenvironmental record from Gransmoor, East Yorkshire, England. *Quaternary Science Reviews* **12**, 659–680.

Wells, N.C., Gould, W.J. & Kemp, A.E.S. (1996) The role of ocean circulation in the changing climate. In: *Oceanography* (eds C.P. Summerhayes & S.A. Thorpe), pp. 41–58. Manson Publishing, London.

Whitlock, M.C. & McCauley, D.E. (1998) Indirect measures of gene flow and migration: $F_{ST} \neq 1/(4Nm+1)$. *Heredity* **82**, 117–125.

Wiens, J.A. (2001) The landscape context of dispersal. In: *Dispersal* (eds J. Clobert, E. Danchin, A.A. Dhondt & J.D. Nichols), pp. 96–109. Oxford University Press, Oxford.

Wilkinson, D.M. (1997) Plant colonization: are wind dispersed seeds really dispersed by birds at larger spatial and temporal scales? *Journal of Biogeography* **24**, 61–66.

Williamson, M. (1981) *Island Populations.* Oxford University Press, Oxford.

Williamson, M. (1991) Oceanic disjunctions. *Nature* **351**, 106.

Williamson, M. (1996) *Biological Invasions.* Chapman & Hall, London.

Wright, S. (1931) Evolution in Mendelian populations. *Genetics* **16**, 97–159.

Wright, S.D., Yong, C.G., Dawson, J.W., Whittaker, D.J. & Gardner, R.C. (2000) Riding the ice age El Niño? Pacific biogeography and evolution of *Metrosideros* subg. *Metrosideros* (Myrtaceae) inferred from nuclear ribosomal DNA. *Proceedings of the National Academy of Sciences of the USA* **97**, 4118–4123.

Index

Page numbers in **bold** refer to tables and in *italic* to figures

Lightning Source UK Ltd.
Milton Keynes UK
02 October 2010

160700UK00001B/1/P

9 780521 549318